Lecture Notes in Artificial Intelligence 10091

Subseries of Lecture Notes in Computer Science

More information about this series at http://www.springer.com/series/1244

Mihoko Otake · Setsuya Kurahashi
Yuiko Ota · Ken Satoh
Daisuke Bekki (Eds.)

New Frontiers in Artificial Intelligence

JSAI-isAI 2015 Workshops, LENLS, JURISIN,
AAA, HAT-MASH, TSDAA, ASD-HR, and SKL
Kanagawa, Japan, November 16–18, 2015
Revised Selected Papers

 Springer

Editors
Mihoko Otake
Chiba University Graduate School
 of Engineering
Chiba
Japan

Setsuya Kurahashi
University of Tsukuba
Tokyo
Japan

Yuiko Ota
Fujitsu Laboratories
Kanagawa
Japan

Ken Satoh
National Institute of Informatics
Tokyo
Japan

Daisuke Bekki
Ochanomizu University
Tokyo
Japan

ISSN 0302-9743   ISSN 1611-3349   (electronic)
Lecture Notes in Artificial Intelligence
ISBN 978-3-319-50952-5   ISBN 978-3-319-50953-2   (eBook)
DOI 10.1007/978-3-319-50953-2

Library of Congress Control Number: 2017936043

LNCS Sublibrary: SL7 – Artificial Intelligence

Printed on acid-free paper

This Springer imprint is published by Springer Nature
The registered company is Springer International Publishing AG
The registered company address is: Gewerbestrasse 11, 6330 Cham, Switzerland

Preface

JSAI-isAI 2015 was the 7th International Symposium on Artificial Intelligence supported by the Japanese Society of Artificial Intelligence (JSAI). JSAI-isAI 2015 was successfully held during November 16–18 at Keio University in Kanagawa, Japan. In total, 246 people from 16 countries participated. The symposium took place after the JSAI SIG joint meeting. As the total number of participants for these two co-located events was 794, it was the second-largest JSAI event in 2015 after the JSAI annual meeting.

JSAI-isAI 2015 included eight workshops, where 27 invited talks and 124 papers were presented. This volume, *New Frontiers in Artificial Intelligence: JSAI-isAI 2015 Workshops*, comprises the proceedings of JSAI-isAI 2015. From the seven of the eight workshops (LENLS 12, JURISIN 9, AAA 2015, HAT-MASH 2015, TSDAA 2015, ASD-HR2015 and SKL 2015), 38 papers were carefully selected and revised according to the comments of the workshop Program Committee. About 33% of the total submissions were selected for inclusion in the conference proceedings.

LENLS (Logic and Engineering of Natural Language Semantics) is an annual international workshop on formal semantics and pragmatics. This year's workshop was the 12th LENLS in the series, and its theme featured talks on a wide range of topics, including discourse particles, disjunction, truth, copredication, expressive content, categorial grammar, dependent-type semantics, sequent calculus, and various aspects of formal pragmatics.

JURISIN (Juris-informatics) 2015 was the ninth event in the series, organized by Satoshi Tojo (Japan Advanced Institute of Science and Technology). The purpose of this workshop was to discuss both the fundamental and practical issues among people from various backgrounds such as law, social science, information and intelligent technology, logic and philosophy, including the conventional "AI and law" area.

AAA (Argument for Agreement and Assurance) has the goal of deepening a mutual understanding and exploring a new research field involving researchers/practitioners in formal and informal logic, artificial intelligence, and safety engineering working on agreement and assurance through argument. The Second Workshop on Argument for Agreement and Assurance took place at Keio University, 2015. The general sessions included a variety of presentations covering material ranging from theoretical work and methodology to demonstrations of practical tools.

HAT-MASH 2015 (Healthy Aging Tech Mashup Service, Data and People) was the first international workshop bringing people together from the fields of healthy aging and elderly care technology, information technology, and service engineering. The main objective of this workshop was to provide a forum for discusing important research questions and practical challenges in healthy aging and elderly care support and to promote transdisciplinary approaches.

TSDAA 2015 (Workshop on Time Series Data Analysis and Its Applications) aimed at providing an interdisciplinary forum for discussion of different approaches and techniques of time series data analysis and their implementation in various real-life applications. As time series data are abundant in nature, a unifying approach is needed to bridge the gap between traditional multivariate time series data analysis with state-of-the-art methodologies of data mining from real-life time series data (numerical and text) in various applications ranging from medical and health related, biometrics, or process industry to finance or economic data analysis or weather prediction.

ASD-HR 2015 (Autism Spectrum Disorders Using a Humanoid Robot) was the first international workshop on interventions for children with autism spectrum disorders using a humanoid robot. As shown in the statistics of the recent reports, the necessity for treatment and education for children and adolescents with autism spectrum disorders (ASD) has been widely recognized. Researchers have recently considered using humanoid robots to treat ASD-associated deficits in social communication. In all, 16 oral presentations including five invited talk were given at the workshop, presenting studies in the interdisciplinary field of research on this topic including both engineering and medical sides.

SKL 2015 (Skill Science) invited researchers who investigate human skills. Human skills involve well-attuned perception and fine motor control, often accompanied by thoughtful planning. The involvement of the body, environment, and tools mediating them makes the study of skills unique among researches of human intelligence. The study of skills requires various disciplines to collaborate with each other because the value of skills is not determined solely in terms of efficiency, but calls for consideration of quality. Participants discussed the theoretical foundations of skill science as well as practical and engineering issues.

It is our great pleasure to be able to share some highlights of these fascinating workshops in this volume. We hope this book introduces readers to the state-of-the-art research outcomes of JSAI-isAI 2015, and motivates them to participate in future JSAI-isAI events.

June 2016

Mihoko Otake
Setsuya Kurahashi
Yuiko Ota
Ken Satoh
Daisuke Bekki

Organization

JSAI-isAI 2015

Chair

Mihoko Otake   Chiba University, Japan

Co-chairs

Setsuya Kurahashi   University of Tsukuba, Japan
Yuiko Ota   Fujitsu Laboratories Ltd., Japan

Publication Co-chairs

Daisuke Bekki   Ochanomizu University, Japan
Ken Satoh   National Institute of Informatics, Japan

Advisory Committee

Daisuke Bekki   Ochanomizu University, Japan
Ho Tu Bao   JAIST, Japan
Ee-Peng Lim   Singapore Management University, Singapore
Thanaruk Theeramunkong   SIIT, Thammasat University, Thailand
Tsuyoshi Murata   Tokyo Institute of Technology, Japan
Yukiko Nakano   Seikei University, Japan
Ken Satoh   National Institute of Informatics, Japan
Vincent S. Tseng   National Cheng Kung University, Taiwan
Takashi Washio   Osaka University, Japan

LENLS 12

Workshop Chair

Eric McCready   Aoyama Gakuin University, Japan

Workshop Co-chairs

Daisuke Bekki   Ochanomizu University, Japan
Koji Mineshima   Ochanomizu University, Japan

Program Committee

Eric McCready   Aoyama Gakuin University, Japan
Daisuke Bekki   Ochanomizu University, Japan
Koji Mineshima   Ochanomizu University, Japan
Alastair Butler   Tohoku University, Japan
Richard Dietz   University of Tokyo, Japan

Yoshiki Mori	University of Tokyo, Japan
Yasuo Nakayama	Osaka University, Japan
Katsuhiko Sano	JAIST, Japan
Katsuhiko Yabushita	Naruto University of Education, Japan
Tomoyuki Yamada	Hokkaido University, Japan
Shunsuke Yatabe	Kyoto University, Japan
Kei Yoshimoto	Tohoku University, Japan

JURISIN 2015

Workshop Chair

Takehiko Kasahara	Toin Yokohama University, Japan
Ken Satoh	National Institute of Informatics and Sokendai, Japan

Steering Committee

Takehiko Kasahara	Toin Yokohama University, Japan
Makoto Nakamura	Nagoya University, Japan
Katsumi Nitta	Tokyo Institute of Technology, Japan
Seiichiro Sakurai	Meiji Gakuin University, Japan
Ken Satoh	National Institute of Informatics and Sokendai, Japan
Satoshi Tojo	JAIST, Japan
Katsuhiko Toyama	Nagoya University, Japan

Advisory Committee

Trevor Bench-Capon	University of Liverpool, UK
Tomas Gordon	Fraunfoher FOKUS, Germany
Henry Prakken	University of Utrecht and Groningen, The Netherlands
John Zeleznikow	Victoria University, Australia
Robert Kowalski	Imperial College London, UK
Kevin Ashley	University of Pittsburgh, USA

Program Committee

Thomas Agotnes	University of Bergen, Norway
Katie Atkinson	University of Liverpool, UK
Marina De Vos	University of Bath, UK
Randy Goebel	University of Alberta, Canada
Guido Governatori	NICTA, Australia
Tokuyasu Kakuta	Nagoya University, Japan
Yoshinobu Kano	Shizuoka University, Japan
Takehiko Kasahara	Toin Yokohama University, Japan
Mi-Young Kim	University of Alberta, Canada
Robert Kowalski	Imperial College London, UK
Masahiro Kozuka	Okayama University, Japan
Nguyen Le Minh	JAIST, Japan

Beishui Liao	Zhejiang University, China
Minghui Ma	Southwest University, China
Makoto Nakamura	Nagoya University, Japan
Katumi Nitta	Tokyo Institute of Technology, Japan
Paulo Novais	University of Minho, Portugal
Antonino Rotolo	University of Bologna, Italy
Seiichiro Sakurai	Meiji Gakuin University, Japan
Katsuhiko Sano	JAIST, Japan
Ken Satoh	National Institute of Informatics and Sokendai, Japan
Akira Shimazu	JAIST, Japan
Hirotoshi Taira	Osaka Institute of Technology, Japan
Fumihiko Takahashi	Meiji Gakuin University, Japan
Satoshi Tojo	JAIST, Japan
Katsuhiko Toyama	Nagoya University, Japan
Minghui Xiong	Sun Yat-sen University, China
Baosheng Zhang	China University of Political Science and Law, China

AAA 2015

Program Committee

Katarzyna Budzynska	Polish Academy of Sciences, University of Dundee, Poland
Martin Caminada	University of Aberdeen, UK
Federico Cerruti	University of Aberdeen, UK
Ewen Denney	SGT/NASA Ames, USA
Juergen Dix	Clausthal University of Technology, Germany
Phan Minh Dung	Asian Institute of Technology, Thailand
C. Michael Holloway	NASA Langley Research Center, USA
Antonis Kakas	University of Cyprus, Cyprus
Tim Kelly	University of York, UK
Hiroyuki Kido	The University of Tokyo, Japan
Yoshiki Kinoshita	Kanagawa University, Japan
John Knight	University of Virginia, USA
Yutaka Matsuno	Department of Computer Engineering, College of Science and Technology, Nihon University, USA
John Rushby	SRI International, USA
Chiaki Sakama	Wakayama University, Japan
Ken Satoh	National Institute of Informatics and Sokendai, Japan
Guillermo Simari	Dep of Computer Science and Engineering, Universidad Nacional del Sur in Bahia Blanca, Argentine
Kenji Taguchi	National Institute of Advanced Industrial Science and Technology, Japan
Kazuko Takahashi	Kwansei Gakuin University, Japan
Toshinori Takai	NAIST, Japan

Makoto Takeyama	Dept. Information Science, Faculty of Science, Kanagawa University, Japan
Paolo Torroni	University of Bologna, Italy
Charles Weinstock	Software Engineering Institute, USA
Stefan Woltran	TU Wien, Austria
Shuichiro Yamamoto	Nagoya University, Japan

Organizing Committee

Kazuko Takahashi	Kwansei Gakuin University, Japan
Kenji Taguchi	AIST, Japan
Tim Kelly	University of York, UK
Hiroyuki Kido	The University of Tokyo, Japan

HAT-MASH 2015

Workshop Chair

Ken Fukuda	AIST, Japan

Advisory Committee

Masaaki Mochimaru	AIST, Japan
Takeshi Sunaga	Kyoto University, Japan
Hideaki Takeda	National Institute of Informatics, Japan

Organizing Committee

Taku Nishimura	AIST, Japan
Hiroyasu Miwa	AIST, Japan
Kentaro Watanabe	AIST, Japan

Program Committee

Ken Fukuda	AIST, Japan
Tom Hope	Tokyo Institute of Technology, Japan
Shuichi Ino	AIST, Japan
Takahiro Kawamura	JST, Japan
Kouji Kozaki	Osaka University, Japan
Noriaki Kuwahara	Kyoto Institute of Technology, Japan
Hiroyasu Miwa	AIST, Japan
Taketoshi Mori	The University of Tokyo, Japan
Yoichi Motomura	AIST, Japan
Marketta Niemela	VTT, Finland
Yoshifumi Nishida	AIST, Japan
Taku Nishimura	AIST, Japan
Yuko Ohno	Osaka University, Japan
Jun Ohta	The University of Tokyo, Japan
Hiroshi Sato	AIST, Japan

Atsushi Shinjo	Keio University, Japan
Marja Toivonen	VTT, Finland
Kentaro Watanabe	AIST, Japan
Naoshi Uchihira	JAIST, Japan

TSDAA 2015

Workshop Chairs

Basabi Chakraborty	Iwate Prefectural University, Japan
Goutam Chakraborty	Iwate Prefectural University, Japan
Tetsuji Kuboyama	Gakushuin University, Japan
Masafumi Matsuhara	Iwate Prefectural University, Japan

Program Committee

Cedric Bornand	HES-SO/HEIG-VD/MiS Institute, Switzerland
Smarajit Bose	Indian Statistical Institute, India
Basabi Chakraborty	Iwate Prefectural University, Japan
Goutam Chakraborty	Iwate Prefectural University, Japan
Bhabatosh Chanda	Indian Statistical Institute, India
Rung Ching Chen	Chaoyang University of Technology, Taiwan
Takako Hashimoto	Chiba University of Commerce, Japan
Tzung-Pei Hong	National University of Kaohsiung, Taiwan
Cheng-Hsiung Hsieh	Chaoyang University of Technology, Taiwan
Yo-Ping Huang	National Taipei University of Technology, Taiwan
Maciej Huk	Wroclaw University of Technology, Poland
Tetsuji Kuboyama	Gakushuin University, Japan
Hsien-Chou Liao	Chaoyang University of Technology, Taiwan
Yusuke Manabe	Chiba Institute of Technology, Japan
Masafumi Matsuhara	Iwate Prefectural University, Japan
Subhas Mukhopadhyay	Massey University, New Zealand
C.A. Murthy	Indian Statistical Institute, India
Amita Pal	Indian Statistical Institute, India
Jagdish Patra	Swinburne University of Technology, Australia
Vinod A. Prasad	Nanyang Technological University, Singapore
David Ramamonjisoa	Iwate Prefectural University, Japan
Keun Ho Ryu	Chungbuk National University, Korea
Kilho Shin	University of Hyogo, Japan
Yukari Shirota	Gakushuin University, Japan
Hideyuki Takahashi	Tohoku University, Japan
Leon S.L. Wang	National University of Kaohsiung, Taiwan
Qiangfu Zhao	University of Aizu, Japan

ASD-HR 2015

Organizers

Hirokazu Kumazaki   Fukui University, Japan
Yoshio Matsumoto   AIST, Japan
Yukie Nagai   Osaka University, Japan
Yuichiro Yoshikawa   Osaka University, Japan

SKL 2015

Workshop Chair

Tsutomu Fujinami   JAIST, Japan

Steering Committee

Masaki Suwa   Keio University, Japan
Ken Hashizume   Osaka University, Japan
Mihoko Otake   Chiba University, Japan
Yoshifusa Matsuura   Yokohama National University, Japan
Keisuke Okuno   Freelance

Advisory Committee

Koichi Furukawa   Keio University, Japan
Alan Robinson   Syracuse University, USA
Jacques Cohen   Brandeis University, USA

Program Committee

Mizue Kayama   Shinshu University, Japan

Sponsored By

The Japan Society for Artificial Intelligence (JSAI)

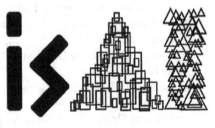

Contents

LENLS 12

Towards a Probabilistic Analysis for Conditionals and Unconditionals   3
  Stefan Kaufmann

What Do Proper Names Refer to? The Simple Sentence Puzzle
and Identity Statements   15
  Tomohiro Sakai

Particles of (Un)expectedness: Cantonese *Wo* and *Lo*   27
  Yurie Hara and Eric McCready

Rhetorical Structure and QUDs   41
  Julie Hunter and Márta Abrusán

An Inference Problem Set for Evaluating Semantic Theories
and Semantic Processing Systems for Japanese   58
  Ai Kawazoe, Ribeka Tanaka, Koji Mineshima, and Daisuke Bekki

Applicative Abstract Categorial Grammars in Full Swing   66
  Oleg Kiselyov

Scope Parallelism in Coordination in Dependent Type Semantics   79
  Yusuke Kubota and Robert Levine

Discourse Particles as CCP-modifiers: German *doch* and *ja*
as Context Filters   93
  Lukas Rieser

Tracking Down Disjunction 109
  Uli Sauerland, Ayaka Tamura, Masatoshi Koizumi,
  and John M. Tomlinson Jr.

The Projection of Not-at-issue Meaning via Modal Support:
The Meaning and Use of the Japanese Counter-Expectational Adverbs 122
  Osamu Sawada

Evaluative Predicates and Evaluative Uses of Ordinary Predicates 138
  Isidora Stojanovic

Strong Permission in Prescriptive Causal Models 151
  Linton Wang

Truth as a Logical Connective 166
  Shunsuke Yatabe

JURISIN 9

Abductive Logic Programming for Normative Reasoning and Ontologies.... 187
  Marco Gavanelli, Evelina Lamma, Fabrizio Riguzzi, Elena Bellodi,
  Zese Riccardo, and Giuseppe Cota

A Belief Revision Technique to Model Civil Code Updates 204
  Ryuta Arisaka

Combining Input/Output Logic and Reification for Representing
Real-World Obligations 217
  Livio Robaldo, Llio Humphreys, Xin Sun, Loredana Cupi,
  Cristiana Santos, and Robert Muthuri

Using Ontologies to Model Data Protection Requirements in Workflows 233
  Cesare Bartolini, Robert Muthuri, and Cristiana Santos

Utilization of Multi-word Expressions to Improve Statistical Machine
Translation of Statutory Sentences............................. 249
  Satomi Sakamoto, Yasuhiro Ogawa, Makoto Nakamura,
  Tomohiro Ohno, and Katsuhiko Toyama

Argumentation Support Tool with Reliability-Based
Argumentation Framework...................................... 265
  Kei Nishina, Yuki Katsura, Shogo Okada, and Katsumi Nitta

Applying a Convolutional Neural Network to Legal Question Answering.... 282
  Mi-Young Kim, Ying Xu, and Randy Goebel

Lexical-Morphological Modeling for Legal Text Analysis 295
  Danilo S. Carvalho, Minh-Tien Nguyen, Chien-Xuan Tran,
  and Minh-Le Nguyen

AAA 2015

On the Issue of Argumentation and Informedness 317
  Martin Caminada and Chiaki Sakama

On the Interpretation of Assurance Case Arguments 331
  John Rushby

Learning Argument Acceptability from Abstract
Argumentation Frameworks 348
  Hiroyuki Kido

HAT-MASH 2015

Designing Intelligent Sleep Analysis Systems for Automated Contextual
Exploration on Personal Sleep-Tracking Data   367
  Zilu Liang, Wanyu Liu, Bernd Ploderer, James Bailey, Lars Kulik,
  and Yuxuan Li

Axis Visualizer: Enjoy Core Torsion and Be Healthy for Health
Promotion Community Support.   380
  Takuichi Nishimura, Zilu Liang, Satoshi Nishimura, Tomoka Nagao,
  Satoko Okubo, Yasuyuki Yoshida, Kazuya Imaizumi, Hisae Konosu,
  Hiroyasu Miwa, Kanako Nakajima, and Ken Fukuda

TSDAA 2015

A Comparative Study of Similarity Measures for Time Series Classification . . .   397
  Sho Yoshida and Basabi Chakraborty

Extracting Propagation Patterns from Bacterial Culture Data
in Medical Facility   409
  Kazuki Nagayama, Kouichi Hirata, Shigeki Yokoyama,
  and Kimiko Matsuoka

Real-Time Anomaly Detection of Continuously Monitored Periodic
Bio-Signals Like ECG.   418
  Takuya Kamiyama and Goutam Chakraborty

Aggregating and Analyzing Articles and Comments on a News Website   428
  David Ramamonjisoa

ASD-HR 2015

Positive Bias of Gaze-Following to Android Robot in Adolescents
with Autism Spectrum Disorders.   447
  Yuichiro Yoshikawa, Yoshio Matsumoto, Hirokazu Kumazaki,
  Yujin Wakita, Sakiko Nemoto, Hiroshi Ishiguro, Masaru Mimura,
  and Masutomo Miyao

Feasibility of Collaborative Learning and Work Between Robots
and Children with Autism Spectrum Disorders   454
  Felix Jimenez, Tomohiro Yoshikawa, Takeshi Furuhashi,
  Masayoshi Kanoh, and Tsuyoshi Nakamura

Teachers' Impressions on Robots for Therapeutic Applications   462
  Hidenobu Sumioka, Yuichiro Yoshikawa, Yasuo Wada,
  and Hiroshi Ishiguro

An Intervention for Children with Social Anxiety and Autism Spectrum
Disorders Using an Android Robot   470
    Hirokazu Kumazaki, Yuichiro Yoshikawa, Yoshio Matsumoto,
    Masutomo Miyao, Hiroshi Ishiguro, Taro Muramatsu,
    and Masaru Mimura

Usefulness of Animal Type Robot Assisted Therapy for Autism Spectrum
Disorder in the Child and Adolescent Psychiatric Ward   478
    Yoshihiro Nakadoi

SKL 2015

The Trend in the Frontal Area Activity Shift with Embodied Knowledge
Acquisition During Imitation Learning of Assembly Work   485
    Yusuke Asaka, Keiichi Watanuki, and Lei Hou

The Cognitive Role of Analogical Abduction in Skill Acquisition   499
    Koichi Furukawa, Keita Kinjo, Tomonobu Ozaki,
    and Makotoc Haraguchi

Identifying Context-Dependent Modes of Reading   514
    Miho Fuyama and Shohei Hidaka

Whole-Body Coordination Skill for Dynamic Balancing on a Slackline   528
    Kentaro Kodama, Yusuke Kikuchi, and Hideo Yamagiwa

Author Index   547

LENLS 12

Logic and Engineering of Natural Language Semantics (LENLS) 12

Eric McCready

Department of English, Aoyama Gakuin University, Tokyo, Japan

1 The Workshop

This year's workshop was the twelfth LENLS and was held at the Hiyoshi campus of Keio University in Kawasaki in November of 2015 as part of the JSAI International Symposia on AI program of the Japanese Society for Artificial Intelligence. The workshop featured invited talks by Nicholas Asher, on the analysis of conversation via a special kind of repeated game, Robert Henderson, on the semantics of mimetics, Magdalena Kaufmann, on free choice, Stefan Kaufmann, on probabilistic approaches to the semantics of conditionals, and Tomohiro Sakai, on reference. Papers based on the talks by Stefan Kaufmann and Tomohiro Sakai appear in the present volume. In addition, 21 papers were selected by the program committee (see Acknowledgements) from the submitted abstracts for presentation and as alternates.

As always with the LENLS workshops, the content of the presented papers was rich and varied. This year's theme featured talks on a wide range of topics, including discourse particles, disjunction, truth, copredication, expressive content, categorial grammar, dependent type semantics, sequent calculus, and various aspects of formal pragmatics. In addition, the first day of the workshop included a workshop on politeness with invited talks by Eric McCready, on honorification, and Daisuke Bekki, on the composition of expressive content in dependent type semantics. As the reader will notice, the range of topics addressed by the contributed papers is very wide. As a result, the workshop was very stimulating; the participants, with their different perspectives, gave useful and interesting comments on each paper. From the perspective of the organizers at least, the result was very successful. We hope (and believe) that the other participants shared this impression. The papers in the present volume represent a selection of the papers presented at the workshop and give a sense (we think) of the breadth of the content presented there.

2 Acknowledgements

Let me finally acknowledge some of those who have helped with the workshop. The program committee and organizers, in addition to myself, were Daisuke Bekki, Koji Mineshima, who were co-chairs of the workshop, and Alastair Butler, Richard Dietz, Yoshiki Mori, Yasuo Nakayama, Katsuhiko Sano, Katsuhiko Yabushita, Tomoyuki Yamada, Shunsuke Yatabe, and Kei Yoshimoto. I would also like to acknowledge the external reviewers for the workshop. Finally, the organizers would like to thank JSAI for giving us the opportunity to hold the workshop.

Towards a Probabilistic Analysis
for Conditionals and Unconditionals

Stefan Kaufmann[✉]

Department of Linguistics, University of Connecticut,
365 Fairfield Way, Unit 1145, Storrs Mansfield, CT 06268, USA
stefan.kaufmann@uconn.edu
http://stefan-kaufmann.uconn.edu/

Abstract. The thesis that the probability of a conditional 'if A, C' is the corresponding conditional probability of C, given A, enjoys wide currency among philosophers and growing empirical support in psychology. In this paper I ask how a probabilisitic account of conditionals along these lines could be extended to *unconditional* sentences, i.e., conditionals with interrogative antecedents. Such sentences are typically interpreted as equivalent to conjunctions of conditionals. This raises a number of challenges for a probabilistic account, chief among them the question of what the probability of a conjunction of conditionals should be. I offer an analysis which addresses these issues by extending the interpretation of conditonals in *Bernoulli models* to the case of unconditionals.

Keywords: Conditionals · Unconditionals · Probability · Bernoulli models

1   Background

1.1   Conditionals

For the purposes of this paper, an English conditional is a complex sentence of the form 'if antecedent, (then) consequent', relating two declarative clauses to each other, here called *antecedent* and *consequent*.[1] Throughout this paper I use the sentential connective '>' for a concise formal representation of this construction, writing '$A > C$' for the conditional 'if A, (then) C'.

The class of actual linguistic forms subsumed under this label is rather diverse. The list in (1) gives examples of the three most prototypical forms conditionals can take, along with labels they frequently receive in the literature.

(1)   a.   If she throws an even number, it will be a six.   *[predictive]*
      b.   If she threw an even number, it was a six.   *[non-predictive]*
      c.   If she had thrown an even number, it was a six.   *[counterfactual]*

---

[1] A competing usage in the descriptive linguistic literature calls the constituents *protasis* and *apodosis*, but this usage is not widespread in formal semantics or philosophy.

© Springer International Publishing AG 2017
M. Otake et al. (Eds.): JSAI-isAI 2015 Workshops, LNAI 10091, pp. 3–14, 2017.
DOI: 10.1007/978-3-319-50953-2_1

Clearly there are semantic differences between the three forms: each of the sentences in (1) can be true while the others are false. Nor is there general agreement on the labels for the sub-classes, or for that matter on the status of the taxonomical divisions themselves [2,15]. However, authors do agree that all three share a common semantic core. Roughly speaking, this common core of "conditional" meaning can be paraphrased as follows: The antecedent A and the consequent C denote propositions, and the conditional is true if and only if C is true *on the supposition that* A is true. In devising a semantic analysis on this basis, much hinges on the way in which such suppositional reasoning is modeled.

This paper is concerned with this common conditional meaning, glossing over the fine distinctions in tense, aspect, and mood exhibited in (1), which drive and constrain the ways in which the core conditional meaning is applied in particular cases. Additionally, although all three examples in (1) have antecedents introduced by the word 'if', conditional meaning also arises in constructions which are not so marked. Those are beyond the scope of this paper, as their existence does not add much to the main goal of clarifying the nature of "conditional meaning" as an abstract semantic category.

One fact about conditionals that does motivate some of the technical work underlying this paper is that they can themselves be compounded and embedded rather freely, as illustrated in the examples in (2).

(2)   a.   If this match is wet, it won't light if you strike it. $A > B$

      b.   If this switch will fail if it is submerged in water, it will be discarded.

$$(A > B) > C$$

      c.   If this vase will crack if it is dropped on wood, it will shatter if it is dropped on marble. $(A > B) > (C > D)$

      d.   If she drew a prime number, it is even, and if she drew an odd number, it is prime. $(A > B) \land (C > D)$

A semantic theory of conditionals ought to have some way of accounting for these facts. What exactly that means depends in part of the semantic framework. For instance, a theory which assigns probabilities to conditionals ought to be extendable to one which assigns probabilities to complex sentences containing them as constituents. This turns out to be a non-trivial goal.

1.2   Unconditionals

Unconditionals are sentences of the form 'Wh-antecedent,consequent' whose distinctive property vis-à-vis conditionals is that their antecedents are not declarative, as in (1) above, but interrogative. Examples are given in (3), along with the labels commonly applied to the respective sub-types of unconditionals in the linguistic literature.

(3)   a.   Whether Mary comes or not, we will have fun.   *[alternative uc]*

      b.   Whether John or Mary comes, we will have fun.   *[alternative uc]*

      c.   Whoever comes, we will have fun.   *[constituent uc]*

      d.   No matter who comes, we will have fun.   *[headed uc]*

In formal semantics, interrogatives are interpreted as denoting sets of propositions, each element representing an answer.[2] In this paper I use the form '$A?$' to refer to interrogative clauses denoting sets of propositions $\{A_1, A_2, \ldots\}$.[3] Intuitively, an unconditional $A? > C$ is true just in case C is true *regardless of which* A_i among the alternatives is true.

There is ample evidence that conditionals and unconditionals are closely related not only semantically, but also grammatically. For instance, declarative and interrogative antecedents can be interleaved fairly freely in either order (4). Furthermore, declarative and interrogative antecedents can make each other redundant, in either order, as shown in (5). Finally, there are counterfactual unconditionals (6).

(4)   a.   If John sings, then whether he is drunk or not, it will be fun.
       b.   Whether John is drunk or not, if he sings, it will be fun.

(5)   a. #If John sings, then whether he sings or not, it will be fun.
       b. #Whether John sings or not, if he sings, it will be fun.

(6)   a.   Whether he had come or not, we would have had fun.
       b.   Whatever John had bought, we would have had fun.

Most importantly for present purposes, there is a tight semantic relation between unconditionals on the one hand, and conjunctions of conditionals, on the other. Thus in each of the three following examples, the unconditional in (a) means the same as the conjunction in (b):

(7)   a.   Whether John comes or not, it will be fun.
       b.   If John comes it will be fun, and if he doesn't come it will be fun.

(8)   a.   Whether John or Mary comes, it will be fun.
       b.   If John comes it will be fun, and if Mary comes it will be fun.

(9)   a.   {Whoever / No matter who} comes, it will be fun.
       b.   If John comes it will be fun, and if Mary comes it will be fun, and if Kim comes it will be fun, ...

Schematically, these examples instantiate a pattern that is generally taken to hold the key to the meaning of unconditionals:

(10)   **Distribution over antecedents.**
       $A? > C \Leftrightarrow (A_1 > C) \wedge (A_2 > C) \wedge \ldots$

Some analyses of unconditionals are explicitly designed as implementations of this pattern. Among them is the approach of Rawlins [25, 26], the most

---

[2] This glosses over important points of variation, for instance as to whether the set of propositions is taken to be the set of possible answers, of true answers, and whether its members are required to cover or partition the set of all possibilities [9, 11, 13]. These are important issues, but they are not crucial for the purposes of this paper.

[3] The typographical similarity with the conventions of Inquisitive Semantics [3, 8], is intended, although in this paper I do not pursue an in-depth study of this connection.

thoroughly worked-out proposal on the topic in recent years. Simply put, Rawlins treats unconditionals within a general theory of conditionals by postulating that when the antecedent has an interrogative denotation, a covert universal quantifier is introduced by default which quantifies over all elements of that denotation. This way, in effect, unconditionals are defined in terms of the right-hand side of (10). Other proposals, including an earlier one of my own [17], may not invoke explicit quantification over alternatives; but they, too, typically have the consequence that the truth conditions of unconditionals come out equivalent to the corresponding conjunction of conditionals.

The present paper does not depart from this general approach. I take the equivalence in (10) as a desideratum without further argument. The question I will be concerned with for the remainder of this paper is what the equivalence amounts to in a probabilistic setting, and how to implement a probabilistic account that derives it as a theorem.

2   Conditionals and Probability

Approaches to conditionals in terms of subjective probability have a long history, going back at least to Jeffrey [12] and Adams [1], and arguably as far as Ramsey [24]. The central idea underlying almost all proposals in this vein is that the probability of a conditional 'if A, (then) C' is the corresponding conditional probability of the consequent C, given the antecedent A.

This idea has strong intuitive appeal and wide currency among philosophers, as well as strong and growing empirical support in psychology [6,22,23].[4] Its implementation faces a number of challenges, however, which doubtless have stood in the way of its widespread acceptance. This paper is not the place to discuss these challenges in great detail. For more discussion, see [18] and the references therein. Instead, I just rehearse the main points by way of introducing the technical apparatus I am going to apply in the remainder of the paper.

2.1   Simple Probability Models

I start with a standard possible-worlds model as typically used to model the logical and epistemological aspects of language meaning and communication: sentences denote propositions, which are represented as sets of possible worlds, and truth-functional connectives are mapped to Boolean operations on propositions in the familiar fashion. There is a straightforward way to add probabilities to such a model: a probability measure is defined on an algebra of propositions, assuming that it has the appropriate properties.

---

[4] Certain counterexamples have been discussed in the literature, but there is hope that those can be explained as systematic deviations that do not undermine the general idea, but rather help to fine-tune its application in particular cases [14,19,29].

Definition 1 (Probability model). *A probability model for the language of propositional logic is a structure* $\langle \Omega, \mathcal{F}, \mathrm{Pr}, V \rangle$, *where:*[5]

a. Ω *is a non-empty set (of possible worlds);*
b. $\mathcal{F}$ *is a σ-algebra of subsets of Ω (propositions);*
c. Pr *is a probability measure on $\mathcal{F}$;*
d.   *V is a valuation function mapping sentences to characteristic functions of propositions, subject to the following constraints:*

$$V(\neg\varphi)(\omega) = 1 - V(\varphi)(\omega)$$
$$V(\varphi \wedge \psi)(\omega) = V(\varphi)(\omega) \cdot V(\psi)(\omega)$$

No generality is lost, and some simplicity is gained, if we assume that $\mathcal{F}$ is the powerset of Ω. I will make this assumption throughout. The immediate benefit is that no special provisions are required to ensure that all propositions denoted by sentences are in the domain of the probability function.

In a simple probability model, sentences and their their truth-functional compounds can be given probabilities in a straightforward manner: The probability of a sentence is the *expectation* of its truth value. In statistical jargon, the function $V(\varphi)$ is a *random variable*.[6] The expectation of a random variable with finite range is defined as the weighted sum of its values, where the weights are the probabilities that it takes those values:[7]

(11)   **Expectation and conditional expectation.**
$$E[\zeta] = \sum_{x \in range(\zeta)} x \cdot Pr(\zeta = x)$$
$$E[\zeta|\eta] = \sum_{x \in range(\zeta)} x \cdot Pr(\zeta = x | \eta = 1)$$

We may then define a function P from sentences to their expectations and speak of $P(\varphi)$ as "the probability of φ", even though P is strictly speaking not a measure.

(12)   **Probability and conditional probability of sentences.**
$$P(\varphi) = E[V(\varphi)]$$
$$P(\psi|\varphi) = E[V(\psi)|V(\varphi) = 1]$$

---

[5] $\mathcal{F} \subseteq \wp(\Omega)$ is σ-algebra iff it contains Ω and is closed under complement and countable union. $\mathrm{Pr} : \mathcal{F} \mapsto [0,1]$ is a probability measure iff $\mathrm{Pr}(\Omega) = 1$ and for any countable set of pairwise disjoint $X_i \in \mathcal{F}, \mathrm{Pr}(\bigcup_i X_i) = \sum_i \mathrm{Pr}(X_i)$.
[6] Random variables with range $\{0, 1\}$ are also called *indicator functions*.
[7] For continuous variables, the summation is replaced by integration, but the basic idea is the same. I write '$Pr(\zeta = x)$' as shorthand for '$Pr(\{\omega \in \Omega | \zeta(\omega) = x\})$'.

Thus we arrive at the following straightforward disquotational slogan: The probability of a sentence φ is the probability that φ is true. For instance:

(13)   **Probabilistic disquotation.**
$P(\text{"she tossed an even number"}) = Pr(\text{she tossed an even number})$

Now, the desired treatment of conditionals is not yet supported by the account. Boolean operations allow us to represent the material conditional $\neg(\varphi \wedge \neg\psi)$, which is however not the correct rendering of $'\text{if } \varphi, \text{ then } \psi'$, since its probability (i.e., the expectation of its truth value) is not equivalent to the conditional probability of ψ given φ. Instead, we are looking for a way to extend the interpretation function V to sentences of the form $\varphi > \psi$ in such a way that the equivalence is ensured, i.e., that the unconditional probability $P(\varphi > \psi)$ equals the conditional probability $P(\psi|\varphi)$ whenever both are defined.

This extension of V is not at all straightforward, however. Starting with [20], a long string of increasingly sophisticated *triviality results* have established that there is no way, except in certain very limited cases, to interpret conditionals alongside atomic sentences and Boolean compounds in such a way that the equivalence holds for all probability distributions. I will not expand on this problem any further here.[8] Suffice it to say these results show, at the very least, that the existence of a solution to the triviality problem within the confines of the framework just outlined is sufficiently unlikely to warrant a search for an alternative.

2.2   Bernoulli Models

The framework I choose to adopt was first introduced by van Fraassen and further developed by Stalnaker and Jeffrey [7,28]; for further discussion and applications see also [16,18].

The main conceptual innovation is the following basic intuition about the evaluation of a conditional $\varphi > \psi$ at a world ω: If the antecedent φ is true at ω, then the truth value of the conditional is that of the consequent ψ at ω. Thus where the antecedent is true, the conditional is equivalent to the corresponding material conditional. If the antecedent is false at ω, the truth value of the conditional is that of the consequent at an *arbitrarily* chosen antecedent-world. More precisely, the choice of an antecedent-world is modeled as a process of repeatedly selecting worlds from Ω (with replacement) according to the probability distribution Pr, until an antecedent-world is found. Formally, van Fraassen proposed to represent this process in a Bernoulli model:[9]

Definition 2 (Bernoulli model).   *Given a probability model $\langle \Omega, \mathcal{F}, \mathrm{Pr}, V \rangle$, the corresponding Bernoulli model is the structure $\langle \Omega^*, \mathcal{F}^*, \mathrm{Pr}^*, V^* \rangle$, where*

---

[8] See [10] for a survey of the status quo in 1994. My own attempt at a readable exposition is found in [15], but see [4] for more.

[9] Van Fraassen called the construction "Stalnaker Bernoulli model" since he saw in it the probabilistic analog of a Stalnaker-style selection function.

a. Ω^* *is the set of all countable sequences of worlds in* Ω. *Notation:*

$\:'\omega^*[n]\:'$ *is the n-th world in* ω^*, $n \geq 1$;

$\:'\omega^*(n)\:'$ *is the "tail" of* ω^* *starting with the n-th world.*

b. $\mathcal{F}^*$ *is the set of all Cartesian products* $X_1 \times \ldots \times X_n \times \Omega^*$ *for* $X_i \in \mathcal{F}$;

c. Pr^* *is a product measure on* $\mathcal{F}^*$ *defined as follows:*

$$\mathrm{Pr}^*(X_1 \times \ldots \times X_n \times \Omega^*) = \mathrm{Pr}(X_1) \times \ldots \times \mathrm{Pr}(X_n)$$

d. V^* *is a function from pairs of sentences and sequences in* Ω^* *to truth values, defined as follows:*

$$V^*(p)(\omega^*) = V(p)(\omega^*[1]) \text{ for atomic } p$$

$$V^*(\neg\varphi)(\omega^*) = 1 - V^*(\varphi)(\omega^*)$$

$$V^*(\varphi \wedge \psi)(\omega^*) = V^*(\varphi)(\omega^*) \cdot V^*(\psi)(\omega^*)$$

$$V^*(\varphi > \psi)(\omega^*) = V^*(\psi)(\omega^*{\uparrow}\varphi) \text{ where } \omega^*{\uparrow}\varphi = \omega^*(n)$$

$$\text{for the least } n \text{ s.t. } V^*(\varphi)(\omega^*(n)) = 1$$

According to the last line in the definition of V^*, the truth value of a conditional at a sequence ω^* is determined by "skipping" forward in ω^*, disregarding a finite initial sub-sequence of $n-1$ worlds and inspecting the first "tail" $\omega^*(n)$ at which the antecedent is true. (If the antecedent is true at ω^*, the value of the conditional is that of the consequent at ω^*.) An alternative way to think about this operation offers itself once we realize that the tail $\omega^*(n)$ is itself a sequence in Ω^*. Thus we can think of the operation of skipping forward in ω^* alternatively as skipping "sideways" to an alternative sequence at which the antecedent is true. Viewed this way, the operation bears a close resemblance to Stalnaker's semantics in terms of a selection function [27]. The crucial difference is, of course, that the move to the alternative sequence is not deterministic, but a series of random choices.

Notice that this means that the truth value of the conditional is only defined at sequences which contain a sub-sequence at which the antecedent is true. It can be shown that if the antecedent has non-zero probability, then the probability that an antecedent-world is found at *some* point (after an arbitrarily long but finite sequence of trials), and hence that the truth value of the conditional is defined, is one. Thus whenever the antecedent has non-zero probability, the set of sequences at which the conditional's value is undefined can be neglected.

3 Bernoulli Models for (Un)conditionals

The last section introduced Bernoulli models, the main formal tool on which I build my proposal for the interpretation of unconditionals. The above discussion was brief and the following subsection on conditionals will likewise be no more than a sketch. The reader is referred to [16, 18] for further details and discussion.

3.1 Bernoulli Models for Conditionals

The most obvious and immediate benefit of Bernoulli models for the interpretation of conditionals is that they deliver the desired equivalence between the probability of a conditional $P^*(\varphi > \psi)$ (recall that this was defined as the expectation of the sentence's truth value under V^*) and the corresponding conditional probability $P^*(\psi|\varphi)$ (i.e., the conditional expectation of ψ's truth value, given that φ is true), without facing the triviality problem. For detailed arguments on why this is the case, see [7,16].

Another, no less important benefit of Bernoulli models is that they assign models to compounded and embedded conditionals. Moreover, those probabilities of conditional sentences in the Bernoulli model can be calculated from the probabilities of their non-conditional constituents in the underlying simple probability model. To illustrate, the following can be shown for arbitrary atomic sentences A, B, C, D.[10]

(14)   If $P(A) > 0$, then $P^*(A > B) = P^*(B|A) = P(B|A)$

(15)   If $P(B) > 0$ and $P(C) > 0$, then
$$P^*(B > (C > D)) = P(C \wedge D|B) + P(D|C)P(\neg C|B)$$

(16)   If $P(A) > 0$ and $P(B|A) > 0$, then
$$P^*((A > B) > D) = P(D|A \wedge B)P(A) + P(D \wedge \neg A)$$

(17)   If $P(A) > 0$ and $P(C) > 0$, then
$$P^*((A > B) \wedge (C > D)) = \\ \frac{P(A \wedge B \wedge C \wedge D) + P(D|C)P(A \wedge B \wedge \neg C) + P(B|A)P(\neg A \wedge C \wedge D)}{P(A \vee C)}$$

(18)   If $P(A) > 0, P(C) > 0$, and $P(B|A) > 0$, then
$$P^*((A > B) > (C > D)) = \frac{P^*((A > B) \wedge (C > D))}{P^*(A > B)}$$

These are significant improvements in the formal apparatus at our disposal. One may still find fault on empirical grounds with the probabilities thus assigned (indeed, a good portion of [16] is dedicated to a modification of the truth definitions to address certain counter-intuitive predictions), but a falsifiable hypothesis is certainly better than none.

Of particular interest for the purposes of the present paper is of course the formula in (17) for conjunctions of conditionals. Recall that what we are aiming to account for in the logical behavior of unconditionals is what I called distribution over antecedents in (10) above.

3.2 Bernoulli Models for Unconditionals

The developments outlined in the previous subsection marked progress on two fronts: First, a probabilistic interpretation of conditionals became possible in

---

[10] More complex compounds also receive truth values and probabilities under the approach, but those are hard to evaluate because intuitive judgments are not easy to come by.

the first place; and second, the new framework also opened up the possibility of extending the treatment to compounds of conditionals, including conjunctions of conditionals. I should re-emphasize that this is a prerequisite for a probabilistic account of unconditionals to even come within reach. This is because, recalling the formulation of distribution over alternatives in (10) above (repeated here for convenience), we now have a systematic way of filling in the right-hand side.

(10)   **Distribution over antecedents.**
$A? > C \Leftrightarrow (A_1 > C) \wedge (A_2 > C) \wedge \ldots$

The remaining question, then, is how to extend the assignment function still further to conditionals with interrogative antecedents, thus filling in the left-hand side, in such a way that the equivalence falls out.

Unconditional Connectives. Recall that the main difference between a conditional $\varphi > \psi$ and an unconditional $\varphi? > \psi$ is that the antecedent of the latter denotes a set of propositions, rather than a single proposition. The goal now is to extend the valuation function V^* to deal with this case.

The way in which I propose to approach this problem is through a modification of the shift from a sequence ω^* to the corresponding alternative φ-sequence. Definition 2 above invoked an mapping $\cdot \uparrow \cdot$ between sequences: $\omega^* \uparrow \varphi$ is the sub-sequence $\omega^*(n)$ for the least n such that $V^*(\varphi)(\omega^*(n)) = 1$. I extend this to a "multiselection" function mapping sequences ω^* and sets of propositions $\Phi = \{X, Y, \ldots\}$ to sets of sequences:

(19)   **Multiselection.**
$\omega^* \uparrow \Phi = \{\omega^* \uparrow X \mid X \in \Phi$ and $\omega^* \uparrow X$ is defined$\}$

Based on this set-valued selection, there are now multiple options for defining the truth conditions for unconditional sentences. In (20) through (22) I list three choices which strike me as plausible candidates.

(20)   **Universal unconditional.**
$V^*(\Phi >_\forall \psi)(\omega^*) = 1$ iff $V^*(\psi)(\omega^{*\prime}) = 1$ for all $\omega^{*\prime}$ in $\omega^* \uparrow \Phi$

(21)   **Existential unconditional.**
$V^*(\Phi >_\exists \psi)(\omega^*) = 1$ iff $V^*(\psi)(\omega^{*\prime}) = 1$ for some $\omega^{*\prime}$ in $\omega^* \uparrow \Phi$

(22)   **Minimal unconditional.**
$V^*(\Phi >_{\min} \psi)(\omega^*) = 1$ iff $V^*(\psi)(\omega^{*\prime}) = 1$ for the "first" $\omega^{*\prime}$ in $\omega^* \uparrow \Phi$

All three of these (and perhaps more) are more or less plausible choices for an interpretation rule for unconditionals. In the next subsection I show that the first choice, which I dub the "universal unconditional", delivers the desired predictions.

The Main Result. Each of the conditional operators in (20) through (22) above is worth careful consideration in its own right, but in the interest of brevity I only state their crucial properties without discussion. I do give further arguments for one of them, viz. the universal unconditional, which turns out to be the one needed to account for the semantic behavior of unconditional sentences. The crucial consequences of each of the connectives are listed in (23) through (25). Each gives an equation which holds for all probability distributions and arbitrary constituents $A?$ and C (provided that all the relevant conditional probabilities are defined). For simplicity, I assume that neither the alternatives in $A?$ nor C are or contain conditionals. This assumption is not essential for the validity of the equations in (23) through (25), but it simplifies the exposition and allows us to calculate the probabilities of the formulas in terms of probabilities under P from the underlying simple model.

(23) **Universal uc.**
$$P^*(A? >_\forall C) = P^*((A_1 > C) \wedge (A_2 > C) \wedge \ldots)$$

(24) **Existential uc.**
$$P^*(A? >_\exists C) = P^*((A_1 > C) \vee (A_2 > C) \vee \ldots)$$

(25) **Minimal uc.**
$$P^*(A? >_{\min} C) = P^*((A_1 \vee A_2 \vee \ldots) > C)$$

It can be shown that the equality in (23) holds for arbitrary unconditionals (subject to the simplification mentioned above, i.e., assuming that all constituents involved are non-conditional). It is useful to start with he following lemma.

Lemma 1 (Fraction lemma $-$[7]**).** *If* $Pr(X) > 0$, *then* $\sum_{n \in \mathbb{N}} Pr(\overline{X})^n = 1/Pr(X)$

Proof. $\sum_{n \in \mathbb{N}} Pr(\overline{X})^n \cdot Pr(X) = \sum_{n \in \mathbb{N}} (Pr(\overline{X})^n \cdot Pr(X))$
$= \sum_{n \in \mathbb{N}} Pr^* (\overline{X}^n \times X \times \Omega^*) = Pr^* (\bigcup_{n \in \mathbb{N}} (\overline{X}^n \times X \times \Omega^*))$
$= Pr^* (\{\omega^* \in \Omega^* | \exists n \in \mathbb{N}. \omega^*[n] \in X\})$
$= 1 - Pr^* (\{\omega^* \in \Omega^* | \forall n \in \mathbb{N}. \omega^*[n] \in \overline{X}\})$
$= 1 - \lim_{n \to \infty} Pr(\overline{X}^n) = 1 - 0$ since $Pr(\overline{X}) < 1$ by assumption.   □

With this, I proceed to show that the equality in (23) holds.

Proof. I only discuss the case of an antecedent with two alternatives $A? = \{A_1, A_2\}$; the extension to more alternatives is straightforward.

a.   If A_1, A_2 are both true at ω^*, then $V^*(A? >_\forall C)(\omega^*) = V^*(C)(\omega^*)$. Thus $P^*((A? > C) \wedge A_1 \wedge A_2) = P^*(A_1 \wedge A_2 \wedge C) = P(A_1 \wedge A_2 \wedge C)$ since A_1, A_2, C do not contain conditionals.

b.   If A_1 is true and A_2 is false at ω^*, then $A? > C$ is true iff C is true at both ω^* and $\omega^* \uparrow A_2$. Thus $P^*((A? > C) \wedge A_1 \wedge \neg A_2)$
$= P(A_1 \wedge \neg A_2 \wedge C)P(A_2 \wedge C) + P(A_1 \wedge \neg A_2 \wedge C)P(\neg A_2)P(A_2 \wedge C)$
$+ P(A_1 \wedge \neg A_2 \wedge C)P(\neg A_2)^2 P(A_2 \wedge C) + \ldots$

$$= P(A_1 \wedge \neg A_2 \wedge C)P(A_2 \wedge C)/P(A_2) \text{ by Lemma 1}$$
$$= P(A_1 \wedge \neg A_2 \wedge C)P(C|A_2).$$

c. By a similar argument, $P^*((A? > C) \wedge \neg A_1 \wedge A_2)$
$$= P(\neg A_1 \wedge A_2 \wedge C)P(C|A_1).$$

d. By (a-c), $P^*((A? > C) \wedge (A_1 \vee A_2))$
$$= P(A_1 \wedge A_2 \wedge C) + P(A_1 \wedge \neg A_2 \wedge C)P(C|A_2)$$
$$+ P(\neg A_1 \wedge A_2 \wedge C)P(C|A_1).$$

e. If $A_1 \vee A_2$ is false at ω^*, then the truth value of $A? > C$ is determined by the first $\omega^*(n)$ at which $A_1 \vee A_2$ is true. Thus by (a-d), $P^*((A? > C))$
$$= P^*((A? > C) \wedge (A_1 \vee A_2))$$
$$+ P(\overline{A_1 \vee A_2})\text{Pr}^*((A? > C) \wedge (A_1 \vee A_2))$$
$$+ P(\overline{A_1 \vee A_2})^2 P^*((A? > C) \wedge (A_1 \vee A_2)) + \dots$$
$$= P^*((A? > C) \wedge (A_1 \vee A_2))/P(A_1 \vee A_2) \text{ by Lemma 1}$$
$$= \frac{P(A_1 \wedge A_2 \wedge C) + P(A_1 \wedge \neg A_2 \wedge C)P(C|A_2) + P(\neg A_1 \wedge A_2 \wedge C)P(C|A_1)}{P(A_1 \vee A_2)}$$

f. (e) instantiates the formula for the conjunction $(A_1 > C) \wedge (A_2 > C)$ from (17) above. □

4 Conclusions and Outlook

To sum up, the universal unconditional operator $>_\forall$ defined in (20) above delivers a suitable rendering of our intuitions about unconditionals. Specifically, it satisfies distribution over alternatives in its probabilistic form. The formal tool of Bernoulli models (or something matching its expressivity) turns out to be necessary to state the condition of distribution over alternatives (which involves conjunctions of conditionals), as well as to define truth conditions for unconditionals.

Acknowledgments. I would like to thank the organizers of LENLS 12 for the opportunity to present this work. Parts of this material were previously presented at the "Work in Progress" seminar in the Philosophy Department at MIT. I am grateful to the audiences at both events for valuable feedback. Thanks also to Yukinori Takubo and Kyoto University for an invitation to a one-semester guest professorship in the fall of 2015, during which some of this work was carried out.

References

1. Adams, E.: The logic of conditionals. Inquiry **8**, 166–197 (1965)
2. Bennett, J.: A Philosophical Guide to Conditionals. Oxford University Press, Oxford (2003)
3. Ciardelli, I., Roelofsen, F.: Generalized inquisitive semantics and logic (2009). http://sites.google.com/site/inquisitivesemantics/. Accessed Nov 2009
4. Edgington, D.: On conditionals. Mind **104**(414), 235–329 (1995)
5. Eells, E., Skyrms, B. (eds.): Probabilities and Conditionals: Belief Revision and Rational Decision. Cambridge University Press, Cambridge (1994)
6. Evans, J.S., Over, D.E.: If. Oxford University Press, Oxford (2004)

7. van Fraassen, B.C.: Probabilities of conditionals. In: Harper, W.L., Stalnaker, R., Pearce, G. (eds.) Foundations of Probability Theory, Statistical Inference, and Statistical Theories of Science. The University of Western Ontario Series in Philosophy of Science, vol. 1, pp. 261–308. D. Reidel, Dordrecht (1976)

8. Groenendijk, J., Roelofsen, F.: Inquisitive semantics and pragmatics (2009). http://sites.google.com/site/inquisitivesemantics/. Accessed Nov 2009

9. Groenendijk, J., Stokhof, M.: Studies in the semantics of questions and the pragmatics of answers. Ph.D. thesis, University of Amsterdam (1984)

10. Hájek, A., Hall, N.: The hypothesis of the conditional construal of conditional probability. In: Eells and Skyrms [5], pp. 75–110

11. Hamblin, C.L.: Questions. Australas. J. Philos. **36**(3), 159–168 (1958)

12. Jeffrey, R.C.: If. J. Philos. **61**, 702–703 (1964)

13. Karttunen, L.: Syntax and semantics of questions. Linguist. Philos. **1**, 3–44 (1977)

14. Kaufmann, S.: Conditioning against the grain: abduction and indicative conditionals. J. Philos. Logic **33**(6), 583–606 (2004)

15. Kaufmann, S.: Conditional predictions: a probabilistic account. Linguist. Philos. **28**(2), 181–231 (2005)

16. Kaufmann, S.: Conditionals right and left: probabilities for the whole family. J. Philos. Logic **38**, 1–53 (2009)

17. Kaufmann, S.: Unconditionals are conditionals. Handout, DIP Colloquium, University of Amsterdam. http://stefan-kaufmann.uconn.edu/Papers/Amsterdam2010_hout.pdf

18. Kaufmann, S.: Conditionals, conditional probabilities, and conditionalization. In: Zeevat, H., Schmitz, H.-C. (eds.) Bayesian Natural Language Semantics and Pragmatics. LCM, vol. 2, pp. 71–94. Springer, Cham (2015). doi:10.1007/978-3-319-17064-0_4

19. Khoo, J.: Probabilities of conditionals in context. Linguist. Philos. **39**, 1–43 (2016)

20. Lewis, D.: Probabilities of conditionals and conditional probabilities. Philos. Rev. **85**, 297–315 (1976)

21. Mellor, D.H. (ed.): Philosophical Papers: F. P. Ramsey. Cambridge University Press, New York (1990)

22. Oaksford, M., Chater, N.: Conditional probability and the cognitive science of conditional reasoning. Mind Lang. **18**(4), 359–379 (2003)

23. Oaksford, M., Chater, N.: Bayesian Rationality: The Probabilistic Approach to Human Reasoning. Oxford University Press, Oxford (2007)

24. Ramsey, F.P.: General propositions and causality. Printed in [21], pp. 145–163 (1929)

25. Rawlins, K.: (Un)conditionals: an investigation in the syntax and semantics of conditional structures. Ph.D. thesis, UCSC (2008)

26. Rawlins, K.: (Un)conditionals. Nat. Lang. Seman. **21**, 111–178 (2013)

27. Stalnaker, R.C.: A theory of conditionals. In: Harper, W.L., Stalnaker, R., Pearce, G. (eds.) IFS. Conditionals, Belief, Decision, Chance and Time. The Western Ontario Series in Philosophy of Science, vol. 15, pp. 41–55. Blackwell, Oxford (1968). doi:10.1007/978-94-009-9117-0_2

28. Stalnaker, R., Jeffrey, R.: Conditionals as random variables. In: Eells and Skyrms [5], pp. 31–46

29. Zhao, M.: Intervention and the probabilities of indicative conditionals. J. Philos. **112**, 477–503 (2016)

What Do Proper Names Refer to?

The Simple Sentence Puzzle and Identity Statements

Tomohiro Sakai[(✉)]

Waseda University, Tokyo, Japan
t-sakai@waseda.jp

Abstract. The purpose of this paper is to solve the simple sentence puzzle about proper names. (1) Superman leaps more tall buildings than Clark Kent. (2) Superman = Clark Kent. (3) Superman leaps more tall buildings than Superman. Even when (1) and (2) are true, (3) is false. It will be shown that this is not a real puzzle, because (i) (1) and (3) do not express singular propositions, and (ii) the identity statement in (2) only concerns singular propositions. In (1) and (3), the proper names refer to aspects of an individual at the level of explicature, while identity statements of the form $X = Y$ mean that Y can be substituted for X *salva veritate*, only in singular propositions about X /Y. Given this difference in reference between (1)/(3) and (2), the conjunction of (1) and (2) does not entail (3), in accordance with our intuition.

Keywords: Simple sentence puzzle · Identity statement · Substitution · Singular proposition · Proposition about aspects

1 The Simple Sentence Puzzle

Even when (1) and (2) are true, (3) seems to be false.

(1) Superman leaps more tall buildings than Clark Kent.
(2) Superman = Clark Kent
(3) Superman leaps more tall buildings than Superman.

Since (2) is true, (1) and (3) should have the same truth value. Intuitively, however, this is not the case. A similar point can be made for (4)–(6). The truth of (4) and (5) does not seem to entail that of (6).

(4) Clark Kent went into a phone booth, and Superman came out.
(5) Superman = Clark Kent
(6) Clark Kent went into a phone booth, and Clark Kent came out.

It is well known that such substitution failure can occur in intensional or opaque contexts, containing verbs of belief like *believe* or modal verbs like *want*.[1] Yet, in the above examples, there are no such problematic expressions; they constitute genuine

---

[1] Even if (i) and (ii) are true, (iii) can be false. (i) John believes that Cicero was a great orator. (ii) Cicero = Tully (iii) John believes that Tully was a great orator. Similarly, the truth of (ii) and (iv) does not entail that of (v). (iv) John wants to meet Cicero. (v) John wants to meet Tully.

© Springer International Publishing AG 2017
M. Otake et al. (Eds.): JSAI-isAI 2015 Workshops, LNAI 10091, pp. 15–26, 2017.
DOI: 10.1007/978-3-319-50953-2_2

extensional contexts. This is called the simple sentence puzzle about proper names [25–27].

In this paper, it will be shown that the simple sentence puzzle is not a real puzzle, because (i) the sentences in (1) and (4) do not express propositions about individuals i.e. singular propositions, but propositions about aspects of an individual, and (ii) identity statements of the form X = Y such as (2)/(4) only concern singular propositions.[2]

2 Basic Assumption

The semantics of proper names is an extremely controversial issue in the philosophy of language. In this paper, assuming that descriptivism as defended by Russell [20] has definitively been rejected, I will stick to the Millian view, the simplest view on proper names:

> [...] whenever the names given to objects convey any information, that is, whenever they have properly any meaning, the meaning resides not in what they denote, but in what they connote. The only names of objects which connote nothing are proper names; and these have, strictly speaking, no signification. ([15]: 42–43)

> A proper name is but an unmeaning mark which we connect in our minds with the idea of the object, in order that whenever the mark meets our eyes or occurs to our thoughts, we may think of that individual object. ([15]: 43)

According to the Millian view, proper names do not carry any "Sinn" in Frege's sense [8], nor are they disguised descriptions as claimed by Russell [20]; they only have "Bedeutung".[3] This view can be summarized as in the following textbook descriptions [14].

> [Names] have their meanings simply by designating the particular things they designate, and introducing those designata into discourse. (Let us call such an expression Millian name, since John Stuart Mill (1843/1973) seemed to defend the view that proper names are merely labels for individual persons or objects and contribute no more than those individuals themselves to the meanings of sentences in which they occur.) ([14]: 31–32)

3 Four Solutions

So far, four solutions have been proposed for the simple sentence puzzle: semantic solution [11], implicature-based solution [1], mistaken evaluation solution [3] and explicature-based solution [9, 16, 17]. All these solutions are compatible with the Millian view assumed here.

---

[2] "Aspects" are also called "(temporal) phases" or "modes of personification".

[3] This is consistent with Wittgenstein's remark ([28]: 3.3): "Nur der Satz hat Sinn: nur im Zuzammenhange des Satzes hat ein Name Bedeutung." (Only propositions have sense. Only in the nexus of a proposition does a name have sense.) This view sharply contrasts with descriptivism whereby the name *Romulus*, for example, is interpreted as a truncated description such as "a person who did such-and-such things, who killed Remus, and founded Rome" ([21]: 79).

3.1 Semantic Solution

In the semantic solution, it is assumed that such names as *Superman* or *Clark Kent* are ambiguous co-referential names [11]. By disambiguating them, we get different truth-conditions for sentences containing these names, depending on the meaning of the names chosen in context (Fig. 1).

ambiguous co-referential name
↓
disambiguation
↓
different truth-conditions

Fig. 1. Semantic solution

Koyama and Nakayama give the following definitions ([11]: 32).

(7) a. "A" and "B" are ambiguous co-referential names when there are weak names and standard co-referential names for A and B such that in some contexts it is appropriate to interpret A = A(−) or B = B(−), and in some contexts it is appropriate to interpret A = A(+) or B = B(+), where "A(−)" and "B(−)" are weak names and "A(+)" and "B(+)" are standard names.
b. A name is a standard name if and only if it stands for an individual and not for its temporal phase.
c. A name is a weak name for an individual object if and only if the name stands not for the object but for one of its temporal phases.

Given the definitions, *Superman* and *Clark Kent* are ambiguous co-referential names to the extent they are standard names in some contexts and weak names in others. Thus, in (1) and in (4), they are used as weak names as in "Superman (−) leaps more tall buildings than Clark Kent (−)" or "Clark Kent (−) went into a phone booth, and Superman (−) came out", whereas in (2)/(5), they are used as standard names as in "Superman (+) = Clark Kent (+)". Since the meanings of the names are not the same in (1)/(4) and (2)/(5), substituting *Clark Kent* for *Superman* or vice versa in these sentences leads to a fallacy of equivocation.

The semantic solution is at odds with the Modified Occam's Razor ([10]: 47), according to which senses are not to be multiplied beyond necessity [9]. Each time we encounter a puzzle like (1)–(3) or (4)–(6), the semantic solution posits ambiguous lexical items. However, the phenomena cannot be attributed to the idiosyncrasy of certain names, because, in principle, any name can exhibit the alleged ambiguity. Other things being equal, the phenomena should be accounted for by a general theory of proper names, rather than by the idiosyncrasy of certain words.

3.2 Implicature-Based Solution

The implicature-based approach assumes that (1) and (3) are both false and that (4) and (6) cannot differ in truth-condition, contrary to out intuition [1]. What differentiates (1)/(4) from (3)/(6) is their pragmatic implicature (Fig. 2).

linguistic meaning
↓
a unique truth-condition
↓
different implicatures

Fig. 2. Implicature-based solution

This solution goes against the Scope Principle ([19]: 271), according to which "[a] pragmatically determined aspect of meaning is part of what is said (and, therefore, not a conversational implicature) if -and, perhaps, only if - it falls within the scope of logical operators such as negation and conditionals" [9]. The differences between (1)/(4) and (3)/(6) fall within the scope of a conditional or a negation, as illustrated by (8)–(9).

(8) If Superman leaps more building than {<u>Clark Kent</u> /# <u>Superman</u>}, Lois will be happy.

(9) What happened was not that Clark Kent went into a phone booth, and <u>Clark Kent</u> came out, but that Clark Kent went into a phone booth, and <u>Superman</u> came out.

This suggests that the differences in questions should be accounted for at the level of explicature rather than at the level of implicature.

Before discussing the explicature-based solution, let us take a look at the mistaken evaluation solution.

3.3 Mistaken Evaluation Solution

The mistaken evaluation solution assumes, with the implicature-based solution, that (1) and (3) are both false and that (4) and (6) cannot differ in truth-value [3]. It denies, as against the implicature-based solution, that the intuitions come from the implicatures of the utterances. According to this position, our evaluations of those propositions for truth-value, and possible differences in truth-value, are mistaken. It explicitly denies what is called the Matching Proposition Principle, generally accepted in the literature, explicitly or implicitly.

(10) (MP) The Matching Proposition Principle:
Suppose that a competent, rational, relevantly well-informed speaker hears and understands an utterance U of a sentence, and judges U to have a (possible) truthvalue T. Then there is some proposition P such that:
(a) U either semantically expresses P or conversationally implicates P; and
(b) P has (possible) truth-value T." ([3]: 18)

Here are some of Braun and Saul's comments on their striking position.

If you had entertained [(2)], and considered its logical relations with (1) and [(3)], you might have realized that you made a mistake. But, for the reasons discussed above [= because you (quite reasonably) don't usually do so when you entertain sentences containing 'Superman' or 'Clark'], you didn't. ([3]: 27)

[...] you did not consider the fact that Superman is Clark, which might have given you pause." ([3]: 28)

[...] you may not have considered the consequences of combining (1) and [(3)] with the identity [=(2)], simply because you were already confident of your answer. Similar points could hold for [(4)] and [(6)]. We do not think that this would have been irrational on your part. We simply cannot take the time to draw out many of the logical consequences of the propositions we believe and entertain before judging whether an English sentence is true. Since in most other cases you quite reasonably do not make the identity substitutions, you quite reasonably failed to do so in this case. ([3]: 29)

To those who are not convinced by their argument, they go so far as to say:

Some ordinary speakers, however, might refuse to alter their initial judgments. They might persist in thinking that (1) is true and [(3)] is false. [...] Do such stubborn ordinary speakers lend any support to the semantic or implicature explanations? We think not. These stubborn ordinary speakers claim that they "meant" something about aspects when they uttered '(1) is true and [(3)] is false'. So they are making claims about their own utterances. (Braun & Saul 2002 ([3]: 26), emphases in the original).

Furthermore, such after-the-fact claims about what one "meant" by past utterances are often unreliable (as are many after-the-fact judgments about one's states of mind). So, we should not take for granted that these speakers' claims about their utterances are correct. ([3]: 34)

The problem with this approach is that it has only theory-internal grounds for calling certain speakers "stubborn ordinary speakers". Having discarded MP, Braun and Saul could equally call those who judge (11) or (12) to be true "stubborn ordinary speakers". But they don't. They only say that those who judge (1) above to be true are "stubborn ordinary speakers".

(11)   The earth is round.
(12)   The capital of Japan is Tokyo.

This is presumably because they just want the assumption that the conjunction of (1) and (2) entails (3) to be preserved. The adequacy of this assumption, however, should first be examined carefully. Otherwise, they would be forced to make "after-the-fact claims" on their part. The solution that we will examine in the next section questions the very assumption.

3.4 Explicature-Based Solution

In the explicature-based solution, it is assumed that proper names sometimes undergo a metonymic meaning shift and refer to aspects of an individual [9, 16, 17] (Fig. 3).

The metonymic meaning shift is a free pragmatic process [19, 20], and it affects the truth-condition of the utterance, even though the lexical meanings of the nouns remain constant.

linguistic meaning

↓

metonymic meaning shift

↓

different truth-conditions

Fig. 3. Explicature-based solution

Under this perspective, (1) and (4) express propositions about aspects of Superman/Clark as in (13) and (14) respectively.

(13) Clark/Superman's Superman-aspect leaps more tall buildings than Clark/Superman's Clark-aspect.

(14) Clark/Superman's Clark aspect went into the phone booth and his Superman-aspect came out.

Note that the meaning shift is an optional process. This makes it possible for the proper names in question to literally refer to individuals in some contexts.[4] For example, (15) and (16) constitute such contexts.[5]

(15) Superman is identical with Clark Kent.

(16) Wow, so sometimes Clark Kent wears a cape and leaps tall buildings.

Braun and Saul contend that the explicature solution raises a problem in connection with the distinction between enlightened and unenlightened contexts in which (1) and (4) are uttered, illustrated in Table 1 [3].

Given this distinction, it is predicted that an utterance of (1) semantically expresses a proposition about aspects only if (a) the speaker is enlightened and (b) the speaker is thinking about aspects. Yet, this prediction is clearly not borne out. Even unenlightened speakers who are not thinking about aspects can fully entertain the propositions expressed by (1) and (4).[6]

---

[4] This is an application of the Optionality Criterion given by Recanati (2004: 101): "Whenever a contextual ingredient of content is provided through a pragmatic process of the optional variety, we can imagine another possible context of utterance in which no such ingredient is provided yet the utterance expresses a complete proposition."

[5] The semantic approach we have seen in 3.1 above interprets *Superman* and *Clark Kent* in (15–16) as standard names, that is, as Superman (+) and Clark Kent (+).

[6] Yoshiki Nishimura (University of Tokyo) and Sayaka Hasegawa (Seikei University) object to this observation by saying that (1) and (4) express different propositions depending on whether the interpreters are enlightened or not, to the extent that unenlightened people's construal of *Superman* is different from enlightened people's (p.c., 2015). Here the term 'construal' is taken from Cognitive Linguistics [12]. This view, however, is incompatible with the Millian view on proper names as assumed in this paper, because it forces us to consider that the truth-conditional meaning of a proper name consists of an individual as well as its construal. If, on the contrary, the construal is taken to be external to the truth-condition of the proposition in which the name occurs, enlightened and unenlightened speakers can, as against Nishimura and Hasegawa's claim, fully entertain one and the same proposition, no matter how different their construal of the name may be.

Table 1. Enlightened context and unenlightened context

Enlightened context	Unenlightened context
The conversational participants are aware of the relevant double lives.	The conversational participants are not aware of such facts.
The conversational participants are in a position to make reference to aspects or modes of personification, and if their focus is on these rather than individuals, the propositions expressed by their utterances will involve aspects or modes of personification.	The conversational participants do not know that reference to aspects or modes of personification might be called for, and so utterances of the names refer only to individuals.

In what follows, we will propose another explicature-based approach by solving the problem raised by the enlightened and unenlightened distinction.

4 Yet Another Solution to the Puzzle

In this section, it will be shown that the simple sentence puzzle about proper names is not a real puzzle, because (i) the sentences in (1) and (4) do not express singular propositions (=propositions about an individual/individuals) but propositions about aspects of an individual as claimed by the explicature-based approach, and (ii) identity statements of the form X = Y such as (2)/(4) only concern singular propositions. Let us begin by identity statements.

4.1 Identity Statements

The semantics of identity statements can be defined as in (17)[7].

(17) X = Y (X is Y, X is identical with Y): In singular propositions about X /Y, Y can be substituted for X and vice versa *salva veritate*.

According to (17), identity statements speak about a relation between the two names X and Y, rather than about the individuals they denote. This kind of metalinguistic characterization goes back to Frege [7] and Wittgenstein ([28]: 6.24). Frege [8] rejected the meta-linguistic analysis of X = Y that he had once advocated [7] in favor of the well-known sense/reference analysis. Frege thought that the metalinguistic analysis could not account for the cognitive significance exhibited by X = Y.

---

[7] Wittgenstein summarizes the nature of the puzzle raised by identity statements as follows ([28]: 5.5303, emphases in the original): "Beiläufig gesprochen: Von *zwei* Dingen zu sagen, sie seien identisch, ist ein Unsinn, und von *Einem* zu sagen, es sei identisch mit sich selbst, sagt gar nichts." (Roughly speaking, to say of *two* things that they are identical is nonsense, and to say of *one* thing that it is identical with itself is to say nothing at all.) As will be shown below, the definition given in (17) enables us to solve the puzzle.

Is it [= equality] a relation? a relation between objects? or between names or signs for objects? I had assumed the latter in my *Begriffsschrift*. [...] What one wants to say by a = b seems to be that the signs or names "a" and "b" mean the same, so that one would be talking precisely about those signs, and asserting a relation between them. Thus a sentence a = b would no longer concern the issue itself, but only our way of using signs; we would not express any proper knowledge with it. However, in many cases that is precisely what we want to do. ([8]: 25–26)

Contrary to what Frege thought, however, X = Y does not have to express any proper knowledge on its own. It just serves as a generator of knowledge, rather than as a bearer of knowledge [22]. Thus, given X = Y as defined in (17), you know F (Y) if you know F (X), you know G (Y) if you know G (X), and so on, insofar as all these propositions are interpreted as singular propositions. A generation of knowledge of this kind can be seen in (16) above, uttered by Lois Lane.

What should be noted about (17) is that it restricts the substitutivity to singular propositions. *Y* can be substituted for *X* only when *X* refers to an individual, and not an aspect of that individual. Suppose that (2) is true. Then, we can get (3) from (1) only when *Superman* in (1) refers to an individual, and not an aspect of that individual. As the explicature-based approach claims [16, 17], however, *Superman* in (1) does not refer to any individual. This is exactly the reason why the substitution fails in (1) and (3), even if (2) is true.

4.2 Primary Reference to Aspects

The restriction of substitutivity to singular propositions indicated in (17) is a natural consequence of the fact that X and Y are names of the same individual. Without the notion of individual, identity statements of the form X = Y would be totally meaningless. This does not imply, however, that the primary reference of a proper name is an individual. As Evans says, "[a]ny producer can introduce another person into the name-using practice as a producer by an introduction ('This is NN')" ([4]: 376). It should be emphasized here that there is no *a priori* reason to restrict the reference of "this" to an individual, given the fact that the use of the name of an object is triggered by a particular interest in that object ([2]: Chap. 5, [4]: 379, [13]: Book 3, Ch.III, [18, 24]). It can then happen that you are interested only in certain aspects of an individual. This is the origin of the names given to aspects such as *Superman* or *Clark Kent*.

The notion of aspect is extensively discussed by Wittgenstein [29], especially in connection with the duck- rabbit figure below (Fig. 4).

Wittgenstein introduces the notion of aspect in the following passage.

I contemplate a face, and then suddenly notice its likeliness to another. I see that it has not changed; and yet I see it differently. I call this experience "noticing of an aspect." ([29]: II, xi; emphasis in the original)

In the duck-rabbit figure, seeing a duck constitutes a process of "noticing an aspect" different from seeing a rabbit. Accordingly, we can give each of these aspects a different name as in (18).

(18)   a. I name the duck John.

      b. I name the rabbit Mary.

Fig. 4. Duck-rabbit Figure

When I have performed the speech acts in (18), I can have the following intuitions.

(19) a. John is {OK a duck /# a rabbit /# a figure}.
      b. Mary is {# a duck /OK a rabbit /# a figure}.

What is crucial here is that the recognition of (20), or the relation illustrated in Table 2, does not affect the intuitions.

Table 2. The relation of the duck-rabbit figure, John and Mary

Duck-rabbit Figure	John
	Mary

(20) John = Mary

This suggests that it is not the case that first we refer to a neutral figure and then refer to a duck or a rabbit; we rather refer to a duck or a rabbit directly, without passing through a neutral figure. There is no metonymic meaning shift involved here; *John* and *Mary* refer directly to aspects of the figure.

This leads to a slight but important modification of the relation between Recanati's two levels of identification [19].

> An object is first identified – or rather localized – as a space-occupier, and then identified as a certain type of object. To think of an object, and to dub it, only the first level of identification is required. Thus if we associate the proper name 'Bozo' with the temporary file corresponding to the first level of identification, we may discover that Bozo was not a plane, after all, but a bird. ([19]: 171–172)

Recanati's observation is basically correct. It is equally true, however, that to think of an object, and to dub it, only the second level of identification is required, as shown by the example of the duck-rabbit figure. Looking at a bird, you can as well say "This is not Bozo". In the use of proper names, reference to aspects of an individual is a rule rather than an exception. We can freely associate a name with anything, an individual or an aspect, following our own interest.

We sometimes even reinterpret a name of an individual as a name of an aspect of that individual as in (21), or reinterpret a name of an aspect A as a name of another aspect B as in (22).

(21)   In short, Ichiro is not Ichiro, and it's fair to argue that his age has caught up with him, meaning little hope of a significant rebound.

http://m.espn.go.com/wireless/story?storyId=6850373&wjb=&pg=2&lang=ES

(22)   An essential part of Superman is hope. Without hope, Superman is not Superman. He is the ideal we aspire to, the man we should all strive to be, and he shows us the way. (http://coolaman4.rssing.com/chan-3588963/all_p114.html)

In (21), the speakers stops associating the name *Ichiro* to the individual in question and reinterpret the name as a name of an aspect of that individual [23]. In (22), the speaker stops associating the name *Superman* with an aspect of an individual and reinterpret the name as a name of another aspect (an aspect that has a hope) of that individual. If a proper name N were always associated with an individual or an aspect, utterances of the form N *is not* N would never be possible.[8]

4.3   Braun and Saul's Mistake

We can now return to the problem on the enlightened/unenlightened distinction addressed by Braun and Saul [3]. Contrary to their claim, the fact that the conversational participants are not aware of the relevant double lives does not entail that the conversational participants do not know that reference to aspects or modes of personification might be called for. In Table 3 below, we can refer directly to A or B, without recognizing that A and B are names of the same individual C, that is, without applying any metonymic meaning shift as assumed by the explicature-based approach.

Table 3. Direct reference to Superman/Clark Kent

Individual C	Aspect A = Superman
	Aspect B = Clark Kent

Braun & Saul believe that the recognition of (2) is required for competent speakers to entertain the propositions about aspects in (1) and (4). This is not the case, however. Competent speakers can perfectly entertain (1) and (4) without recognizing (2). The recognition of (2) is rather required for speakers to interpret (1) and (4) as singular propositions.

---

[8] In the approach defended here, sentences of the form X *is* Y as in (2) and sentences of the form X *is (not)* X as in (21–22) receive different interpretations. While the latter present the speakers' subjective view on extra-linguistic states of affairs, the former correspond to metalinguistic comments on the substitutivity of the terms X and Y as articulated in (17).

5 Conclusion

Since the conjunction of (1) and (2) /(4) and (5) does not entail (3) /(6), there is no puzzle about intuitions.

(1) Superman leaps more tall buildings than <u>Clark Kent</u>.
(2) Superman = Clark Kent
(3) Superman leaps more tall buildings than <u>Superman</u>.
(4) Clark Kent went into a phone booth, and <u>Superman</u> came
(5) Superman = Clark Kent
(6) Clark Kent went into a phone booth, and <u>Clark Kent</u> came out.

Presumably, confusion of the linguistic and social factors has complicated the issue on the reference of proper names. From a linguistic point of view, reference to aspects of an individual is a rule rather than an exception in the uses of proper names. This is what Braun and Saul overlooked when they criticized the explicature-based solution [3]. The preference for singular reference is only supported by the social convention whereby proper names are supposed to refer to individuals. Linguistically, however, a proper name can refer to anything, depending on the interest of its user. The only constraint is that X and Y be interpreted as names of individuals in identity statements of the form $X = Y$, which presuppose the notion of individual and hence of singular propositions. Only in social contexts can identity statements be interpreted properly, to the extent that the notion of individual is highly social by nature.

Acknowledgments. I would like to thank Daisuke Bekki (Ochanomizu University/JST CREST/National Institute of Informatics), who kindly gave me the opportunity to present this paper at LENLS 12, held at Keio University, Yokohama, Japan, on November 15–17 2015. This work was supported by JSPS KAKENHI Grant Number 25370437.

References

1. Barbar, A.: A pragmatic treatment of simple sentences. Analysis **60**, 300–308 (2000)
2. Bloom, P.: How Children Learn the Meanings of Words. The MIT Press, Cambridge (2000)
3. Braun, D., Saul, J.: Simple sentences, substitutions, and mistaken evaluation. Philos. Stud. **111**, 1–41 (2002)
4. Evans, G.: The Varieties of Reference. Oxford University Press, Oxford (1982)
5. Forbes, G.: How much substitutivity? Analysis **57**, 109–113 (1997)
6. Forbes, G.: Enlightened semantics for simple sentences. Analysis **59**, 86–91 (1999)
7. Frege, G.: Begriffsschrift. Nebert, Halle (1879)
8. Frege, G.: Über Sinn und Bedeutung. Zeitschrift für Philosophische Kritik 100, 25–50 (1892)
9. Fujikawa, N.: Namae ni Nan no Imi ga Arunoka (What's in a Name?: Philosophy of Proper Names). Keisoushobou (in Japanese) (2014)
10. Grice, P.: Studies in the Way of Words. Harvard University Press, Cambridge (1989)
11. Koyama, T., Nakayama, Y.: The simple sentence puzzle and ambiguous co-referential names. Ann. Jpn. Assoc. Philos. Sci. **10**(3), 127–138 (2001)

12. Langacker, R.W.: Cognitive Grammar. Oxford University Press, Oxford (2008)
13. Locke, J.: An Essay Concerning Human Understanding. Penguin Classics (1997 [1689])
14. Lycan, W.G.: Philosophy of Language: A Contemporary Introduction, 2nd edn. Routledge, New York (2000)
15. Mill, J.S.: A System of Logic. Longmans, London (1843)
16. Moore, J.: Saving substitutivity in simple sentences. Analysis **59**, 91–105 (1999)
17. Moore, J.: Did clinton lie? Analysis **60**, 250–254 (2000)
18. Noya, S.: Katarienu Mono Wo Kataru (Speaking the Unspeakable). Koudansha (in Japanese) (2011)
19. Recanati, F.: Direct Reference: From Language to Thought. Blackwell Publishers, Oxford (1993)
20. Recanati, F.: Literal Meaning. Cambridge University Press, Cambridge (2004)
21. Russell, B.: The Philosophy of Logical Atomism and Other Essays 1914-19 (The collected papers of Bertrand Russell 8). Allen & Unwin Ltd, London (1956)
22. Sakai, T.: L'énigme des énoncés d'identité de type « a = b » : Une solution grammaticale. Langue et littérature françaises **101**, 23–37 (2012)
23. Sakai, T.: Contextualizing tautologies: from radical pragmatics to meaning eliminativism. Engl. Linguist. **29–1**, 38–68 (2012)
24. Sato, N.: Retorikku No Kigouron (Semiotics of Rhetoric). Koudansha (in Japanese) (1993)
25. Saul, J.: Substitutions and simple sentences. Analysis **57**, 102–108 (1997)
26. Saul, J.: Substitution, simple sentence, and sex scandals. Analysis **59**, 106–112 (1999)
27. Saul, J.: Simple Sentences, Substitution, and Intuitions. Oxford University Press, Oxford (2007)
28. Wittgenstein, L.: Tractatus Logico-Philosophicus. Routledge & Kegan Paul, Side-by-side-by side edition at https://people.umass.edu/klement/tlp/ (1922)
29. Wittgenstein, L.: Philosophische Untersuchungen. In: Anscombe, G.E.M. (ed.) Philosophical Investigations. Basil Blackwell, Oxford (1953)

Particles of (Un)expectedness: Cantonese *Wo* and *Lo*

Yurie Hara[1(✉)] and Eric McCready[2]

[1] School of Creative Science and Engineering, Waseda University, Tokyo, Japan
yuriehara@aoni.waseda.jp
[2] Department of English, Aoyama Gakuin University, Shibuya, Japan
mccready@cl.aoyama.ac.jp

Abstract. Cantonese has a number of sentence-final particles which serve various communicative functions. This paper looks into two of the most frequently used particles, *wo3* and *lo1*. We propose that *wo3* and *lo1* are expressive items: *Wo3* indicates unexpectedness of the propositional content or the current discourse move, while *lo1* indicates expectedness of the propositional content or the current discourse move. We employ Default Logic to characterize the notion of (un)expectedness by normality conditionals. The analysis has a further implication on the Gricean Cooperative Principle in that the use of *wo3* and *lo1* makes reference to the general world knowledge which includes conditions on how the discourse should normally proceed.

Keywords: Particle · Expressive · Conventional implicature · Default logic · Expectedness · Discourse relation · Grice · Cooperative principle

1 Introduction

Cantonese has a number of sentence-final particles which serve various communicative functions. This paper looks into two of the most frequently used particles, *wo3*, *wo* with Tone 3 and *lo1*, *lo* with Tone 1 (simply *wo* and *lo* hereafter). In the framework of Conversation Analysis, Luke (1990) shows that the meaning of *wo* involves the violation of expectations. In (1), given that both she and her husband have straight personalities, C finds her son's behaviour puzzling and unexpected.

(1)   C: Me and my husband are both very straight and we don't like to lie and cheat others.

The research is partly supported by City University of Hong Kong Strategic Research Grant (7004334) awarded to the first author and by JSPS Kiban C Grant #25370441 awarded to the second. We would like to thank our student helpers, Peggy Pui Chi Cheng and Phoebe Cheuk Man Lam, Jerry Hok Ming So and Agnes Nga Ting Tam and the audience at LENLS2015 for helpful comments. We are also grateful to Shinichiro Ishihara, Kazuya Wada and Mutsuyo Wada for their help in statistics.

© Springer International Publishing AG 2017
M. Otake et al. (Eds.): JSAI-isAI 2015 Workshops, LNAI 10091, pp. 27–40, 2017.
DOI: 10.1007/978-3-319-50953-2_3

C: daan6hai6 koei5 le1   zau6 hou2 zung1ji3 gong3daai6waa6 go3 *wo3*
      but   he   PRT PRT very like   lie   PRT PRT
      'But he likes to lie very much.'
C: I don't know why so I want to ask for your advice how should I teach
      him?

(Luke 1990, p. 201)

In contrast, *lo* is the inverse of *wo*; it indicates expectedness rather than unexpectedness. In (2), if it is a common knowledge that Jimmy lies all the time, then it is obvious why A is so unhappy: Jimmy likes to lie very much.

(2)   Context: A is bothered by his son Jimmy lying all the time. A's hus-
       band, who also knows that Jimmy lies very often, asks A why she
       is so unhappy. A thinks it should be obvious to him, but A answers
       anyway:
A: Jimmy hou2 zung1ji3 gong3daai6waa6 *lo1*.
       Jimmy very like   lie   PRT
       'Jimmy likes to lie very much.'

Wo and *lo* can indicate not only the (un)expectedness of the content but also that of the current discourse move that the speaker makes. Our first case is the following simple discourse translated from Asher and Lascarides (2003, p. 412). (3) illustrates meta-level unexpectedness encoded by *wo*. A asks a question which presupposes that John has a job. In response to this, since B knows that the presupposition is false, B's assertion denies the presupposition rather than answering the question, "John's job is x." This is an unexpected response: A anticipates that B will answer the question rather than undermining its basis.

(3)   A: John zou6 mat1je5 je5   aa3?
            John do   what   things PRT
            'What's John's job?'
       B: koei5 mou5   zou6 je5 wo3
            he   not.have do   job PRT
            'He doesn't have a job.'

Conversely, if the addressee provides an answer to the questioner's question straightforwardly, the addressee's discourse move is an expected one. Thus, it is predicted that *lo* is suffixed in those cases. This prediction is indeed attested. Luke (1990) shows that *lo* is most commonly used in the second position in a Question-Answer sequence.

(4)   L: mm jau5-gei2daai6 aa3
            mm how-old   PRT
            'mm how old is he?'
       C: keoi5 gam1nin4 ee duk6 form-one sap6sei3 seoi3 *lo1*
            he   this-year um study First-form fourteen years PRT
            'He's in First Form this year he's fourteen.'   (Luke 1990, p. 121)

To recapitulate, *wo* indicates unexpectedness of the propositional content or the current discourse move, while *lo* indicates expectedness of the propositional content or the current discourse move. This introspection-based generalization is further supported by the two experiments summarized in the appendix. We propose that *wo* and *lo* are expressive items and employ Default Logic to characterize the notion of (un)expectedness by normality conditionals. The analysis has a further implication on the Gricean Cooperative Principle in that the use of *wo* and *lo* makes reference to the general world knowledge which includes conditions on how the discourse should normally proceed.

2   Proposal: Default Logic

This section formally characterizes the notion of (un)expectedness using normality conditionals and provides a default-logic analysis of *wo* and *lo*. In default logics (Reiter 1980), '>' is a so-called 'normality conditional' which indicates conclusions that can be drawn under normal circumstances, i.e. in the absence of defeaters. The semantics of '>' is given in (5). $*_M$ is a function mapping a world and a proposition into a proposition, thus $*_M(w, \llbracket A \rrbracket^M)$ indicates a set of worlds in M where things are normal with respect to A according to the defaults in w.

(5) $\llbracket A > B \rrbracket^M(w)$ is true iff $*_M(w, \llbracket A \rrbracket^M) \subseteq \llbracket B \rrbracket^M$

(Asher and Lascarides 2003)

The conditional '$\varphi > \psi$' can be read 'if φ, then normally it follows that ψ'.

The use of default logic to capture expectedness and unexpectedness in interpretation has a long history going back to the analysis of generics (e.g., Pelletier and Asher 1997) and progressive aspect (e.g., Asher 1992).[1] Generic sentences have exceptions, but they are taken to hold of ordinary class-exemplars.

(6)   a.   Birds fly.
       b. $Bx > Fx$

Progressive sentences give rise to the 'progressive paradox' on which the event being described is not completed, but they can be thought of as indicating that the event currently underway will culminate in a complete event of the relevant type.

(7)   a.   John is crossing the street.
       b.   (John is engaged in an activity a) &
             (a finishes $>$ a yields a street-crossing)

We characterize (un)expectedness using normality conditionals (for both content-level and metalevel unexpectedness), and the sets of them that we can view as making up our body of world knowledge. In this setting, an expected

---

[1] See also (Nute 1994; Horty 2014) for more on the properties of default logics themselves.

fact is one which corresponds to the consequent of a normality conditional in the knowledge base such that the antecedent of that conditional holds, while an unexpected fact is one which corresponds to the negation of the consequent:

(8)   Suppose that $\phi > \psi$ is part of world knowledge and ϕ is known. Then,

    a. $\neg\psi$ is unexpected, **Unexpected**$(\neg\psi)$ and

    b. ψ is expected, **Expected**(ψ).

Thus, $\neg\psi$ is unexpected if it conflicts with expectations about the normal course of events and ψ is expected if it is a natural consequence given the background knowledge. Using this notion of (un)expectedness (8), we propose that *wo* and *lo* are conventional implicature (CI) inducers (Potts 2005; McCready 2010) which project a pair of independent meanings. One is an at-issue meaning which is the input of the suffixed utterance passed on unmodified, the prejacent proposition or discourse move. The other is an expressive component or CI meaning which marks the prejacent proposition or discourse move as (un)expected.

(9)   a. $[\![p\text{-}\mathrm{WO}]\!] = \langle p, \mathbf{Unexpected}(\neg q)\rangle$

    b. $[\![p\text{-}\mathrm{LO}]\!] = \langle p, \mathbf{Expected}(q)\rangle$

In case of content-level (un)expectedness, q is the consequent of the normality conditional $r > p$. Thus, $q = p$. In case of meta-level (un)expectedness, q is the consequent of the normality conditional regarding discourse relations, e.g., $(\mathcal{M}_1$ is a question$) > Answer(\mathcal{M}_1, \mathcal{M}_2)$ (see below for illustrations).

2.1   Content-level (Un)expectedness

Consider (1), repeated here as (10), and suppose that our world-knowledge assumptions include $straight(m) \wedge straight(f) \wedge son.of(s, m \oplus f) > straight(s)$.

(10)   C: Me and my husband are both very straight and we don't like to lie and cheat others.

        C: daan6hai6 koei5 le1   zau6 hou2 zung1ji3 gong3daai6waa6 go3

             but   he   PRT PRT very like   lie   PRT

             wo3

             PRT

             'But he likes to lie very much.'

        C: I don't know why so I want to ask for your advice how should I teach him?

                                                  (Luke 1990, p. 201)

C's first utterance sets up that the antecedent $straight(m) \wedge straight(f) \wedge son.of(s, m \oplus f)$ is known. Thus, because the antecedent is satisfied, but nonetheless the consequent is denied, $(\neg straight(s))$, (8-a) applies and C's son's actual behaviour described by the *wo*-suffixed utterance is unexpected, **Unexpected**$(\neg straight(s))$. The role of *wo* is then to mark this fact and that the speaker recognizes it. Thus, the *wo*-suffixed utterance has two discourse effects.

One is a plain assertion providing a piece of information that C's son likes to lie very much, while the other is the introduction of an expressive meaning which indicates that the content of the assertion is unexpected given the extant background knowledge.

Similarly, in (2), repeated here as (11), it is a reasonable assumption that people who lie all the time like to lie ($p > q$).

(11)   Context: A is bothered by his son Jimmy lying all the time. A's husband, who also knows that Jimmy lies very often, asks A why she is so unhappy. A thinks it should be obvious to him, but A answers anyway:

   A: Jimmy hou2 zung1ji3 gong3daai6waa6 *lo1*.
        Jimmy very like   lie   PRT
        'Jimmy likes to lie very much-LO.'

The context says that it is a common knowledge that Jimmy lies all the time (p). By (8-b), thus, it is expected that Jimmy likes to lie, **Expected**(q).

2.2   Meta-level (Un)expectedness

In defining discourse-level normality conditionals, we employ the notion of discourse relations in Asher and Lascarides' (2003) Segmented Discourse Representation Theory (SDRT). The second discourse move is expected when it establishes an expected discourse relation, while it is unexpected when there is no expected discourse relation as such. Let us go back to the example in (3), repeated here as (12), in which a question is followed by a non-answer.

(12)   A: John zou6 mat1je5 je5   aa3?
             John do   what   things PRT
             'What's John's job?'

        B: koei5 mou5   zou6 je5 *wo3*
             he   not.have do   job PRT
             'He doesn't have a job.'

In understanding the use of *wo* in B's utterance, it is a reasonable assumption that if the speaker asks a question, then normally the hearer should provide an answer as in (13). Thus, by *wo*-suffixing, B is indicating that his discourse move violates (13).

(13)   Let $\mathcal{M}_1$ and $\mathcal{M}_2$ be discourse moves by agents A and B, respectively.
        ($\mathcal{M}_1$ is a question) $> Answer(\mathcal{M}_1, \mathcal{M}_2)$

The same discourse normality conditional is deployed in the frequent *lo*-suffixing in Question-Answer pairs as in (4), repeated here as (14).

(14)   L: mm jau5-gei2daai6 aa3
             mm how-old   PRT
             'mm how old is he?'

> C: keoi5 gam1nin4 ee duk6 form-one   sap6sei3 seoi3 *lo1*
>   he   this-year um study First-form fourteen years PRT
>   'He's in First Form this year he's fourteen.'   (Luke 1990, p. 121)

Since L's move is a question and C is straightforwardly answering the question, there is an expected *Answer* relation between the two moves.

Let us look at other discourse-level normality conditionals. *Wo* can mark a discourse move as unexpected when an assertion on the part of one conversational participant is followed by a non-acceptance of the assertion by the other. In (15), C challenges the proposal made by L rather than accepting it.

(15)   L: Especially boys, do tend to be more active and energetic.
        C: koei5 jau6 m4 hai6 hou2 wut6joek3 *wo3*
             he   really not be very energetic PRT
             'well but he isn't really very energetic.'   (Luke 1990, p. 217)

Since Stalnaker (1978), an assertion is analyzed as a proposal made by the speaker to add a proposition to the Common Ground. The proposition can enter the Common Ground only after it is accepted by the hearer. Given this fact, for the speaker to gain any potential benefit from an assertion, the hearer must choose to believe the content the speaker has proffered. Based on these considerations, it is a reasonable assumption that if the speaker makes an assertion, then normally the hearer should accept it, as codified in (16). The use of *wo* indicates that the speaker is aware of the unexpectedness of her own discourse move.

(16) $(\mathcal{M}_1$ is an assertion$) > Rel(\mathcal{M}_2, \mathcal{M}_1)$ such that *Rel* is veridical in its second argument.

Conversely, if the respondent assents to the content of the first move, the second move is expected and *lo* is suffixed as seen in (17).

(17)   M: About going for a walk tomorrow, R said, "God I hate walking what's the point".
        J: gam2 mai6 m4-hou2 haang4 *lo1*
             so   then don't   walk   PRT
             'Don't walk then.'
        M: gam2 mai6 m4-hou2 haang4 *lo1*
             so   then don't   walk   PRT
             'Don't walk then.'   (Luke 1990, pp. 141–142)

Furthermore, Luke (1990) claims that *lo* can be characterized as a completion proposal as it is often suffixed at the end of storytelling as follows:

(18)   J: And we thought yesterday there would still be a bus running I mean after the meal.
        M: There wouldn't be.

J: And then we came out, we realized (when) we went to the stop for
     service 40, it was already-

M: There weren't buses anymore.

J: There weren't any service 40M anymore.

M: So that the best thing to do is a taxi, right?

J: aam1aam1   zau6-jau5 dik1si2 mai6 zit6-zo2 dik1si2 ceot1-lei4
     there-happen to-be   taxi   so   stopped taxi   come-out
     lo1
     PRT
     'There happened to be a taxi so we took a taxi and came out.

M: I see.   (Luke 1990, p. 165)

We can include this function as one kind of expectedness. Suppose that a
storytelling discourse is like a script: There is a preset sequence of events or
activities, s which should culminate with some particular event, which would be
the final bit of the narration. If we have a script of length n and events up to
$n - 1$ have all been completed, event n is expected as a completion of the story.
This could be codified in a default rule:

(19)   Suppose: $\exists s[|s| = n \land \exists e_1 \ldots e_{n-1}[e_{n-1}$ realizes step $n - 1$ of $s]]$
       and $\mathcal{M}_{n-2} = \text{ASSERT}(e_{n-2})$ and $\mathcal{M}_{n-1} = \text{ASSERT}(e_{n-1})$
       Then, $Narration(s, \mathcal{M}_{n-2}, \mathcal{M}_{n-1})$ $>$ $Conclude(s, \mathcal{M}_n)\&\mathcal{M}_n$ $=$
       $\text{ASSERT}(e_n)$

Asher and Lascarides (2003) defines *Narration* as follows: We infer *Narration*
when there is a sequence such that the first event realized by α occasions the
event realized by β; here, λ is the larger discourse constituent in which β is to
be embedded.

(20)   Suppose: ? is a skolem constant, and α, β are discourse moves.
       $(?(\lambda, \alpha, \beta) \land occasion(\alpha, \beta)) > Narration(\lambda, \alpha, \beta)$
                        (Adapted from Asher and Lascarides 2003, p. 201)

We infer $occasion(\alpha, \beta)$, if two events described have certain properties, ϕ
and ψ, which are related to each other:

(21)   Suppose: ? is a skolem constant, α, β are discourse moves and ϕ and ψ
       are event predicates.
       Then, $(?(\alpha, \beta) \land \phi(\alpha) \land \psi(\beta)) > occasion(\alpha, \beta)$
                        (Adapted from Asher and Lascarides 2003, p. 201)

We define *Conclude* as follows: If α describes the final bit of an event
sequence, α concludes the sequence:

(22)   Suppose: s is a preset sequence of events or activities and $|s| = n$
       Then, $(?(s, \alpha) \land \alpha = \text{ASSERT}(e_n)) > Conclude(s, \alpha)$

As an anonymous reviewer pointed out, our current definitions presuppose that the interlocutors know beforehand the size of the event sequence, i.e., when the narration of the sequence will finish. Justification of this presupposition is left for future research.

3   Gricean Cooperative Principle

We relate the discourse normality conditionals given above to the Gricean (1975) Cooperativity principle, "make your contribution such as it is required, at the stage at which it occurs, by the accepted purpose or direction of the talk exchange in which you are engaged". Asher and Lascarides provide a default logic-based statement of Cooperativity as follows:

(23)   Asher and Lascarides's (2003) `Cooperativity`
     a.   Agent B normally adopts agent A's goals,
     b.   If B doesn't for whatever reason, then he normally indicates this to A.

Note that the original Gricean principle is entirely based on the speaker's actions and intentions, thus the hearer only played an indirect role, while in (23) both A and B's intentions are taken into account. Indeed, as we have already seen above, there must be Cooperativity constraints on hearer behavior as well: if asked a question, she should answer (13). If proposed a statement, she should accept (16).

Suppose now that a language has an expression which carries an expressive meaning of (un)expectedness. If a speaker chooses not to use this expression, it will implicate that she does not believe that what she is saying is (un)expected (cf., MAXIMIZE PRESUPPOSITION in Schlenker 2012). In the case of metalevel (un)expectedness, where expected behavior is imposed by social and conversational norms, this will implicate that she is a defective speaker, which is a priori undesirable as the speaker is likely to accrue a reputation for bad linguistic behavior (c.f., McCready 2015). Thus, the use of *wo* and *lo* in metalevel-(un)expected contexts is dictated by general Gricean considerations about cooperativity.

4   Conclusion

4.1   Summary

We proposed analyses for the content-level and metalevel (un)expectedness expressively marked by *wo* and *lo*, on the basis of default expectations about normal courses of events and the interaction of such expectations with Gricean cooperativity.

4.2   Future Directions

There are several future directions for research related to this analysis. First, it would be fruitful to investigate the lexical encoding of (un)expectedness cross-linguistically. Some languages are claimed to have mirativity markers which indicate that the prejacent proposition is surprising (DeLancey 1997). Also, Japanese has the sentence-final particle *yo* which indicates the informativeness of the utterance (McCready 2009; Davis 2009). As pointed out by an anonymous reviewer, however, we do not know any other languages which have a single lexical object that expresses (un)expectedness at the level of both propositional content and discourse moves.

Second, what would happen when the expectedness of the propositional content and the discourse move conflicts? Incidentally, it is possible to combine the two particles, but only in the *lo-wo* order (note also that the tone of *lo* is different, Tone 3 not 1):

(24)   lei5 nam2-haa2 seng4   saa1aa6 jan4   dou1 mei6 git3   fan1 *lo3wo3*
        you consider   almost 30   person still not   marry PRT
        'Just think, someone who's almost thirty but is not married yet.'

(Luke 1990, p. 213)

Further investigations of the patterns of felicity and preference in cases of 'conflict' will shed new light on possible linguistic encodings of defaults in the world knowledge and discourse structures.

A   Experiments

A.1   Experiment I: Naturalness Rating

The predictions for the distribution of particles and context are as follows:

(25)   a.   *Lo*-utterances should be rated more natural than *wo*-utterances in EXPECTED contexts.
        b.   *Wo*-utterances should be rated more natural than *lo*-utterances in UNEXPECTED contexts.

The purpose of Experiment I is to verify these predictions.

Method

Stimuli. The stimuli had two fully-crossed factors—contexts (COMMON GROUND/EXPECTED/UNEXPECTED) and sentence-final particles (*laa/lo/wo*), which resulted in nine conditions.

(26)   Contexts:

a.   COMMON GROUND context
A gok3dak1 keoi5 go3 zai2 Jimmy sing4jat6 gong2daai6waa6
hou2 kwan3jiu2 A soeng2 man6 keoi5 go3 pang4jau5 B ji3gin3
tung4maai4 check haa5 B hai6mai6 dou1 zi1dou3 Jimmy zung1ji3
gong2daai6waa6
'A is bothered by his son Jimmy lying all the time. A wants to have
some advice from her friend B and wants to check if B also knows
that Jimmy likes to lie:'

b.   EXPECTED context
A gok3dak1 keoi5 go3 zai2 Jimmy sing4jat6 gong2daai6waa6
hou2 kwan3jiu2 A ge3 lou5gung1 dou1 zi1dou3 Jimmy sing4jat6
gong2daai6waa6 keoi5 man6 A dim2gaai2 gam3 m4 hoi1sam1 A
gok3dak1 gam2 deoi3 keoi5 lai4 gong2 jing1goi1 hou2 ming4hin2
daan6hai6 A jing4jin4 wui4daap3 waa6
'A is bothered by his son Jimmy lying all the time. A's husband,
who also knows that Jimmy lies very often, asks A why she is so
unhappy. A thinks it should be obvious to him, but A answers anyway:'

c.   UNEXPECTED context
A waa6 bei2 keoi5 go3 pang4jau5 B zi1 A tung4 A go3
lou5gung1 hai6 sing4sat6taan2baak6 ge3 jan4 so2ji5 keoi5 m4
ming4baak6 dim2gaai2 keoi5dei6 go3 zai2 Jimmy gam3 zung1ji3
gong2daai6waa6
'A tells her friend B that A and her husband are straight shooters so
she doesn't understand why their son Jimmy likes to lie very much.'

(27)   Target Sentences:

a.   *laa*-utterance

Jimmy	hou2	zung1ji3	gong2daai6waa6	laa1	pei3jyu4	kam4jat6
Jimmy	very	like	lie	PRT	for-example	yesterday

keoi5	mou5	faan1	hok6	daan6hai6	keoi5	waa6	ngo5	zi1	keoi5
he	no	return	school	but	he	say	me	know	he

jau5	faan1dou3
have	return

'Jimmy likes to lie very much. For example, he didn't go to school
yesterday but he told me that he did.'

b.   *lo*-utterance

Jimmy	hou2	zung1ji3	gong2daai6waa6	lo1
Jimmy	very	like	lie	PRT

'Jimmy likes to lie very much.'

c.   *wo*-utterance

Jimmy	hou2	zung1ji3	gong2daai6waa6	wo3
Jimmy	very	like	lie	PRT

'Jimmy likes to lie very much.'

Each of the nine conditions had 12 items, resulting in 108 target sentences (12 items * 9 conditions). 36 questions from another experiment were also included.

Procedure. The rating experiment was conducted in a quiet meeting room at City University of Hong Kong. The stimuli were presented in Chinese characters by Qualtrics.[2] The first page of the test showed the instructions.

In the main section, the participants were asked to read each stimulus, and then judge the naturalness of the stimuli on a 7-point scale (provided in Chinese characters): from "7: very natural" to "1: very unnatural".

The main experiment was organized into 12 blocks. Each block contained 9 items. None of the stimuli were repeated. To avid minimal pair sentences from appearing next to each other, the order of the blocks and the stimuli within each block were randomized by the Qualtrics software.

Participants. Ten native speakers of Cantonese participated in the rating experiment. They were undergraduate students recruited from City University of Hong Kong and received 80 Hong Kong dollars as compensation.

Statistics. The responses were recorded as numerical values: from very natural=7 to very unnatural=1. Context types and particle types were fixed factors. To analyze the results, a general linear mixed model (Baayen 2008; Baayen et al. 2008; Bates 2005 was run using the lmerTest package (Kuznetsova et al. 2015) implemented in R (R Core Team 2015). Context types and particle types were the fixed factors. Speakers and items were the random factors. The p-values were calculated by the Markov chain Monte Carlo method using the LanguageR package (Baayen 2013).

If the naturalness of the particles depends on the type of context, then the dependency is expected to result in a significant interaction between contexts and particles.

Result. Figure 1 shows the average naturalness ratings in each condition. The discussion above leads to the prediction that *lo*-utterances are more natural in EXPECTED contexts than *wo*-utterances. This prediction was confirmed ($t = -2.695$, $p < 0.001$). In UNEXPECTED contexts, *wo*-utterances are more natural in EXPECTED contexts than *lo*-utterances ($t = -1.941$, $p < 0.1$).

---

[2] Qualtrics is a web-based system that conducts online surveys. Version 45634 of the Qualtrics Research Suite. Copyright©2013 Qualtrics. Qualtrics and all other Qualtrics product or service names are registered trademarks or trademarks of Qualtrics, Provo, UT, USA. http://www.qualtrics.com.

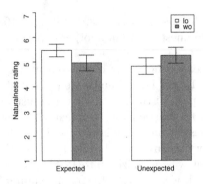

Fig. 1. Average naturalness ratings

A.2   Experiment II: Force-Choice

In Experiment II, Predictions parallel to Experiment I are attested in a force-choice experiment.

(28)   a.   *Lo*-utterances should be selected in EXPECTED contexts more than in UNEXPECTED contexts.
       b.   *Wo*-utterances should be selected in UNEXPECTED contexts more than in EXPECTED contexts.

Method

Stimuli. The same contexts and sentences as Experiment I are used. There were 12 items and each question had 3 contexts (COMMON GROUND/EXPECTED/UNEXPECTED), resulting in 36 questions (12 items * 3 contexts). 108 questions from another experiment were also included.

Procedure. In the main section, the participants were asked to read each context, and then select the most natural utterance among the three choices, utterances suffixed with *laa/lo/wo*.

The main experiment was organized into 12 blocks. Each block contained 3 items. The other aspect of the procedure was the same as Experiment I.

Participants. Ten native speakers of Cantonese who did not participate in Experiment I participated in the force-choice experiment. The other aspect of the procedure was the same as Experiment I.

Statistics. The responses were recorded as categorical data. To analyze the results, `chisq.test()` was run implemented in R (R Core Team 2015). If the naturalness of particle depends on the type of context, then the dependency is expected to result in a significant interaction between contexts and particles.

Table 1. Total of Force-choice test

	lo	wo
Expected	74	17
Unexpected	37	40

Result. Table 1 shows the total of responses to each condition. *Lo*-utterances were selected in EXPECTED contexts more than in UNEXPECTED contexts ($\texttt{X-squared} = 12.333, p < 0.001$). *Wo*-utterances were selected in UNEXPECTED contexts more than in EXPECTED contexts ($\texttt{X-squared} = 9.2807, p < 0.01$).

References

Asher, N.: A default, truth-conditional semantics for the progressive. Linguist. Philos. **15**, 463–508 (1992)

Asher, N., Lascarides, A.: Logics of Conversation. Cambridge University Press, Cambridge (2003)

Baayen, H.R.: Analyzing Linguistic Data: A Practical Introduction to Statistics Using R. Cambridge University Press, Cambridge (2008)

Baayen, H.R., Davidson, D., Bates, D.M.: Mixed-effects modeling with crossed random effects for subjects and items. J. Mem. Lang. **59**, 390–412 (2008)

Baayen, R.H.: LanguageR: data sets and functions with "Analyzing Linguistic Data: a practical introduction to statistics" (2013). http://CRAN.R-project.org/package=languageR. r package version 1.4.1

Bates, D.: Fitting linear mixed models in R. R News **5**, 27–30 (2005)

Davis, C.: Decisions, dynamics and the Japanese particle *yo*. J. Semant. **26**, 329–366 (2009)

DeLancey, S.: Mirativity: The grammatical marking of unexpected information. In: Plank, F. (ed.) Linguistic Typology, vol. 1, pp. 33–52. De Gruyter, New York (1997)

Grice, H.P.: Logic and conversation. In: Cole, P., Morgan, J. (eds.) Syntax and Semantics. Speech Acts, vol. 3, pp. 43–58. Academic Press, New York (1975)

Horty, J.: Reasons as Defaults. Oxford University Press, Oxford (2014)

Kuznetsova, A., Brockhoff, P.B., Christensen, R.H.B.: lmerTest: tests in linear mixed effects models (2015). http://CRAN.R-project.org/package=lmerTest. r package version 2.0-29

Luke, K.K.: Utterance particles in Cantonese Conversation. J. Benjamins Pub. Co., Amsterdam (1990)

McCready, E.: What man does. Linguist. Philos. **31**, 671–724 (2009)

McCready, E.: Varieties of conventional implicature. Semant. Pragmatics **3**(8), 1–57 (2010)

McCready, E.: Reliability in Pragmatics. Oxford University Press, Oxford (2015)

Nute, D.: Defeasible logic. In: Gabbay, D., Hogger, C. (eds.) Handbook of Logic for Artificial Intelligence and Logic Programming, vol. III, pp. 353–395. Oxford University Press, Oxford (1994)

Pelletier, J., Asher, N.: Generics and defaults. In: van Benthem, J., ter Meulen, A. (eds.) Handbook of Logic and Language, pp. 1125–1177. MIT Press, Cambridge (1997)

Potts, C.: The Logic of Conventional Implicatures. Oxford Studies in Theoretical Linguistics. Oxford University Press, Oxford (Revised 2003 UC Santa Cruz PhD thesis) (2005)

R Core Team: R: A Language and Environment for Statistical Computing. R Foundation for Statistical Computing, Vienna, Austria (2015). http://www.R-project.org/

Reiter, R.: A logic for default reasoning. Artif. Intell. **13**, 81–132 (1980)

Schlenker, P.: Maximize presupposition and Gricean reasoning. Nat. Lang. Semant. **20**, 391–429 (2012)

Stalnaker, R.: Assertion. In: Cole, P. (ed.) Syntax and Semantics, pp. 315–322. Academic Press, New York (1978)

Rhetorical Structure and QUDs

Julie Hunter[1][(✉)] and Márta Abrusán[2]

[1] IRIT, Université Paul Sabatier, Toulouse, France
juliehunter@gmail.com
[2] IRIT, CNRS, Toulouse, France

Abstract. We consider two hypotheses about how rhetorical structure and QUD structure might come together to provide a more general pragmatic theory. Taking SDRT ([2]) and some basic principles from [18]'s QUD framework as starting points, we first consider the possibility that rhetorical relations can be modelled as QUDs, and vice versa. We ultimately reject this hypothesis in favor of the possibility that QUDs correspond to topics that bind together the members of complex discourse units.

Theories of rhetorical structure [2,13] and theories of discourse structure centered around a Question Under Discussion or QUD [8,18] share many of the same principles.[1] Both approaches hold that the interpretation of a given sentence or *elementary discourse unit* (EDU) depends in part on that EDU's relation to other moves that have been made in the same discourse.[2] Both also hold that a discourse context must therefore keep track of not only (some subset of the) prior discourse moves but also certain structural relations between these moves. These structural relations are believed to play an integral role in discourse coherence and the relevance of individual discourse moves, and as such, to influence various semantic and pragmatic phenomena, including ellipsis of various sorts, anaphora (rhetorical theories), and prosody (QUD theories).

To clarify their potential contribution to analyses of semantic and pragmatic phenomena, we need a better understanding of how rhetorical and QUD frameworks are related: are they fundamentally distinct, but complementary theories, or do they aim to model the same phenomena? If the latter, do they end up describing two sides of the same coin or are they in conflict? The goal of this paper is to propose and evaluate two hypotheses about how the two frameworks might correspond, so that we can eventually come to a better understanding of the respective roles that these frameworks play with regard to phenomena from the semantics-pragmatics interface. The first hypothesis, which we develop and ultimately reject in Sect. 2.1, is that there is a direct correspondence between

---

We gratefully acknowledge support from the ERC grant 269427 and Marie Curie FP7 Grant, PCIG13-GA-2013-618550.

[1] See also [21] for a framework that combines QUDs and rhetorical relations.

[2] How a discourse is broken down into basic units can vary from theory to theory, but all rhetorical theories and QUD theories must take a stand on what constitutes a basic discourse move.

M. Otake et al. (Eds.): JSAI-isAI 2015 Workshops, LNAI 10091, pp. 41–57, 2017.
DOI: 10.1007/978-3-319-50953-2_4

instances of discourse relations and QUDs in a discourse. The second hypothesis, which we present in Sect. 2.2, is that QUDs correspond to *complex discourse units* in a discourse graph. We judge this hypothesis to be more promising.

Our discussion focuses on two particular theories. On the side of rhetorical structure, we adopt *Segmented Discourse Representation Theory* [2], which we briefly introduce in Sect. 1.1. Of all of the rhetorical theories, SDRT is the most developed from a semantic point of view; it assigns semantics to each of its discourse relations and posits semantic constraints on how its hierarchical discourse representations, which capture the contents of full discourses, can be constructed. These choices are fuelled by concerns about anaphora resolution, presupposition, temporal interpretation and other phenomena relevant to the semantics-pragmatics interface, putting SDRT in an ideal position to be compared to alternative semantic/pragmatic theories of discourse structure. On the side of QUDs, we take inspiration from [18] whose proposal, we think, is closest in spirit to SDRT. Like SDRT it aims to capture relations between discourse moves and provide hierarchical discourse structures that model global features of the discourse context. However, because the account outlined in [18] less thoroughly developed than SDRT, the possible formulations of QUD frameworks that we consider in Sect. 1.2 sometimes move beyond basic principles laid out in [18] in directions that we think deserve consideration, but which may not be endorsed by Roberts.

1   Background on SDRT and QUD

1.1   A Very Brief Introduction to SDRT

A fundamental principle of rhetorical theories, including SDRT, is that the relations that utterances stand in to one another affect on the one hand the interpretation of the discourse in which the utterances figure and on the other, the interpretation of the utterance contents themselves. Consider (1):

(1)   I missed my meeting this morning. My car broke down.

Seeking a connection between the two parts of (1), an addressee will naturally understand the content of the second sentence as providing an explanation for the content of the first. This interpretation of the discourse comes with its own truth conditions: it is true just in case the speaker was late for her meeting, her car broke down, and the latter event was the cause of the former. At the same time, the inferred causal connection affects the interpretation of the two sentences in (1). Both sentences are in the past tense, which indicates that the events that they describe occurred before the speech time. The inferred causal relation between the two events entails that in addition, the second event must have occurred before the first—a cause must occur before its effect. This sequence of events does not follow from the tense of the verbs and the order in which the events are described in the discourse, but only from understanding how the two sentences are coherently or rhetorically related. Of course, in the absence of an explicit marker for causality (e.g. *because*), this causal connection is at best an implicature; nevertheless, speakers regularly accept and act on such implicatures. SDRT aims to model such connections by developing rules for constructing logical forms for discourses and by providing models to interpret these logical forms.

To capture the rhetorical connections in a discourse, rhetorical theories must accomplish three tasks. First, the discourse must be segmented into EDUs according to rules set by the theory. In SDRT, the aim is to have each EDU denote a single eventuality. The next task is to attach each EDU to some other part of the discourse. SDRT maintains that a new move can only be a coherent extension of a given discourse if it is relevant to some other move that has been made previously; thus, each EDU will be attached to at least one other EDU that was discourse prior to it, with the exception of the initial EDU, which will only have attachments to later moves. In this way, a rhetorical relation is a kind of *anaphoric* relation, but one that holds between discourse units rather than referring expressions and discourse referents. The final task to be accomplished, which is in practice accomplished in tandem with labelling and even segmentation at times, is the labelling of each discourse attachment with a rhetorical relation (Explanation, Elaboration, Contrast, Narration, etc.).

Attachment and labelling involve default reasoning about the contents of the EDUs involved and world knowledge. Importantly, however, they also take into account *global* features from the discourse. For example, because discourse relations are often implicatures, the inference of an attachment/label that might otherwise be justified can be blocked by information carried by other EDUs in the discourse. For this reason, SDRT provides defeasible rules for inferring discourse relations and stresses the need for global constraints on the development of a discourse structure.

One such global constraint, which affects EDU attachment, is the Right Frontier Constraint (RFC). SDRT, along with other discourse theories [15,16], posits that even if the content of a new EDU satisfies the semantic conditions for being attached to some other EDU in a discourse, this connection is only coherent if the prior EDU is *accessible*[3], that is, if it is on the Right Frontier (RF). Compare (2-a)-(2-c):

(2)   a.   [John speaks German.]$_{\pi_1}$ [He can translate for you while you're in Berlin.]$_{\pi_2}$

       b.   [John speaks German]$_{\pi_{1'}}$ [and his sister speaks French.]$_{\pi_{2'}}$?? [He can translate for you while you're in Berlin.]$_{\pi_{3'}}$

       c.   [John speaks German.]$_{\pi_{1''}}$ [(because) He lived in Stuttgart for 10 years.]$_{\pi_{2''}}$ [He can translate for you while you're in Berlin.]$_{\pi_{3''}}$

While π_1 and π_2 in (2-a) are related by Result—a connection supported by the contents of the EDUs and world knowledge about what being able to translate

---

[3] SDRT does allow for violations of the RFC in cases that [1] calls *discourse subordination,* but such violations need to be explicitly signalled, e.g. *Let's go back to your first point.*

in Berlin would normally require—the same connection is blocked for $\pi_{1'}$ and $\pi_{3'}$ in (2-b). (2-c), however, shows that it is not blocked by just any intervening EDUs.

SDRT defines the RF so as to reflect facts about accessibility like those illustrated in (2-a)–(2-c). We start by introducing SDRT's discourse structures, which are connected graphs rather than trees. Graphs are needed to model certain facts about discourse, two of which are relevant for this paper: (i) some units can have incoming links from two different EDUs, and (ii) multiple DUs can work together to form a *complex discourse unit* (CDU), which serves as a single argument to a discourse relation.

(3)　　a.　　Sam is being punished.$_{\pi_1}$ She took her parents' car without permission,$_{\pi_2}$ so they've grounded her for 2 weeks.$_{\pi_3}$

　　　　b.　　*Explanation* (π_1, π_2), *Result* (π_2, π_3), *Elaboration* (π_1, π_3)

(4)　　a.　　$\pi_{1'} + \pi_{2'}$, but their parents don't speak any foreign languages.$_{\pi_{3'}}$

　　　　b.　　*Continuation*$(\pi_{1'}, \pi_{2'})$, *Contrast* $([\pi_{1'}, \pi_{2'}], \pi_{3'})$

In (3), π_3 is the second argument for an instance of Result and an instance of Elaboration, whose first arguments are distinct. In (4), which builds on $\pi_{1'}$ and $\pi_{2'}$ from (2-b), not only are both $\pi_{1'}$ and $\pi_{2'}$ needed to provide the necessary antecedent for *their* in $\pi_{3'}$, but these units are attached to each other and both satisfy the conditions needed to support the Contrast with $\pi_{3'}$. SDRT would therefore group them into a single, though internally complex, argument for Contrast.

SDRT's graphs thus contain two types of edges: (i) edges that are labelled with discourse relations, and (ii) edges that relate each CDU to each DU that it contains. Edges of type (i) can be further subdivided into *subordinating* and *coordinating* edges, governed by the semantics of the relations that label these edges. When a DU is attached to another DU via a subordinating relation, both arguments remain accessible for further attachments. In (2-c), for example, the role of $\pi_{2''}$ is to provide background information that explains how John came to speak German ($\pi_{1''}$) or to simply back up that claim. $\pi_{1''}$ therefore remains central to the discussion and salient enough to be accessible to $\pi_{3''}$ despite the intervening $\pi_{2''}$. Subordinating relations include Explanation, Elaboration, Background, and Question-Answer Pair, among others. When a DU is attached to another DU via a coordinating relation, the RF is pushed dynamically forward so that the second DU is on the RF, but the first is knocked off. In (2-b), for instance, $\pi_{2'}$ goes on to tell us that John's sister speaks French, so the discourse is no longer centered on John's German speaking abilities when it comes time to add $\pi_{3'}$. An attachment from $\pi_{3'}$ to $\pi_{1'}$ is thus difficult to achieve. Coordinating relations include Contrast, Continuation, Narration (Sequence), Result, and Conditional, among others.

We conclude with a description of SDRT's RF. A node π_x is on the RF of a graph G, i.e. $\text{RF}_G(\pi_x)$, just in case (a) π_x is *Last*, i.e. π_x is the EDU introduced most recently into G following the textual order of the EDUs in G, or (b) $\exists \pi_y(\text{RF}_G(\pi_y))$ such that (i) $e_s(\pi_x, \pi_y)$ for a subordinating edge e_s or (ii) $\pi_y \in \pi_x$

(i.e., π_x is a CDU). The Right Frontier Constraint (RFC) simply states that an incoming EDU that needs to be attached to a graph G should be attached to a node along the RF.

1.2 QUD

The concept of a QUD is not a homogenous one; we will aim in this section to clarify two ways of understanding this concept from the literature. Perhaps the widest application of QUD has been in the context of focus phenomena ([5,6,18]). Consider (5) and (6) (where the capital letters indicate the desired intonational emphasis):

(5)   a.   John gave the cake to TOM.
          b.   John gave the CAKE to Tom.

(5-a) and (5-b) have the same syntactic structure and the same truth conditions: each is true just in case John gave the cake in question to Tom. Yet utterances of (5-a) and (5-b) are not appropriate in the same contexts. (5-a), for example, would be an appropriate answer to the question (Q1) *To whom did John give the cake?* but not to (Q2) *What did John give to Tom?*; (5-b) would be an appropriate answer to (Q2) but not to (Q1).

[19] proposed that focus is licensed if there is a suitable set of anaphorically available alternative propositions in the context. Since questions are standardly analysed as introducing a set of alternative questions (*Alt-Q*) (cf. [11]), the anaphoric requirement of focus can be satisfied by a *congruent question*, a question whose alternative set *Alt-Q* is a subset of *Alt-F*. The *Alt-F* sets for (5-a) and (5-b) are as follows:

(6)   a.   *Alt-F* of (5-a)=$\{x \in D_e$: John gave the cake to $x\}$
          b.   *Alt-F* of (5-b)=$\{x \in D_e$: John gave x to Tom$\}$

(Q1) is a suitable antecedent for the alternatives introduced by (5-a) but not by (5-b) because *Alt-(Q1)*=$\{x \in D_e$: John gave the cake to $x\}$ is a subset of *Alt-F* of (5-a) but not *Alt-F* of (5-b). Conversely, (Q2) is a suitable antecedent for (5-b) but not (5-a).

Sometimes the QUD for a discourse can be a sub-question of a larger question that needs to be addressed, as illustrated in (7). Such hierarchical structures or *stacks* of questions have been proposed by [6,17,18,21], among others.

(7)   John gave the CAKE to [Tom]$_{CT}$ and the ICE cream to [Linda]$_{CT}$.

[6] would analyze (7) as follows. The first unit can be taken to answer the question (Q3) *What did John give to Tom?*, and the second, the question (Q4) *What did John give to Linda?*. The presence of contrastive topic marking on *Tom* and *Linda* indicates that the two questions should be understood as addressing the super-question (Q5), *John gave what to whom?*. In this case, (Q3) and (Q4) form a *strategy* for answering (Q5). The QUD for an utterance u is always the open question that is the most salient at the time u is made. Thus (Q3) is the QUD

for *John gave the CAKE to Tom*, but once this question has been answered, the next subquestion of (Q5), namely (Q4), is addressed.

While the above notion of QUD is helpful for the analysis of focus, contrastive topic and other focus-related phenomena, it is a very local notion in the sense that it cannot tell us what accounts for the incoherence of the following mini-discourse:

(8)   JOHN ate the beans. My sister is a NURSE. Rose was gone for 10 MINutes!

We can retrieve a QUD for each sentence in (8) based on its focus structure: *Who ate the beans?*, *What does your sister do?*, and *How long was Rose gone?*, respectively. Moreover, each sentence provides a complete answer to its QUD, so moving from one sentence to the next should be a perfectly fine strategy. Yet (8) is not a good discourse.

[18] aims to push the notion of QUD further in order to account for the felicity of the focus of utterances in the contexts in which they are actually used. In a nutshell: the sentences in (8) do not hang together well because we do not understand them as participating in a strategy to answer the same, larger question. In more detail, [18] posits that the structure of open questions in a discourse can be represented with a *stack*. In (7), for example, when (Q3) is the QUD, it is on the top of the stack, just above (Q5). Once (Q3) is answered, it is popped from the stack and (Q4) is pushed on to the top. Once all sub-questions of (Q5) have been addressed, and (Q5) has therefore been answered, (Q5) is also pushed from the stack. The heart of Roberts's idea is to push the stack paradigm further: don't assume in a discourse like (7) that (Q5) is the question on the bottom. In a real discourse, (Q5) would in turn figure in a strategy to answer a yet larger question, and so on, all leading up (or down) to the Big Question: *What is the way things are?*. Of course, the Big Question is far too large to answer in one discourse, so interlocutors adopt the strategy of breaking it down into smaller questions. Exactly which sub-questions are relevant in a given discourse are determined by the *conversational goals* of the interlocutors; the sub-questions chosen signify a strategy for achieving this goal.[4] What is wrong with (8), then, is that it's not clear what sub-question of the Big Question is at issue; the individual segments do not form a clear strategy for answering an obvious question.

The rules of the language game constrain how different types of linguistic structures update the discourse context, with the following principal effects: (a) If an assertion is accepted by the interlocutors in a discourse, it is added to the common ground. (b) If a question is accepted by the interlocutors in a discourse, then it is added to the set of questions under discussion, and it becomes the immediate topic of the discussion. This in turn commits the interlocutors

---

[4] That the conversational goals and intentions of speakers are relevant for computing the pragmatic meaning of an utterance goes back at least to [9]. See also [7,10, 12,20]. See [8] for a QUD-based theory that shares many features with [18], but is importantly different in ways we cannot consider in this paper.

to a common goal, namely, finding the answer. By the principle of Relevance (described in (9) below), the interlocutors should attempt to answer the question as soon as it is asked. A member of the set of questions under discussion in a discourse is removed from that set iff its answer is entailed by the common ground or it is determined to be unanswerable.

(9)   A move m is **Relevant** to the question under discussion q, i.e., to last(QUD(m)), iff m either introduces a partial answer to q (m is an assertion) or is part of a strategy to answer q (m is a question). ([18], p. 21)

As a result of the above rules, all discourse moves must be relevant to the QUD, though they need not all answer (completely or partially) the QUD: it is enough if they are part of the strategy for answering the QUD.

2   Locating the Correspondence

[18] has hypothesized that rhetorical theories and QUD-based theories are compatible and that understanding how they work together might give us a more complete pragmatic theory. The goal of this paper is to determine, on the assumption that both theories are correct, how the theories might be related. We assume that they cannot be independent: if a discourse move m_n, say, explains the content of a prior discourse move m_k, and both m_n and m_k are relevant to the QUD, then the fact that m_n explains m_k should also be relevant to the QUD—relating relevant discourse moves in irrelevant or off-topic ways would presumably lead to a very odd discourse. In other words, we assume that if the two are compatible, the discourse structures predicted by QUD-theories should correspond in some way to the structures posited by SDRT, such that the discourse graph for a whole discourse should shed light on the question structure for that discourse.

In what follows we explore two hypotheses about how the theories might be related: the first, which we will reject, assumes that rhetorical relations correspond to QUDs; the second, which will be shown to be more promising, assumes that QUDs can be associated with complex discourse units in SDRT graphs.

2.1   Rhetorical Relations as QUDs

An intuitive starting point for the comparison of SDRT and QUD is the hypothesis that rhetorical relations can be analyzed as QUDs, and vice versa. For example, when two discourse units, π_i and π_j, are related by Narration, the QUD associated with π_i, to which π_j would provide an answer, would be *What happened next?*. Similarly, if π_i and π_j are related by Explanation, the QUD associated with π_i, to which π_j would provide an answer, would be *Why?*. This hypothesis, which casts SDRT and QUD as two sides of the same coin, is frequently raised in our discussions with colleagues on this topic, and it appears to inform the work of [14,21].

We call the hypothesis that there is a one-to-one correspondence between discourse relation tokens and QUDs (or, more precisely, that each relation instance can be analyzed as a QUD at the point that the relation is added to the graph) **R-QUD**:

(10)   **R-QUD**: For discourse units π_i and π_j and rhetorical relation R in an SDRT graph, if $R(\pi_i,\pi_j)$, then there is some question $QUD(\pi_i)$ that π_i gives rise to and that π_j answers fully or partially.

Structural correspondence. R-QUD also suggests a structural correspondence between the hierarchical structures of SDRT and those of QUD. The hierarchical structures in SDRT, for example, are fully determined by the instances of rhetorical relations in the discourse, so if each rhetorical relation in an SDRT graph corresponds to a question in the QUD structure for the same discourse, then we might expect the set of questions to display a hierarchical structure that echoes that of the SDRT graph. This expectation is reinforced by the fact that SDRT's Right Frontier (RF) and Roberts's QUD stacks reflect (a part of) the hierarchical structure of a discourse and are designed to perform the same task, namely that of tracking live and salient issues in a discourse. Indeed, the two structures are similar, at least at first glance. The RF of a graph G consists of the last EDU introduced in G as well as any DU that is super-ordinate to a node on the RF via a chain of subordinating relations. Likewise, a QUD stack will include the most recent open question introduced in the discourse as well as any question that is a super-question of a question on the stack via a chain ordered by the sub-question relation.[5] In SDRT, a new DU in a graph G that attaches to a node m along the RF of G will knock any other node that is subordinate to m off of the RF. Similarly, a new question Q in a QUD stack S that is a sub-question of another question Q_m on the stack, entails that any question Q_n, such that Q_n is higher on S than Q_m be popped from the stack (unless Q is a sub-question of Q_n as well). Finally, both SDRT and QUD posit constraints that require new moves to attach to/address a move on the RF/stack.

Despite these apparent similarities, there are also some obvious, and deep, dissimilarities between the RF and stacks. Most notably, the RF orders DUs, not relations, while a QUD stack orders questions, not answers. Given that R-QUD associates relations and questions (and the arguments to relations with answers), R-QUD will inevitably limit the role that either QUD stacks or the RF can play in determining the coherence and relevance of discourse moves. From the point of view of QUD, a new move, m, in a discourse d must be relevant to one of the open questions on the QUD-stack for d at the time that m is made. Yet if we assume R-QUD and SDRT's RF, then the nature and order of the open questions on a QUD stack for a discourse d would themselves be derived through rhetorical reasoning—there are no independent principles that would lead to exactly this set of questions for d in exactly the desired order. From the perspective of SDRT, a new EDU e in a discourse d must be attachable

---

[5] This is the ordering adopted by [18] on page 15, clause (g.iii).

to some node along the RF of the graph for d at the time that the utterance that introduces e into the discourse is made. Yet if we assume R-QUD and QUD stacks, then coherence and relevance cannot be driven by the arguments to the rhetorical relations, for these correspond to *answers* not questions on the QUD stack. Given R-QUD, then, QUD and SDRT must be understood as giving two very different and incompatible models of what drives coherence and relevance in discourse.

Another structural consequence of assuming SDRT and R-QUD would be that the stack derived from the RF of a given discourse graph would not necessarily be ordered by the sub-question relation. (And conversely, a consequence of assuming QUD and R-QUD would be that the relations connecting nodes along the RF would be severely limited by the sub-question ordering on the QUD stacks.) Again, it is in part the QUD and the need to answer that QUD that is supposed to determine whether a given discourse move is coherent or not in a QUD framework; the sub-question relation is important because it keeps conversation on topic by simply breaking down the QUD into smaller QUDs. Examples like (11) show, however, that a combination of R-QUD and SDRT exclude the possibility of a QUD stack ordered by the sub-question relation.

(11)   We had so much fun in London!$_{\pi_1}$ We got to see the Lion King!$_{\pi_2}$ I've been wanting to go for a really long time$_{\pi_3}$ and my mom finally gave me tickets for my birthday.$_{\pi_4}$ We also got to ride on the big Ferris wheel$_{\pi_5}$...

SDRT would predict the following structure for (11):

(11)'   Elaboration(π_1, π_2), Background($\pi_2, [\pi_3, \pi_4]$), Continuation(π_3, π_4), Elaboration(π_1, π_5), Continuation(π_2, π_5).

From the first two relation instances, R-QUD would yield questions: (q1) *What did you do?* and (q2) *What makes that so exciting?* (or something like that). Suppose we stack these questions so that (q2), the more discourse-recent question, is on top of (q1). The fact that we (eventually) have Elaboration(π_1, π_5) and Continuation(π_2, π_5) shows that (q1), *What did you do?*, has not yet been fully answered when π_3 and π_4 are uttered, so it must still be on the stack of open questions. (q2), however, is not a subquestion of (q1). Still, the utterances of π_3 and π_4 are coherent and relevant in the discourse.

A third consequence of R-QUD is that it would result in a loss of information for SDRT. A node in an SDRT graph, as noted in Sect. 1.1, can have incoming links from two different nodes, though it is unclear how such dependencies could be modelled using stacks. In (3), repeated here, π_2 appears to explain π_1, and so could be taken to answer the question (q1) *Why?* (or *Why is Sam being punished?*), while π_3 appears to elaborate on π_1, and so could be taken to answer the question (q2) *How is Sam being punished?*. At the same time, π_3 also describes a result of π_2 and so could be taken to answer the question (q3) *what happened as a result?* or something like that.

(3)   Sam is being punished.$_{\pi_1}$ She took her parents' car without permission,$_{\pi_2}$
       so they've grounded her for 2 weeks.$_{\pi_3}$

The stack architecture does not tell us where (q3) fits in to the stack; stacks
are more naturally compatible with a tree-based structure, not the graph-based
structure of SDRT. Nor does the stack architecture predict that reversing the
order of (q1) and (q2) (and so that of π_2 and π_3) would make a difference to
the discourse content, though clearly such a reversal would make a significant
impact.

Another source of information loss for SDRT, given R-QUD, relates to CDUs.
The need for CDUs is illustrated in (12):

(12)   I finally figured out the right baking process for cannelés.$_{\pi_1}$ I left them
        at 210 for 20 min,$_{\pi_2}$ then I turned them down to 190 for 30 min,$_{\pi_3}$ and
        then I finished at 180 for the last 15 min.$_{\pi_4}$

In (12), π_2-π_4 together describe the correct baking process introduced in π_1; no
single unit alone provides this information. In terms of questions, no single unit
answers the question: *What is the right baking process?*. In other words, π_2-π_4
form a CDU that serves as the second argument to the Elaboration relation that
intuitively holds between π_1 and π_2-π_4. At the same time, the CDU composed of
π_2-π_4 has its own internal structure. Its component EDUs are related by Narration
or Sequence, which imposes temporal constraints on the EDUs—the cooking steps
cannot be followed in just any order. QUD stacks derived through R-QUD (or
otherwise) do not provide the structure needed to capture the layers of discourse
graphs that result from CDUs.

Relation instances. Even if we abandon a correspondence between the RF
and stacks (which would render R-QUD far less interesting), the correspondence
posited in R-QUD breaks down for several reasons. First, for certain rhetorical
relations at least, we run into a problem of circularity in which the intuitive
question has to mention the associated discourse relation or marker directly.
For example, SDRT would posit the relation Contrast(π_1, π_2) for (13), but it's
unclear what question could replace the Contrast relation that wouldn't itself
presuppose the same rhetorical relation.

(13)   [Pat]$_F$ came to the party$_{\pi_1}$ but [Mel]$_F$ didn't.$_{\pi_2}$

The most suitable implicit question to posit in this case would be *What does
π_1 contrast with?*. Similar remarks can be made for instances of Parallel,
Result (*What happened as a result?/And so?*), and Narration (*What happened
then/next/after that?*). This calls into question the possibility of truly translating relations into independent questions.

An alternative question for (13), which would be more in line with the ideas
presented in [6,17], is that π_1 and π_2 in (13) could be taken as addressing the
same question *Who came to the party?*. Yet the fact that π_1 and π_2 both answer
the same question doesn't justify the use of the contrastive marker *but*. In fact,
a variant of (13) in which *but* is replaced by *and*, and π_1 and π_2 are related

with Parallel, is acceptable as well, though the intuitive background question for such a variant would be the same: *Who came to the party?*. Similar remarks can be made for certain examples of Contrast that have possible Continuation-based variants. For instance, (14-a) and (14-b) could both be understood as answering the question: *What's going on with John?*, but this doesn't capture the intuitive difference between the examples, brought about by the presence of the contrastive marker in (14-a).

(14)   a.   John called but he said he's running late.
        b.   John called and he said he's running late.

In sum, if we assume R-QUD, the QUD framework runs into a problem because it does not easily extend to handle most co-ordinating relations.

Summary. The forgoing discussion reveals that a correspondence like that assumed by R-QUD would lead to a loss of information and an abandonment of basic principles from both SDRT and QUD. From the point of view of SDRT, if we assume R-QUD and QUD, then we must abandon the role that rhetorical reasoning plays in determining the coherence and relevance of individual discourse moves. We also lose information because QUD is not designed to handle triangular discourse structures (like that underlying (3)), CDUs, or coordinating relations. Nor is the correspondence posited by R-QUD favorable for QUD. If we assume SDRT + R-QUD, we must abandon the hypothesis that QUD-stacks of the sort posited in [18] guide the coherence and relevance of individual discourse moves. Moreover, the QUDs predicted from R-QUD would not look like the QUDs that a QUD-framework would predict otherwise. In [18] and elsewhere, it is suggested that focus marking indicates the QUD. In the case of (15), the second sentence has focus marking, but the question that can be derived from this marking (*What did John buy?*) is much more fine-grained than the question type that seems to correspond to the rhetorical relation Elaboration (e.g., *Can you tell me more?*). In fact, the question generated from focus is compatible with many rhetorical relations.

(15)   John went to the store (π_1). He bought [apples]$_F$ (π_2).

(16)   Mary is mad at John (π_1). He bought [apples]$_F$ (π_2). (instead of pears.)

SDRT would posit Elaboration(π_1, π_2) for (15), but Explanation(π_1, π_2) in (16). Yet in both cases, the question generated from the focus structure of π_2 is: *What did John buy?* Thus, in adopting R-QUD, we lose information from QUD as well.

2.2   Complex Discourse Units as QUDs

In this section, we take a step back and consider another hypothesis about how SDRT and QUD might interact. One of the intuitions behind [18]'s QUD account is that QUD stacks model the plan structure of a discourse. The rough idea is that given a specific discourse goal, which either triggers a QUD or is itself modelled as a QUD, a speaker or set of interlocutors will aim to answer this QUD by breaking it down in smaller and smaller QUDs until the QUDs become more manageable. As these smaller QUDs are answered, their collective answers work together to provide answers to larger QUDs and so on until the QUD triggered by the original discourse goal is answered. The question that we pose in this section is whether there are features of an SDRT graph that might reveal the planning structure of a discourse, yielding a correspondence more in touch with the planning aspect of [18]'s QUD stacks. The hypothesis that we consider is that CDUs and the relations between them yield the desired planning structure. In short, the idea is that CDUs manifest a certain topical cohesion that allows us to treat their contents as complex answers to implicit questions.

CDUs. A CDU, as briefly described in Sect. 1.1, is a collection of discourse units (EDUs, CDUs or a mixture of both) that work together to function as an individual argument to a discourse relation. (4) provides an example of a CDU in a Contrast relation; (12) provides an example in an Elaboration relation. Further empirical motivation for CDUs comes from the behavior of anaphora:

(17)   One plaintiff was passed over for promotion three times (π_1). Another didn't get a raise for five years (π_2). A third plaintiff was given a lower wage compared to males who were doing the same work (π_3). But the jury did not believe this (π_4). ([2], p. 15)

As noted in [2], the anaphor *this* in π_4 can be resolved to π_3 or to the set $\{\pi_1, \pi_2, \pi_3\}$, but not to, say, π_2 alone. The second interpretation of (17), represented in Fig. 1, requires the construction of a CDU $\pi_{t(1-3)}$ formed from $\{\pi_1, \pi_2, \pi_3\}$.

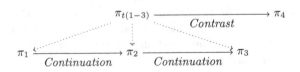

Fig. 1. SDRT graph for (17)

Thus CDUs are constructed when called for by the demands of rhetorical relations or anaphora, but we can also think of CDUs in a more general way as collections of DUs that share a thematic or rhetorical coherence. Suppose that two people are discussing a recent democratic presidential debate in the U.S. and one speaker is arguing that Bernie Sanders won the debate. She might provide a multi-step argument for her position, providing numerous justifications for it and elaborating on each justification in turn. We can think of her whole argument as yielding one large CDU whose members work together to provide support for the claim that Bernie Sanders won. Within this larger CDU, each justification for her main position will also yield a CDU whose members are the set of DUs that participate in the argument for this justification. For example, the speaker might offer as justification of her main claim the argument that Sanders provided a superior plan for attacking Wall Street. The members of the CDU associated with this justification would be the set of DUs that work together to support the claim that Sanders provided a superior plan for Wall Street. Within this CDU, the speaker might go on to provide multiple reasons for why Sanders' plan was superior, thereby giving rise to smaller CDUs and so on.

In this way, we can think of each CDU in a discourse as having a topic that glues its members together,[6] although the topic of a given CDU is not (necessarily) derivable from the content of any explicit move in the discourse; it is rather a topic that emerges from considering what the discourse moves work together to accomplish. In addition, because CDUs are collections of *connected* DUs—i.e. not just arbitrary sets of DUs from the discourse—the set of CDUs in a discourse will naturally give rise to a partial order based on the subset relation: if a CDU π_i includes a CDU π_j, the set of members of π_j will be a strict subset of the set of members of π_i. A CDU cannot merely overlap another CDU without one being entirely included in the other.

CDU-QUD. We summarize the hypothesis that the content of a CDU fully answers an implicit QUD as follows (Although what counts as a 'full' answer can depend on the context and interlocutors' interests as in [18]):

(18)   **CDU-QUD$_a$:** For every CDU π_i in an SDRT graph there is some question $\text{QUD}(\pi_i)$ that the the discourse units in π_i answer fully.

To illustrate the correspondence, consider the textbook example (19) and the associated SDRT graph in Fig. 2.

(19)   John had a great evening (π_1). He had a great meal (π_2). He ate salmon (π_3). He devoured lots of cheese (π_4). Then he won a dancing competition (π_5).

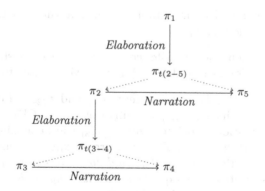

Fig. 2. SDRT graph for (19)

---

[6] Though see [3] for a discussion of the difficulties of defining discourse topic in SDRT.

Putting aside deeper questions about how exactly to predict the construction of CDUs and their associated implicit topics, which would be an interesting topic for future research, let us assume that there are three CDUs in this structure: $\pi_{t(1-5)}$[7], i.e., the discourse as a whole, $\pi_{t(2-5)}$, and $\pi_{t(3-4)}$. We derive $\pi_{t(2-5)}$ from the fact that if π_2 participates in the second argument of an Elaboration relation off of π_1, then $\pi_{t(3-4)}$ does as well because $\pi_{t(3-4)}$ is simply an Elaboration of π_2. These three CDUs can each be associated with a question as follows:

1. $\pi_{t(1-5)}$ is associated with a Q_1, e.g.: *What was John's evening like?*
2. $\pi_{t(2-5)}$ is associated with a Q_2, e.g.: *What did he do in the evening?*
3. $\pi_{t(3-4)}$ is associated with a Q_3, e.g.: *What did he eat?*

Note that the CDUs posited for (19) are ordered by the subset relation (if $\pi_i \in \pi_{t(3-4)}$, then $\pi_i \in \pi_{t(2-5)}$, etc.). This relation is moreover reflected in what appears to be a sub-question relation that holds between Q_1, Q_2, and Q_3 (if an assertion answers Q_3, it partially answers Q_2, etc.).

Given the correspondence between the CDU structure of (19) and the proposed associated QUD structure, a reasonable hypothesis is that we can extend the correspondence posited by CDU-QUD$_a$ with CDU-QUD$_b$.

(20)   **CDU-QUD$_b$:** If a CDU π_j is a member of another CDU π_i then QUD(π_j) is a subquestion of QUD(π_i)

CDU-QUD$_b$ runs into problems when we consider a wider range of rhetorical relations than the ones in the above example, however. As mentioned above, $\pi_{t(3-4)}$ figures in $\pi_{t(2-5)}$ because $\pi_{t(3-4)}$ is related to π_2 via an Elaboration relation. This Elaboration relation in turn naturally gives rise to a set of questions ordered by the sub-question relation: even once we get down to $\pi_{t(3-4)}$, we're still talking about John's evening. The situation is less clear for CDUs figuring in other subordinating relations, e.g. Explanation, Consequence, Commentary, and so on.

Consider (21) and its discourse graph in Fig. 3, in which the topic of the CDU provides an Explanation for π_1:

(21)   Yesterday John and his wife went to the fanciest restaurant in Paris (π_1). It was John's birthday (π_2) and his wife wanted to spoil him (π_3).

There are two CDUs in the above structure: $\pi_{t(1-3)}$ and $\pi_{t(2-3)}$. Because $\pi_{t(1-3)}$ is simply the graph as a whole, $\pi_{t(2-3)}$ is a subset of $\pi_{t(1-3)}$. CDU-QUD$_b$ therefore predicts that the question associated with $\pi_{t(2-3)}$ will be a sub-question of the question associated with $\pi_{t(1-3)}$. This is incorrect: $\pi_{t(1-3)}$ would intuitively be associated with a question such as *What did John do yesterday?*, while $\pi_{t(2-3)}$ would likely be associated with a question such as *Why did he do this?*. Yet the latter question is not a sub-question of the former.

(22) illustrates the same point but with a different discourse relation:

---

[7] The CDU that represents the discourse as a whole is identical to the discourse graph as a whole, so we do not use special notation to label it in Fig. 2.

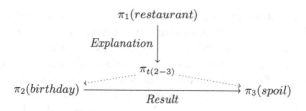

Fig. 3. SDRT graph for (21)

(22)   The Fed lowered the prime interest again today for the third time in a month (π_1). Most economists greeted the move with skepticism (π_2) but were afraid to express this publicly (π_3). (Modified version of example (5) from [4])

Structurally (22) is like (21); only the rhetorical relations are different. Again, the question intuitively associated with $\pi_{t(2-3)}$ (e.g. *What was the economists reaction to this?*) does not seem to be a sub-question of the question associated with the whole graph (e.g. *What did the FED do today?*) (Fig. 4).

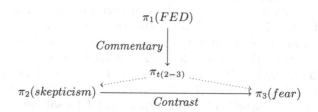

Fig. 4. SDRT graph for (22)

Because the structure of the attachment between $\pi_{t(1-5)}$ and $\pi_{t(2-5)}$ in (19) is identical to the structure of the attachments in (21) and (22), the sub-question relation posited by CDU-QUD$_b$ to hold between QUDs is not entailed by the structural relations in the discourse graph. The nature of the rhetorical relations, which are represented by the labels of the structural relations, must also be taken into account. As a result, CDU-QUD$_b$ is probably only applicable when the CDU π_j (from the definition of CDU-QUD$_b$) attaches to the top node of the CDU π_i with a relation that implies subeventhood. The semantics of the Elaboration relation, for example, entails that the eventuality described by the second argument of an Elaboration is a sub-eventuality of the eventuality described by the first.

Future directions. If the applicability of CDU-QUD$_b$ is indeed as restricted as the preceding discussion suggests, then this hypothesis entails that many QUDs derived from CDU-QUD$_a$ will not be linked to other QUDs semantically, though QUDs might still be ordered by a precedence relation. This is an unsatisfying result because it potentially prevents many QUDs in a discourse from working with other QUDs to form a plan or strategy for achieving a discourse goal.

One way to save the idea of a CDU-QUD correspondence, which we ultimately find promising for at least certain kinds of discourse, is to loosen the requirement on the QUD side that QUDs be ordered by the sub-question relation and allow strategies for achieving discourse goals to be more complex than sequences of sub-questions. That is, we should abandon CDU-QUD$_b$ and replace is with a more general hypothesis. One idea, given the proposed correspondence in CDU-QUD$_a$, is that a QUD might be retrieved from the rhetorical structure of a discourse much like a question can be retrieved from focus structure in frameworks like [5,6,18]. After all, CDUs are sets of connected nodes from a discourse graph, so the possible CDUs for a given discourse must be determined by the rhetorical connections at work in the discourse.

Associating QUDs with CDUs could have a value for SDRT as well. In particular, QUDs could be used to capture the topical cohesion of CDUs. This would give us a means for abstracting over the details of an SDRT graph to look at the higher-level strategy that a speaker, or group of interlocutors, has adopted to achieve her/their discourse goal. We assume that this strategy will be apparent at least in cooperative, information-seeking discourses.

This high-level approach to thinking about the relation between QUDs, strategies for achieving discourse goals, and rhetorical structure reflects, we think, the spirit of [17] (last paragraph before the conclusion) and [18] (pp. 62–63) where the idea that each explicit move in a discourse should address a QUD, as suggested by R-QUD, is de-emphasized or outright rejected, and it is suggested that the connection between the rhetorical structure of larger chunks of discourse and strategies for achieving discourse goals should be further explored. Future work on the connection between CDUs and QUDs will depend, however, on these notions being more clearly defined in their respective frameworks than they currently are. What we suggest is that these aspects of QUD and SDRT might most beneficially be developed in tandem.

3   Conclusion

Adopting SDRT and some basic notions from [18]'s QUD framework as starting points, we have explored two hypotheses about how rhetorical structure and QUD structure might come together to provide a more general pragmatic theory. The first hypothesis, which posits a correspondence between rhetorical relations and QUDs, was rejected. The second hypothesis posits a correspondence between QUDs and more global features of an SDRT graph, namely how discourse units are thematically grouped together into larger, complex chunks of discourse. This hypothesis is, we have argued, more promising.

References

1. Asher, N.: Reference to Abstract Objects in Discourse. Kluwer Academic Publishers, Dordrecht (1993)
2. Asher, N., Lascarides, A.: Logics of Conversation. Cambridge University Press, Cambridge (2003)
3. Asher, N.: Discourse topic. Theor. Linguist. **30**, 163–201 (2004)
4. Asher, N., Vieu, L.: Subordinating and coordinating discourse relations. Lingua **115**, 591–610 (2005)
5. Beaver, D., Clark, B.: Sense and Sensitivity. Blackwell, Oxford (2008)
6. Büring, D.: On D-trees, beans, and B-accents. Linguist. Philos. **26**, 511–545 (2003)
7. Carlson, L.: Dialogue Games: An Approach to Discourse Analysis (Studies in Linguistics and Philosophy), vol. 17. Springer, Netherlands (1985)
8. Ginzburg, J.: The Interactive Stance. Oxford University Press, Oxford (2012)
9. Grice, H.P.: Studies in the Way of Words. Harvard University Press, Cambridge (1989)
10. Grosz, B.J., Sidner, C.L.: Attention, intentions, and the structure of discourse. Comput. Linguist. **12**(3), 175–204 (1986)
11. Hamblin, C.: Questions in Montague English. Found. Lang. **1**, 41–53 (1973)
12. Lewis, D.: Scorekeeping in a language game. J. Philos. Logic **8**(1), 339–359 (1979)
13. Mann, W., Thompson, S.: Rhetorical structure theory: towards a functional theory of text organization. Text **8**, 243–281 (1988)
14. Onea, E.: Potential Questions in Discourse and Grammar. University of Göttingen (2013). Habilitation Thesis
15. Polanyi, L.: A theory of discourse structure and discourse coherence. In: Eilfort, W.H., Kroeber, P.D., Peterson, K.L. (eds.) Papers from the General Session at the 21st Regional Meeting of the Chicago Linguistic Society. Chicago Linguistic Society (1985)
16. Polanyi, L., Scha, R.: A syntactic approach to discourse semantics. In: Proceedings of the 10th International Conference on Computational Linguistics (COLING84), Stanford, pp. 413–419 (1984)
17. Roberts, C.: Context in dynamic interpretation. In: Horn, L., Ward, G. (eds.) Handbook of Contemporary Pragmatic Theory, pp. 197–220. Blackwell, Oxford (2004)
18. Roberts, C.: Information structure in discourse: towards an integrated formal theory of pragmatics. Semant. Pragmatics **5**, 1–69 (2012)
19. Rooth, M.: A theory of focus interpretation. Nat. Lang. Semant. **1**(1), 75–116 (1992)
20. Thomason, R.H.: Accommodation, meaning, and implicature: interdisciplinary foundations for pragmatics. In: Cohen, P.R., Morgan, J., Pollack, M.E. (eds.) Intentions in Communication. System Development Foundation Benchmark, pp. 325–363. The MIT Press, Cambridge (1990)
21. Van Kuppevelt, J.: Directionality in discourse: prominence differences in subordination relations. J. Semant. **13**(4), 363–395 (1996)

An Inference Problem Set for Evaluating Semantic Theories and Semantic Processing Systems for Japanese

Ai Kawazoe[1]([✉]), Ribeka Tanaka[2], Koji Mineshima[2], and Daisuke Bekki[2]

[1] National Institute of Informatics, 2-1-2 Hitotsubashi Chiyoda-ku, Tokyo, Japan
zoeai@nii.ac.jp
[2] Ochanomizu University, 2-1-1 Ohtsuka Bunkyo-ku, Tokyo, Japan

Abstract. This paper introduces a collection of inference problems intended for use in evaluation of semantic theories and semantic processing systems for Japanese. The problem set categorizes inference problems according to semantic phenomena that they involve, following the general policy of the FraCaS test suite. It consists of multilingual and Japanese subsets, which together cover both universal semantic phenomena and Japanese-specific ones. This paper outlines the design policy used in constructing the problem set and the contents of a beta version, currently available online.

1 Introduction

Explaining the validity (or invalidity) of inference (e.g. entailment, presupposition, and implicature) among sentences is one of the main objectives of studies of meaning. Since the start of the PASCAL RTE challenge (Dagan et al. 2006), the recognition of such inference relations has been a core component of NLP tasks, and the necessity and importance of inference problem sets is widely recognized. Aiming to contribute to the development and evaluation of semantic theories and semantic processing systems for Japanese, we are constructing a data set which comprises inference problems involving Japanese semantic phenomena. For English, the FraCaS test suite (Cooper et al. 1996), which covers major semantic phenomena, has been used for textual entailment (TE) recognition tasks, but no such data set has previously been constructed for Japanese. Our problem set consists of two parts: a multilingual subset and a Japanese subset. The multilingual subset includes Japanese counterparts of FraCaS problems. The Japanese subset covers some universal phenomena not included in FraCaS and some specific to Japanese, such as *toritate* particles and *wa–ga* constructions. In this paper, we outline the design policy used in constructing the problem set and describe a beta version that we have released online.

Each inference problem in the original FraCaS test suite is a triplet: a premise or set of premises; a yes/no question; and the answer to that question. A "hypothesis" sentence, a declarative counterpart of the yes/no question, have since been

© Springer International Publishing AG 2017
M. Otake et al. (Eds.): JSAI-isAI 2015 Workshops, LNAI 10091, pp. 58–65, 2017.
DOI: 10.1007/978-3-319-50953-2_5

added in a machine-readable version by Bill MacCartney.[1] In the following sections, we simply omit questions and show only the premises (P), hypotheses (H), and answers when illustrating inference problems, exemplified as follows.

(1) fracas-141 answer:unknown

P1 John said Bill had hurt himself.
_____
H Someone said John had been hurt.

2 Background

When evaluating linguistic theories, the primary requirement for evaluation data is that each data item represents only the target phenomena for that item to the greatest possible extent. For inference relations, it is ideal that the data involve only the target phenomena, and do not include other factors that could affect speakers' judgments of the validity of an inference between a premise set and a hypothesis. This requirement is also important for data to be used in NLP tasks such as textual entailment. Recently, several researchers have pointed out the necessity of data that can allow measuring system performance on specific phenomena (e.g., Bos 2008; Sammons et al. 2010; Bentivogli et al. 2010; Cabrio et al. 2013).

The FraCaS test suite was created by the FraCaS Consortium as a benchmark by which to measure the semantic competence of NLP systems. It contains 346 inference problems that collectively demonstrate basic linguistic phenomena for which formal semantics should account; these include quantification, plurality, anaphora, ellipsis, tense, comparatives, and propositional attitudes. Each problem is designed to include exactly one target phenomenon, to exclude other phenomena, and to be independent of background knowledge. A machine-readable version of FraCaS has been used for evaluation and error analyses of several TE models (e.g., MacCartney and Manning 2007, 2008; Lewis and Steedman 2013; Tian et al. 2014; Abzianidze 2015; Mineshima et al. 2015). Currently, the MultiFraCaS project, headed by Robin Cooper, is working to create a multilingual FraCaS test suite.[2]

One strength of FraCaS-type data sets is that they are based on the outcomes of standard linguistic studies and thus represent reliable observations of phenomena. This means that the quality of the data set (e.g., the accuracy of judgements about the validity of inference, and the validity of analyses) is ensured by the community of linguists. In addition, FraCaS-type data sets have enough generality that the validity of inference will usually not be changed if we replace content words in the sentences and the situation of utterance. This is because each problem represents a generalization about relevant semantic phenomena.

For the Japanese language, some existing inference problem sets have been designed so as to restrict the number of phenomena or factors that affect possibilities of inference. For example, NTCIR RITE provides the UnitTest data set, following the methodologies in Sammons et al. (2010) and Bentivogli et al. (2010).

_____

[1] http://www-nlp.stanford.edu/~wcmac/downloads/fracas.xml.
[2] http://www.ling.gu.se/~cooper/multifracas/.

The Kyodai Textual Entailment Evaluation Data, which was created by Kotani et al. (2008), contains premise–hypothesis pairs with only one or two factors relevant to the validity of inference. However, many of the problems in these sets involve lexical, syntactic, or world knowledge; further, basic semantic phenomena, such as those in FraCaS, are not fully represented. A set with a wider variety of inference problems should be created to cover major semantic phenomena of Japanese.

3   Overview of the Problem Set

3.1   Multilingual and Japanese Subsets

Our problem set categorizes inference problems according to the semantic phenomena they involve, following the design policy of the FraCaS test suite. The content of the problem set is shown in Table 1. In that table, phenomena marked with "⋆" are those covered by the beta version.

Table 1. Sections of our problem set. Those covered by the beta version as of April 2015 are denoted by "⋆".

Subsets	Descriptions	Sections	Num.
Multilingual subset	Japanese counterparts of FraCaS problems	⋆Generalized quantifier, ⋆plurality, ⋆anaphora, ⋆ellipsis, ⋆adjectives, ⋆comparatives, ⋆tense, ⋆verbs, ⋆propositional attitude	624
Japanese subset	Problems with universal phenomena not covered by FraCaS	Modality, conditionals, negation, ⋆adverbs, focus, and more phenomena with ⋆adjectives, ⋆verbs, ⋆comparatives and ⋆propositional attitudes, etc	166
	Problems with Japanese-specific phenomena	⋆*Toritate* particles, *wa–ga* construction, etc	

The multilingual subset of our problem set contains Japanese counterparts of FraCaS problems, but we have not adhered to the principle of one-to-one correspondence that is followed for the MultiFraCaS test sets. As a result, 90 of the FraCaS problems correspond to more than one of the problems in our data set. In particular, those FraCaS problems with generalized quantifiers have many Japanese counterparts, which was done because there are many Japanese expressions and word patterns that are truth-conditionally equivalent to quantificational NPs in English (but may introduce different presuppositions and/or implicatures).

Not all problems in the multilingual subset are literal translations of FraCaS problems. For those FraCaS items that have no natural translation, we created Japanese problems that target similar phenomena, albeit with different syntactic structures or lexical items. The majority of the Japanese subset is still being developed. We have covered more phenomena involved with adjectives and verbs which are not covered by FraCaS and some phenomena with adverbs and toritate particles.

3.2 Format

We adopted the following format for our problem set.

- `problem`: an inference test
  - `jsem_id` attribute: an unique ID
  - `answer` attribute: validity of inference (yes, no, unknown, or undef)
  - `phenomena` attribute: type of phenomena (multiple entries allowed)
  - `inference_type` attribute: type of inference
- `link`: a link to a resource in other languages
  - `resource` attribute: the name of the linked resource
  - `link_id` attribute: the ID of the corresponding test in the linked resource
  - `translation` attribute: specifies whether the inference test is a literal translation of the linked test or not. (allowed: yes, no, unknown)
  - `same_phenomena` attribute: specifies whether the inference test represents the same phenomena as the linked test or not (allowed: yes, no, unknown)
- `p`: premise
- `h`: hypothesis
- `note`: comments

For example, the Japanese equivalent of fracas-141 (shown above as (1)) is described as follows:

problem	id: 449
	answer: yes
	phenomena: Nominal anaphora, intra-sentential anaphora, zibun
	inference_type: entailment
link	resource: fracas
	link_id: 141
	translation: yes
	same_phenomena: no
p1	ジョンは、ビルが自分を傷つけたと言った。
English	John said Bill had hurt himself.
h	誰かが、ジョンが傷つけられたと言った。
English	Someone said John had been hurt.
note	As is well-known, unlike "himself," "zibun" allows long-distance anaphora (Kuno 1978). This makes it possible to interpret "John" as the antecedent of "zibun" in p1, which is in contrast to the English counterpart.

Some elements—such as `problem`, `p`, `h`, `note`, and the attribute `answer`—are based on the FraCaS and MultiFraCaS representation. We added some new elements and attributes, as described below.

Links to the FraCaS Problems. The `link` element is added to show information about a linked resource. The attributes `translation` and `same_phenomena` are introduced to specify translation- and phenomena-level similarities and differences between multi-language problem pairs. As shown in the example above, we can see that the Japanese problem is a possible translation of an English counterpart by looking at the value of the `translation` attribute. The value of the `same_phenomena` attribute shows that they involve disparate phenomena, leading to a different answer ("yes") for the Japanese version than for the English version (which has answer "unknown"). Details of relevant phenomena and references to relevant literature are given in the `note` element.

Categories of Phenomena. Phenomena involved in an inference problem are concisely described by `phenomena` attributes. We allow multiple value entries for this attribute and recommend creating new values as necessary in the process of constructing the problem set. In our construction, created values have been collected afterwards and edited.

Currently the values for `phenomena` in our problem set are classified into three types: section titles, universal phenomena, and Japanese-specific phenomena. Section titles are taken from the nine sections of the original FraCaS (Generalized Quantifiers, Plurals, Nominal Anaphora, etc.). Some of the universal phenomena values are also taken from subsections or problem descriptions in FraCaS, and others were newly created by us (e.g., factive/non-factive/counterfactive for propositional attitude problems; some other types of anaphora, such as coreference and bound variable anaphora). Values that indicate Japanese-specific phenomena include several anaphora types (*no* anaphora, *soo su* anaphora), elliptic constructions (stripping with or without case markers), word order patterns of a quantifier and an NP it modifies (pre- or post-nominal quantifiers, floating quantifiers), and key functional words (anaphoric expressions such as *zibun*, *kare/kanozyo* and *so*-series demonstratives, various conjunctive particles, etc.).

Inference Types. Inference is a complex phenomenon that typically involves various linguistic and contextual factors when we judge the validity or invalidity of an inference relation among sentences. In our problem set, we specify the type of inference for each premise–hypothesis pair, using the `inference_type` attribute for this purpose, in addition to specifying the type of phenomenon via the `phenomena` attribute. This enables evaluation of TE models according to inference type.

The major values for the `inference_type` attribute are `entailment` and `pre-supposition`. Distinction between the two inference types is based on well-known inference classifications in formal semantics and pragmatics (e.g., Chierchia and McConnell-Ginet 2000; Levinson 2000; Kadmon 2001; Potts 2005). Entailment is an at-issue content of utterance, also called "asserted content", "What is Said", (Grice 1989) or "semantic entailment", as distinguished from "entailment" in the broader sense. The problem (2), one of the counterparts of fracas-017, shows a typical example of entailment.

(2)   id:117 answer:yes

P1 一人のアイルランド人がノーベル文学賞を受賞した。
One-CL-GEN Irishman-NOM Nobel literature prize-ACC win-PAST
"An Irishman won the Nobel Prize in Literature."

H 一人のアイルランド人がノーベル賞を受賞した。
One-CL-GEN Irishman-NOM Nobel prize-ACC win-PAST
"An Irishman won a Nobel Prize."

Presupposition, in contrast, acts as background content for an utterance, and is often indicated by specific lexical items or expressions. The problem (3) is an example of presupposition, signified by the factive predicate *koto-o uresiku omou* (lit., be pleased to know that).

(3)   id:737 answer:yes

P1 太郎は花子が高校を卒業したことを嬉しく思った。
Taroo-TOP Hanako-NOM high school-ACC graduate-PAST-COMP-ACC pleased think-PAST
"Taro was pleased to know that Hanako graduated from high school."

H 花子は高校を卒業した。
Hanako-TOP high school-ACC graduate-PAST
"Hanako graduated from high school."

Neither entailment nor presupposition can be cancelled by subsequent contexts, but the former disappears and the latter survives when the premise appears in a modal or negated context and when it appears in the antecedent of a conditional.

Conventional and conversational implicature are two other major types of inference, and are important data for TE recognition tasks. Although the current beta version of our problem set covers only problems with entailment and presupposition, we plan to expand it to include these implicature cases.

3.3   Creation of the Problem Set

Four linguists constructed the problem set. In principle, one linguist constructed inference problems for each section and another reviewed them, revising as necessary. We strongly recommended referring to the relevant literature when introducing new problems. In the review process, we checked for the presence or absence of factors other than targeted factors, for ambiguity and naturalness of sentences, and for cross-rater reliability of inference judgments.

4   Concluding Remarks

We have introduced a FraCaS-type inference problem set that covers semantic phenomena in Japanese. We have shown some new features that may contribute to a cross-linguistic evaluation of TE models according to phenomenon or inference type. The problem set is now being expanded to cover more phenomena, both universal and Japanese-specific ones. We encourage linguists to become collaborators for the data set by contributing inference problems with their specialized knowledge and findings.

References

Abzianidze, L.: Towards a wide-coverage tableau method for natural logic. In: Murata, T., Mineshima, K., Bekki, D. (eds.) JSAI-isAI 2014. LNCS (LNAI), vol. 9067, pp. 66–82. Springer, Heidelberg (2015). doi:10.1007/978-3-662-48119-6_6

Bentivogli, L., Cabrio, E., Dagan, I., Giampiccolo, D., Leggio, M.L., Magnini, B.: A methodology for isolating linguistic phenomena relevant to inference. In: Proceedings of LREC 2010, pp. 3544–3549, Valletta, Malta (2010)

Bos, J.: Let's not argue about semantics. In: Proceedings of the 6th Language Resources and Evaluation Conference (LREC 2008), pp. 2835–2840. Marrakech, Morocco (2008)

Chierchia, G., McConnell-Ginet, S.: Meaning and Grammar: An Introduction to Semantics. MIT Press, Cambridge (2000)

Cooper, R., Crouch, D., van Eijck, J., Fox, C., van Genabith, J., Jan, J., Kamp, H., Milward, D., Pinkal, M., Poesio, M., Pulman, S., Briscoe, T., Maier, H., Konrad, K.: Using the framework. Technical report, FraCaS: A Framework for Computational Semantics. FraCaS Deliverable D16 (1996)

Cabrio, E., Magnini, B.: Decomposing semantic inferences. In: Condoravi, C., Zaenen, A. (eds.) Linguistic Issues in Language Technology (LiLT), vol. 9(1). Special Issue on The Semantics of Entailment (2013)

Dagan, I., Glickman, O., Magnini, B.: The PASCAL recognising textual entailment challenge. In: Quiñonero-Candela, J., Dagan, I., Magnini, B., d'Alché-Buc, F. (eds.) MLCW 2005. LNCS (LNAI), vol. 3944, pp. 177–190. Springer, Heidelberg (2006). doi:10.1007/11736790_9

Grice, P.: Studies in the Way of Words. Harvard University Press, Cambridge (1989)

Kadmon, N.: Formal Pragmatics. Blackwell, Oxford (2001)

Kotani, M., Shibata, T., Nakata, T., Kurohashi, S.: Building textual entailment Japanese data sets and recognizing reasoning relations based on synonymy acquired automatically. In: Proceedings of the 14th Annual Meeting of the Association for Natural Language Processing, Tokyo, Japan (2008)

Kuno, S.: Danwa no Bunpoo [grammar of discourse]. Taishukan, Tokyo (1978)

Levinson, S.C.: Presumptive Meanings: The theory of generalized conversational Implicature. MIT Press, Cambridge (2000)

Lewis, M., Steedman, M.: Combining distributional and logical semantics. Trans. Assoc. Comput. Linguist. 1, 179–192 (2013)

MacCartney, B., Manning, C.D.: Natural logic for textual inference. In: Proceedings of the ACL-PASCAL Workshop on Textual Entailment and Paraphrasing, pp. 193–200 (2007)

MacCartney, B., Manning, C.: Modeling semantic containment and exclusion in natural language inference. In: The 22nd International Conference on Computational Linguistics (Coling 2008), Manchester, UK (2008)

Mineshima, K., Martínez-Gómez, P., Miyao, Y., Bekki, D.: Higher-order logical inference with compositional semantics. Proceedings of the 2015 Conference on Empirical Methods in Natural Language Processing (EMNLP 2015), Lisbon, Portugal, pp. 2055–2061 (2015)

Potts, C.: The Logic of Conventional Implicatures. Oxford University Press (2005). Sammons 2010 Sammons, M., Vinod Vydiswaran, V.G., Roth, D.: Ask not what textual entailment can do for you.... In: Proceedings of the 48th Annual Meeting of the Association for Computational Linguistics:1199–1208, Uppsala, Sweden (2010)

Tian, R., Miyao, Y., Matsuzaki, T.: Logical inference on dependency-based compositional semantics. In: Proceedings of ACL, pp. 79–89 (2014)

Applicative Abstract Categorial Grammars in Full Swing

Oleg Kiselyov[✉]

Tohoku University, Sendai, Japan
oleg@okmij.org

Abstract. Recently introduced Applicative Abstract Categorial Grammars (AACG) extend the Abstract Categorial Grammar (ACG) formalism to make it more suitable for semantic analyses while preserving all of its benefits for syntactic analyses. The surface form of a sentence, the abstract (tecto-) form, as well as the meaning are all uniformly represented in AACG as typed terms, or trees. The meaning of a sentence is obtained by applying to the abstract form a composition of elementary, deterministic but generally partial transformations. These term tree transformations are specified *precisely* and can be carried out mechanically. The rigor of AACG facilitates its straightforward implementation, in Coq, Haskell, Grammatical Framework, etc.

We put AACG through its paces, illustrating its expressive power and positive as well as negative predictions on the wide range of analyses: gender-marked pronouns, quantifier ambiguity, scoping islands and binding, crossover, topicalization, and inverse linking. Most of these analyses have not been attempted for ACG.

AACG offers a different perspective on linguistic side-effects, demonstrating compositionality not just of meanings but of transformations.

1 Introduction

One of the goals of the semantic analysis is comprehension: determining the meaning of an utterance. The subject of the analysis is usually not the raw utterance but rather more or less abstract form thereof, abstracting away peculiarities of pronunciation and writing and often declination, conjugation, etc. One may as well call this 'logical form'; we will stay however with 'abstract form'. Finding the right level of abstraction is one of the challenges.

One may discern two broad perspectives of the analysis. One is proof search: building a logical system derivation whose conclusion is directly related to the abstract form of an utterance. The successfully obtained derivation is taken as the evidence that the utterance is grammatical. The meaning is then read off the derivation. The proof search perspective is manifest in type-logical grammars; it also underlies general categorial grammars and Minimalist Grammars. Parsing as deduction is truly compelling and logically insightful. Yet it is often hard to characterize the space of possible derivations, to see that everything derivable will be judged by the native speakers as grammatical, and everything that

© Springer International Publishing AG 2017
M. Otake et al. (Eds.): JSAI-isAI 2015 Workshops, LNAI 10091, pp. 66–78, 2017.
DOI: 10.1007/978-3-319-50953-2_6

judged as grammatical can be derived. In particular, it is difficult to get negative predictions: to prove that there really cannot be a derivation for a particular
sentence, no matter how hard we try to build one.

Another perspective is most clearly described by Chung-chieh Shan in his
papers on linguistic side effects [6]. (Discourse representation theory (DRT) is
also in the same vein.) It is based not on proof but on computation: the source
sentence is taken to be a program that computes the meaning as a logical formula.
As any program, it may fail: raise an 'exception' or stop before computing any
result. Thus in the computational perspective, the meaning is obtained as the
result of a mostly deterministic, inherently partial, usually precisely specified
and often mechanized process. The algorithmic nature of this process lets it
lay a credible claim to 'real life', physiological relevance. On the downside, the
computational process is rigid. It is also too easy to get bogged down to details
(control operators, implementation of monads, etc.) Taken the computational
perspective as the starting point, the present work aims to raise its level of
abstraction, eliding implementation details and adding the 'right' amount of
flexibility.

Applicative Abstract Categorial Grammars (AACG) recently introduced in
[4] is a reformulation of Abstract Categorial Grammars (ACG) [1]. AACG supports syntactic analyses within the well-understood second-order ACG. In addition and in contrast to ACG, it lets us do semantic analyses without compro-
mising syntactic ones. The meaning of a sentence is obtained from its abstract
(parsed) form through a precisely specified process – so precise that it can be
carried out mechanically, by a computer. The process is a composition of simple
tree transformations. It may fail to deliver the meaning formula, thus predicting
the unacceptability of the sentence.

AACG is formally introduced and compared to ACG in [4]. That paper however used rather simple illustrations. We now apply AACG to more interesting
analyses, of phenomena exhibited in the following examples.

(1)   Every girl$_i$'s father loves her$_i$ mother.
(1a) *Every girl$_i$'s father loves its$_i$ mother.
(1b) *Her$_i$ father loves every girl$_i$'s mother.
(1c) A girl$_i$ met every boy who liked her$_i$.
(2a) That John$_i$ left upset his$_i$ teacher.
(2b) *That every boy$_i$ left upset his$_i$ teacher.
(3a) Alice's present for him$_i$, every boy$_i$ saw.
(3b) *Every boy$_i$, his$_i$ mother likes.
(4a) At least two senators on every committee voted against the bill.
(4b) Two politicians spy on someone from every city.
(4c) Some man from every city secretly despises it.

We thus analyze, in Sect. 3, binding (1) with gender-marked pronouns (1a)
and crossover (1b), scoping islands and binding (2), and cataphora and anaphora
in topicalization (3). Section 4 deals with inverse linking (4). The examples have
many subtleties: (1c) has in reality no reading with "every boy" outscoping

"a girl". Likewise, in (4b), "two politicians" may scope either wider than "someone from every city" or narrower. There is however no reading in which "two politicians" scope between "someone" and "every city". The inadmissibility and the absence of the readings are reproduced in our analyses. These phenomena have not been dealt with, or even considered, within the original ACG. We drew most of the examples from the pioneering paper on linguistic side-effects [6], to contrast with its approach to direct compositionality. The inverse linking examples come from [5].

2   AACG Background and Quantifier Ambiguity

For reference we very briefly recall AACG, on the example of quantifier ambiguity; see [4] for all details.

In AACG, the surface form of a sentence, its various abstract (parsed) forms and the logical formula expressing the meaning – all are uniformly represented as typed terms (trees), in a T-language, Fig. 1.

$$\begin{array}{llll}
\text{Base types} & v & \text{T-Constants} & c \\
\text{T-Types} & \sigma ::= v \mid \sigma \rightarrowtail \sigma & \text{T-Terms} & d ::= c \mid d \centerdot d
\end{array}$$

$$\frac{}{c: \sigma} \qquad \frac{d_1: \sigma_1 \rightarrowtail \sigma_2 \quad d_2: \sigma_1}{d_1 \centerdot d_2: \sigma_2} \; Tapp$$

Fig. 1. T-languages

T-types σ are formed from base types v and the binary connective $- \rightarrowtail -$. T-terms d are formed from constants c and the left-associative binary connective $- \centerdot -$. Each constant is assigned a T-type; the T-type of a composite term, if any, is determined by the inference rule (Tapp). A T-language is the set of all well-typed terms – or, a set of finite trees. On the latter view, the constants with their assigned types constitute a tree grammar. Different T-languages differ in their set of base types and their constants.

The following table shows three T-languages with the sample of constants, to be used throughout. T_S is the surface language of strings, with many string constants and one binary operation (written in infix) for string concatenation. T_A is for abstract forms: Curry's tecto-grammar. Its types are familiar categories. The silent constant cl combines an NP and a VP into a clause. These T-languages are first-order.

v	c
T_A S, NP, N, VP	$John$: NP man: N $like$: $NP \rightarrowtail VP$ cl: $NP \rightarrowtail VP \rightarrowtail S$
T_S string	$\cdot$: string $\rightarrowtail$ string $\rightarrowtail$ string "John", "every", $\ldots$: string
T_L e, t	conj, disj, $\ldots : t \rightarrowtail t \rightarrowtail t$ john: e man: $e \rightarrowtail t$ $\forall, \exists$: $(e \rightarrowtail t) \rightarrowtail t$ $x, y, z, \ldots : e$ $\hat{x}, \hat{y}, \hat{z}, \ldots : t \rightarrowtail (e \rightarrowtail t)$

The language T_L is to express meaning, as a logic formula. In addition to the standard logical constants it has an infinite supply of distinct *constants* $x, y, z, \ldots$ of the type e and the corresponding set of constants $\hat{x}$, etc., intended as binders. For example, the meaning of the sentence "a man left" is written in T_L as

$$\exists \centerdot (\hat{x} \centerdot (\text{conj} \centerdot (\text{man} \centerdot x) \centerdot (\text{left} \centerdot x)))$$

We will often informally use the conventional logic notation however: $\exists x. \text{man } x \wedge$ left x. Although T_L is sort of higher-order, it has no notion of reduction or substitution; its terms are just trees, but with bindings.

Determining the meaning of a sentence is transforming its abstract form T_A to the logical formula T_L. The transformation is easier to grasp if done in small steps. We will be using a variety of T_A languages to express the intermediate abstract forms. Each language adds to the core T_A described above new constants, whose set is summarized in the following table:

$$
\begin{aligned}
every &: N \rightarrowtail NP \\
a &: N \rightarrowtail NP \\
var_x, var_y, \ldots &: NP \\
U_x, U_y, \ldots &: N \rightarrowtail S \rightarrowtail S \\
E_x, E_y, \ldots &: N \rightarrowtail S \rightarrowtail S \\
he, she, it &: NP
\end{aligned}
$$

We will refer to the set $var_x, var_y, \ldots$ as just *var*, and similarly for U and E. These sets of constants are analogous to $x, y, \ldots, \hat{x}, \hat{y}, \ldots$ of the T_L language and represent (to be) bound variables and their binders. They are *distinct* from lambda-bound variables and are not subject to substitution, α-conversion or capture-avoidance. They are the variables and binders that Kobele [5] wished for and had to emulate in a complicated way.

2.1   Quantification and Quantifier Ambiguity

As an example, the sentence "every boy likes a girl" has the abstract form

$$ex_{bg} \stackrel{\text{def}}{=} cl \cdot (every \cdot boy) \cdot (like \cdot (a \cdot girl))$$

written in the language $T_A \cup \{a, every\}$. One can easily imagine a transformation $\mathcal{L}_{syn}$ converting ex_{bg} to the T_S term

$$\texttt{"every"} \cdot \texttt{"boy"} \cdot \texttt{"like"} \cdot \texttt{"a"} \cdot \texttt{"girl"}$$

taken to be the surface form of the sentence. The transformation is a set of simple re-writing rules for converting T_A terms to T_S, as specified in Table 1. Alternatively, we can implement it as the mapping of T_A terms to the terms of linear simply-typed lambda-calculus with T_S terms as constants: Table 2. Applying such $\mathcal{L}_{syn}$ to the original ex_{bg} gives a lambda-expression, which, upon normalization, becomes the desired T_s term. The lambda-calculus view of $\mathcal{L}_{syn}$ goes back to the original ACG [1]. It is a good way of mechanically implementing the transformation, and used in our semantic calculator. After all, lambda-calculus is a term-rewriting system. Nevertheless, this view is "too concrete", showing implementation details like the normalization step. The term re-writing–system view, shown in Table 1, offers the right amount of abstraction. Therefore, it will be often used throughout the paper.

Table 1. $\mathcal{L}_{syn}$ as a term re-writing system

$$
\begin{array}{lcl}
\mathcal{L}_{syn}[boy] & \mapsto & \texttt{"boy"} \\
\mathcal{L}_{syn}[girl] & \mapsto & \texttt{"girl"} \\
\mathcal{L}_{syn}[a \cdot d] & \mapsto & \texttt{"a"} \cdot \mathcal{L}_{syn}[d] \\
\mathcal{L}_{syn}[every \cdot d] & \mapsto & \texttt{"every"} \cdot \mathcal{L}_{syn}[d] \\
\mathcal{L}_{syn}[like \cdot d] & \mapsto & \texttt{"like"} \cdot \mathcal{L}_{syn}[d] \\
\mathcal{L}_{syn}[cl \cdot d_1 \cdot d_2] & \mapsto & \mathcal{L}_{syn}[d_1] \cdot \mathcal{L}_{syn}[d_2]
\end{array}
$$

Table 2. $\mathcal{L}_{syn}$ as a mapping to lambda-calculus

$$
\begin{array}{lcl}
\mathcal{L}_{syn}[boy] & \mapsto & \texttt{"boy"} \\
\mathcal{L}_{syn}[girl] & \mapsto & \texttt{"girl"} \\
\mathcal{L}_{syn}[a] & \mapsto & \lambda d.\,\texttt{"a"} \cdot d \\
\mathcal{L}_{syn}[every] & \mapsto & \lambda d.\,\texttt{"every"} \cdot d \\
\mathcal{L}_{syn}[like] & \mapsto & \lambda d.\,\texttt{"like"} \cdot d \\
\mathcal{L}_{syn}[cl] & \mapsto & \lambda d_1 d_2.\, d_1 \cdot d_2
\end{array}
$$

There is yet another way to look at $\mathcal{L}_{syn}$: Table 3 is derived from Table 1 by replacing each $\mathcal{L}_{syn}[d]$ expression with the type of the T_A term d. The result looks like a context-free grammar. Therefore, the abstract term ex_{bg} can be viewed as a parsed tree of the grammar, and $\mathcal{L}_{syn}$ as computing its yield. There is clearly

Table 3. $\mathcal{L}_{syn}$ as a context-free grammar

$$
\begin{array}{rcl}
N & \mapsto & \texttt{"boy"} \\
N & \mapsto & \texttt{"girl"} \\
NP & \mapsto & \texttt{"a"} \cdot N \\
NP & \mapsto & \texttt{"every"} \cdot N \\
VP & \mapsto & \texttt{"like"} \cdot NP \\
S & \mapsto & NP \cdot VP
\end{array}
$$

an inverse transformation $\mathcal{L}_{syn}^{-1}$ – parsing from the surface form T_S to the parse tree T_A. The grammar in Table 3 is rather simple, thanks to the simple, unlifted types of the quantifying determiners like *every*.

The meaning of ex_{bg} is derived by applying a sequence of other transformations to it. The first transformation $\mathcal{L}_U$ turns ex_{bg} to a term in a language $T_A \cup \{a, var, U\}$, with new constants in place of *every*.

$$
\begin{array}{rcl}
\mathcal{L}_U[cl \cdot C[every \cdot d_r] \cdot d] & \mapsto & (U_x \cdot d_r) \cdot (cl \cdot C[var_x] \cdot d) \\
\mathcal{L}_U[cl \cdot d \cdot C[every \cdot d_r]] & \mapsto & (U_x \cdot d_r) \cdot (cl \cdot d \cdot C[var_x])
\end{array}
$$

where $C[]$ is a context (a term with a hole) such that the hole is not a subterm of a cl term. Section 4 will show a few more rules for $\mathcal{L}_U$. If none of them apply, $\mathcal{L}_U$ is the identity. The result of $\mathcal{L}_U[ex_{bg}]$

$$
(U_x \cdot boy) \cdot (cl \cdot var_x \cdot (like \cdot (a \cdot girl)))
$$

is in effect the Quantifier Raising (QR) of "every boy", but in a rigorous, deterministic way. The intent of the new constants should become clear: U is to represent the raised quantifier, and *var* its trace. Unlike QR, the raised quantifier $(U_x \cdot boy)$ lands not just on any suitable place. $\mathcal{L}_U$ puts it at the closest boundary marked by the clause-forming constant cl. It should also be mentioned that $\mathcal{L}_U$, as $\mathcal{L}_{sym}$, is type-preserving: it maps a well-typed term to also a well-typed term. The type preservation is the necessary condition for the correctness of the transformations. Again unlike QR, we specify the correctness conditions precisely.

The analogous $\mathcal{L}_E$ transformation raises "a girl", turning the above $T_A \cup \{a, var, U\}$ term into a term in the language $T_A \cup \{var, U, E\}$, without the constant a and with new constants E:

$$
(U_x \cdot boy) \cdot ((E_y \cdot girl) \cdot (cl \cdot var_x \cdot (like \cdot var_y)))
$$

There are no longer any of the original quantifiers. The raised existential is placed at the boundary marked by cl.

The final, straightforward transformation $\mathcal{L}_{sem}$, Table 4, turns the above term into the T_L logical formula $\forall x.\, boy\ x \implies \exists y.\, girl\ y \wedge like\ y\ x$ representing the meaning of the original sentence. If $\mathcal{L}_E$ is applied first and $\mathcal{L}_U$ second, the resulting logical formula will have the opposite order of quantifiers, denoting the inverse reading of the sentence. The quantifier ambiguity hence comes from the order of applying individual transformations.

Table 4. $\mathcal{L}_{sem}$ as a term re-writing system

$$\mathcal{L}_{sem}[boy] \qquad\qquad \mapsto \quad boy$$
$$\mathcal{L}_{sem}[girl] \qquad\qquad \mapsto \quad girl$$
$$\mathcal{L}_{sem}[var_x] \qquad\qquad \mapsto \quad x$$
$$\mathcal{L}_{sem}[(U_x \centerdot d_r) \centerdot d] \quad \mapsto \quad \forall x.\, \mathcal{L}_{sem}[d_r] \Rightarrow \mathcal{L}_{sem}[d]$$
$$\mathcal{L}_{sem}[(E_x \centerdot d_r) \centerdot d] \quad \mapsto \quad \exists x.\, \mathcal{L}_{sem}[d_r] \wedge \mathcal{L}_{sem}[d]$$
$$\mathcal{L}_{sem}[like \centerdot d] \qquad \mapsto \quad like\ \mathcal{L}_{sem}[d]$$
$$\mathcal{L}_{sem}[cl \centerdot d_1 \centerdot d_2] \quad\ \mapsto \quad \mathcal{L}_{sem}[d_2]\ \mathcal{L}_{sem}[d_1]$$

When a sentence has several quantifying determiners, each of them can be associated with its own transformation $\mathcal{L}$ that will eliminate, or raise them. Although there can be many ways to order such transformations, not every order will give a distinct reading, as we shall see soon.

The quantifier ambiguity was also treated in ACG [3], but differently, using quantifier lowering, which required guessing the (parsed) abstract form with the raised quantifiers. On our analysis, the parsed form has quantifiers in-situ, represented by the simple second-order ACG. No guessing of abstract forms is required.

2.2 Scope Islands

In the case of a scope island, like in "That every boy left upset a teacher", the abstract form

$$cl \centerdot (that \centerdot (cl \centerdot (every \centerdot boy) \centerdot left)) \centerdot (upset \centerdot (a \centerdot teacher))$$

has two cl constants corresponding to the subordinate and matrix clauses. The former is the closest to "every boy" and becomes the landing place for the raised universal, which is hence confined to the subordinate clause. As the result, changing the order of $\mathcal{L}_U$ and $\mathcal{L}_E$ transformations does not change the resulting logic formula: the original sentence does not have the quantifier ambiguity.

2.3 Lexicon: Term-Language Transformation

The transformations like $\mathcal{L}_U$, $\mathcal{L}_{syn}$, etc. (called 'lexicon') have been specified as type-preserving and confluent term-rewriting systems. As we have already mentioned for $\mathcal{L}_{syn}$, lexicon can also be programmed in linear lambda-calculus. After all, lambda-calculus is a type-preserving and confluent term-rewriting system. To transform a T_1 term into a term in the language T_2, we replace each constant in the original T_1 term by a λ-expression $\Lambda(T_2)$ and normalize the result. Here, $\Lambda(T_2)$ means a typed lambda-calculus with sums, products and the fixpoint whose constants are terms of T_2. If the normalization succeeds, we end up with the term T_2, the result of the transformation. Because of the fixpoint, there may be no normal form: the term transformations are inherently partial.

The transformations $\mathcal{L}_{syn}$ and $\mathcal{L}_{sem}$ are clearly simple homomorphisms. $\mathcal{L}_U$ and $\mathcal{L}_E$ are intuitively understood as movements, to a precisely-defined boundary. The exact details of the lambda-calculus implementation, fully described in [4], are not needed to understand the present paper or to use AACG. Often the term-rewriting system view is the right level of abstraction.

3 Anaphora and the Modeling of Dynamic Semantics

This section describes AACG analyses of binding and its interaction with other phenomena, in particular, quantification and scoping islands. We start with the trivial example "Mary loves her mother", whose abstract form is as follows:

$$cl \cdot Mary \cdot (love \cdot (possess \cdot she \cdot mother))$$

It is written in the language $T_A \cup \{pronoun\}$; the constant $possess$ has the type $NP \rightarrowtail N \rightarrowtail NP$.

We now describe in detail the transformation $\mathcal{L}_{dyn}$ that will eliminate the pronoun, replacing it with its referent. The transformation is performed in two phases. The preliminary phase adds annotations to the abstract form, to be eliminated, along with the pronouns, by the $\mathcal{L}_{dyn}$ proper. The annotation phase $\mathcal{L}_{ann}$ is a lexicon transformation like before, into a T_A language enriched with two additional constants

$$update \colon Gender \rightarrowtail NP \rightarrowtail \sigma \rightarrowtail \sigma$$
$$post \colon NP \rightarrowtail NP$$

The annotation transformation is

$$
\begin{aligned}
\mathcal{L}_{ann}[Mary] &\longmapsto & post \cdot (update \cdot Fem \cdot Mary \cdot Mary) \\
\mathcal{L}_{ann}[var_x] &\longmapsto & post \cdot var_x \\
\mathcal{L}_{ann}[(U_x \cdot girl) \cdot d] &\longmapsto & (U_x \cdot (update \cdot Fem \cdot var_x \cdot girl)) \cdot \mathcal{L}_{ann}[d]
\end{aligned}
$$

It is the identity otherwise. Applying $\mathcal{L}_{ann}$ to or example gives

$$cl \cdot (post \cdot (update \cdot Fem \cdot Mary \cdot Mary)) \cdot$$
$$(love \cdot (possess \cdot she \cdot mother))$$

The transformation $\mathcal{L}_{dyn}$ will try to eliminate the pronoun such as she, replacing it with a post-ed term. (And the follow-up trivial transformation will erase the no-longer needed $post$ and $update$ annotations). It is easier to explain the context-sensitive re-writing $\mathcal{L}_{dyn}$ using the notion of a "discourse context" – which is the global state maintained by $\mathcal{L}_{dyn}$ as it traverses the term in-order: left-to-right, depth-first. When $\mathcal{L}_{dyn}$ encounters $post \cdot d$, it posts d into the discourse context. The annotation $update \cdot Fem \cdot d_1 \cdot d_2$ represents d_2 while recording a constraint in the context: d_1 has feminine gender. The $update$ annotation does not post the term; it merely records the constraint. $\mathcal{L}_{dyn}[she]$ searches the context for a posted term associated with the feminine gender, and returns that term. In our example, $(post \cdot (update \cdot Fem \cdot Mary \cdot Mary))$ records the constraint that *Mary* has feminine gender, and then posts *Mary* into the context, where *she* will find it. In the result, $\mathcal{L}_{dyn}$ gives

$$cl \cdot Mary \cdot (love \cdot (possess \cdot Mary \cdot mother))$$

which no longer has any pronouns.

If we replace "she" with "he" in the original sentence, $\mathcal{L}_{dyn}[he]$ will fail to find the suitable referent (assuming the initially empty discourse context). The failure means the $\mathcal{L}_{dyn}$ transformation was unsuccessful and hence the sentence corresponding to the source term is unacceptable. This negative prediction is not delivered in [6] or ACG analyses since they not make gender distinctions.

$\mathcal{L}_{dyn}$ may be implemented using a global mutable state, state monad, delimited continuations, and so on – or as a context-sensitive term re-writing system. Either way, the implementation details should not matter. What matters is that $\mathcal{L}_{dyn}[she]$ examines the parents and the left siblings of *she*, looking for a posted term of the feminine gender.

The transformation $\mathcal{L}_{dyn}$ easily composes with $\mathcal{L}_U$ and $\mathcal{L}_E$, giving an account of quantification and binding, as in

(1) Every girl$_i$'s father loves her$_i$ mother.
(1b) *Her$_i$ father loves every girl$_i$'s mother.
(2a) That John$_i$ left upset his$_i$ teacher.
(2b) *That every boy$_i$ left upset his$_i$ teacher.

To wit, the abstract form of (1), written in the language $T_A \cup \{pronoun, every\}$ is

$$cl \cdot (possess \cdot (every \cdot girl) \cdot father) \cdot (love \cdot (possess \cdot she \cdot mother))$$

The transformation $\mathcal{L}_{sem}$ that produces the logic formula does not apply since that lexicon has no mapping for pronouns and *every*. They have to be eliminated first, by applying $\mathcal{L}_U$ followed by $\mathcal{L}_{dyn}$. First, $\mathcal{L}_U$ with $\mathcal{L}_{ann}$ produce

$$(U_x \cdot (update \cdot Fem \cdot var_x \cdot girl)) \cdot$$
$$(cl \cdot (possess \cdot (post \cdot var_x) \cdot father) \cdot$$
$$(love \cdot (possess \cdot she \cdot mother)))$$

Therefore, $\mathcal{L}_{dyn}$ will first record the constraint that var_x is feminine, then post var_x, and then find it when encountering *she*. The result is

$$(U_x \cdot girl) \cdot (cl \cdot (possess \cdot var_x \cdot father) \cdot (love \cdot (possess \cdot var_x \cdot mother)))$$

with the eliminated (resolved) pronoun. $\mathcal{L}_{sem}$ can now derive the logical formula representing the meaning of the sentence.

From the left-to-right, depth-first traversal mode of $\mathcal{L}_{dyn}$ we predict that a referent for a pronoun in an acceptable sentence must be located to the 'left' of the pronoun, i.e., earlier in the in-order tree traversal. For example, we predict that (1b) is unacceptable. The scoping island condition for the universal then explains the unacceptability of (2b).

Our prediction may seem at odds with (3a) and (3b):

(3a) Alice's present for him_i, every boy_i saw.
(3b) *Every boy_i, his_i mother likes.

Looking at the abstract form of (3b), for example,

$$(frontNP \cdot (every \cdot boy)) \cdot (cl \cdot (possess \cdot he \cdot mother) \cdot (like \cdot \rule{1em}{0.4pt}))$$

we see the constant $frontNP$ that corresponds to the comma that sets off the fronted NP. $\mathcal{L}_{sem}$ has no mapping for that constant and for the trace $\rule{1em}{0.4pt}$, therefore, they have to be transformed away first. Our transformation stack will hence begin with $\mathcal{L}_{lower}$ that lowers the fronted phrase into the position indicated by $\rule{1em}{0.4pt}$. The other transformations run afterwards. By the time of $\mathcal{L}_{dyn}$, the variable representing the moved out $every \cdot boy$ will be to the right of he and so the pronoun cannot refer to it.

We finish with the interesting example

(1c) A $girl_i$ met every boy who $liked_i$ her.

Although it has an existential and a universal QNPs, it exhibits no quantifier ambiguity. To understand why, consider its abstract form

$$cl \cdot (a \cdot girl) \cdot (met \cdot (every \cdot (who \cdot boy \cdot (liked \cdot she))))$$

where the constant who has the type $N \rightarrowtail VP \rightarrowtail N$. Applying $\mathcal{L}_E$ with $\mathcal{L}_{ann}$ (and omitting $update$ for clarity) gives

$$(E_x \cdot girl) \cdot (cl \cdot (post \cdot var_x) \cdot (met \cdot (every \cdot (who \cdot boy \cdot (liked \cdot she)))))$$

If the universal is raised first, the result

$$(U_y \cdot (who \cdot boy \cdot (liked \cdot she)))((E_x \cdot girl) \cdot (cl \cdot (post \cdot var_x) \cdot (met \cdot var_y)))$$

clearly has the pronoun she to the right of its posted referent and hence cannot be bound by it.

4   Inverse Linking

This section describes the analyses of inverse linking (4a–4c):

(4a) At least two senators on every committee voted against the bill.
(4b) Two politicians spy on someone from every city.
(4c) Some man from every city secretly despises it.

The examples, borrowed from [5] demonstrate three characteristic features of this phenomenon. First, a QNP embedded into a prepositional phrase attached to a noun of another QNP takes scope over that outer noun phrase: (4a) has a reading with "every committee" taking scope over "the two senators".

The second feature of inverse linking is that an external QNP like "two politicians" in (4b) cannot scope between the inversely linked "every" and "someone". Lastly, the embedded QNP on the inverse linking reading may bind pronouns in other parts of the sentence.

Previously, the restrictor of a QNP was rather simple, often merely a common noun. Inverse linking is about QNPs whose restrictor itself contains a QNP, which requires generalization of our $\mathcal{L}_U$ and $\mathcal{L}_E$ transformations. To see why, consider what happens if we apply the existing $\mathcal{L}_E$ to (4b):

$$(E_x \cdot (from \cdot person \cdot (every \cdot city))) \cdot$$
$$(cl \cdot (two \cdot politician) \cdot (spyOn \cdot var_x))$$

We have to eliminate $every$ since $\mathcal{L}_{sym}$ has no mapping for it. However, $\mathcal{L}_U$ cannot apply to $(every \cdot city)$ since that subterm does not appear within the cl context.

We have to generalize the quantifier movement transformations, and take restrictors seriously. First, we re-define U_x and E_x constants and introduce the explicit constant for restrictors:

$$U_x, U_y, \ldots : S \rightarrowtail S \rightarrowtail S$$
$$E_x, E_y, \ldots : S \rightarrowtail S \rightarrowtail S$$
$$restr : N \rightarrowtail NP \rightarrowtail S$$

The new $\mathcal{L}_U$ is as follows:

(1) $\mathcal{L}_U[cl \cdot C[every \cdot d_r] \cdot d] \qquad\qquad \mapsto$
$\qquad (U_x \cdot (restr \cdot d_r \cdot var_x)) \cdot (cl \cdot C[var_x] \cdot d)$

(2) $\mathcal{L}_U[cl \cdot d \cdot C[every \cdot d_r]] \qquad\qquad \mapsto$
$\qquad (U_x \cdot (restr \cdot d_r \cdot var_x)) \cdot (cl \cdot d \cdot C[var_x])$

(3) $\mathcal{L}_U[restr \cdot C[every \cdot d_r] \cdot var_x] \qquad \mapsto$
$\qquad (U_y \cdot (restr \cdot d_r \cdot var_y)) \cdot (restr \cdot C[var_y] \cdot var_x)$

(4) $\mathcal{L}_U[U_x \cdot (restr \cdot C[every \cdot d_r] \cdot var_x)] \mapsto$
$\qquad (U_y \cdot (restr \cdot d_r \cdot var_y)) \cdot (U_x \cdot (restr \cdot C[var_y] \cdot var_x))$

(5) $\mathcal{L}_U[E_x \cdot (restr \cdot C[every \cdot d_r] \cdot var_x)] \mapsto$
$\qquad (U_y \cdot (restr \cdot d_r \cdot var_y)) \cdot (E_x \cdot (restr \cdot C[var_y] \cdot var_x))$

The hole in the context $C[]$ should not appear as a sub-term of cl, $restr$, or within a QNP.

We now show the analysis of (4b) with the generalized quantifier movements. First we observe that we cannot move out $every \cdot city$ from the original term since this subterm is part of larger QNP. Recall that the hole in the context $C[]$ must not appear within a QNP. Suppose we first apply $\mathcal{L}_E$, producing

$$(E_x \cdot (restr \cdot (from \cdot person \cdot (every \cdot city)) \cdot var_x)) \cdot$$
$$(cl \cdot (two \cdot politician) \cdot (spyOn \cdot var_x))$$

There are now two choices of applying the $\mathcal{L}_U$ transformation. The first choice, using rule (3) of the new $\mathcal{L}_U$ transformation, gives

$(E_x$.

$\quad ((U_y \cdot (restr \cdot city \cdot var_y)) \cdot (restr \cdot (from \cdot person \cdot var_y) \cdot var_x)))$.

$\quad (cl \cdot (two \cdot politician) \cdot (spyOn \cdot var_x))$

with "someone" scoping over "every city". The other choice, using rule (5) of $\mathcal{L}_U$, produces

$\quad\quad ((U_y \cdot (restr \cdot city \cdot var_y))$.

$\quad\quad\quad (E_x \cdot (restr \cdot (from \cdot person \cdot var_y) \cdot var_x)))$.

$\quad\quad\quad (cl \cdot (two \cdot politician) \cdot (spyOn \cdot var_x))$

with the wide-scoping "every city". This is the case of inverse linking. In either case, "two politicians" takes the narrowest scope. If "two politicians" is moved out first, it takes the sentence-wide scope. In no case this QNP can scope between "someone" and "every city", reproducing the empirical restriction.

5   Conclusions

We have demonstrated AACG as the framework for uniform analyses of variety of phenomena related to quantification and binding, including the rarely treated (outside dynamic semantics) the gender marking of pronouns. The key idea is successive transformations of an abstraction of the syntactic form until we obtain the logic formula representing its meaning. We specifically used the examples of [6] to contrast the AACG's take on linguistic side-effects, effect interactions and the notion of evaluation. Whereas in [6] all linguistic side-effects (movements) occur during the single tree traversal (evaluation), we decompose them into separate simple traversals. AACG takes compositionality to a new level, not only of meanings but transformations.

Linguistic side-effects use the single delimited control operator shift for everything: to implement the discourse context and the movements similar to $\mathcal{L}_U$. The versatility had the price of rigidity. Quantification ambiguity could only be realized by changing the global evaluation order, which affects all other predictions. Delimited control is also a low-level implementation detail. In our approach, the transformations are specified rather abstractly, as type-preserving term rewriting. Although each individual lexicon transformation is confluent, there is a choice in their ordering, giving enough flexibility for individual movements – with rigidity to reproduce the scoping restrictions of inverse linking.

The present paper talked entirely about semantic interpretation of sentences, or parsing a sentence to a logical form, so to speak. We have said nothing about the converse, finding a sentence whose meaning matches the given logical form. That is, we have been solving the problem of comprehension and have not at all investigated generation. Although using arbitrary applicative functors clearly makes the generation problem intractable, there could be constraints imposed on the applicatives that make the 'inversion' of their transformations tractable, like the almost linear constraint of [2]. Viewing generation as logic programming problem, as did [2], seems promising.

References

1. de Groote, P.: Towards abstract categorial grammars. In: Proceedings of the 40th Annual Meeting of the Association for Computational Linguistics, pp. 148–155. Morgan Kaufmann, San Francisco, July 2002. http://www.aclweb.org/anthology/P01-1033
2. Kanazawa, M.: Parsing and generation as Datalog query evaluation (2011). http://research.nii.ac.jp/~kanazawa/publications/pagadqe.pdf
3. Kanazawa, M., Pogodalla, S.: Advances in Abstract Categorial Grammars: Language theory and linguistic modeling. Lecture notes, ESSLLI 09, Part 2, July 2009. http://www.loria.fr/equipes/calligramme/acg/publications/esslli-09/2009-esslli-acg-week-2-part-2.pdf
4. Kiselyov, O.: Applicative abstract categorial grammar. In: Kanazawa, M., Moss, L.S., de Paiva, V. (eds.) NLCS 2015. Third Workshop on Natural Language and Computer Science, EasyChair Proceedings in Computing, vol. 32, pp. 29–38. EasyChair (2015)
5. Kobele, G.M.: Inverse linking via function composition. Nat. Lang. Seman. **18**(2), 183–196 (2010)
6. Shan, C.-c.: Linguistic side effects. In: Barker, C., Jacobson, P. (eds.) Direct Compositionality, pp. 132–163. Oxford University Press, New York (2007)

Scope Parallelism in Coordination in Dependent Type Semantics

Yusuke Kubota[1]([✉]) and Robert Levine[2]

[1] University of Tsukuba, Tsukuba, Japan
kubota.yusuke.fn@u.tsukuba.ac.jp
[2] Ohio State University, Columbus, USA
levine@ling.ohio-state.edu

Abstract. The scope parallelism in the so-called Geach sentences in right-node raising (*Every boy admires, and every girl detests, some saxophonist*) poses a difficult challenge to many analyses of right-node raising, including ones formulated in the type-logical variants of categorial grammar (e.g. Kubota and Levine (2015)). In this paper, we first discuss Steedman's (2012) solution to this problem in Combinatory Categorial Grammar, and point out some empirical problems for it. We then propose a novel analysis of the Geach problem within Hybrid Type-Logical Categorial Grammar (Kubota and Levine 2015), by incorporating Dependent Type Semantics (Bekki 2014) as the semantic component of the theory. The key solution for the puzzle consists in linking quantifiers to the argument positions that they correspond to via an anaphoric process. Independently motivated mechanisms for anaphora resolution in DTS then automatically predicts the scope parallelism in Geach sentences as a consequence of binding parallelism independently observed in right-node raising sentences.

Keywords: Scope parallelism · Coordination · Right-node raising · Hybrid Type-Logical Categorial Grammar · Dependent Type Semantics

1 Scope Parallelism in Coordination and Type-Logical Categorial Grammar

In his recent book, Steedman (2012) offers what at first sight appears to be an elegant account of the scope parallelism in examples like (1), first noted by Geach (1972):

(1) Every boy admires, and every girl detests, some saxophonist.
$$(\forall > \exists \wedge \forall > \exists; \exists > \forall \wedge \exists > \forall)$$

(1) has only the two readings shown above, and crucially, lacks readings in which the right-node raised (RNR'ed) existential scopes above the universal in one conjunct but below it in the other conjunct.

Steedman's own account of this problem hinges on his treatment of indefinites as underspecified Skolem functions. In Steedman's approach, indefinites are

© Springer International Publishing AG 2017
M. Otake et al. (Eds.): JSAI-isAI 2015 Workshops, LNAI 10091, pp. 79–92, 2017.
DOI: 10.1007/978-3-319-50953-2_7

translated as Skolem functions whose values can be fixed via an operation called 'Skolem specification' at different points in the derivation, potentially affecting their interpretations. For example, in a simple non-RNR sentence like (2), we obtain different interpretations based on whether Skolem specification takes place at ♣ or ♠.

(2)

$$
\cfrac{
\cfrac{
\cfrac{\underset{S/(NP\backslash S)}{\text{Every farmer}} \quad \underset{(NP\backslash S)/NP}{\text{owns}}}{S/NP : \lambda x.\forall y[farmer'y \to own'xy]}\ {}^{>B} \quad
\cfrac{\underset{(S/NP)\backslash S}{\text{a donkey}}\ {}^{\clubsuit}}{: \lambda q.q(skolem'donkey')}
}{S : \forall y[farmer'y \to own'(skolem'donkey')y]}\ {}^{\spadesuit}
}{S : \forall y[farmer'y \to own'sk^{(y)}_{donkey'}y]}
$$

with

$$: \lambda p.\forall y[farmer'y \to py] \qquad : \lambda x.\lambda y.own'xy$$

In the derivation illustrated in (2), Skolem specification takes place at ♠. This has the effect that the interpretation of the Skolem function depends on the value of the variable bound by the outscoping universal (as indicated by the superscript y). This yields the $\forall > \exists$ reading, where there can be different donkeys corresponding to different farmers. In an alternative derivation in which Skolem specification takes place at an earlier point in the derivation ♣, the Skolem function is not under the scope of the universal. In this case the Skolem function is assigned a constant value, corresponding to wide scope for the existential.

In the case of a more complex derivation involving RNR, the same early specification of the Skolem function has the effect of giving the existential interpretation associated with the resulting skolem constant wide scope over the entire conjunction within which the universal quantifiers take scope. As a result, the existential inevitably winds up outscoping these universals—a possibility illustrated in the following derivation (Steedman 2012, p. 166):

(3)

$$
\cfrac{
\cfrac{\text{Every boy admires and every girl detests}}{S/NP}\quad
\cfrac{\underset{S\backslash(S/NP)}{\text{some saxophonist}}}{\begin{array}{c}: \lambda q.q(skolem'sax')\\ \cdots\cdots\cdots\cdots\\ : \lambda q.q(sk_{sax'})\end{array}}
}{S : \forall y.[boy'y \to admires'sk_{sax'}y] \land \forall z[girl'z \to detests'sk_{sax'}z]}\ {}^{<}
$$

with

$$: \lambda x.\forall y.[boy'y \to admires'xy] \land \forall z[girl'z \to detests'xz]$$

The identical Skolem constant $sk_{sax'}$ appearing in each conjunct entails the existence of a single constant saxophonist that is the target of admiration in the first conjunct and detestation in the second, and hence corresponds to an existential scoping widely over the conjunction. For an RNR sentence, another possibility is to wait to state the Skolem specification till after the RNRed existential is β-converted into the position of the x variable in each conjunct. At that point, the Skolem term is under the scope of the universal in each conjunct, and hence receives an interpretation dependent on the universally bound y and z variables, corresponding to the $\forall > \exists$ reading in both conjuncts.

The CCG analysis of RNR thus ensures that the source of scope ambiguity in RNR sentences is exactly the same as in simpler sentences like (2) (with positions in the derivation corresponding to ♣ and ♠ yielding two distinct readings). Crucially, there is no way to 'split' the derivational step at which Skolem specification takes place in the two conjuncts in an RNR sentence. It then follows that 'mixed scope' readings are ruled out automatically.

However, a closer look suggests that the same aspects of Steedman's approach which blocks the mixed-scope reading in Geach sentences lead to some serious undergeneration problems. As already noted, in Steedman's analysis, there are only two possibilities: either Skolem specification occurs after the Skolem function combines with (and falls under the scope of) the universal quantifier, or it occurs at an earlier step, in which case we obtain the wide-scope reading for the existential along the lines illustrated in (3). In other words, the only way for the Skolem function to distribute over the conjunction—i.e., for the latter to outscope it—is for the function to be specified after it falls under the scope of the universal in both conjuncts, in which case the universal will always outscope it.

But it is not difficult to find instances of RNR with a reading in which the conjunction outscopes the Skolem function but where the latter takes wide scope over the universal in each conjunct. Consider the example in (4).

(4)   Every American respects, and every Japanese admires, some novelist—namely, their respective most recent Nobel Prize winner.

The relevant reading can be paraphrased as 'There is some American novelist such that every American respects that novelist and there is some Japanese novelist such that every Japanese respects that novelist'. The existential distributes over the conjunction, but within each conjunct it takes widest scope.

The Skolem function analysis of existentials cannot capture the salient reading of (4). CCG could perhaps be extended to license cases such as (4) by taking existentials to have lexical entries not only as Skolem functions but as standard generalized quantifiers, and assuming a higher-order polymorphic category for RNR (e.g., conjunction of $S/((S/NP)\backslash S)$). However, at least the first of these assumptions is severely at odds with the core premises of Steedman's own analysis of indefinites.

Another problem comes from examples such as (5):

(5)   Every boy in that prep school started going out steadily with, and every one of his relatives ended up having serious reservations about his marrying, some totally unsophisticated rural girl.

This sentence is most naturally understood on the $\forall_{\textbf{boy}} > \exists_{\textbf{girl}} > \forall_{\textbf{relative}}$ reading, which requires having the universal quantifier in the first conjunct scope over the whole coordinate structure, thereby binding the pronoun in the restriction of the universal in the second conjunct. But since in typical cases universals scope only within their conjuncts, Steedman (2012) builds supposedly syntactic island restrictions including the Coordinate Structure Constraint into his combinatorics for quantifier interpretation. This prevents quantifiers from scoping out of the conjuncts they occur in. A reading of this sentence in which the quantifier in the first conjunct scopes over the whole coordinate structure is therefore underivable.[1]

Unlike CCG, Type-Logical Categorical Grammar (TLCG) does not suffer from undergeneration, but it suffers from overgeneration. Specifically, in TLCG, without constraining quantifier scope relations, the mixed readings are licensed. Here, we illustrate this problem with the treatment of quantifier/RNR interactions in Hybrid TLCG (Kubota and Levine 2015; K&L). In what follows, we assume basic familiarity with Hybrid TLCG. For further details, see Kubota and Levine (2015).

The overgeneration issue becomes clear once we see how the existential narrow scope reading is obtained in K&L's fragment. This part of their proof rests on the 'slanting' lemma, from which type specifications for quantifiers involving directional (i.e. forward and backward) slashes—including those separately listed in the lexicon in CCG—all follow. For example, we can derive a directionally slashed version for an object quantified NP in English as follows:

---

[1] In connection to this point, note that Steedman (2012) cites the following example from Fox (2000, pp. [51–54]):

(i)   Some student likes every professor and hates her assistant.

which (apparently) has a reading in which *her* is bound by *every professor*. Steedman suggests that this may not be a real instance of bound anaphora and that perhaps 'something other than compositional semantics is at work' (Steedman 2012, p. 173) here, since the possessive can be replaced with an epithet such as *the old dear's* (which according to Steedman (2012) normally resists bound variable construals) with the same reading preserved. If compositional semantics is not responsible for the relevant interpretation of (i), it is not clear what else is, and Steedman (2012) does not elaborate on this point any further anywhere in the whole book.

Moreover, the assumption that epithets in general cannot support real bound anaphoric interpretation seems questionable, even though such a claim is pervasive in the literature, based on the alleged evidence that such construals are unavailable in c-commanded positions in examples such as the following (see Déchaine and Wiltschko (2014) for a similar example)?:

(ii)   *Every dictator$_i$ was outraged that the bastard$_i$ was criticized by the press.

But note that the following example is fine on the bound variable construal even with the epithet in a clearly c-commanded position:

(iii)   Every [two-bit drug dealer we pull in]$_i$ is going to hear it from me that [the son-of-a-bitch]$_i$ is going to prison when this is all over.

We take it that the contrast between (iii) and (ii) should be explained in terms of difference in perspectives and that the unacceptability of (ii) does not provide any evidence for the assumption that epithets are different from pronouns in their ability to induce bound variable readings.

In any event, Steedman's (2012) account of (i), whatever one makes of it, does not seem to extend to the case of (5) in any event, since at least for the second author of the present paper, replacing *his relatives* with *the lad's relatives* in (5) does not preserve the bound-variable reading.

(6)

$$\cfrac{\cfrac{[\varphi_2; P; S/NP]^2 \quad [\varphi_1; x; NP]^1}{\cfrac{\varphi_2 \circ \varphi_1; P(x); S}{\lambda\varphi_1 . \varphi_2 \circ \varphi_1; \lambda x.P(x); S{\upharpoonright}NP}\ {\upharpoonright}I^1}}{\cfrac{\lambda\sigma . \sigma(\textsf{someone}); \exists_{\textbf{person}}; S{\upharpoonright}(S{\upharpoonright}NP) \quad \lambda\varphi_1 . \varphi_2 \circ \varphi_1; \lambda x.P(x); S{\upharpoonright}NP}{\cfrac{\varphi_2 \circ \textsf{someone}; \exists_{\textbf{person}}(\lambda x.P(x)); S}{\textsf{someone}; \lambda P.\exists_{\textbf{person}}(\lambda x.P(x)); (S/NP)\backslash S}\ \backslash I^2}\ {\upharpoonright}E}$$

The last line is identical to the lexical entry that Steedman posits for a quantifier in direct object position, but which, in Hybrid TLCG, is simply a consequence of the $S{\upharpoonright}(S{\upharpoonright}NP)$ type under the inference rules of the hybrid calculus.

We can then derive an expression subcategorizing for a 'slanted' quantifier (e.g. (6) in the object position:

(7)

$$\cfrac{\cfrac{\cfrac{\varphi_2 \circ \textsf{admires}; \lambda w.\textbf{admire}(w)(v); S/NP \quad [\varphi_1; cV; (S/NP)\backslash S]^1}{\cfrac{\varphi_2 \circ \textsf{admires} \circ \varphi_1; \mathscr{V}(\lambda w.\textbf{admire}(w)(v)); S}{\lambda\varphi_2 . \varphi_2 \circ \textsf{admires} \circ \varphi_1; \lambda v.\mathscr{V}(\lambda w.\textbf{admire}(w)(v)); S{\upharpoonright}NP}} \quad \cfrac{\lambda\sigma_1 . \sigma_1(\textsf{every} \circ \textsf{boy});}{\textbf{V}_{\textbf{boy}}; S{\upharpoonright}(S{\upharpoonright}NP)}}{\textsf{every} \circ \textsf{boy} \circ \textsf{admires} \circ \varphi_1; \textbf{V}_{\textbf{boy}}(\lambda v.\mathscr{V}(\lambda w.\textbf{admire}(w)(v))); S}}{\textsf{every} \circ \textsf{boy} \circ \textsf{admires}; \lambda\mathscr{V}.\textbf{V}_{\textbf{boy}}(\lambda v.\mathscr{V}(\lambda w.\textbf{admire}(w)(v))); S/((S/NP)\backslash S)}$$

Two or more such signs can be conjoined to produce functor of the same type, which will take the directionally slashed quantifier term as its argument, as for example in (8):

(8) $\textbf{V}_{\textbf{boy}}(\lambda z.\exists_{\textbf{saxist}}(\lambda x.\textbf{admire}(x)(z))) \ \wedge\ \textbf{V}_{\textbf{girl}}(\lambda z.\exists_{\textbf{saxist}}(\lambda x.\textbf{detest}(x)(z)))$

However, this analysis also overgenerates the mixed readings since the type $S/((S/NP)\backslash S)$ sign for the conjunct can be associated with a different scoping relation between the two quantifiers:

(9)

$$\cfrac{\cfrac{\cfrac{\textsf{every} \circ \textsf{girl} \circ \textsf{detests} \circ \varphi_1; \textbf{V}_{\textbf{girl}}(\lambda y.\textbf{detest}(u)(y)); S}{\textsf{every} \circ \textsf{girl} \circ \textsf{detests}; \lambda u.\textbf{V}_{\textbf{girl}}(\lambda y.\textbf{detest}(u)(y)); S/NP} \quad [\varphi_2; \mathscr{U}; (S/NP)\backslash S]^1}{\textsf{every} \circ \textsf{girl} \circ \textsf{detests} \circ \varphi_2; \mathscr{U}(\lambda u.\textbf{V}_{\textbf{girl}}(\lambda y.\textbf{detest}(u)(y))); S}}{\textsf{every} \circ \textsf{girl} \circ \textsf{detests}; \lambda\mathscr{U}.\mathscr{U}(\lambda u.\textbf{V}_{\textbf{girl}}(\lambda y.\textbf{detest}(u)(y))); S/((S/NP)\backslash S)}$$

Conjoining (9) and (7) yields a functor taking (6) as an argument to give rise to a mixed reading.

We are thus left in the unsatisfactory situation of either blocking mixed readings for Geach sentences but undergenerating (4) and (5) via the CCG analysis, or licensing (4) and (5) while overgenerating the mixed Geach reading via Hybrid TLCG. In the latter case, the source of the problem is that in its current formulation, Hybrid TLCG has no way to ensure that the bound higher order variables in each conjunct, corresponding to the RNR'ed generalized quantifier argument of the conjunction, have parallel scope with regard to the other quantifier term in their respective conjunct. As things stand, given the inherent lack of correlation between how hypothetical reasoning is carried out in two different parts of a proof, it is difficult to see how this parallelism could be enforced. In the following section, therefore, we argue for an analysis of quantifier interpretation which is quite different from the standard assumptions common across the various different versions of TLCG.

2   Scope Parallelism as Binding Parallelism in Dependent Type Semantics

We adopt Dependent Type Semantics (DTS; Bekki (2014)) as the compositional semantic theory for Hybrid TLCG to solve the overgeneration problem. DTS is a proof-theoretic compositional dynamic semantics based on Dependent Type Theory (Martin-Löf 1984).

2.1   Anaphora in DTS

We start by illustrating the analysis of anaphora resolution in DTS with (10):

(10)   A man entered. He sat down.

DTS assigns the following semantic translation for the above mini-discourse:

(11)
$$\lambda c. \begin{bmatrix} v : \begin{bmatrix} u : \begin{bmatrix} x : \textbf{Ent} \\ \textbf{Man}(x) \end{bmatrix} \\ \textbf{Enter}(\pi_1 u) \end{bmatrix} \\ \textbf{Sit\text{-}down}(@_1(c, v)) \end{bmatrix}$$

Here u is a proof term of the dependent sum type $\begin{bmatrix} x : \textbf{Ent} \\ \textbf{Man}(x) \end{bmatrix}$ (roughly corresponding to the existential quantification $\exists x.\textbf{Man}(x)$; we sometimes abbreviate this as $[(x : \textbf{Ent}) \times \textbf{Man}(x)]$ below). The existence of this proof term means that there is a proof that x is of type **Ent** (entity) and that x is a man. Unlike existential quantification, since u is technically just a pair, its components can be referenced by projection functions. Thus, $\pi_1 u$ corresponds to the variable x, and this means that v is (roughly) a proof that the proposition $\exists x.[\textbf{Man}(x) \wedge \textbf{Enter}(x)]$ is true. Finally, pronouns encode underspecified proof terms with the @ operator (which comes with indices to impose certain identity conditions; see below for its use). Resolving this underspecification amounts to resolving anaphora. In (11), a context (formally modelled as a pair) consisting of the previous discourse context c and the whole proof term for the immediately preceding sentence (v) is passed on to $@_1$ as an argument. By instantiating this underspecified term as $@_1 = \lambda c.\pi_1\pi_1\pi_2 c$, we obtain the intended reading for (10), where the individual that sat down is the man who entered (note that with this resolution, $@_1(c, v)$ corresponds to x).

2.2 Mixed Binding Problem Solved

DTS solves an important overgeneration issue in a previous approach to anaphora resolution in a proof theoretic setup by Krahmer and Piwek (Krahmer and Piwek 1999, Piwek and Krahmer 2000), which incorrectly licenses the mixed binding reading ('John loves his own father and Bill loves somebody else's father') for the following example:

(12)   Each of John and Bill loves his father.

Specifically, the following representation is assigned as the meaning for (12) in Bekki's (2014) approach:

(13)
$$\lambda c.\mathbf{L}(\mathbf{j}, \pi_1((@_1 : \gamma_1 \rightarrow \begin{bmatrix} x : \mathbf{Ent} \\ \mathbf{FatherOf}(x, (@_0 : \gamma_0 \rightarrow e)(c, \mathbf{j})) \end{bmatrix})(c, \mathbf{j})))$$
$$\wedge \mathbf{L}(\mathbf{b}, \pi_1((@_1 : \gamma_1 \rightarrow \begin{bmatrix} x : \mathbf{Ent} \\ \mathbf{FatherOf}(x, (@_0 : \gamma_0 \rightarrow e)(c, \mathbf{b})) \end{bmatrix})(c, \mathbf{b})))$$

The crucial assumption here is that the $@_0$ operator corresponding to the pronoun *his* is resolved in the same way. By resolving it as $@_0 = \pi_2$ we obtain the parallel sloppy reading. If, by contrast, $@_0$ is resolved in some other way to pick up some antecedent from the discourse (represented by the context variable c), then a parallel strict reading is obtained. Crucially, because of the coindexing (by the subscript 0) of the $@_0$ operator, $@_0$ is to be instantiated in exactly the same way in its two occurrences in the two conjuncts. Thus, mixed binding readings are ruled out.

As we show below, our own solution for the Geach problem below crucially relies on DTS's solution for the mixed binding problem.

2.3 HTLCG+DTS

We now present a fragment adopting DTS as the dynamic compositional semantics theory for Hybrid TLCG (HTLCG+DTS). HTLCG+DTS is modelled after Bekki's (2014) fragment combining CCG and DTS (CCG+DTS) and mostly involves straightforward translation of Bekki's CCG lexicon to Hybrid TLCG. The only major difference is that the lexical assignments of quantifiers and pronouns with directional slashes are replaced by ones with the vertical slash (of type $S{\upharpoonright}(S{\upharpoonright}NP)$):

(14)   a. $\lambda \varphi \lambda \sigma . \sigma(\mathsf{every} \circ \varphi); \lambda P \lambda Q \lambda c.(u : [(x : \mathbf{Ent}) \times Pxc]) \rightarrow Q(\pi_1 u)(c, u);$ $S{\upharpoonright}(S{\upharpoonright}NP){\upharpoonright}N$

   b. $\lambda \varphi \lambda \sigma . \sigma(\mathsf{some} \circ \varphi); \lambda P \lambda Q \lambda c. \begin{bmatrix} u : [(x : \mathbf{Ent}) \times Pxc] \\ Q(\pi_1 u)(c, u) \end{bmatrix}; S{\upharpoonright}(S{\upharpoonright}NP){\upharpoonright}N$

   c. $\lambda \sigma . \sigma(\mathsf{it}); \lambda P \lambda c.P(@_1 c)(c, @_1 c); S{\upharpoonright}(S{\upharpoonright}NP)$

We can now analyze the donkey sentence *Every farmer who owns a donkey beats it* as in (15), which yields exactly the same translation as Bekki's CCG+DTS fragment.

(15)

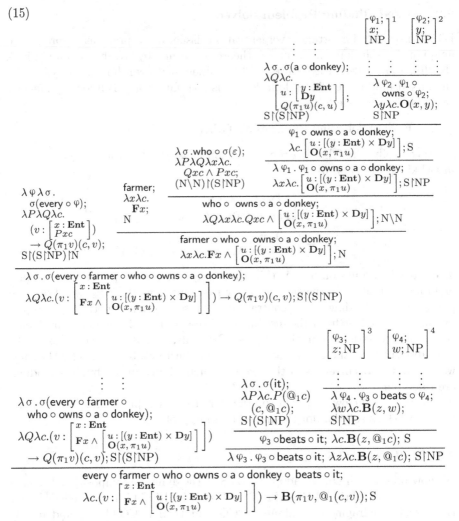

As in Bekki's (2014) analysis, by instantiating the $@_1$ operator as $@_1 = \lambda c.\pi_1\pi_1\pi_2\pi_2\pi_2 c$, we obtain the reading in which the pronoun refers to the donkey.

2.4   Geach Problem Solved

The intuition behind our analysis of Geach sentences is that the derivation for the mixed reading for (1) fails for the same reason that mixed readings for pronouns is ruled out in coordination contexts, as illustrated above for the case of ordinary constituent coordination. This was in fact one of the key motivations for the use of the @ operator for anaphora resolution in DTS. The same binding parallelism extends to RNR contexts, as noted by Jacobson (1999):

(16)   Every Englishman respects, and every American loves, his mother.

The following translation is assigned to (16) by assuming the direct constituent coordination analysis of RNR standard in many variants of CG:

(17)
$$\lambda c.(u : \begin{bmatrix} x : \mathbf{Ent} \\ \mathbf{E}(x) \end{bmatrix}) \to \mathbf{R}(\pi_1 u, \pi_1((@_1 : \gamma_1 \to \begin{bmatrix} y : \mathbf{Ent} \\ \mathbf{M}(y, (@_0 : \gamma_0 \to e)(c, u)) \end{bmatrix})(c, u)))$$

$$\wedge (u : \begin{bmatrix} x : \mathbf{Ent} \\ \mathbf{A}(x) \end{bmatrix}) \to \mathbf{L}(\pi_1 u, \pi_1((@_1 : \gamma_1 \to \begin{bmatrix} y : \mathbf{Ent} \\ \mathbf{M}(y, (@_0 : \gamma_0 \to e)(c, u)) \end{bmatrix})(c, u)))$$

The situation is completely parallel to the binding parallelism in ordinary constituent coordination. If the $@_0$ operator is instantiated as $@_0 = \lambda c.\pi_1 \pi_2 c$, then we obtain the parallel sloppy reading and if it is instantiated to pick some other individual available in the context c, then the parallel strict reading is obtained. Since these are the only two possibilities for instantiating $@_0$, mixed readings are correctly blocked.

To capture the parallel behavior of pronominal binding and quantificational scope in Geach sentences, we take quantifiers to leave an 'invisible pronoun' in the trace position. The intuition behind this proposal is that sentences like (1) receive interpretations that can roughly be paraphrased as follows:

(18)   Every boy$_i$ is such that there is a saxophonist$_j$ such that he$_i$ admires him$_j$, and every girl$_k$ is such that there is a saxophonist$_l$ such that she$_k$ admires him$_l$. (on the $\forall > \exists \wedge \forall > \exists$ reading)

The problem of mixed scope readings for quantifiers then reduces to the problem of mixed binding readings, since the mixed scope readings for (1) require the 'hidden pronoun' corresponding to the shared object quantifier to resolve anaphora differently in the two conjuncts. To put it informally, although the antecedent existential quantifier is 'visible' as antecedent to the pronoun in the trace position in both scope configurations, the relative scope between the subject and object quantifiers differ in the two clauses in the mixed reading cases since the 'target' antecedent quantifier is located in different positions in the context passed to the @ operator, thus making the mixed construal unavailable.

Below, we formalize an analysis that implements the above analytic idea in HTLCG+DTS and show how it solves the problem of mixed scope readings in Geach sentences. Since the 'hidden pronoun' that is involved in the interpretation of quantifiers plays a nontrivial role only for the shared object quantifier in RNR sentences like (1), for ease of exposition, we illustrate the analysis with a simplified translation that involves a hidden pronoun only for the object quantifier. To simplify the analysis further, we treat pronouns in syntactic category NP_p as in (19a), and assume that verbs can undergo lexical operations to take pronouns in any argument position, as in (19b):

(19)   a.   it; $\lambda c.@_i c$; NP_p

      b.   admires; $\lambda f \lambda x \lambda c. \mathbf{A}(x, fc)$; $(NP \backslash S)/NP_p$

A quantifier entry that introduces a hidden pronoun in the trace position can then be written as follows:[2]

(20) $\lambda \varphi \lambda \sigma . \sigma(\mathsf{some} \circ \varphi); \lambda P \lambda Q \lambda c. \begin{bmatrix} u : [(x : \mathbf{Ent}) \times Pxc] \\ Q(\lambda d.@_1^{\pi_1 u} d)(c, u) \end{bmatrix}; \mathrm{S} {\upharpoonright} (\mathrm{S}{\upharpoonright} \mathrm{NP}_p) {\upharpoonright} \mathrm{N}$

Here, $@_1^{\pi_1 u}$ is an abbreviation of a *constrained* @ operator such that, for any c, $@_1^{\pi_1 u} c$ is well-defined only if $@_1^{\pi_1 u} c = \pi_1 u$.

With the entry for the existential quantifier in (20), we obtain the following analysis for the surface scope reading of the sentence *Every boy admires some saxophonist*:

(21)

$$
\begin{array}{c}
\begin{array}{cc} \vdots & \vdots \end{array} \qquad \begin{bmatrix} \varphi_1; \\ z; \mathrm{NP} \end{bmatrix}^1 \quad \begin{bmatrix} \varphi_2; \\ f; \mathrm{NP}_p \end{bmatrix}^2
\end{array}
$$

$\lambda \sigma . \sigma(\mathsf{some} \circ \mathsf{saxophonist});$
$\lambda Q \lambda c. \begin{bmatrix} u : [(x : \mathbf{Ent}) \times \mathbf{S}x] \\ Q(\lambda d.@_1^{\pi_1 u} d)(c, u) \end{bmatrix};$
$\mathrm{S}{\upharpoonright}(\mathrm{S}{\upharpoonright} \mathrm{NP}_p)$

$\begin{array}{cc} \vdots & \vdots \end{array} \qquad \begin{array}{cc} \vdots & \vdots \end{array}$
$\underline{\hspace{4cm}}$
$\lambda \varphi_2 . \varphi_1 \circ \mathsf{admires} \circ \varphi_2;$
$\lambda f \lambda c. \mathbf{A}(z, fc); \mathrm{S}{\upharpoonright} \mathrm{NP}_p$

$\underline{\hspace{10cm}}$

$\varphi_1 \circ \mathsf{admires} \circ \mathsf{some} \circ \mathsf{saxophonist};$
$\lambda c. \begin{bmatrix} u : [(x : \mathbf{Ent}) \times \mathbf{S}x] \\ \mathbf{A}(z, @_1^{\pi_1 u}(c, u)) \end{bmatrix}; \mathrm{S}$

$\begin{array}{cc} \vdots & \vdots \end{array}$
$\lambda \varphi \lambda \sigma . \sigma(\mathsf{every} \circ \mathsf{boy});$
$\lambda Q \lambda c.(v : [(x : \mathbf{Ent}) \times \mathbf{B}x])$
$\quad \to Q(\pi_1 v)(c, v);$
$\mathrm{S}{\upharpoonright}(\mathrm{S}{\upharpoonright} \mathrm{NP})$

$\underline{\hspace{6cm}}$
$\lambda \varphi_1 . \varphi_1 \circ \mathsf{admires} \circ \mathsf{some} \circ \mathsf{saxophonist};$
$\lambda z \lambda c. \begin{bmatrix} u : [(x : \mathbf{Ent}) \times \mathbf{S}x] \\ \mathbf{A}(z, @_1^{\pi_1 u}(c, u)) \end{bmatrix}; \mathrm{S}{\upharpoonright} \mathrm{NP}$

$\underline{\hspace{12cm}}$

$\mathsf{every} \circ \mathsf{boy} \circ \mathsf{admires} \circ \mathsf{some} \circ \mathsf{saxophonist};$
$\lambda c.(v : [(x : \mathbf{Ent}) \times \mathbf{B}x]) \to \begin{bmatrix} u : [(x : \mathbf{Ent}) \times \mathbf{S}x] \\ \mathbf{A}(\pi_1 v, @_1^{\pi_1 u}((c, v), u)) \end{bmatrix}; \mathrm{S}$

The $@_1^{\pi_1 u}$ operator is constrained to be well-defined only if it returns the term $\pi_1 u$ for any given c as an argument. Thus, the only possible instantiation that yields an interpretable result for (21) is $@_1^{\pi_1 u} = \lambda c. \pi_1 \pi_2 c$. With this instantiation of $@_1^{\pi_1 u}$, we obtain the following as the final translation of (21) after anaphora resolution has taken place:

(22)

$$
\lambda c.(v : [(x : \mathbf{Ent}) \times \mathbf{B}x]) \to \begin{bmatrix} u : [(x : \mathbf{Ent}) \times \mathbf{S}x] \\ \mathbf{A}(\pi_1 v, \pi_1 u) \end{bmatrix}
$$

Moving on to the Geach sentences, corresponding to (7) and (9), we now obtain the following signs:

(23) $\mathsf{every} \circ \mathsf{boy} \circ \mathsf{admires}; \lambda \mathscr{P} \lambda c.(w : [(x{:}\mathbf{Ent}) \times \mathbf{B}x]) \to \mathscr{P}(\lambda f \lambda c. \mathbf{A}(\pi_1 w, fc))(c, w);$
$\mathrm{S}/((\mathrm{S}/\mathrm{NP}_p)\backslash \mathrm{S})$

---

[2] The idea that 'traces' of movement contain more contentful material than just bound variables seems to have something in common with various proposals made in the minimalist literature in the context of the so-called 'copy theory' of movement (cf., e.g., Fox (1999; 2002)).

(24) every $\circ$ girl $\circ$ detests; $\lambda\mathscr{P}.\mathscr{P}(\lambda f\lambda c.(w{:}[(x : \mathbf{Ent}) \times \mathbf{G}x]) \rightarrow \mathbf{D}(\pi_1 w, f(c, w)));$
S/((S/NP$_p$)\S)

By conjoining two conjuncts of the form in (24) via dynamic generalized conjunction and giving a slanted existential (which has the same semantic translation as (20)) as an argument to that functor, we obtain the following translation for the whole sentence:

(25)
$$\lambda c. \left[t: \begin{bmatrix} w:[(x:\mathbf{Ent}) \times \mathbf{S}x] \\ (u:[(x:\mathbf{Ent}) \times \mathbf{B}x]) \rightarrow \mathbf{A}(@_1^{\pi_1 w}((c,w),u)) \end{bmatrix} \\ \begin{bmatrix} s:[(x:\mathbf{Ent}) \times \mathbf{S}x] \\ (v:[(x:\mathbf{Ent}) \times \mathbf{G}x]) \rightarrow \mathbf{D}(@_1^{\pi_1 s}(((c,t),s),v)) \end{bmatrix} \right]$$

By resolving the $@_1$ operator as $@_1 = \lambda c.\pi_1\pi_2\pi_1 c$, we obtain a parallel wide scope interpretation for the existential suitable for examples like (4).

Conjoining (23) and (24) and feeding the slanted existential to it yields the following translation:

(26)
$$\lambda c. \left[t: \left((w:[(x:\mathbf{Ent}) \times \mathbf{B}x]) \rightarrow \begin{bmatrix} u:[(x:\mathbf{Ent}) \times \mathbf{S}x] \\ \mathbf{A}(@_1^{\pi_1 u}((c,w),u)) \end{bmatrix} \right) \\ \begin{bmatrix} s:[(x:\mathbf{Ent}) \times \mathbf{S}x] \\ (v:[(x:\mathbf{Ent}) \times \mathbf{G}x]) \rightarrow \mathbf{D}(\pi_1 v, (@_1^{\pi_1 s}(((c,t),s),v))) \end{bmatrix} \right]$$

But this does not correspond to any coherent interpretation of the sentence since there is no coherent way to instantiate the $@_1$ operator. Thus, the mixed readings are correctly ruled out.

(5), which was problematic for Steedman's (2012) account, can be dealt with easily in the present approach. For expository convenience, we analyze a slightly different example (27), which exhibits essentially the same scopal relation between the two subject quantifiers and the RNR'ed quantifier:

(27) A famous professor$_i$ in our department agreed to fix, and (therefore) a student of his$_i$ will wind up eliminating, every remaining problem in Taking Scope.

To derive the relevant reading ($\exists_{\mathbf{prof}} > \forall_{\mathbf{problem}} > \exists_{\mathbf{student}}$) for this example, we need the following pronominalization operator that turns a missing NP into a pronominal one:

(28) $\lambda\sigma.\sigma$; $\lambda P\lambda f\lambda c.P(fc)c$; (S↾NP$_p$)↾(S↾NP)

The derivation then goes as follows (here, we simplify the meaning of the existential quantifier instead of the universal quantifier; the last step which introduces the subject quantifier for the first conjunct is omitted):

(29)

$$[\varphi_1; x; \mathrm{NP}]^1 \quad [\varphi_2; y; \mathrm{NP}]^2$$

$$\begin{bmatrix} \varphi_3; \\ z; \\ \mathrm{NP} \end{bmatrix}^3 \quad \begin{array}{c} \vdots \quad \vdots \\ \varphi_1 \circ \text{ agreed} \circ \text{to} \circ \text{fix} \circ \text{and} \circ \varphi_2 \circ \text{eliminates}; \\ \lambda z \lambda c.[(\mathbf{F}(x,z)) \times \mathbf{E}(y,z)]; \mathrm{S/NP} \end{array}$$

$$\begin{array}{c} \lambda \sigma \,.\, \sigma(\mathrm{a} \circ \text{student}); \\ \lambda Q \lambda c. \begin{bmatrix} u : \begin{bmatrix} y : \mathbf{Ent} \\ \mathbf{S}y \end{bmatrix} \\ Q(\pi_1 u)(c, u) \end{bmatrix}; \\ \mathrm{S} {\restriction} (\mathrm{S} {\restriction} \mathrm{NP}) \end{array} \quad \begin{array}{c} \varphi_1 \circ \text{ agreed} \circ \text{to} \circ \text{fix} \circ \text{and} \circ \varphi_2 \circ \text{eliminates} \circ \varphi_3; \\ \lambda c.[(\mathbf{F}(x,z)) \times \mathbf{E}(y,z)]; \mathrm{S} \\ \hline \lambda \varphi_2 \,.\, \varphi_1 \circ \text{ agreed} \circ \text{to} \circ \text{fix} \circ \text{and} \circ \varphi_2 \circ \text{eliminates} \circ \varphi_3; \\ \lambda y \lambda c.[(\mathbf{F}(x,z)) \times \mathbf{E}(y,z)]; \mathrm{S} {\restriction} \mathrm{NP} \end{array}$$

$$\begin{array}{c} \varphi_1 \circ \text{ agreed} \circ \text{to} \circ \text{fix} \circ \text{and} \circ \mathrm{a} \circ \text{student} \circ \text{eliminates} \circ \varphi_3; \\ \lambda c. \begin{bmatrix} u : [(y : \mathbf{Ent}) \times \mathbf{S}y] \\ [(\mathbf{F}(x,z)) \times \mathbf{E}(\pi_1 u, z)] \end{bmatrix}; \mathrm{S} \end{array}$$

$$\begin{array}{c} \lambda \varphi_3 \,.\, \varphi_1 \circ \text{ agreed} \circ \text{to} \circ \text{fix} \circ \text{and} \circ \mathrm{a} \circ \text{student} \circ \text{eliminates} \circ \varphi_3; \\ \lambda z \lambda c. \begin{bmatrix} u : [(y : \mathbf{Ent}) \times \mathbf{S}y] \\ [(\mathbf{F}(x,z)) \times \mathbf{E}(\pi_1 u, z)] \end{bmatrix}; \mathrm{S} {\restriction} \mathrm{NP} \end{array}$$

$$\begin{array}{c} \vdots \quad \vdots \\ \lambda \sigma \,.\, \sigma; \\ \lambda P \lambda f \lambda c. P(fc)c; \\ (\mathrm{S} {\restriction} \mathrm{NP}_p) {\restriction} (\mathrm{S} {\restriction} \mathrm{NP}) \end{array} \quad \begin{array}{c} \lambda \varphi_3 \,.\, \varphi_1 \circ \text{ agreed} \circ \text{to} \circ \text{fix} \circ \\ \text{and} \circ \mathrm{a} \circ \text{student} \circ \text{eliminates} \circ \varphi_3; \\ \lambda z \lambda c. \begin{bmatrix} u : [(y : \mathbf{Ent}) \times \mathbf{S}y] \\ [(\mathbf{F}(x,z)) \times \mathbf{E}(\pi_1 u, z)] \end{bmatrix}; \mathrm{S} {\restriction} \mathrm{NP} \end{array}$$

$$\begin{array}{c} \vdots \quad \vdots \\ \lambda \varphi \lambda \sigma \,.\, \sigma(\text{every} \circ \text{problem}); \\ \lambda Q \lambda c.(v : [(x : \mathbf{Ent}) \times \mathbf{B}x]) \\ \to Q(\lambda d. @_1^{\pi_1 v} d)(c, v); \\ \mathrm{S} {\restriction} (\mathrm{S} {\restriction} \mathrm{NP}_p) \end{array} \quad \begin{array}{c} \lambda \varphi_3 \,.\, \varphi_1 \circ \text{ agreed} \circ \text{to} \circ \text{fix} \circ \\ \text{and} \circ \mathrm{a} \circ \text{student} \circ \text{eliminates} \circ \varphi_3; \\ \lambda f \lambda c. \begin{bmatrix} u : [(y : \mathbf{Ent}) \times \mathbf{S}y] \\ [(\mathbf{F}(x, fc)) \times \mathbf{E}(\pi_1 u, fc)] \end{bmatrix}; \mathrm{S} {\restriction} \mathrm{NP}_p \end{array}$$

$$\begin{array}{c} \varphi_1 \circ \text{ agreed} \circ \text{to} \circ \text{fix} \circ \text{and} \circ \mathrm{a} \circ \text{student} \circ \text{eliminates} \circ \text{every} \circ \text{problem}; \\ \lambda c.(v : [(x : \mathbf{Ent}) \times \mathbf{B}x]) \to \begin{bmatrix} u : [(y : \mathbf{Ent}) \times \mathbf{S}y] \\ [(\mathbf{F}(x, @_1^{\pi_1 v}(c, v))) \times \mathbf{E}(\pi_1 u, @_1^{\pi_1 v}(c, v))] \end{bmatrix}; \mathrm{S} \end{array}$$

The point here is that the parallel scope requirement for the right node-raised quantifier is lifted since the anaphora resolution process for the hidden pronoun corresponding to the quantifier takes place outside of the conjunction. Thus, an asymmetrical $\exists_{\mathbf{prof}} > \forall_{\mathbf{problem}} > \exists_{\mathbf{student}}$ reading does not pose any problem as long as the right node-raised quantifier takes scope outside of the conjunction.

3   Open Questions

A reviewer for the present paper notes that the constraint we impose on anaphora resolution for the covert pronoun in the interpretation of quantifiers may be too strong. In this section, we discuss some examples which may turn out to be problematic for the current formulation of our account for this reason. The key readings of the relevant examples are somewhat complex, and it is difficult to determine whether they are available readings. For this reason, we leave it as an open issue to decide whether these examples ultimately provide counterexamples for our approach.

The first type of potential problem for the present approach comes from examples such as the following:

(30)   John admires, and every girl detests, some saxophonist.

The $\forall > \exists$ reading for the second conjunct is unproblematic, since in that case, the existential is the outermost 'discourse referent' in the context passed on to the hidden pronoun in the two conjuncts. Thus, $@_1 = \lambda c.\pi_1\pi_2 c$ suffices to resolve the anaphoric reference correctly. The reading in which the existential outscopes the whole conjunction is similarly unproblematic. In this case, just as in (27), the hidden pronoun itself scopes over the conjunction, and thus, the index on the @ operator does not introduce any further constraint on its interpretation. However, the present approach does not license the reading in which the existential scopes over the universal but *within* the second conjunct, unless proper names such as *John* are treated on a par with quantifiers. To see this, note that our analysis assigns the following representation for the relevant reading of the sentence:

$$(31) \quad \lambda c. \left[\begin{array}{l} t : \left[\begin{array}{l} u : [(x : \mathbf{Ent}) \times \mathbf{S}x] \\ \mathbf{A}(\mathbf{j}, @_1^{\pi_1 u}(c, t)) \end{array} \right] \\ \left[\begin{array}{l} s : [(x : \mathbf{Ent}) \times \mathbf{S}x] \\ (v : [(x : \mathbf{Ent}) \times \mathbf{G}x]) \to \mathbf{D}(\pi_1 v, (@_1^{\pi_1 s}(((c, t), s), v))) \end{array} \right] \end{array} \right]$$

Just as in (27), there is no way of instantiating the @ operator so that it yields a coherent interpretation.

It is currently unclear to us whether (30) has the relevant reading. (For the second author of the present paper, that reading does not seem to be available.) Obviously, more work needs to be done to first clarify the empirical issue, and then, to make any necessary theoretical adjustments, but we have to leave this task for future study.

More generally, when different numbers of quantifiers are present in the two conjuncts, a parallel in-conjunct wide scope reading is blocked on the present account. Thus, the following example is predicted to lack the reading parallel to (4) where the right node-raised existential scopes above the subject universal separately in the two conjuncts:

(32) Every American detests, and every Japanese has some serious reservations for, some Nobel Prize winner—namely, their respective most recent Literature Prize winners.

Again, it is difficult to tell whether the intended reading is available for this sentence.

Acknowledgments. We thank two anonymous reviewers for LENLS 12 for their insightful comments. Discussions with Daisuke Bekki and Koji Mineshima have also been very useful. The first author acknowledges the financial support from the Japan Society for the Promotion of Science (KAKENHI; Grant number 15K16732).

References

Bekki, D.: Representing anaphora with dependent types. In: Asher, N., Soloviev, S. (eds.) Logical Aspects of Computational Linguistics 2014. Lecture Notes in Computer Science, vol. 8535, pp. 14–29. Springer, Heidelberg (2014)

Déchaine, R-M., Martina W.: Bound variable anaphora (2014). http://ling.auf.net/lingbuzz/002280

Fox, D.: Reconstruction, binding theory, and the interpretation of chains. Linguist. Inq. **30**, 157–196 (1999)

Fox, D.: Economy and Semantic Interpretation. MIT Press, Cambridge (2000)

Fox, D.: Antecedent-contained deletion and the copy theory of movement. Linguist. Inq. **33**(1), 63–96 (2002)

Geach, P.T.: A program for syntax. In: Davidson, D., Harman, G.H. (eds.) Semantics of Natural Language, pp. 483–497. D. Reidel Publishing Co., Dordrecht (1972)

Jacobson, P.: Towards a variable-free semantics. Linguist. Philos. **22**(2), 117–184 (1999)

Krahmer, E., Piwek, P.: Presupposition projection as proof construction. In: Bunt, H., Muskens, R. (eds.) Computing Meaning, vol. 1, pp. 281–300. Kluwer, Dordrecht (1999)

Kubota, Y., Levine, R.: Against ellipsis: arguments for the direct licensing of 'non-canonical' coordinations. Linguist. Philos. **38**(6), 521–576 (2015)

Martin-Löf, P.: Intuitionistic Type Theory. Bibliopolis, Naples (1984)

Piwek, P., Krahmer, E.: Presuppositions in context: constructing bridges. In: Bonzon, P., Cavalcanti, M., Nossum, R. (eds.) Formal Aspects of Context, pp. 85–106. Kluwer, Dordrecht (2000)

Steedman, M.: Taking Scope. MIT Press, Cambridge (2012)

Discourse Particles as CCP-modifiers:
German *doch* and *ja* as Context Filters

Lukas Rieser[(✉)]

Graduate School of Letters, Kyoto University, Kyoto, Japan
lukasjrieser@gmail.com

Abstract. This paper proposes an analysis of declaratives with the German discourse particles *doch* and *ja* as context change potentials (CCPs), imposing restrictions on input and output contexts consisting of public beliefs of the discourse participants. The analysis accounts for a wider range of data than previous approaches and makes novel predictions on the distribution of *doch* and *ja* by defining public beliefs as independent of fist-order beliefs, and modeling the difference between *ja* and *doch* in terms of whether a public belief of the addressee is presupposed (*ja*), or not (*doch*).

Keywords: German · Discourse particles · Context change potentials

The paper is structured as follows: Section one gives an overview of the data, followed by a discussion of previous approaches in section two. The third section lays out the semantics of particle-utterances as CCPs based on higher-order beliefs, section four discusses how this analysis accounts for the data. Section five discusses directions for further research.

1 German *doch* and *ja* in Discourse

This section illustrates four uses of *doch*-declaratives with different communicative effects within the discourse, in some of which *ja*-declaratives are also felicitous. Three of these uses have been discussed more or less extensively in the literature, one additional use is presented here as it can not be straightforwardly accounted for under many analyses, but falls out naturally from the one proposed here. Examples are limited to declaratives, as is the analysis.[1]

First, *doch*-, but not *ja*-declaratives can be used in *rejections*, which reject, or call into question, the truth of some contextually salient proposition. In our example, the rejection is uttered in reaction to an assertion of this proposition.

[In reaction to: "Mary is coming, too!"]

(1) (Aber) sie ist { doch / #ja } verreist.
    "(But) she is *doch* / *ja* traveling."

---

[1] Possible expansion to other illocutionary forces is briefly discussed in Sect. 5.

© Springer International Publishing AG 2017
M. Otake et al. (Eds.): JSAI-isAI 2015 Workshops, LNAI 10091, pp. 93–108, 2017.
DOI: 10.1007/978-3-319-50953-2_8

The communicative effect of asserting the prejacent φ = "[Mary] is traveling" is to reject or call into question the antecedent ψ = "Mary is coming".[2] This requires a premise "if φ holds, then (normally) ψ does not". Whether or not *doch* is involved in introducing or marking this premise will be discussed in Sect. 2.2. For now, note that the conjunction *aber* ("also") encodes contrast based on this premise, henceforth "propositional contrast". Also note that *doch* is not required for a felicitous rejection, as an alternative utterance "No, she is traveling" has a similar effect. The intuitive difference to the *doch*-declarative is twofold: first, with *doch* the rejection comes across as negotiable, and second, *doch* marks the prejacent as given in the sense of not being unknown to the addressee. One goal of this paper is to account for these intuitions.

Next, both *doch*- and *ja*-declaratives can be used in *acceptances*, which are reactions to an antecedent utterance without propositional contrast.[3]

[In reaction to "Peter looks bad."]

(2) Er war { doch / ?ja } (auch) lange krank.
   "He was *doch* / *ja* sick for a long time (after all)."

With the prejacent φ = "[Peter] has been sick" and the antecedent ψ = "Peter looks bad", the premise here is "if φ holds, the (normally) ψ holds", *i.e.* there is no contrast between the two propositions. Acceptances essentially elaborate on the antecedent, providing an explanation for it. The sentential adverb *aber* encodes this relation parallel to *auch* in rejections.[4] The *doch*-declarative intuitively conveys that the antecedent utterance is somehow degraded in terms of felicity, as the addressee is not (sufficiently) taking the prejacent into consideration, a nuance which *ja* lacks. The *ja*-declarative conveys that the prejacent is not new information, and is otherwise an elaboration, which are slightly worse without *auch* (thus the ? mark).

The third use are *reminders*, which differ from both rejections and acceptances in that they lack an antecedent and can occur discourse-initially.

(3) Du bist { doch / ja } Linguist (?, oder?)
   "You are *doch*/*ja* a linguist (, right?)"

Reminders can function as preparatory utterances for a subsequent turn-holding move of the speaker[5] and are compatible with a question tag "... *oder?*" which

---

[2] I write φ for the "prejacent" (the proposition of the particle utterance), ψ for the "antecedent" (the proposition of the antecedent utterance).

[3] For space, I do not mention uses where there is propositional contrast between φ and ψ, but both hold, which can be understood as cases of blocked defeasible inference. As they allow *ja*-declaratives, I consider them a sub-category of acceptances.

[4] As a sentential adverb/discourse connective, the additive particle *auch* indicates that the prejacent is (additional) justification for believing some proposition asserted, similar to English "after all".

[5] Typical continuations include questions, in (3) for example "Tell me, how many languages are there?", to which the addressee is likely to know the answer if the prejacent holds, and elaborations on the prejacent on part of the speaker.

is worse with the *ja* than with the *doch*-declarative (thus the $^?$ mark). This illustrates the intuition that the *doch*-declarative has more of a confirming nuance.

The common denominator of all uses discussed so far is that the *ja*-declaratives mark the prejacent as non-new information, and the *doch*-declaratives convey that the addressee is not, or may not be, taking the (non-new) prejacent into consideration. Many previous analyses are based on this observation and assume common belief marking to be at the core of both *ja* and *doch*.

The final use of *doch*- but not *ja*-declaratives challenge this view. I call this the *manipulative*[6] use based on its uncooperative, pragmatically marked status in previous analyses, but will argue that it is not, in fact, "manipulative" in the sense of being uncooperative.

[In reaction to an assertion the speaker does not agree with:]

(4) Das ist { doch / #ja } nicht wahr.
   "That's *doch* / *ja* not true."

In this example, the speaker reacts to the assertion of an antecedent ψ by asserting the prejacent $\varphi =$ "ψ is not true". If *doch* marks the prejacent as a common belief, it should not be felicitous here, as it is hardly a reasonable assumption that the addressee, having just asserted ψ, shares the belief φ. One possible explanation is that the utterance is a manipulative move in the sense of insinuating that the addressee has asserted something they know to be false, for example in order to make them take their commitment to ψ back. I will propose an alternative analysis which predicts the felicity of *doch*, but not *ja*, in (4).

2   Previous Analyses, Issues

Lindner (1991) [11] provides an early empirical generalization covering all but manipulative uses of *doch* and *ja*, here in Zimmermann's (2011) wording:[7]

(5) $[\![doch]\!](p) =$ "[The] speaker assumes φ not to be activated at the current stage in the discourse."[14, 2017]
(6) $[\![ja]\!](p) =$ "[The] speaker believes φ uncontroversial."[14, 2016]

There are two major challenges to making these paraphrases formally explicit. First, how to define "uncontroversial" and "inactive", and second, how to connect the meanings of *doch* and *ja*.[8]

In order to capture the intuitions regarding *doch* and *ja*, a number of later analyses use a notion of contrast conveyed by *doch* in addition to some notion of

---

[6] This label is borrowed from Karagjosova (2004), who however includes reminders in case the speaker has no reason to assume that the prejacent is an inactive belief.

[7] Zimmermann includes the truth of the prejacent in the paraphrases which I leave out as it is not relevant for the expressive meaning component I am concerned with.

[8] Other than perceived similarities, the fact that their contributions become indistinguishable when they co-occur in the same utterance supports such a connection.

givenness, the latter being shared with *ja*. Abstracting away from concrete implementations, this can be simplified as some use-conditional[9] function GIVEN(φ) shared by both particles, and, based mainly on rejections, an additional use-conditional function CONTRAST(φ, ψ). This prevalent view is schematically represented below.

(7) $[\![ja(\varphi)]\!] = [\![\varphi]\!], [\![\text{GIVEN}(\varphi)]\!]$

(8) $[\![doch(\varphi)]\!] = [\![\varphi]\!], [\![\text{GIVEN}(\varphi)]\!], [\![\text{CONTRAST}(\varphi, \psi)]\!]$

Below, I discuss complications necessary for proponents of propositional or other forms of contextual contrast in order to account for uses of *doch*-declaratives other than rejections, and an alternative view closer to Lindner's original paraphrase in which contrast is defined non-contextually, effectively resulting in two different notions of givenness for *ja* and *doch*, which can not straightforwardly account for manipulative uses, and to some extent reminders.

2.1 Contextual Contrast at the Core of *doch*

The function CONTRAST can be understood as adding a use-condition that there be a relation of defeasible entailment between an element pertaining to the doch-utterance and another, contextually salient element. An early proponent is Ormelius-Sandblom (1997) [12], who proposes that *doch* gives rise to an implicature that there is a (salient) proposition which entails the falsity of the prejacent. Bárány (2009) [2], arguing that reference to the negation of the prejacent is too strong a claim, reverses the direction of entailment so that the truth of the prejacent entails the falsity of another proposition. Karagjosova (2004) [9] proposes a defeasible version based on Asher and Lascarides (2003) [1].[10] Writing defeasible entailment as '>', this is represented in (9), where φ is the element related to the particle-utterance, ψ is some contextually salient element.

(9) *Contextual contrast*: $[\![\text{CONTRAST}(\varphi, \psi)]\!] = \varphi > \neg\psi$

In basic examples of rejections, φ and ψ are the prejacent and the proposition of an antecedent utterance, and the condition in (9) is satisfied. In other cases, some salient ψ needs to be identified. Below, I briefly discuss how this is realized in two previous analyses.

Grosz (2010) [6] proposes that *doch* associates with focus, requiring that there be a focus alternative ψ of the prejacent φ and that the context entail $\neg(\varphi \wedge \psi)$.[11] The identification of ψ is straightforward when there is narrow focus on a constituent, but otherwise somewhat difficult. In one case, Grosz suggests that *doch* associates with "wide sentential focus", alternatives of which can be

---

[9] This wording is possibly biased, but reflects the widely accepted observation that *doch* and *ja* do not alter the truth conditions of utterances, but add felicity conditions which depend on the utterance's propositional content.

[10] She does not, however, take this relation to be part of *doch*'s meaning, but makes use of it to explain the interpretation of rejections with *doch*.

[11] This is logically equivalent to $\varphi \rightarrow \neg\psi$ and thus a variant of contextual contrast.

characterized by a question "What is the case?". This appears to be rather unrestricted—for example, he identifies ψ for a reminder as "I am mistaken and $\neg\varphi$". Furthermore, cases without propositional contrast, such as acceptances, can not straightforwardly be accounted for in this analysis.

Egg (2010) [4] modifies contextual contrast by allowing not only propositions, but also felicity and preparatory conditions of utterances to fill the slots of φ and ψ.[12] For instance, Egg takes the acceptance in (2) to convey that the surprise conveyed by the antecedent utterance is not felicitous, as if the prejacent holds (and is previously known to the addressee), the antecedent is not surprising. Reminders are analyzed as marking a violation of the condition that the prejacent be new to the hearer, this being the salient ψ, while φ is the fact $doch$-utterance has been made.

Thus, relying on contextual contrast at the core of $doch$ requires considerable complications for all but rejections. Before moving on to discussing an alternative proposal circumventing these problems by making no reference to contextual elements in the meaning of $doch$, I briefly discuss independent evidence suggesting that $doch$ does, in fact, not encode propositional contrast.

2.2   Propositional Contrast Is Independent of *doch*

Propositional contrast at the core of *doch* is mainly motivated by rejections, as in our example repeated below.

[In reaction to: "Mary is coming, too!"]

(10) (Aber) sie ist { doch / #ja } verreist.
       "(But) she is *doch / ja* traveling."

There are two arguments as follows speaking against *doch* as the source of propositional contrast in this example.

First, the conjunction *aber* in (10) encodes precisely the type of relation of defeasible entailment that is modeled as CONTRAST in the prevalent view. While *aber* is optional in the *doch*-declarative, provided that its status as a rejection is sufficiently clear from the utterance situation, the addition of *doch* to a plain declarative does not mark propositional contrast. Rather, it conveys that the proposition is not new to the hearer, independent of whether or not they (appear to) entertain a conflicting belief.

Second, there is a specific intonational pattern associated with rejections: the late-peak contour, which Grice and Baumann (2002) [5] describe as a low tone on the nuclear phrase accent, followed by a high tone and a final low phrase boundary tone ($L^*+H-L\%$ in *GToBI* notation), which is a marked intonational pattern compared to the neutral $H^*-L\%$. While the exact contribution of the

---

[12] Egg takes both the defeasible entailment relation and the salient element to be defeasibly deducible from the context set.

$L*+H$ nuclear accent is a complicated matter,[13] rejections are to my intuition degraded without it which leads me to assume that it plays a part in conveying propositional contrast, independently of *doch*.

2.3   Two Variants of Givenness for *doch* and *ja*

Karagjosova (2004) develops a version of CONTRAST closer to Lindner's early paraphrase, in which both *ja* and *doch* convey a speaker belief that the prejacent is a common belief, but only *doch* conveys that this common belief may not be an active belief. The formalization is based on Wassermann's (2000) [13] AGM-based model of resource-bounded belief revision which distinguishes between active and inactive beliefs. Simplified versions of Karagjosova's *ja* and *doch* are given in (11) and (12), where $B_{AS}(\varphi)$ stands for a belief φ entertained by both participants, speaker (S) and addressee (A), *B.act* marking active beliefs, *B.expl* explicit beliefs:[14]

(11) $[\![ja(\varphi)]\!] = [\![\varphi]\!], B_S(B.expl_{AS}(\varphi))$
(12) $[\![doch(\varphi)]\!] = [\![\varphi]\!], B_S(B.expl_{AS}(\varphi)) \wedge \neg B_S(B.act_{AS}(\varphi))$

Karagjosova's analysis can be subsumed under the prevalent view in that there is a common meaning component encoding givenness, in addition to which *doch* encodes contrast in the form of possible non-activation. There is, however, no propositional or other contextual contrast involved, as what corresponds to CONTRAST here takes only one argument: the prejacent. In this way, what *doch* contributes is essentially a modified version of what *ja* does, which can be interpreted as two different versions of givenness. On this analysis, both particles convey the speaker's belief that the prejacent is a common belief, and *doch*, in addition to this, the speaker's assumption that this common belief may not be active. This predicts two uses of *doch*-declaratives to be "manipulative", or uncooperative in the sense that a speaker belief is not justified. First, cases of reminders in which the speaker does not have good reason to assume that common belief is inactive (*e.g.* because of the addressee forgetting about it); Second, the use labeled manipulative above, where the speaker's belief that the prejacent is given, *i.e.* a common belief, is not justified, *cf.* the discussion of (4).

There are two parallels to the analysis proposed in this paper. First, *ja* and *doch* differ only in the kind of givenness they convey. While Karagjosova's notion of common belief in *doch* is stronger[15] than that conveyed by *ja*, my analysis has it that *doch*'s givenness is weaker than that of *ja*, as the former, but not

---

[13] Grice and Baumann give two examples for its use which they paraphrase "self-evident statement", and "involved or sarcastic statement" [5, 295, translation my own], which I assume represents typical uses of the contour rather than its meaning.

[14] Explicit beliefs are the core of an agents set of beliefs in Wassermann's model in the sense that they form the basis for reasoning.

[15] In Wassermann's articulated model of belief states, explicit beliefs can be either active or inactive, so that the set the speaker believes φ to be part of in (12) is a subset of the corresponding set in (11).

the latter, is underspecified in terms of addressee beliefs. This covers both cases problematic for Karagjosova. Second, *doch* is formulated in terms of higher-order beliefs in both analyses, but makes use of public belief in a version excluding first-order beliefs. While the meaning of *doch* being formulated in terms of speaker's beliefs can be taken to mean that *doch*-utterances make (higher-order) beliefs public, this is not made explicit in Karagjosova's proposal.

3   Discourse Particles in Declaratives as CCPs

The analysis put forward here is based on a model of declaratives as context change potentials (CCPs) operating on contexts from higher-order public beliefs. Restrictions on admissible input and output contexts are used to model the contributions (pre- and postsuppositions) of *doch* and *ja*.

3.1   Non-first Order Public Beliefs

The model of CCPs used here is based on Gunlogson (2003), who defines contexts in terms of participant-relative *commitment sets*, which are in turn defined as sets of worlds in which each participants' *public beliefs* hold [7, 41–43]. For the purposes of this paper, only public beliefs, but not commitments, are of interest.[16] Gunlogson's definition of public belief is as follows, where φ is a proposition, X is the set of discourse participants, and x is an individual participant [7, 42].[17]

(13) φ is a *public belief* of x iff 'x believes φ' is a mutual belief of all $x \in X$.

For the purposes of the present analysis, I define *mutual belief* as follows, where ϕ is a belief state or a proposition.

(14) ϕ is a *mutual belief* of participants x and y iff $B_x B_y \phi$ and $B_y B_x \phi$

In this definition, ϕ is a mutual belief iff each participants believes that *everyone else* believes ϕ. This is different from *common belief*, a belief which all participants entertain, defined as follows.

(15) ϕ is a *common belief* of participants x and y iff $B_x \phi$ and $B_y \phi$

Substituting "mutual belief" in the definition in (13) with (14) yields the following version of public belief, where PB_x is the set of participant x's public beliefs, the only participants being x and y.

---

[16] I will remain neutral in regard to the necessity of valuation of public beliefs against worlds and thus in regard to both the question of whether a Stalnakerian analysis of the common ground is preferable for what *doch* and *ja* operate on.

[17] The symbols have been modified from the original; Gunlogson refers to speaker and addressee as the two discourse participants. Furthermore, she uses the propositions representing these beliefs to construct *discourse commitments*, which will not be necessary for the present purposes as no validation of propositions takes place.

(16) $\varphi \in PB_x$ iff $B_y(B_x(B_x\varphi)) \wedge B_x(B_y(B_x\varphi))$

On this definition, φ is a public belief of x if "x believes φ" is a belief x attributes to y and vice versa. Assuming that a belief $B_x B_x\varphi$ is equivalent to $B_x\varphi$,[18] this can be simplified as below.

(17) $\varphi \in PB_x$ iff $B_y(B_x\varphi) \wedge B_x(B_y(B_x\varphi))$

Another simplification is possible assuming that when ϕ is a public belief, all participants are aware of its existence. This can be formulated as *mutual introspection*, by which for each belief of a given participant, all other participants are aware of that belief, *i.e.* have a higher-order belief ascribing the original belief to the participant.

(18) *Mutual introspection:* $\forall B_x \; \forall x' \in X \cap x \; (\exists B_{x'}(B_x))$

This makes stating $B_y B_x B_y\varphi$ in (17) superfluous, as it follows from $B_x B_y\varphi$, so that the minimal version of mutual public belief is as follows.

(19) *Public belief (mutual), introspective:* $\varphi \in PB_x$ iff $B_y B_x\varphi$ and (18) holds.

On the other hand, substituting common rather than mutual belief in (13) yields the stronger version below, labeled CPB for common public belief here.

(20) *Public belief (common), introspective:* $\varphi \in CPB_x$ iff $B_x\varphi$ and (18) holds.

As private, first order beliefs of the participants are not at stake in the mutual version of public belief (which is used in the present analysis), it is weaker than a version with first-order beliefs (which is used in Gunlogson's analysis). This can easily be confirmed, as the condition $B_y B_x\varphi$ in (19) is entailed by the condition $B_x\varphi$ in (20) under mutual introspection (18), but not the other way around.

3.2   Mutual Public Beliefs and Update Dynamics

As a final preliminary before the analysis proper, this section discusses how mutual introspection reflects reasoning on belief states of cooperative agents, and how a non first-order version of the common ground comes about.

First, assume that participant x makes a belief $B_x\varphi$[19] public by means of a linguistic signal observable to the (only) other participant y (according to the proposal to follow, this is what a falling declarative does). When y observes this signal, and assumes that x is cooperative, y will believe that $B_x\varphi$, thus $B_y B_x\varphi$. When x in turn has made sure that y has perceived the signal or has no reason to believe otherwise and can anticipate y's move, $B_x B_y B_x\varphi$, etc. Thus, mutual introspection reflects a successful exchange of information, or the default effect

---

[18] This assumption may not be uncontroversial, parallel to positive introspection for knowledge, but I do not see any obstacle to make this simplification for the present analysis.

[19] I use the propositional symbol φ here for simplicity, but the same goes for belief states.

of making a private belief public is. Note that even if x, the originator of the initial signal, does not actually hold a belief that φ (but, for example, wants to deceive y) this is how things go if y deems x cooperative.

Next, assume that x ascribes a belief that φ to y, for example because y does not disagree with x's assertion of φ. This results in an additional belief $B_x B_y \varphi$ is aware of this belief, we get $B_y B_x B_y \varphi$ and so on by mutual introspection in case the exchange is successful in the sense of being transparent. The resulting state is one of *shared public belief*, as shown below.

(21) *Shared public belief:* $\varphi \in PB_{x,y} = B_x B_y \varphi \wedge B_y B_x \varphi$

This is as close as my version of public beliefs gets to a notion of φ being part of the common ground, with the crucial difference that it need not be a common belief. In the proposal below, I model both *doch* and *ja* as mapping to an output context with a shared public belief. In the case of *ja* this just means carrying over the presupposition, in the case of *doch* adding a speaker belief ascribing a belief φ to the addressee to the context. The default move, then, is for the addressee to acknowledge this speaker belief, and shared public belief comes about as outlined above.

3.3   Declaratives as CCPs

In order to develop the semantics of particle-declaratives, I build on Gunlogson's compositional analysis of declaratives with final rising and falling intonation [7, 52] in Davis's (2011) relational version. The crucial feature of the version proposed here is that contexts consist of higher-order beliefs only.[20] In this model, utterances are CCPs mapping contexts to sets of contexts consisting of the public beliefs, as defined above, of all participants. Utterances differ in the conditions they impose on public beliefs in the input and output contexts. The meaning of a declarative asserting a proposition φ is given below as a function ASSERT, based on Davis's DECL. The declarative maps input context c to output context c', where φ is in the set of public beliefs of an underspecified discourse participant x in c' [3, 44].

(22) $[\![\text{ASSERT } \varphi]\!] = \lambda x.\{\langle c, c' \rangle | \varphi \in PB_x^{c'}\}$

Final falling or rising intonation resolves x to the speaker S or the addressee A, respectively, which will henceforth be assumed to be the only participants, as shown in the representation below [3, 45].

(23) $[\![\downarrow (\text{ASSERT } \varphi)]\!] = \{\langle c, c' \rangle | \varphi \in PB_S^{c'}\}$
(24) $[\![\uparrow (\text{ASSERT } \varphi)]\!] = \{\langle c, c' \rangle | \varphi \in PB_A^{c'}\}$

To better illustrate the conditions (particle) declaratives impose on input and output contexts, I introduce an alternative notation as follows.

---

[20] This is possibly too weak an assumption for modeling the effect of assertions on contexts, see Sect. 4.4 for discussion.

(25) $\{\langle c, c' \rangle | [\![\varphi]\!] \in PB_S^{c'}\} \equiv \dfrac{PB_S^c : -}{PB_A^c : -} \bigg| \dfrac{PB_S^{c'} : \varphi}{PB_A^{c'} : -}$

In this notation, $PB_x^c : \varphi$ stands for a condition that $\varphi \in PB_x^c$, and $PB_x^c : -$ that there are no restrictions on the public beliefs of x in c.

3.4   What *doch* and *ja* do in Declaratives

Declaratives with *doch* and *ja* are defined by the conditions they impose on input contexts, *i.e.* presuppositions, and the conditions they impose on output contexts, which I call "postsuppositions", following Davis (2011) [3, 48]. In my understanding, as postsuppositions in our model characterize the contextual changes an utterance brings about in terms of higher-order beliefs, they can alternatively be understood as conventional implicatures of their respective utterances.

The CCPs for *doch*- and *ja*-declaratives are given below, for now ignoring the question of whether a compositional analysis is possible. It should be noted here that *doch*- and *ja*-declaratives both occur with final falling intonation only.

(26) $[\![\text{DOCHASSERT } \varphi]\!] = \dfrac{PB_S^c : \varphi}{PB_A^c : -} \bigg| \dfrac{PB_S^{c'} : \varphi}{PB_A^{c'} : \varphi}$

Assuming that elements in the input context carry over to the output context if the presupposition is satisfied, the conditions for (26) are given below.

(27) $[\![\text{DOCHASSERT } \varphi]\!] = \{\langle c, c' \rangle | \varphi \in PB_S^c, PB_A^{c'}\}$

Thus, *doch* presupposes that φ be a public belief of the speaker, and conveys ("postsupposes") that φ be a public belief of the addressee (as well). Next, consider the representation of the CCP for *ja* below.

(28) $[\![\text{JAASSERT } \varphi]\!] = \dfrac{PB_S^c : \varphi}{PB_A^c : \varphi} \bigg| \dfrac{PB_S^{c'} : \varphi}{PB_A^{c'} : \varphi}$

Notice that nothing changes from input to output context, so that the *ja*-declarative is purely presuppositional, as represented below.

(29) $[\![\text{JAASSERT } \varphi]\!] = \{\langle c, c' \rangle | \varphi \in PB_{S,A}^c\}$

The representations in (26) and (28) have the advantage that they bring out the fact that the postsuppositions of *doch* and *ja* are the same, *i.e.* they result in output contexts restricted in the same way in regard to public beliefs on the prejacent. The difference in meaning is solely dependent on whether such an output context is fully presupposed, as in the case of *ja*, or partially presupposed, as in the case of *doch*.

4   Accounting for the Data

This section shows the application of the analysis to the data discussed in the first section, including manipulative uses, and discusses the possibility of reintroducing a stronger version of public belief for bare declaratives.

4.1   Accounting for Rejections

The first example to show the application of the proposal is a rejection repeated below from (1). Recall that there is a contrastive relation of defeasible entailment of the form $\varphi > \neg\psi$ between the prejacent φ = "Mary is traveling" and the antecedent ψ = "Mary is coming, too" which is independent of the presence or absence of particles.

[In reaction to: "Mary is coming, too!"]

(30) (Aber) sie ist { doch / #ja } verreist.
       "(But) she is *doch / ja* traveling."

According to the present proposal, the presupposition of *doch* is a partial type of givenness, namely that φ be a public belief of the speaker. How can this be satisfied? As the model stands, the most simple way is previous assertion of φ by the speaker, as can be easily verified by comparing the output-context side of a bare declarative (25) with the input-context side of a *doch*-declarative (26). Another possibility is that speaker believed φ was a shared belief, but the addressee's asserting ψ has lead them to the assumption that this is no longer the case, similar to an inactive-belief analysis. Both possibilities are plausible scenarios for (30). Furthermore, in case *doch*'s presupposition, represented below, is not satisfied, it is more easily accommodated than a presupposition based on common belief.

(31) *Presupposition of doch:* $B_A B_S \varphi \wedge B_S B_A B_S \varphi \wedge \ldots$

Accommodating this presupposition means for the addressee to ascribe a public belief that φ to the speaker ($B_A B_S \varphi$) rather than acceptance of or commitment to φ. The latter is necessary under analyses according to which *doch* presupposes that (the speaker believes) φ be a common belief, arguably predicting *doch*-declaratives harder to accommodate than they intuitively are. In the present proposal, all that needs to be accommodated is the "publicity" of the speaker's belief, *i.e.* a higher-order belief.

While in in our case, no notion of common belief is made reference to, *ja* presupposes that φ be a shared public belief, which is predicted to be harder to accommodate than *doch*'s presupposition and explains its badness in rejections. Recall that the premise $\varphi > \neg\psi$ together with the addressee's assertion of ψ make it highly unlikely that φ is a public belief of the addressee. However, this is precisely what the presupposition of *ja* requires, as highlighted below.

(32) *Presupposition of ja:* $B_A B_S \varphi \wedge B_S B_A B_S \varphi \wedge \ldots, \underline{B_S B_A \varphi} \wedge \underline{B_A B_S B_A \varphi} \wedge \ldots$

In this way, the badness of *ja* in rejection is predicted as its presupposition is clearly not satisfied, and accommodation would require the addressee to ascribe a public belief that φ to themselves, which would be inconsistent with their previous assertion of ψ.

The contrast between *doch* and *ja* in this example furthermore illustrates a novel prediction on the contexts in which the two particles are felicitous: as can be easily verified by comparing (31) and (32), *ja* imposes an additional condition (here underlined) on input contexts as compared to *doch*, predicting *ja*-felicitous contexts to be a proper subset of *doch*-felicitous ones.

4.2   Revealing Similarities of Acceptances and Reminders

Our example of an acceptance, which has an antecedent utterance but no propositional contrast, is repeated below from (2). There is a relation between antecedent and prejacent of the form $\varphi > \psi$,[21] again independent of the particles.

[In reaction to "Peter looks bad."]

(33) Er war { doch / ?ja } (auch) lange krank.
     "He was *doch* / *ja* sick for a long time (after all)."

Without particles, the prejacent provides an explanation for the addressee's observation that ψ. The *ja*-declarative additionally presupposes that φ be a shared public belief. In contrast to this, the *doch*-declarative is compatible with contexts where only the speaker has such a public belief. I assume that in contexts where both particles are available, the fact that the *doch*-declarative is less informative[22] gives rise to an implicature that the speaker does *not* ascribe a belief that φ to the addressee in the input context. This is compatible with non-activation of a shared belief, but also compatible with cases of acceptances in which the speaker marks the antecedent utterance as not (fully) felicitous, in light of the prejacent, which the addressee ought to take into consideration.

As for our example of a reminder, repeated from (3), under the present analysis the only difference to the acceptance is the absence of an antecedent utterance.

[discourse-initially]

(34) Du bist { doch / ja } Linguist (?, oder?)
     "You are *doch* / *ja* a linguist (, right?)"

Similar to the acceptance, the confirming nuance of the *doch*-declarative follows from choosing it over the *ja*-declarative, implicating that there is a possibility that the addressee does not believe the prejacent to be true. It also explains

---

[21] "If [Peter] has been sick for a while ($= \varphi$), then (normally) he looks bad. ($= \psi$)".

[22] How use-conditional content or presuppositions can be compared parallel to logical strength of truth conditional content is beyond the scope of this paper, but see the comparison of presuppositions in Sect. 4.1 above.

why tag-questions are good with *doch*, but degraded with *ja*, assuming that a question tag introduces the negation of the prejacent as a possibility[23]—if the presupposition of *ja* that the prejacent be a shared belief is satisfied, making the negated prejacent salient is inconsistent.

4.3 "Manipulative" Uses: from Speaker- to Shared Belief

One motivation for the present analysis has been the (in my experience rather rather large) class of "manipulative" examples, our instance of which is repeated below from (4), which can not be directly accounted for in previous proposals. The fact that there is no propositional contrast with the antecedent utterance requires significant complications of approaches with contextual contrast; the low likelihood of the prejacent being a common, but inactive, belief is a problem for characterizing *doch* in terms of non-activation.

[In reaction to an assertion the speaker does not agree with:]

(35) Das ist { doch / #ja } nicht wahr.
   "That's *doch* / *ja* not true."

The current proposal naturally accounts for (35). Parallel to rejections, a speaker assumption that the addressee entertain the prejacent is not well founded, making the *ja*-declarative infelicitous. As for the *doch*-declarative, the postsupposition that the prejacent be a shared public belief and the presupposition requiring the prejacent to be a pubic belief of the speaker (thus known to the addressee) taken together with the high likelihood of the addressee believing it to be false explain the nuance of exasperation the utterance conveys. What is necessary to make (35) felicitous is that the speaker's belief that φ = "ψ is not true" be a public belief in the input context.

At this point, it should be noted that a public belief φ does not necessarily need to come about by assertion of the proposition, but for example can be derived from other public beliefs of the same participant. It also seems plausible that there are uses of (35) in which the speaker insinuates that their belief φ is easily inferable from the utterance context, so that the addressee can be expected to have been aware of this belief, which is facilitated by the relative ease of accommodation of *doch*'s presupposition discussed in Sect. 4.1.

Summing up, the *doch*-declarative in (35) does not need to be considered manipulative in the present analysis. An actually manipulative move would be uttering the *ja*-declarative in the same situation, insinuating that the addressee has made an utterance which they know (and publicly believe) to be false. While marginal, such a use is possible to my intuition, but has an intensely condescending and dismissive feel to it which fits well with the semantics proposed here.

---

[23] While this assumption glosses over a number of possible complications in an analysis of tag questions, the details are orthogonal to the question tag's making the notion of "confirming nuance" more concrete and testable against intuitions.

4.4 Bringing Back Strong Assertions

Recall that for the current proposal, I have defined public beliefs independent of first-order beliefs by replacing common with mutual beliefs. While I maintain that this has welcome results when modeling *doch-* and *ja*-declaratives, it may be too weak for assertions. Thus, I suggest using the the common-belief based version of public belief (17) rather than the mutual-belief version (16) for plain declaratives. The CCP of the strong version of rising and falling bare declaratives is given below (as usual, participant x is resolved to the speaker with final falling, the addressee with final rising intonation).

(36) $[\![\text{ASSERT } \varphi]\!] = \{\langle c, c' \rangle | \varphi \in CPB_x^{c'}\}$

Support for reintroducing the strong version of assertion comes from the "manipulative" example (35) above: a bare declarative in the same context is good with a final fall, but out with a final rise, as shown below.

[In reaction to an assertion the speaker does not agree with:]

(37) Das ist nicht wahr. $\{ \downarrow\ /\ \#\uparrow \}$
       "That's not true."

The final rise is predicted as bad for similar reasons as *ja* is: ascribing a belief that the prejacent holds to the addressee is not well-founded. The difference is that the badness of *ja* is due to presupposition failure, while that of the rising declarative is due to a preparatory condition of commitment not being met. This lines up with the intuition that the rising declarative in (37) is worse than the *ja*-declarative in (35), the latter being clearly uncooperative but with the possibility of a "true manipulative" use as described in Sect. 4.3 above—to my intuition, such a move is all but impossible with a rising declarative, supporting the assumption that stronger than mutual public beliefs are at stake.

5 Conclusion and Outlook

In this paper, I have proposed an analysis of *doch-* and *ja*-declaratives which accounts for a wider range of uses than previous proposals. From this analysis, a novel prediction regarding felicitous contexts for *doch* and *ja* arises, namely that the contexts (or utterance situations, all else being equal) in which *ja* is felicitous are a proper subset of those in which *doch* is. Within the non-exhaustive discussion of uses in this paper, *doch* and *ja* appear to pattern in this way, but showing that this prediction is borne out is left for further research.

Another open topic is expanding the current model to utterances other than declaratives and illocutionary forces other than assertion, a possible starting point for which is the following. One way of satisfying *doch*'s presupposition in declaratives is a previous assertion of the proposition by the speaker. It seems intuitively appealing to extend this to other illocutionary forces so that, for instance, a *doch*-imperative would indicate that the speaker has either uttered the same imperative before, or something with the same effect on the context happened (and the hearer did not act accordingly, thus needs to be "reminded"). While the formal details are far from trivial, this matches the intuitive meaning of a *doch*-imperative. On possible way is shown by Davis (2011), who applies his relational CCPs to imperatives by replacing public beliefs with "public intentions" [3, 147].

Also worth mentioning is the question of whether particle-utterances should be analyzed compositionally, in the sense of consisting of a "regular" speech act and the pre- and postsuppositions from the particle, or whether particle-utterances have a weaker effect on the participants' belief states than such without particles. While a detailed discussion has to be left for future work, I believe that there is some merit to the latter option based on observations regarding the strength of addressee commitment as in (37), and on the strength of speaker commitment as follows. Assuming the stronger version of declaratives given in (36), bare falling declaratives necessarily result in speaker commitment to the prejacent, which is not the case with particle declaratives. In fact, *doch*-declaratives are frequently used to draw attention to their prejacent when its truth is to be decided, or has been called in question, rather than necessarily conveying that the speaker judges the prejacent true. Rejections, for instance, can be used to prompt the addressee to provide whatever additional information they might have for settling the open question of which holds, the prejacent or the antecedent. While other kinds of utterances, such as polar questions with outer negation (paraphraseable as "Isn't it the case that...?") have a similar effect, it is difficult to achieve with plain declaratives.

It should also be noted that the presuppositions of *doch* and *ja* proposed here are not only compatible with the current model of CCPs, but in principle carry over to other approaches. For example, Gutzmann (2015) [8] proposes analyzing *ja* as an element taking a truth-conditional argument and delivering the use-condition that the prejacent be "common knowledge" or "verifiable on the spot" [8, 262]. The condition that the prejacent be a mutual public belief could be easily replaced for this paraphrase and has the advantage of being more formally explicit. Provided that there is a suitable way of differentiating between pre- and postsuppositions, the semantics for *doch* proposed here could also be integrated into a compositional framework of use-conditional meaning.

An alternative route to providing a formalization of the semantics of *doch* and *ja*, which is also able to account for "manipulative" uses but is based on a Kratzerian view of modality, is proposed in Kaufmann and Kaufmann (2012). According to their proposal, *ja* and *doch* both encode "uncontroversiality", presupposing that any rational agent can find out whether the prejacent holds in the current context. They differentiate between the two particles with an additional meaning component of "normalcy": *ja* presupposes that the context is hearer-normal in the sense of the aforementioned presupposition, *doch* that it is not. In other words, *doch* conveys that there is a possibility that the hearer can not find out whether the prejacent holds, even though this is normally the case [10, 212, 220–221]. I forgo a comparison with the analysis in this paper due to space, but note that both are flexible enough to account for a wider range of data than (other) previous analyses.

Finally, an articulated model of belief revision is not included in the present proposal, but is necessary to explicitly show the role of *doch-* and *ja*-declaratives within this process. While a formalization of belief revision is a complex matter far beyond the scope of this paper, the output contexts of particle declaratives as proposed here in principle allow for the prejacent to retain a status like that of "provisional belief" in Wassermann's (2000) system of articulated belief states, which is an additional advantage of leaving first-order beliefs out of the model.

Acknowledgments. I thank Magda and Stefan Kaufmann for thorough discussion of and helpful comments on an earlier version, the audience at LENLS for comments, and Daisuke Bekki and Eric McCready for encouraging me to participate.

References

1. Asher, N., Lascarides, A.: Logics of Conversation. Cambridge University Press, Cambridge (2003)
2. Bárány, A.: Form and interpretation of the German discourse particles ja, doch and wohl. Master's thesis, Universität Wien (2009)
3. Davis, C.: Constraining interpretation: Sentence final particles in Japanese. Ph.D. thesis, University of Massachusets, Amherst (2011)
4. Egg, M.: A unified account of the semantics of discourse particles. In: Proceedings of SIGDIAL 2010, pp. 132–138 (2010)
5. Grice, M., Baumann, S.: Deutsche intonation und GToBI. Linguistische Ber. **191**, 267–298 (2002)
6. Grosz, P.: German doch: an element that triggers a contrast presupposition. Proc. CLS **46**, 163–177 (2010)
7. Gunlogson, C.: True to form: Rising and falling declaratives as questions in English. Ph.D. thesis, UCSC (2003)
8. Gutzmann, D.: Use-conditional meaning: Studies in multidimensional semantics. Oxford Unviersity Press, Oxford (2015)
9. Karagjosova, E.: Meaning and function of German modal particles. Ph.D. thesis, Universität des Saarlandes (2004)
10. Kaufmann, M., Kaufmann, S.: Epistemic particles and performativity. Proc. SALT **22**, 208–225 (2012)
11. Lindner, K.: 'Wir sind ja doch alte Bekannte'. The use of German ja and doch as modal particles. In: Abraham, W. (ed.) Discourse Particles: Descriptive and theoretical investigations on the logical, syntactic and pragmatic properties of discourse particles in German, pp. 163–201. John Benjamins, Amsterdam (1991)
12. Ormelius-Sandblom, E.: Die Modalpartikeln ja, doch und schon. Zu ihrer Syntax, Semantik und Pragmatik. Almqvist & Wiksell International, Stockholm (1997)
13. Wassermann, R.: Resource-Bounded Belief Revision. Ph.D. thesis, University of Amsterdam (2000)
14. Zimmermann, M.: Discourse particles. In: von Maienborn, C., von Heusinger, K., Portner, P. (eds.) Semantics: An International Handbook of Natural Language Meaning, vol. 2, pp. 2012–2038. Mouton de Gruyter, Berlin (2011)

Tracking Down Disjunction

Uli Sauerland[1]([✉]), Ayaka Tamura[2], Masatoshi Koizumi[2],
and John M. Tomlinson Jr.[1]

[1] Center for General Linguistics (ZAS), Schützenstr. 18, 10117 Berlin, Germany
uli@alum.mit.edu
[2] Tohoku University, Aoba-ku, Kawauti 27-1, Sendai 980-8576, Japan

Abstract. Kuno (1973) and others describe the Japanese junctor *ya*
as conjunction. But, Sudo (2014) analyzes *ya* as a disjunction with a
conjunctive implicature. We compare *ya* with other junctors and implicature triggers experimental using mouse-tracking. Our two main results
are: (1) *ya* differs from lexical conjunctions corroborating Sudo's (2014)
proposal. (2) The time-course of the conjunctive implicature of *ya* argues
against the details of Sudo's (2014) implementation, and instead favors
an account similar to other cases of conjunctive implicatures.

Keywords: Implicature · Disjunction · Conjunction · Alternatives ·
Numerals · Mouse-tracking · Japanese

1 Introduction

Propositional logic provides the two propositional junctors $\wedge$ for *and* and $\vee$ for
or. Japanese expresses these too (*to* and *mo* for $\wedge$, *ka* for $\vee$), but also has one
further lexical item for a propositional junctor: *ya*. Descriptive studies classify
ya with *to* as a conjunction (Kuno 1973; Ohori 2004 and others). But recently,
Sudo (2014) proposes to analyze *ya* as a disjunction with a conjunctive implicature. Sudo's proposal is inspired, though different from other recent work on
disjunction that has shown that a conjunctive implicature is possible.

In example (1), *ya* like the other NP-conjunctions of Japanese *mo* and *to* (and
unlike the disjunction *ka*) triggers the conjunctive inference that Taro drank both
coffee **and** tea.

(1)   Tarou-wa kouhii {ya / to   / mo / ka} koucha-o nonda
        Taro-TOP coffee YA / and / and / or   tea-ACC drank

© Springer International Publishing AG 2017
M. Otake et al. (Eds.): JSAI-isAI 2015 Workshops, LNAI 10091, pp. 109–121, 2017.
DOI: 10.1007/978-3-319-50953-2_9

But as Sudo (2014) shows, *ya* behaves differently in (2) and other examples, where *ya* is embedded in an antitone environment.[1] In (2), *ya* unlike *to* and *mo* has a disjunctive interpretation.

(2)   [Tarou-ga kouhii ya koucha-o nom-eba] yoru nemur-e-nai   darou
       [Taro-NOM coffee YA tea-ACC drink-if] night sleep-can-NEG INFER
       'If Taro drinks coffee **or** tea, he won't be able to sleep at night.'

Sudo proposes that the conventional meaning of *ya* is disjunction ∨, and that *ya* triggers a conjunctive implicature in (1). In antitone environments where implicatures are blocked, the disjunctive interpretation is apparent.

In this paper, we first review recent work on disjunction including Sudo's proposal for *ya* in Sect. 2. In the same section, we formulate three possible analyses for *ya* that make different processing predictions. In Sect. 3, we present a mouse-tracking experiment that we conducted to tease apart the predictions of the three different theories. Section 4 is the conclusion.

2   The Pragmatics of Disjunction

Markers of disjunction are interesting from a pragmatic perspective because a disjunction in entailment preserving, *isotone*, contexts is logically less informative than either of its two disjuncts. Hence disjunctive statements are generally expected to give rise to implicatures. Recent work on disjunction has argued that at least two different situations can arise depending on the alternatives a listener associates with a marker of disjuction. We refer to the two different alternative spaces that can arise as the *diamond* and the *substring space*. The difference is whether only the domain alternatives of Chierchia (2013) are present or also the scalar alternative ∧. The two spaces of alternatives are shown schematically in (3), where the arrows indicate entailment relationships.

(3)   *diamond space*   *domain space*

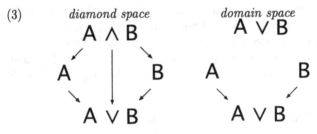

---

[1] In this paper, we use the terms *isotone* for contexts f that preserve the order by subset or entailment (i.e. $a \subset b \rightarrow f(a) \subset f(b)$), and *antitone* for those the reverse this order (i.e. $a \subset b \rightarrow f(b) \subset f(a)$) following e.g. Birkhoff (1940). In linguistics, the terms ±affective, up-/downward entailing (UE/DE), or up-/downward monotone are also used for the same concepts. But we find these terms less convenient because of their association with elementary calculus and intuitions that don't apply in the algebraic framework of semantics.

The diamond space for disjunction was introduced independently by Lee (1995) and Sauerland (2004). The substring space was more recently put to use by Meyer (2012; 2016), Singh et al. (2016), Nicolae (2015), and others. The difference between the two spaces is whether the conjunction $A \wedge B$ is associated as an alternative with disjunction. To our knowledge, there are three cases where disjunction is argued to be associated only with the substring alternatives: (1) *or-else*-disjunction in adult English (Meyer 2012; 2016), (2) disjunction in child language (Singh et al. 2016; Tieu et al. 2017) and (3) adult Warlpiri (Bowler 2015). In all three cases, disjunction acquires a conjunctive implicature in isotone environments. For both alternative spaces, alternatives don't affect interpretation in antitone environments where implicatures are reversed so in the such environments the weakest meaning – i.e. logical (inclusive) disjunction – emerges. The most straightforward case of a disjunction associated with only a substring space is Warlpiri *manu*. *Manu* can be translated as *or* into English if it occurs in the scope of negation (an antitone environment) in (4a), but in an isotone environment as in (4b) a translation with *and* is appropriate.

(4) a. Kula=rna   yunparnu *manu* wurntija   jalangu. Lawa.
         NEG=1SG.SUBJ sing.PST *manu* dance.PST today   nothing
         '*I didn't sing* or *dance today. I did nothing.*'
     b. Ngapa ka   wantimi   *manu* warlpa ka   wangkami.
         water AUX fall.NPST *manu* wind   AUX speak.NPST
         '*Rain is falling* and *wind is blowing.*'

Bowler (2015) argues that the lexical meaning of *manu* should be analyzed as disjunction. This explains the interpretation of (4a) directly since no implicature is predicted to arise in the antitone environment. To explain the conjunctive inference of (4b), Bowler follows the account of Singh et al. (2016) and Meyer (2012; 2016) according to which $A \vee B$ acquires a conjunctive meaning by addition the anti-exhaustive implicatures $\neg(A \wedge \neg B)$ and $\neg(\neg A \wedge B)$.

The derivation of the conjunctive implicature depends on the non-availability of conjunction as an alternative as the following reasoning shows. The conjunctive inference assumes double exhaustification following Fox (2007), i.e. the logical form $\mathbf{exh}(\mathbf{exh}(A \vee B))$. $\mathbf{exh}$ following Fox (2007) is defined with appeal to innocent exclusion as in (5) (see also Spector 2016).

(5) $\mathbf{exh}_A(p) = p \wedge \bigwedge\{\neg q \mid q \in \bigcap\{S \subseteq A \mid \bot \not\Leftarrow p \wedge \bigwedge\{\neg q' \mid q' \in S\}\}\}$

Because the negations of the two alternatives A and B combined are inconsistent with the asserted disjunction $A \vee B$, A or B don't result in implicatures. Therefore, $\mathbf{exh}(A \vee B)$ is only stronger than $A \vee B$ in the diamond space, and then the anti-conjunctive implicature is added as shown in (6). For the second level of **exh**, we need to also exhaustify the alternatives of $A \vee B$ once, resulting in the two alternative spaces shown in (6).

(6)   *diamond space, one* **exh**   *substring space, one* **exh**

$$A \wedge B$$

$$A\wedge\neg B \qquad B\wedge\neg A$$
$$\searmark \qquad \swarrow$$
$$(A\vee B)\wedge\neg(A\wedge B)$$

$$A\wedge\neg B \qquad B\wedge\neg A$$
$$\searrow \qquad \swarrow$$
$$A \vee B$$

Innocent exclusion predicts that the second level of **exh** in the diamond space is vacuous because the negations of $A\wedge\neg B$ and $B\wedge\neg A$ together are inconsistent with $A\vee B\wedge\neg(A\wedge B)$. In case of the substring space, however, this inconsistency doesn't arise since $A \vee B$ didn't acquire an anti-conjunctive inference from the lower **exh**. Therefore the higher **exh** gives rise to the conjunctive inference $A\wedge B$ because the higher **exh** can exclude the two alternative $A \wedge \neg B$ and $B \wedge \neg A$, and combined with $A \vee B$ this entails $A \wedge B$.

The substring space predicts for Warlpiri exactly the pattern in (4). Furthermore, it is clear why in Warlpiri only the substring space can be available at least if we assume that the alternatives available are constrained by the lexical material of the language in question: Because Warlpiri has no conjunction morpheme by assumption, the diamond space just cannot be possible. Similarly in two other cases where the substring scenario has been invoked are cases where it is plausible that conjunction isn't available as an alternative to disjunction: Meyer (2012; 2016) argues that when disjunction is followed by *else*, the alternative with *and* is ungrammatical for English adults. She derives from this proposal the conjunctive inference of *A or else B* to A. Singh et al. (2016) and Tieu et al. (2017) show that children at ages 4 to 6 generally interpret disjunction conjunctively, i.e. reject *A or B* unless both *A* and *B* hold. This follows if children at this age don't yet associate *or* with the lexical alternative *and* as suggested by Chierchia et al. (2001) and Barner et al. (2011), and therefore can only access the substring space of alternatives.

2.1   Japanese *ya*

Sudo (2014) observes that Japanese *ya* exhibits a similar distribution of conjunctive/disjunctive inferences as Warlpiri *manu*. The data in (7) (repeated from (1) and (2)) show that in isotone environments, *ya* seems to have conjunctive interpretation, while in antitone environments, *ya* has a disjunctive interpretation.

(7)   a.   Tarou-wa [kouhii ya koucha]-o nonda
             Taro-top [coffee ya tea]-ACC drank
             'Taro drank things like coffee and tea.'
       b.   [Tarou-ga [kouhii ya koucha]-o nom-eba] yoru nemur-e-nai
             [Taro-nom [coffee ya tea]-ACC drink-if] night sleep-can-NEG
             darou
             infer
             'If Taro drinks things like coffee or tea, he won't be able to sleep at
             night.'

Sudo's (7b) is difficult to reconcile with an analysis of *ya* as conjunction of
Kuno (1973). But Sudo discusses one possibility based on the observation that
frequently junctors are positive polarity items (Spector 2014). In fact, he points
out that the conjunction *to* is a positive polarity item in Japanese, and therefore
when it occurs in an antitone environment such as the restrictor of a universal
as in (8), it receives an interpretation that must be translated by disjunction *or*.

(8)   kouhii to koucha-o nonda dono gakusei-mo kibun-ga waruk unatta
       coffee and tea-ACC drink which student-all got sick
       'Every student who drank coffee or tea got sick.'

If *to* is a positive polarity conjunction, though, we expect that it must take
scope outside of the restrictor of *mo* in (8), and therefore (8) is consistent with
an analysis of *to* as logical conjunction. The more appropriate paraphrase of (8)
is *'Every student who drank coffee and every student who drank tea got sick.'*,
where *and* is indeed translating *to*. Sudo argues that a similar analysis for (7b) is
less plausible because if *ya* is replaced by *to* in (7b) as in (9), only the conjunctive
interpretation is available.

(9)   [Tarou-ga [kouhii to koucha]-o nom-eba] yoru nemur-e-nai   darou
       [Taro-nom [coffee to tea]-ACC drink-if] night sleep-can-NEG infer
       'If Taro drinks coffee and tea, he won't be able to sleep at night.'

Sudo's argument is well taken, but could also be overcome by a stipulation
that *to* is only a local PPI while *ya* is a global one (Homer 2011).[2]
    Sudo instead proposes a different account based on the assumption that *ya*
is lexically a disjunction, that is strengthened by an implicature to a conjunctive
interpretation. However, he doesn't assume the substring alternative space, but
instead assumes that *ya* is associated with the *bona fide* disjunction *ka* as an
alternative, but not directly with conjunction. Furthermore, *ka* associates the
diamond space of alternatives and therefore acquires the anti-conjunctive inference in isotone environments. Assuming double exhaustification for *ya*, Sudo

---

[2] Sudo (personal communication) points out that there is a further argument to be
made in favor of the disjunctive analysis of *ya* based on non-monotonic environments.
Since Sudo's work on this is in progress, I refrain from presenting his argument at
this point.

shows that the higher **exh** leads to the inference the anti-conjunctive inference of *ka* must be false. This amounts to a conjunctive inference for *ya* in isotone environments.

A third possible analysis of *ya* is completely analogous to that of Warlpiri. Since Japanese unlike Warlpiri has a morpheme expressing lexical conjunction, though, this requires the stipulation that only the disjunction *ka* can associate with conjunction as an alternative. For *ya*, then only the substring alternatives are available and therefore the conjunctive implicature arises. These three analyses are the background for the following experiment we conducted.

3   Experiment

Our experiment addresses the distinction between the three alternative analysis. We compare *ya* on the one hand with the conjunction *to* and *mo* and on the other hand with the disjunction *ka* using mouse-tracking. Mouse-tracking was used by Tomlinson Jr. et al. (2013) to investigate implicature processing for scalar *some* in English. Specifically Tomlinson Jr. et al. (2013) show that mouse-tracking can be used to diagnose implicatures vs logical content. If *ya* was also a conjunction, we expect it to pattern with *to* and *mo* like logical content, while the two implicature accounts predict a difference. To distinguish between the two implicature accounts–competition with disjunction *ka* and competition with only substring alternatives, we rely here on the emerging consensus that there are two broad classes of implicatures distinguished by psycho-linguistic tests. One class of implicatures is slow in adult processing and late in language acquisition. The other class of implicatures is fast in adult processing and early in language acquisition. The class of fast implicatures are based on what we will call *explicit alternatives* in the following, namely alternatives that were explicitly mentioned, substring alternatives (Katzir 2007; Fox and Katzir 2011), and the alternatives of numerals. Evidence that explicit alternatives trigger early implicatures is given by e.g. Huang and Snedeker (2009), Huang et al. (2013), Chemla and Bott (2014), Tieu et al. (2017), and others, and that it the implicatures of explicit alternatives are fast in adult processing is shown by Huang and Snedeker (2011). Slow and late implicatures, on the other hand, are for example the not-all implicature of *some* and the anti-conjunctive, not-both implicature of *or* (Bott and Noveck 2004; Tomlinson Jr. et al. 2013 and others). The two implicature accounts make different predictions for whether the conjunctive inference of *ya* should be fast or slow. If *ka* is an alternative, it predicts the implicature to be slow because the conjunctive inference of *ya* involves conventional lexical scales $\langle ya, ka, to \rangle$. But if the conjunctive inference derives only from substring scales, the implicature is predicted to be fast in the relevant sense.

Methods: We compared the conjunctive inference of *ya* with logical content and scalar implicatures with mouse-tracking. Tomlinson Jr. et al. (2013) first used mouse-tracking to investigate the *not-all*-implicature of English *some*. The report that for items such as (10), participants don't respond uniformly, but instead split into two groups (also Noveck and Posada 2003; Bott and Noveck 2004 and others). One group respond based on the logical meaning of disjunction only – i.e. give the response 'true' – while the others respond 'false' based on the pragmatically strengthened meaning of disjunction.

(10)   Some elephants are mammals.

Tomlinson Jr. et al. (2013) report that mouse-tracking reveals a further difference between the two groups: The participant that give the logical response move the mouse almost straight from the start position to the 'true' response box in one corner of the screen. This is schematically shown in the Fig. 1 as the *one step–easy* response pattern. But the participants that give the pragmatic response target the corner for the 'false' response on a markedly different trajectory. Namely, they initially move towards the 'true' response and only in the course of their movement change direction and target the opposite corner. This is shown schematically as the *two step* curve in the Fig. 1. Mouse-tracking can therefore be used to distinguish implicature based responses from those based on literal meaning alone.

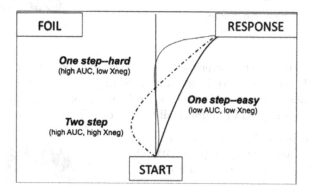

Fig. 1. Classification of some mouse-track patterns by Tomlinson Jr. et al. (2013)

To apply mouse-tracking to Japanese junctors we created an experiment with 200 sentences and corresponding pictures. The experiment included 8 items of condition ya1 like (11a) and 4 each of condition mo1 and to1 like (11b). Participants' task was to indicate how well the picture fit to the sentence with the response being either *ii* ('good') or *warui* ('bad'). For both items in (11), the expected response was 'bad' because of the conjunctive inference of *ya, mo* and *to*.

(11)   a. ya1: kuma-ya gorira-ga   imasu.
              bear-YA gorila-NOM exist

        'There're a bear YA a gorilla.'

     b. mo/to1: kuma-mo gorira-mo imasu. / kuma-to gorira-ga
                 bear-AND gorira-AND exist   / bear-AND gorira-NOM

        imasu.
        exist

        'There're a bear and a gorilla.'

We also compared the conjunctive inference of *ya* with two other implicatures: the anti-conjunctive implicature of the disjunction *ka* and the upper bound implicature of the numeral *one*. 16 items of condition *ka2* in (12a), and 4 item of condition one2 in (12b) tested these.

(12)   a. ka2: budo-ka momo-(ka)-ga arimasu.
              grape-or peach-or-NOM exist

        'There're grapes or a peach.'

     b. one2: hebi-ga   ip-piki imasu.
               snake-NOM one-CL exist

        'There's one snake.'

Controls. In addition, the experiment contained 164 controls and filler items. Two specifically relevant controls are illustrated in (13). The ka1-condition (13a) is a logically and pragmatically true control for the ka2-condition. The two1 condition is a logically and pragmatically false control for the one2-condition.

(13)   a. ka1: budou-ka ringo-(ka)-ga   arimasu.
              grapes-or apple-(or)-NOM exist
              'There're grapes or an apple.'

     b. two1: momo-ga   futatsu arimasu.
               peach-NOM two   exist
               'There're two peaches.'

Presentation. We presented single Japanese sentences such as in (13) on screen followed by picture such as in (13) with a "good"/"bad" decision task. The sentences and all on-screen instructions were shown in the experiment in standard Japanese script. Each participant was shown 200 items including practice items. Each trial consisted of the three phases in (14).

(14)   a.   "Start" Button: Participants clicked the *hajimeru* ('start') button
              at the bottom center of the screen
        b.   Sentence Presentation: A sentence was shown for 2000 ms in the
              screen center. The mouse was frozen during sentence presentation.
        c.   Picture Presentation & Forced Choice Decision: A picture appeared
              in the screen center. Also two response boxes titled *ii* ('good') and
              warui ('bad') appeared in the top left and right corners of the
              screen. Participants had to click one response box before proceed-
              ing to the next item. If participants didn't respond within 2000 ms,
              they saw an on-screen warning message.

The order of presentation was randomized. Before the experimental trials,
subjects saw an instruction screen, then eight practice items, and had an opportunity to ask clarification questions. The instruction was that subject should
indicate how well the picture matched the sentence. The position of good/bad
response boxes was counterbalanced across subjects in a Latin square design.
After half of the experimental trials (about 10 min), subjects were asked to take
a short break. The experiment took on average 20 min per subject. Data were
recorded with the Mousetracker software (Freeman and Ambady 2010). 67 native
Japanese speakers participated; most were undergraduate students at Tohoku
University, Sendai, Japan. Participants were compensated for their participation. Statistical analyses and graphic preparation were conducted in R (R Core
Team 2012).

Results: Overall accuracy on controls and fillers was 97%. Our data show a
clear difference between *ya* and the lexical conjunctions *to* and *mo* (i.e. the
items in (11) above) in response accuracy, reaction times, and mouse-tracks.
For to1 and mo1, accuracy was 95%, but for ya1 significantly lower at 75%,
i.e. 25% of participants judged the sentence with *ya* as acceptable if only one
of the two items was present. Also for to1 and mo1, mean reaction time of
correct responses was 1743 ms, while it was significantly longer (2037 ms) for
ya1. Finally the mousetrack data also confirm that *ya* differs from the lexical
conjunctions. Figure 2 shows the comparison graphically. The mouse paths for
ya1 diverge more from the straight line to the target as shown by a significant
difference in the area-under-the-curve (AUC). The difference between ya1 on the
one hand and mo1 and to1 on the other argues against an analysis of *ya* as a
lexical conjunction and corroborates the implicature proposal of Sudo (2014).
But at the same time the mousetracks don't clearly show the two-step pattern
of response as in Fig. 1 above.

For the further analysis we compare *ya* with other implicature items in our
experiment. Prima facie one possibility may be that implicature items in our setup generally don't exhibit the two-step response because of differences between
the method we used and that of Tomlinson Jr. et al. (2013) (i.e. the use of pictures
and differences in the way we presented the sentences). But the result for the
ordinary disjunction *ka* in the scenario where both disjuncts are true indicates
that our method worked to elicit the two-step response. For this analysis, we
compare the responses to all the either with *ka* (either a single *ka* or two *ka*s)

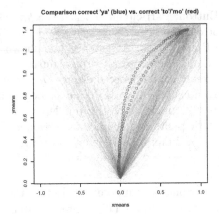

Fig. 2. Comparison of mousetracks for 'ya' (blue) versus lexical conjunctions (red) with one conjunct true (as in (11)), only correct responses (i.e. 'false'). Each light line represents a single response, and the two dotted lines represent the mean response curve. (Color figure online)

when both disjuncts are true into logical (i.e. *good*) and pragmatic responses (i.e. *bad*) in Fig. 3. There is a clear difference: While the logical responses exhibit the 'one step–easy'-pattern of Fig. 1, the pragmatic responses show the 'two step' pattern. In other words, we successfully extended the result of Tomlinson Jr. et al. (2013) for English *some* to the Japanese disjunction *ka*.

What is the comparable result for the implicatures of *ya* and the numeral *one*? The differences between logical and pragmatic responses are much smaller. The comparison between the logical and pragmatic response curves is in Fig. 3 made difficult by the targeting that responses in different corners of the display. For the subtler comparisons, it is therefore better to mirror the responses such that they all seem to target the same corner even though the responses were in fact in different positions. In Fig. 4, the mousetracks of logical and pragmatic responses for the three implicature items (ya1 from (11) and the two from (12)) are compared in this way. Note that the panel on the left in Fig. 4 presents the same data as Fig. 3, but with the red mousetracks mirrored. The results show that we find a clear difference between pragmatic and logical responses only with *ka* ('or').

The pattern is confirmed by the accuracy rates of the three implicature conditions with the corresponding control conditions given in (14). The experiment included also specific non-implicature controls for ka2 and one2 in (13).

(15)   a. | ka2 | ka1 | b. | ya1 | mo/to1 | c. | one2 | two1 |
       acc. 33.9% 92.6%   acc. 75.4% 95.1%   acc. 79.9% 100%

We performed a linear mixed model analysis of the area under the curve with fixed factors condition (ya1, ka1, one2) and response type (logical, pragmatic).

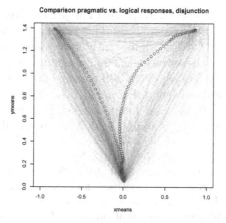

Fig. 3. Comparison of mousetracks for disjunction logical (red) versus pragmatics responses (blue) with both disjuncts true (as in (12)). (Color figure online)

ka ('or')   *ya*   *i* ('one')

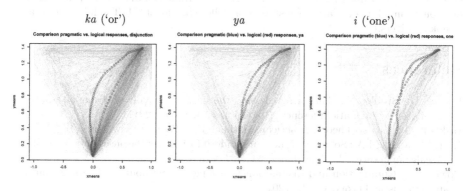

Fig. 4. Comparison of mousetracks for logical (red) vs pragmatic (blue) responses to the three implicature items. (Color figure online)

The analysis confirms that the interaction between condition and response type for condition one2 didn't differ from that for ya1 ($t = 1.3, p = .201$), but the difference to the ya1-interaction is highly significant for conditions 1ka2 ($t = -4.1, p < .0001$) and 2ka2 ($t = -5.0, p < .000001$).

4   Conclusions

Our results confirm Sudo's (2014) proposal that *ya* is different from other conjunctions. The differences we observe between *ya* on the one hand and *mo* and *to* on the other in the items where only one disjunct is true show that *ya* cannot be treated as a logical conjunction.

But our further results don't support the Sudo's implementation of the implicature analysis. His account predicts that item ya1 should pattern similarly to item ka2 or be slower because the response to in *ya1* involves the computation of the implicature of *ka* on Sudo's analysis. This is not what we find: For one, the implicature rates for *ka* (34%) is significantly lower than for ya (75%). Also the comparison of mouse-tracks of the logical and pragmatic responders corroborates this picture: in ka2, there is a significant difference, but not in condition ya1. The implicature rate of cardinals (condition one2) is similar to that of ya1, though, by both accuracy rate (80%) and mouse-tracks. This result corroborates the proposal the implicature of *ya* is similar to that of a numeral, i.e. processed fast.

Acknowledgments. We are grateful to Hye-Jeong Lee, Yasutado Sudo, Kazuko Yatsushiro, four anonymous reviewers, and the audiences of LENLS at the Ochanomizu University, the workshop on Questions and Disjunction at the University of Vienna, and the 39th GLOW colloquium at the University of Göttingen for their comments, and to Ryo Tachibana for technical support. The work reported in this paper was funded primarily by the German research council (DFG project SSI, SA 925/11-1, within SPP 1727 XPrag.de. In addition, it was supported in part by the German Federal Ministry for Education and Research (BMBF, Grant No. 01UG1411), the Alexander von Humboldt foundation (postdoctoral fellowship for J. M. Tomlinson Jr.), and Tohoku University.

References

Barner, D., Brooks, N., Bale, A.: Accessing the unsaid: the role of scalar alternatives in children's pragmatic inference. Cognition **118**, 87–96 (2011)

Birkhoff, G.: Lattice Theory. American Mathematical Society, New York (1940)

Bott, L., Noveck, I.A.: Some utterances are underinformative: the onset and time course of scalar inferences. J. Mem. Lang. **51**, 437–457 (2004)

Bowler, M.: Conjunction and disjunction in a language without 'and'. In: Semantics and Linguistic Theory, vol. 24 (2015)

Chemla, E., Bott, L.: Processing inferences at the semantics/pragmatics frontier: disjunctions and free choice. Cognition **130**, 380–396 (2014)

Chierchia, G.: Logic in Grammar: Polarity, Free Choice, and Intervention. Oxford University Press, Oxford (2013)

Chierchia, G., Crain, S., Guasti, M.T., Gualmini, A., Meroni, L.: Evidence for a grammatical view of scalar implicatures. In: Do, A.-J., et al. (ed.) BUCLD 25 Proceedings (2001)

Fox, D.: Free choice and the theory of scalar implicatures. In: Sauerland, U., Stateva, P. (eds.) Presupposition and Implicature in Compositional Semantics, pp. 71–120. Palgrave Macmillan, Basingstoke (2007)

Fox, D., Katzir, R.: On the characterization of alternatives. Nat. Lang. Seman. **19**, 87–107 (2011)

Freeman, J.B., Ambady, N.: MouseTracker: software for studying real-time mental processing using a computer mouse-tracking method. Behav. Res. Methods **42**, 226–241 (2010)

Homer, V.: Polarity and modality. Ph.D. thesis, UCLA (2011)

Huang, Y.T., Snedeker, J.: Semantic meaning and pragmatic interpretation in 5-year-olds: evidence from real-time spoken language comprehension. Dev. Psychol. **45**, 1723–1739 (2009)

Huang, Y.T., Snedeker, J.: Logic and conversation revisited: evidence for a division between semantic and pragmatic content in real-time language comprehension. Lang. Cogn. Process. **26**, 1161–1172 (2011)

Huang, Y.T., Spelke, E., Snedeker, J.: What exactly do numbers mean? Lang. Learn. Dev. **9**, 105–129 (2013)

Katzir, R.: Structurally-defined alternatives. Linguist. Philos. **30**, 669–690 (2007)

Kuno, S.: The Structure of the Japanese Language. MIT Press, Cambridge (1973)

Lee, Y.: Scales and alternatives: disjunction, exhaustivity, and emphatic particles. Ph.D. thesis, University of Texas, Austin (1995)

Meyer, M.-C.: Missing alternatives and disjunction. In: Beltrama, A., Chatzikonstantinou, T., Lee, J., Pham, M. (eds.) Proceedings of CLS 48. University of Chicago, Chicago, IL (2012)

Meyer, M.-C.: Generalized free choice and missing alternatives. J. Semant. **33**, 703–754 (2016)

Nicolae, A.C.: Simple disjunction PPIs – a case for obligatory epistemic inferences. NELS 46 (2015)

Noveck, I., Posada, A.: Characterizing the time course of an implicature: an evoked potentials study. Brain Lang. **85**, 203–210 (2003)

Ohori, T.: Coordination in mentalese. In: Coordinating Constructions, vol. 58. John Benjamins Publishing, pp. 41–66 (2004)

R Core Team: R: A Language and Environment for Statistical Computing. R Foundation for Statistical Computing, Vienna, Austria (2012). http://www.R-project. org/

Sauerland, U.: Scalar implicatures in complex sentences. Linguist. Philos. **27**, 367–391 (2004)

Singh, R., Wexler, K., Astle- Rahim, A., Kamawar, D., Fox, D.: Children interpret disjunction as conjunction: Conse- quences for theories of implicature and child development. Nat. Lang. Seman. **24**(4), 305–352 (2016)

Spector, B.: Comparing exhaustivity operators. Semant. Pragmatics **9**, 1–33 (2016)

Spector, B.: Global positive polarity items and obligatory exhaustivity. Semant. Pragmat. **7**, 1–61 (2014)

Sudo, Y.: Higher-order scalar implicatures of 'ya' in Japanese. Handout, TEAL 9, University of Nantes (2014)

Tieu, L., Yatsushiro, k., Cremers, A., Romoli, J., Sauerland, U., Chemla, E.: On the role of alternatives in the acquisition of simple and complex disjunctions in French and Japanese. J. Semant. 34, 127–152 (2017)

Tomlinson, J.M., Bailey, T.M., Bott, L.: Possibly all of that and then some: scalar implicatures are understood in two steps. J. Mem. Lang. **69**, 18–35 (2013)

The Projection of Not-at-issue Meaning via Modal Support: The Meaning and Use of the Japanese Counter-Expectational Adverbs

Osamu Sawada[✉]

Department of Humanities, Mie University, Tsu, Japan
sawadao@human.mie-u.ac.jp

Abstract. This paper discusses the phenomenon of what I call the "projection of non-at issue meaning via modal support" shown in the Japanese counter-expectational intensifier *yoppodo* and the counter expectational scale-reversal adverb *kaette*, and considers the variation of projective content from a new perspective. I show that, unlike the typical conventional implicatures (CIs) like appositives and expressives (e.g., Potts [19]), *kaette* and *yoppodo* can project out of the complement of a belief predicate only if there is a modal in the main clause. I argue that *yoppodo* and *kaette* belong to a new class of projective content that requires consistency between an at-issue meaning and a CI meaning in terms of a judge. This paper provides a new perspective for the typology of projective content.

Keywords: Projection · Judge · Modal support · Obligatory local effect · Consistency of a judge

1 Introduction

In this paper, I investigate the meaning and use of the Japanese, counter-expectational scalar modifiers *yoppodo* and *kaette* and reconsider the current classification and theories of projective content from a new perspective.

Recently, important theories and classifications have been proposed for projective content, which include presuppositions and conventional implicature (CI). Particularly well-investigated phenomena are expressives and appositives, such those in (1):

(1) a. That bastard Kresge is famous. (Potts [19], pp. 168)
        (CI/expressive meaning: Kresge is bad in the speaker's opinion.)

O. Sawada—I would like to thank Chris Davis, Ikumi Imani, Robert Henderson, Magdalena Kaufmann, Stefan Kaufmann, Koji Kawahara, Susumu Kubo, Ai Kubota, Yusuke Kubota, Takeo Kurafuji, Lars Larm, Eric McCready, Kenta Mizutani, David Oshima, Harumi Sawada, Jun Sawada, Ayaka Sugawara, Eri Tanaka, Jérémy Zehr and the reviewers of LENLS 12 for their valuable comments and discussions. Parts of this paper were presented at the International Modality Workshop at Kansai Gaidai (2015), the Semantics Workshop in Tokai (2015), and LSA (2016), and I thank the audiences for their valuable comments and discussions. This paper is based upon work supported by JSPS KAKENHI Grant Number 26770140. All remaining errors are of course my own.

© Springer International Publishing AG 2017
M. Otake et al. (Eds.): JSAI-isAI 2015 Workshops, LNAI 10091, pp. 122–137, 2017.
DOI: 10.1007/978-3-319-50953-2_10

    b. Lance, a cyclist, is training. (Potts[21], pp. 97)
       (CI: Lance is a cyclist.)

Potts [19] argues that the meanings of expressives and appositives are a conventional implicature (Grice [4]) and logically and dimensionally independent of "what is said." Potts [19] also contends that CIs are different from presuppositions in terms of projection. In contrast to typical presuppositions, CIs can project even if they are embedded in the complement of an attitude predicate such as *believe* or verbs of saying, which function as a presupposition plug (Karttunen [8]), as in (2):

(2)   a. Sue believes that bastard Kresge should be fired. (♯I think he's a good guy.) (Potts [19], pp. 170)

     b. Sheila believes that the agency interviewed Chuck, a confirmed psychopath, just after his release from prison. (Potts [21], pp. 115)

Potts [19] claims that appositives and expressives are invariably speaker-oriented, regardless of their syntactic environment. However, recent studies have shown that, contrary to Potts' [19] initial claim, CI expressions such as appositives and expressives can have a non-speaker orientation (e.g., Karttunen and Zaenen [9]; Wang, Reese, and McCready [29]; Amaral, Roberts, and Smith [1]; Potts [18]; Sauerland [20]; Haris and Potts [7]). For example, Amaral et al. [1] show that the sentences in (3) have a subject-anchored interpretation:

(3)   a. (Context: Joan is crazy. She's hallucinating that some geniuses in Silicon Valley have invented a new brain chip that's been installed in her left temporal lobe and permits her to speak any of a number of languages she's never studied): Joan believes that her chip, which she had installed last month, has a twelve year guarantee. (Amaral et al. [1], pp. 735f.)

     b. (Context: We know that Bob loves to do yard work and is very proud of his lawn, but also that he has a son Monty who hates to do yard chores. So Bob could say (perhaps in response to his partner's suggestion that Monty be asked to mow the lawn while he is away on business)):
       Well, in fact Monty said to me this very morning that he hates to mow the friggin' lawn. (Amaral et al. [1], pp. 736)

Harris and Potts [7] present both corpus and experimental evidence to indicate that appositives and expressives are generally speaker-oriented, but certain discourse conditions can counteract this preference. A non-speaker orientation becomes the dominant interpretation if certain discourse conditions are met.

In this paper, I present a new perspective on studies of projective behaviors of CIs based on new phenomena, namely the Japanese counter-expectational adverbs *yoppodo* and *kaette*. I show that the projective behaviors of these adverbs are radically different from those of typical CIs (and typical presupposition). More specifically, I argue that the not-at-issue meaning of *yoppodo* and *kaette* can project out of the complement of an attitude predicate only when there is speaker-oriented modality in the main clause. (I will call this phenomenon the "projection of not-at-issue meaning via modal support.") I further argue that *yoppodo* and *kaette* belong to a new class of projective content that requires consistency between an at-issue meaning and a CI meaning in terms of a judge.

2   The Japanese Counter-Expectational Intensifier *yoppodo*

Let us focus first on the meaning and projective properties of *yoppodo*, which has complex semantic/pragmatic characteristics. Consider the example in (4):

(4)   (Context: Taro is looking at a ramen restaurant from outside. He sees a lot of people waiting in front of the restaurant.)

Ano raamen-ya-wa   yoppodo   oishii-nichigainai.
That ramen-restaurant-TOP YOPPODO delicious-must

At-issue: That ramen restaurant must be very delicious.
Not-at-issue: I am inferring the degree via evidence, and the degree is above my expectation.

In (4) the speaker is attempting to explain the unusual situation of there being many people waiting in front of a ramen restaurant by inferring that [the food served at] the restaurant is very delicious. There is also a counter-expectational feeling in (4) that the degree of deliciousness is higher than the speaker's expectation.

Yoppodo has several distinct properties that differ from those of other more regular intensifiers. First, in the adjectival environment, *yoppodo* must co-occur with an evidential modal. As Watanabe [30] also descriptively observes, *yohodo/yoppodo* cannot occur in a simple adjectival sentence, as in (5):[1]

(5)   * Ano raamen-ya-wa   yoppodo   oishii.
          That ramen-restaurant-TOP YOPPODO delicious

'That ramen restaurant is *yoppodo* delicious.'

This suggests that *yoppodo* has a modal concord-like relationship with evidential modality similarly to the case of the German modal particle *ruhig* (Grosz [5]; Kaufmann [10]) which also requires a modal.

---

[1] Note that there are also other uses of *yoppodo*, e.g. comparative-intensifier use, a conditional use, or an eventive/volitive use, and these do not require an evidential modal:

(i)   Okinawa-no   hoo-ga   Tokyo-yori yoppodo   suzushii-desu.
       Okinawa-GENI direction-NOM Tokyo-than YOPPODO cool-PRED.POL

       At-issue: It is much cooler in Okinawa than in Tokyo.(Not-at-issue: Generally, the opposite is true.)

(ii)   Yoppodo   i-te   yar-oo-ka-to   omo-tta.
       YOPPODO say-TE give-volitive-that think-PAST

       At-issue: I had a desire to say a bad word. (Not-at-issue: My degree of willingness to say a bad word is above my expectation, but I didn't say a bad word.)

(iii) Yoppodo   isyoukenmei benkyoo si-nai-to   siken-ni ukara-nai-yo.
       YOPPODO hard   study   do–NEG-COND exam-to pass-NEG-YO

       'You will not be able to pass the exam unless you study *yoppodo* hard.'

In this paper, I mainly focus on the evidential/adjective use of *yoppodo*.

Notice further that non-evidential modals cannot co-occur with the evidential type of *yoppodo*. The epistemic modals *kamoshirenai* 'may' and *daroo* 'possibly' have no evidential component and cannot co-occur with *yoppodo*, as is clear from (6):

(6)   ?? Ano raamen-ya-wa   yoppodo   oishii-{kamoshirenai/daroo}.
       That ramen-restaurant-TOP YOPPODO delicious-may/possibly

      At-issue: That ramen restaurant may be very delicious.

However, if the particle no is added to *kamoshirenai* and *daroo* (*no-kamoshirenai*, *no-daroo*) then (6) becomes natural, as in (7):

(7)   Ano raamen-ya-wa   yoppodo   oishii-no-{kamoshirenai/daroo}.
      That ramen-restaurant-TOP YOPPODO delicious-NODA-may/possibly

      At-issue: That ramen restaurant may be very delicious.
      Not-at-issue: I am inferring the degree via evidence, and the degree is above my expectation.

In (7) *no-kamoshirenai* and *no-daroo* behave with evidential modality, presumably due to the meaning of the discourse particle *no(da)* (see Sawada [21]).

Another distinctive characteristic of *yoppodo* is that it not only semantically intensifies a degree but also conventionally implies that the given degree is above a judge's expectation. Thus, it triggers a counter-expectational feeling as a CI. In the Gricean approach, CI is considered to be independent of "what is said" (the at-issue meaning) (Grice [4]; Potts [19]). Regular intensifiers like the Japanese *totemo* 'very' do not trigger this kind of unexpected meaning, as exemplified in (8):[2]

(8)   (Context: Taro is looking at a ramen restaurant from outside. He sees a lot of people waiting in front of the restaurant.)
      Ano raamen-ya-wa   totemo oishii-nichigainai.
      That ramen-restaurant-TOP very   delicious-must

      At-issue: That ramen restaurant must be very delicious.

The sentence in (8) denotes that the degree of deliciousness of the food served at the ramen restaurant is high, but it does not convey that the degree is unexpected.

The idea that the counter-expectational meaning of *yoppodo* is a CI is supported by the fact that a normal objection—"No, that will be false"—cannot challenge the CI part/not-at-issue part in (8), supporting the idea that it is independent of "what is said."

(9)   A: Ano raamen-ya-wa   yoppodo   oishii-nichigainai.
          That ramen-restaurant-TOP YOPPODO delicious-must

          At-issue: That ramen restaurant must be very delicious. Not-at-issue: I am inferring the degree via evidence, and the degree is above my expectation.

      B: Iya, sore-wa   uso-daroo.
          No that-TOP false-epistemic

          'Well, that will be false.'

---

[2] Notice, however, that there is also an expressive/CI use of *totemo*, which intensifies an unlikelihood/impossibility of a given proposition (Sawada [22]).

Furthermore, the idea that *yoppodo* functions for semantic intensification is supported by the fact that an addressee can challenge the at-issue part (intensification part) of (8) by uttering the sentence in (10):

(10) Iya, sonnan-demo   nai-to   omoun-da-kedo
      No such level-DEMO not-that think-PRED-but

      'Well, I don't think that it is that high ...'

I propose that *yoppodo* (the explanatory use) has mixed content (McCready [14]; Gutzmann [6]); it has an intensified meaning at the at-issue level (the left side of ♦) and inferential/counter-expectational meanings at the CI level (the right side of ♦):

(11) [[*yoppodo_{evidential}*]]:
      $= \lambda G \lambda x \lambda t \lambda w \exists d[d >!!STAND \wedge G(d)(x)(t)(w)]$♦ $d > d'$ for j (where j is a judge [either a speaker or a subject] and d' is a judge's degree of expectation, w is bound by an evidential modal)

Note that the CI meaning is non-quantificational and d in the CI dimension is a free variable, which is anaphoric to the degree in the at-issue component. (See Sudo [26] for the anaphoric approach to the relationship between at-issue and presupposition.) In prose, *yoppodo* in (11) semantically denotes that the degree associated with a gradable predicate is far greater than a contextual standard at the at-issue level. It also conventionally implies that the given degree is above the judge's expectation. Note that there is a restriction that w in the at-issue component must be bound by an evidential modal. This component explains the phenomenon of modal concord/modal matching in the evidential use of *yoppodo*. That is, d is a degree inferred via evidence.

Let us now consider how *yoppodo* is computed in a compositional manner. The important point is that only the at-issue part of *yoppodo* logically interacts with other at-issue (semantic) elements. To ensure that the meaning of mixed content is computed in a compositional fashion, McCready [14] proposes compositional rule(s) for mixed content, which involve(s) the shunting type s, as in (12):

(12) Mixed application (based on McCready [15])

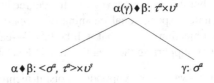

Note that α and β form a single lexical item (mixed content). The crucial point of the rule in (12) is that it is resource sensitive. Note that the rule is different from Potts' [19] CI application, in which the at-issue argument of the CI-inducing element is passed up to the above level —in other words, the application is resource insensitive.[3]

If β is complete (does not have a variable), then the bullet • is introduced to separate the at-issue dimension from the CI dimension as in (13):

---

[3] See McCready [14] for a detailed discussion on the difference between a resource-sensitive CI application(shunting application) and Potts' resource-insensitive one.

(13) Final interpretation rule: Interpret $\alpha \blacklozenge \beta : \sigma^a \times t^s$ as follows: $\alpha : \sigma^a \bullet \beta : t^s$

Let us now analyze the meaning of a sentence with *yoppodo*. Consider the example in (14):

(14) Ano raamen-ya-wa   yoppodo   oishii-nichigainai.
      That ramen-restaurant-TOP YOPPODO delicious-must

      At-issue: That ramen restaurant must be very good.
      Not-at-issue: I am inferring the degree via evidence, and the degree is above my expectation.

Regarding the meaning of gradable predicates, I assume that they represent relationships between individuals and degrees (Seuren [24]; Cresswell [2]; von Stechow [28]; Klein [13]; Kennedy [12]), as in (15):

(15) $[[ooshii]] : \langle d^a, \langle e^a, \langle i^a, \langle s^a, t^a \rangle \rangle \rangle \rangle = \lambda d \lambda x \lambda t \lambda w.delicious(x)(t)(w) = d$

As for the modal *nichigainai* 'must', I posit the lexical meaning reflected in (16):[4]

(16) $[[nichigainai]]^{w,g} : \langle \langle s^a, t^a \rangle, t^a \rangle = \lambda p_{\langle s^a, t^a \rangle}.\forall w'$ compatible with the evidence in $w_0 : p(w') = 1$

Figure (17) shows the logical structure of (14):

(17) The semantic derivation of *yoppodo* (the evidential type)

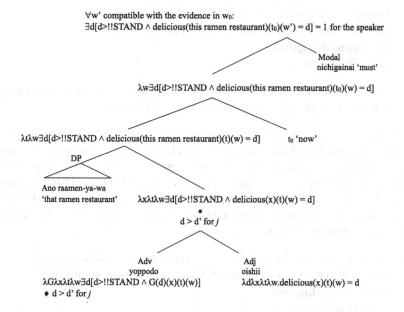

---

[4] There is also a possibilty that the evidential component of *nichigainai* is not-at-issue (CI/presupposition). See, e.g., Portner [16], von Fintel and Gillies [3], and McCready [14] for the discussions on the semantic status of evidentiality.

2.1 The Projective Behavior of *yoppodo*

What is puzzling about the evidential use of *yoppodo* is that it has a complex property of projection. If *yoppodo* is embedded under an attitude predicate and there is an evidential modal inside the embedded clause, then the not-at-issue component of *yoppodo* is always subject-oriented, as in (18):

(18)   (Context: Taro saw a lot of people waiting in front of the ramen restaurant and thought that this situation was unusual.)

    Taro-wa   [ano ramen-ya-wa   yoppodo oishii-**nichigainai**]-to
    Taro-TOP that ramen-restaurant-TOP yoppodo delicious-must-that
    omo-tteiru.
    think-TEIRU

    'Taro thinks that the ramen restaurant must be *yoppodo* delicious.' (The non-at-issue meaning of *yoppodo* = always subject-oriented)

However, if *yoppodo* is embedded under an attitude predicate and there is an evidential modal (a concord element) in the main clause, the non-at-issue component of *yoppodo* will always be speaker-oriented, as in (19):

(19)   (Context:The speaker observes that Taro goes to the ramen restaurant KIKUYA every day.)

    Taro-wa   [ano ramen-ya-wa   yoppodo oishii]-to
    Taro-TOP that ramen-restaurant-TOP yoppodo delicious-that
    omo-tteiru-**nichigainai**.
    think-TEIRU-must

    'Taro must think that that ramen restaurant is *yoppodo* delicious.' (The non-at-issue meaning of *yoppodo* = always speaker-oriented.)

This means that whether or not the not-at-issue meaning of *yoppodo* can project depends on the presence of a modal in the main clause; the modal supports the projection of the not-at-issue meaning. I call this phenomenon the "projection of not-at-issue meaning via modal support." Note that, if there is no modal in (18) or (19), the sentence becomes ill-formed. Notice further that, if two evidential modals are present (one in the main clause and another in the embedded clause), then *yoppodo* can (in principle) match either of them, as in (20):

(20)   Taro-wa   [ano ramen-ya-wa   yoppodo oishii-**nichigainai**]-to
    Taro-TOP that ramen-restaurant-TOP yoppodo delicious-must-that
    omo-tteiru-**nichigainai**.
    think-TEIRU-must
    'Taro must think that ramen restaurant must be *yoppodo* delicious.' (The non-at-issue meaning of *yoppodo* = speaker-oriented/subject-orineted.)

There may be a preference for *yoppodo* to interact with the nearest modal but it seems that it can interact with the modal in the main clause if we posit a context like that in (19).

Let us now try to explain the projective behaviors of *yoppodo*. The question arises as to why (18) does not have a speaker-oriented reading. We might say that *yoppodo* in (18) cannot be speaker-oriented because such a reading will lead to a conflict in terms of modal matching (more specifically, matching by a judge). In (18), as *nichigainai* is embedded under an attitude predicate, the person who evaluates the proposition (based on evidence) has to be the subject (Taro). This makes it impossible for *yoppodo* in (18) to project outside the attitude predicate. However, if there is an evidential modal in the main clause, as in (19), *yoppodo* can project because there is an appropriate match between *yoppodo* and *nichigainai* in terms of the judge. The question then arises as to why (19) does not have a subject-oriented reading. We might say that this is because there is no appropriate match between a judge in the CI and the at-issue domain in terms of the judge. If the judge of *yoppodo* is the speaker, the CI component is not met and the sentence becomes infelicitous (although it is grammatical).

3   The Japanese Scale-Reversal Adverb *kaette*

3.1   The Meaning of *kaette*

Let us now consider the meaning of the scalar reversal adverb *kaette*. Descriptively, this adverb signals that the opposite of the at-issue situation is normally true.[5]

In terms of distribution, it can appear in both adjectival and comparative sentences. Unlike the evidential *yoppodo*, there is no modal matching for *kaette*. Consider the examples in (21):

(21)   a.   Kono basyo-wa   kaette   kiken-da.
            This place-TOP KAETTE dangerous-PRED

         At-issue: This place is dangerous.
         Not-at-issue: Generally, this place is considered safe.

       b.   Konbini-no   koohii-no-hoo-ga   restoran-no
            Convenience store-GEN coffee-GEN-direction-NOM restaurant-GEN
            koohii-yori-mo   kaette   oishii.
            coffee-than-MO KAETTE tasty

         At-issue: A convenience store's coffee is tastier than a restaurant's coffee.
         Not-at-issue: A restaurant's coffee is generally tastier than a convenience store's coffee.

---

[5] Note that *kaette* is a kind of degree adverb. Although it does not intensify a degree, it reverses the polarity of a gradable predicate. As illustrated in (i), if an attached predicate is not gradable, the resulting sentence becomes ill-formed:

(i)   * Taro-wa   *kaette* gakusei-da.
       Taro-TOP *kaette* student-PRED
       'Taro is *kaette* a student.'

In (21a), the speaker conveys that the place is generally safe. In (21b), the speaker indicates that a restaurant's coffee is generally tastier than a convenience store's coffee.

The example in (22) is odd because the CI meaning conflicts with the general geographical knowledge that it is cooler in Tokyo than in Okinawa:

(22) ?? Kaette   Tokyo-no   hoo-ga   Okinawa-yori suzushii.
      KAETTE Tokyo-GENI way-NOM Okinawa-than cool

   At-issue: It is cooler in Tokyo than in Okinawa.

   Not-at-issue: 'Generally, it is cooler in Okinawa than in Tokyo.'

In terms of use, *kaette* is used in abnormal situations. The given proposition is normally assumed to be false. This means that the at-issue part of the sentence with *kaette* is non-generic. (We will come back to this point later.)

Regarding the semantic status of *kaette*, I assume that *kaette* is a CI. This idea is supported by the fact that denial only targets the at-issue part of a sentence with *kaette*, as in (23):

(23) Iya, sore-wa   uso-da.
      No that-TOP false-PRED

   'Well, that's false.'

The meaning triggered by *kaette* is not a semantic presupposition because it is possible to determine the truth value of the given proposition even if the meaning triggered by *kaette* is false. Namely, there is no logical dependency between the at-issue meaning and the meaning triggered by *kaette*.[6]

Based on this argument, I assume the meaning of *kaette* in (21a) as follows:

(24) $[[kaette_{pos}]] = \lambda P_{pos}\lambda x\lambda t\lambda w.P(x)(t)(w) \blacklozenge \lambda P_{pos}\lambda x\lambda t\lambda w.$ the judge j assumes that generally $P(x)(t)(w) = 0$ but admits the possibility that $P(x)(t)(w) = 1$.

In this analysis, *kaette* possesses a special kind of mixed content. The at-issue part of *kaette* behaves as an identity function and has no concrete lexical meaning. In the CI component, it implies that the judge assumes that generally P is not true of x but admits the possibility that P can be true of x. This ensures that the at-issue proposition is always interpreted as a non-generic proposition. Namely, it is true only in an abnormal situation.

---

[6] However, it may be possible to analyze the meaning of *kaette* in terms of 'pragmatic presupposition' (Stalnaker [25]). In the pragmatic presupposition approach, presuppositions are considered to be "the background beliefs of the speaker"— propositions whose truth he takes for granted, or seems to take for granted, in making his statement (Stalnaker [25], pp. 48). It seems that the meaning triggered by *kaette* often corresponds to a common knowledge. Whether the meaning of *kaette* is a CI or a presupposition needs further investigation. I leave this question for future research. In this paper, I will assume that *kaette* is a CI.

In terms of compositionality I assume that, because *kaette* does not modify a degree, it does not directly modify a gradable predicate. Rather, it combines with the constituent that consists of a positive morpheme *pos* and a gradable predicate, *pos*(ADJ). The semantic function of *pos* is to relate the degree argument of the adjectives to an appropriate standard of comparison (Cresswell [2]; von Stechow [28]; Kennedy and McNally [11]; among others), as in (25):

(25) $[[pos]] = \lambda g \lambda x \lambda t \lambda w \exists d[d > STAND \wedge g(d)(x)(t)(w)]$

The figure in (26) shows part of the semantic derivation of the sentence in (21a):

(26)   The semantic derivation of *kaette* (the positive sentence)

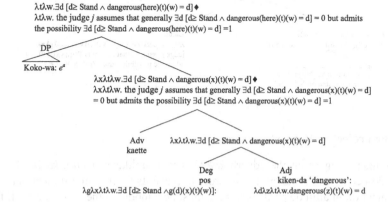

Let us now consider the meaning of *kaette* in a comparative environment. In the case of a comparative sentence with *kaette*, the *pos* morpheme is unnecessary as there is no norm-related meaning in comparison. I assume that *kaette* has a slightly different denotation in the comparative environment, as reflected in (27). Note that P_{comp} corresponds to the comparative phrase "DP-*yori*" and *yori* has the denotation in (28).

(27) $[[kaette_{comp}]] = \lambda P_{comp} \lambda g \lambda x \lambda t \lambda w. P(g)(x)(t)(w) \blacklozenge \lambda P_{comp} \lambda g \lambda x \lambda t \lambda w.$ the judge j assumes that generally $P(g)(x)(t)(w) = 0$ but admits the possibility that $P(g)(x)(t)(w) = 1$.

(28) $[[yori]] = \lambda x \lambda g \lambda y \lambda t \lambda w.max\{d'|g(d')(y)(t)(w)\} > max\{d''|g(d'')(x)(t)(w)\}$

Figure (29) shows the part of the logical structure of (21b):

(29)   The semantic derivation of *kaette* (the comparative sentence)

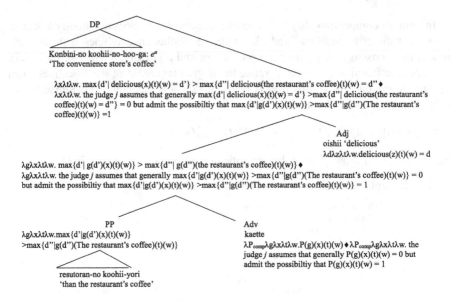

3.2   The Projective Property of *kaette*

What is interesting about the adverb *kaette* is that, unlike *yoppodo*, *kaette* does not have to co-occur with a modal, but in order to make its CI meaning project out of the complement of an attitude predicate, there must be a modal in the main clause. If there is no modal in the main clause, *kaette* can only be anchored to the subject, as in (30). However, if there is a modal in the main clause, *kaette* can be anchored either to the subject or to the speaker, as in (31):

(30)   Taro-wa   [konbini-no   koohii-no-hoo-ga
       Taro-TOP convenience store-GEN coffee-GEN-direction-NOM
       kissaten-no   koohii-yori-mo kaette   oishii]-to omo-tteiru.
       coffee shop-GEN coffee-than-MO KAETTE tasty-that think-TEIRU

       'Taro thinks that a convenience store's coffee is tastier than a coffee shop's coffee.'

       (**Non-at-issue, subject-oriented reading**: A coffee shop's coffee is (generally) tastier than a convenience store's coffee for Taro.)

(31)   Taro-wa   [konbini-no   koohii-no-hoo-ga
       Taro-TOP convenience store-GEN coffee-GEN-direction-NOM
       kissaten-no   koohii-yori-mo kaette   oishii]-to
       coffee shop-GEN coffee-than-MO KAETTE tasty-that
       omo-tteiru-kamoshirenai.
       think-TEIRU-may

       'Taro may think that a convenience store's coffee is tastier than a coffee shop's coffee.'

(a. **Non-at-issue, subject-oriented reading**: A coffee shop's coffee is (generally) tastier than a convenience store's coffee for Taro.)

(b. **Non-at-issue, speaker-oriented reading**: A coffee shop's coffee is (generally) tastier than a convenience store's coffee for me (i.e. the speaker).)

The natural context of the speaker-oriented reading in (31) is that the speaker observes that Taro buys coffee at the convenience store every day and he/she begins to think that the opposite of his/her previous assumption could be true.

One might wonder whether it is possible to get a speaker-oriented reading without a speaker-oriented modality. However, the following example (32) supports the idea that it is difficult to get a speaker-oriented reading if there is no modal in the main clause. (Note that in the following sentence there cannot be a subject-oriented reading because the subject, Taro, is 2 years old and does not have common sense/general assumption):

(32)   (Context: Taro is 2 years old and he does not know that usually, a restaurant's curry is more delicious than a retort-pouch curry.)

?? Taro-wa   retoruto-no   karee-no   hoo-ga   (resutoran-no
   Taro-TOP retort-pouch-GEN curry-GEN direction-NOM restaurant-GEN
karee-yori-mo) kaette   oishii-to   omo-tteiru.
curry-than-MO KAETTE delicious-that think-STATE

At-issue: Taro thinks that a retort-pouch curry is more delicious than a restaurant's curry.
Not-at-issue, speaker-oriented reading (intended): A restaurant's curry is (generally) more delicious than a retort-pouch curry for me.

The above sentence sounds odd as a speaker-oriented reading. This is because the speaker is not actually reversing his/her own assumption. In the at-issue component the sentence merely reports Taro's belief. However, if there is a speaker-oriented modal in the main clause, the speaker can reversing his own established assumption and admits the possibility that the opposite can be true.

One might also question whether there is really a "purely" speaker-oriented reading in an embedded example like (31). Theoretically, there is also a possibility of a "mixed-perspective reading" where a speaker takes a subject's perspective (Yusuke Kubota, personal communication). The following example supports the idea that there is a purely speaker-oriented reading in the environment like (31):

(33)   (Context: A speaker likes Japanese food very much, but Bill does not like Japanese food at all.)

Bill-wa   pasta-no   hoo-ga   udon-yori-mo kaette   oishii-to
Bill-TOP pasta-GEN direction-NOM udon-than-mo KAETTE delicious-than
omo-tteiru-kamosirenai.
think-TEIRU-may

'Bill may think that pasta is kaette more delicious than udon.'
(**Not-at-issue**: Udon is normally tastier than pasta for me.) (= speaker-oriented)

In this sentence it is assumed that the subject Bill does not like Japanese food. Since Bill hates Japanese food, the speaker is the only judge who can assume that Udon is normally tastier than pasta.

Let us now consider the question as to why *kaette* can be speaker-oriented if there is a modal (or speaker-oriented modal like expression) in the main clause. I claim that we can answer this question naturally by considering the lexical meaning of *kaette*. I defined the meaning of *kaette* above as in (34):

(34) $[[kaette_{comp}]]$ $=$ $\lambda P_{comp}\lambda g\lambda x\lambda t\lambda w.P(g)(x)(t)(w)\blacklozenge\lambda P_{comp}\lambda g\lambda x.$ the judge j assumes that generally $P(g)(x)(t)(w) = 0$ but admits the possibility that $P(g)(x)(t)(w) = 1$.

If a modal is added to the main clause, *kaette* can be speaker-oriented because the whole proposition is evaluated by the speaker, enabling him/her to draw a contrast between his/her assumption and the at-issue situation. If there is no modal in the main clause, *kaette* is not speaker-oriented, because there is a strong feeling that there is only one judge, namely a subject. As a speaker's perspective is not salient (or not activated) in the discourse, he/she cannot reverse his/her assumption and admit that the opposite might be true.

4   Theoretical Implications and Conclusion: A New Class of Projective Content

The phenomenon of the counter-expectational (explanational) use of the Japanese intensifiers *yoppodo* and *kaette* provides an important insight for current theory regarding the taxonomy of projective content. This is especially so for the parametric classification based on "obligatory local effect" (Tonhauser et al. [27]), given in (35).

(35)   **OBLIGATORY LOCAL EFFECT:** A projective content m with trigger t has obligatory local effect if and only if, when t is syntactically embedded in the complement of a belief-predicate B, m necessarily is part of the content that is targeted by, and within the scope of, B. (Tonhauser et al. [29], pp.93)

According to this parameter, typical presupposition triggers such as *stop* will be classified as having an obligatory local effect because their presuppositional implications do not project beyond the belief predicate. For example, the possessive expression in (36a) creates the presupposition that "Sam has a kangaroo," but if (36a) is embedded under the attitude predicate believe, the flow of presupposition is blocked, as shown in (36b):

(36)   a.   Sam's kangaroo is sick. (presupposition: Sam has a kangaroo.)

b.   John believes that Sam's kangaroo is sick.

This fact is corroborated by the example in (37):

(37)   Sue believes that Sam's kangaroo is sick, but that's ridiculous—Sam doesn't own a kangaroo. (Potts [20])

However, typical CIs can be classified as not having an obligatory local effect. For example, appositives and expressives, which are often assumed to trigger CIs, do not have an obligatory local effect because they are typically anchored to a speaker (e.g., Potts [17–19]; Amaral et al. [1]; Harris and Potts [7]; Tonhauser et al. [27]).[7]

(38) a. Sue believes that bastard Kresge should be fired. (♯I think he's a good guy.)
          (Potts [19], pp.170)

      b. Sheila believes that the agency interviewed Chuck, a confirmed psychopath,
          just after his release from prison.
          (Potts [21], pp.115)

Although the parameter of obligatory local effect may be useful for distinguishing a typical presupposition trigger (such as *stop*) from a typical CI expression (such as an appositive or expressive) in terms of projection, it does not seem to capture the difference between typical CIs/presuppositions and *yoppodo/kaette*. The parameter would predict that *yoppodo* and *kaette* have a property of non-local effect, but it is difficult to capture the fact that they are "basically" local (i.e., they are opposite to typical CIs in this respect). In the case of *yoppodo* and *kaette*, the not-at-issue meaning can project out of the complement of an attitude predicate only when there is a modal in the main clause. If there is a modal in the main clause, *yoppodo* and *kaette* have a non-local effect (similar to typical CIs; in the case of *yoppodo*, it is obligatorily non-local), but if there is no modal in the main clause, they have an obligatory local effect (similar to typical presuppositions).

In this paper, I have attempted to explain the behavior of *yoppodo* and *kaette* by assuming that they require consistency between an at-issue meaning and a CI meaning in terms of a judge. I also explained how this requirement arises from the lexical meanings of *yoppodo* and *kaette*. I hope to have shown the possibility of a new class of projective content that is neither typical of presuppositions nor typical of CIs.

A number of issues remain for future research. First, it is still theoretically unclear why *yoppodo* and *kaette* lexically require matching between an at-issue dimension and a CI dimension in terms of a judge. Although this is a tentative idea, it may be that the requirement has to do with the compositionality of *yoppodo* and *kaette*. Both *yoppodo* and *kaette* "recycle" the at-issue gradable meaning to convey a CI meaning; the at-issue and the CI meanings of a given sentence are both scalar and relevant (see also Sawada [23]).

Second, I would like to consider the relation between *yoppodo* and evidential modality in more detail and compare it with other related phenomenon (e.g. German discourse particle *ruhig* (Grosz [5]; Kaufmann [10])).

Third, more rigorous empirical investigation is required regarding the interpretation of embedded *kaette*. Since *kaette* does not require the co-occurrence with modal, it may be that speaker-oriented expressions other than a modal can support the projection of *kaette*. Because of space issues, I could not discuss this point in detail, but it seems that discourse particles, belief predicates, or confirmation-seeking questions can also affect the projection of *kaette*.

---

[7] Recall that appositives and expressives can have a non-speaker orientation as well.

Finally, the extent to which the sensitivity of *kaette* and *yoppodo* to modality is connected to the interaction between projective content and external factors in general, requires investigation. McCready and Sudo [15] discuss the meaning of the adverb *sekkaku*, claiming that content external to the local context of the adverbial is relevant in determining felicity/grammaticality. Although the case of *sekkaku* is related to felicity/grammaticality rather than to projection, the proposal also suggests that projective content may be sensitive to external factors.

References

1. Amaral, P., Roberts, C., Smith, A.: Review of the logic of conventional implicatures by chris potts. Linguist. Philos. **30**, 707–749 (2007)
2. Cresswell, M.J.: The semantics of degree. In: Partee, B. (ed.) Montague Grammar, pp. 261–292. Academic Press, New York (1976)
3. von Fintel, K., Gillies, A.: Must … Stay … Strong!. Nat. Lang. Seman. **18**, 351–383 (2010)
4. Grice, P.H.: Logic and conversation. In: Cole, P., Morgan, J. (eds.) Syntax and Semantics, iii: Speech Acts, pp. 43–58. Academic Press, New York (1975)
5. Grosz, P.: Grading modality: a new approach to modal concord and its relatives. In: Proceedings of Sinn und Bedeutung 14, pp. 185–201 (2010)
6. Gutzmann, D.: Use-conditional Meaning: studies in multidimensional semantics. Doctoral Dissertation, University of Frankfurt (2012)
7. Harris, J.A., Potts, C.: Perspective-shifting with appositives and expressive. Linguist. Philos. **32**(6), 523–552 (2009)
8. Karttunen, L.: Presuppositions of compound sentences. Linguist. Inq. **4**(2), 169–193 (1973)
9. Karttunen, L., Zaenen, A.: Veridicity. In: Katz, G., Pustejovsky, J., Schilder, F. (eds.) Annotating, Extracting and Reasoning about Time and Events. Dagstuhl Seminar Proceedings 05151. Springer, Heidelberg (2005)
10. Kaufmann, M.: Discourse particle 'Ruhig': discourse effects, desires, and modality. In: 3rd Cornell Workshop in Linguistics and Philosophy: Modal Talk and Reasoning (2013)
11. Kennedy, C., McNally, L.: Scale structure, degree modification, and the semantics of gradable predicates. Language **81**(2), 345–381 (2005)
12. Kennedy, C.: Vagueness and grammar: the semantics of relative and absolute gradable adjectives. Linguist. Philos. **30**(1), 1–45 (2007)
13. Klein, E.: Comparatives. In: von Stechow, A., Wunderlich, D. (eds.) Semantik: Ein Internationales Handbuch der Zeitgenossischen Forschung, pp. 673–691. Walter de Gruyter, Berlin (1991)
14. McCready, E.: Varieties of conventional implicature. Semant. Pragmatics **3**, 1–57 (2010)
15. McCready, E., Sudo, Y.: Operating on presuppositions: 'Sekkaku' revisited. In: Proceedings of Formal Approaches to Japanese Linguistics 5, pp. 155–166. MA: MITWPL, Cambridge (2012)
16. Portner, P.: Modality. Oxford University Press, Oxford (2009)
17. Potts, C.: Presupposition and implicature. In: Lappin, S., Fox, C. (eds.) The Handbook of Contemporary Semantic Theory, 2nd edn, pp. 168–202. Wiley-Blackwell, Oxford (2015)
18. Potts, C.: The expressive dimension. Theor. Linguist. **33**(2), 165–197 (2007)
19. Potts, C.: The Logic of Conventional Implicatures. Oxford University Press, Oxford (2005)
20. Sauerland, U.: Beyond unpluggability. Theor. Linguist. **33**(2), 231–236 (2007)
21. Sawada, H.: Modaritii (Modality). Kaitakusya, Tokyo (2006)

22. Sawada, O.: Polarity sensitivity and update refusal: the case of the Japanese negative totemo 'very'. In: Proceedings of the 11th International Workshop on Logic and Engineering of Natural Language Semantics (LENLS 11), pp. 313–326 (2014)
23. Sawada, O.: Pragmatic aspects of scalar modifiers. Doctoral Dissertation, University of Chicago (2010)
24. Seuren, P.A.M.: The comparative. In: Kiefer, F., Ruwet, N. (eds.) Generative Grammar in Europe, pp. 528–564. Reidel, Dordrecht (1973)
25. Stalnaker, R.C.: Pragmatic presupposition. In: Munitz, M.K., Unger, P.K. (eds.) Semantics and Philosophy, pp. 197–214. New York University Press, New York (1974)
26. Sudo, Y.: On the semantics of phi features on pronouns. Doctoral Dissertation, MIT (2012)
27. Tonhauser, J., Beaver, D., Roberts, C., Simons, M.: Toward a taxonomy of projective content. Language **89**, 66–109 (2013)
28. von Stechow, A.: Comparing semantic theories of comparison. J. Seman. **3**, 1–77 (1984)
29. Wang, L., Brian, R., McCready, E.: The projection problem of nominal appositives. Snippets **10**, 13–14 (2005)
30. Watanabe, M.: Hikakufukushi yohodo ni tuite (On the Comparative Adverb Yohodo.) Sophia University, Kokubungakka kiyoo 4, 39–52 (1987)

Evaluative Predicates and Evaluative Uses of Ordinary Predicates

Isidora Stojanovic[✉]

Institut Jean-Nicod, CNRS, ENS, PSL Research University, Paris, France
isidora.stojanovic@ens.fr

Abstract. This paper has two aims. The first is to provide a characterization of evaluative predicates ('good', 'horrible', 'beautiful'). The second is to explain how ordinary predicates, such as 'intense' or 'insane', may be used evaluatively, and how they convey sometimes a positive and sometimes a negative evaluation, depending on the context. I propose a semantic account, which, in a nutshell, relies on the fact that evaluative predicates are typically multidimensional adjectives, and that the choice, as well as the respective weights of the relevant dimensions, may vary with the context. Thus in a context in which a negative dimension has been brought to salience, the overall evaluation carried by the use of the predicate will likely be negative; *mutatis mutandis* for the positive case. The paper ends with a comparison between this approach and the pragmatic approach, and suggests that rather than compete, the two complement each other.

Keywords: Evaluative predicates · Predicates of personal taste · Multidimensional adjectives · Value-judgments · Expressive content · Semantics-pragmatics interface

1 Demarcating Evaluative Adjectives from Predicates of Taste

Among semanticists and philosophers of language, there has been a recent outburst of interest in predicates such as 'tasty' and 'fun', called *predicates of personal taste* (PPT for short; see [6, 9, 14, 15]). Somewhat surprisingly, the question of whether PPTs belong in the same class as evaluative predicates such as 'good' and 'bad' and aesthetic predicates such as 'beautiful' has been largely underestimated. In this section, which relies heavily on the discussion and results from [8], I survey some general criteria that linguists have proposed for semantically classifying all kinds of adjectives, and I highlight the criteria that are relevant for distinguishing the class of evaluative predicates. The upshot of this section is to show that the existing accounts of PPTs are not directly applicable to evaluative predicates, as well as to provide some background for the discussion in the remainder of the paper.

1.1 Gradability

One of the most basic characteristics to classify adjectives in semantics is whether they are gradable or not. I start with gradability not only because most evaluative adjectives

© Springer International Publishing AG 2017
M. Otake et al. (Eds.): JSAI-isAI 2015 Workshops, LNAI 10091, pp. 138–150, 2017.
DOI: 10.1007/978-3-319-50953-2_11

are gradable, but also to forestall a possible confusion. In order for a gradable adjective to truthfully apply to some individual, it is typically not enough that the property in question be held to just any degree; rather, it must be held to a degree that passes a certain *threshold*. The choice of threshold typically depends on a contextually determined comparison class ([3–5]). The fact that different speakers may appeal to different comparison classes can lead to disagreement about whether an adjective applies in a given case. Some researchers (e.g. [10]) consider all gradable adjectives whose threshold depends on a context to be *evaluative*. However, in line with a more philosophical tradition, I reserve the term "evaluative" for those cases in which a certain *value-judgment* is being expressed or conveyed.

1.2 Dimensionality

Another characteristic to classify adjectives is whether they are *unidimensional* or *multidimensional* (see [1, 12] for recent discussion). The main criterion to test for multidimensionality is whether the adjective may be felicitously used with constructions such as *in every/some/most respect(s)* or *except for* (see [12]: 336).[1] As shown below, all-purpose evaluative adjectives such as *good* and aesthetic adjectives such as *beautiful* pattern with multidimensional adjectives such as *similar*, while certain predicates of taste such as *salty* pattern with unidimensional adjectives such as *tall*.

These cars are similar in every respect.   (1)

This car is good in every respect.   (2)

Paris is beautiful in every respect.   (3)

?This soup is salty in every respect.   (4)

#She is tall in every respect.   (5)

These cars are similar, except for their speed capacity.   (6)

This car is good, except for its speed capacity.   (7)

Paris is beautiful, except for the filthiness of its streets.   (8)

?This soup is salty, except for the noodles.   (9)

---

[1] While all multidimensional adjectives are felicitous with the construction "in every/most/some respect", only a subclass of multidimensional adjectives – namely, the so-called conjunctive ones – are required to be felicitous with "except for". In this sense, the fact that an adjective cannot be felicitously used with "except for" is not yet evidence that it is not multidimensional (see [12] for discussion). It should also be added that these felicity criteria are indicative rather than conclusive. In particular, "except for" can be understood as referring to a part of the object that fails to instantiate the property, enhancing the felicity of sentences such as sentence (9) below (note though that the part-exception reading fails for (10), because 'tall' does not apply to parts of a body). Also, "in every respect" may be coerced into a metalinguistic reading, giving rise to puns such as "The titles of this newspaper are bold in every respect".

#She is tall, except for the upper part of her body. (10)

One crucial point is that in deciding whether an adjective that denotes a multidimensional property holds of some individual involves not only determining a threshold of applicability, but also determining which dimensions contribute to the property in question, as well as the relative weight of each of these dimensions. This is an additional source of context-dependence; it is one, moreover, that makes room for disagreement between speakers who may weigh the dimensions differently.

1.3   The Presence/Absence of an Experiencer

The third component to adjective meaning relevant for classifying adjectives involves the presence (or absence) of an experiencer, that is, a sentient individual who perceives the property in question. Some adjectives describe properties whose applicability may depend on the way in which they are experienced by some individual. Examples include adjectives such as *tasty*, *fun*, *salty*, or *loud*, as well as many adjectives derived from verbs denoting situations with experiencers, such as *shocking, disgusting, astonishing, enjoyable*, or *boring*. Various diagnostics have been proposed to distinguish adjectives that entail an experiencer from those that do not. When the adjective is deverbal, a clear indicator is the possibility of adding a *to* or *for* prepositional phrase that identifies the experiencer in the event described by the verb root, as in the following transitions:

The idea shocked/ disgusted/bored/astonished/ offended us. (11)

The idea was shocking/disgusting/boring/astonishing/offensive to us. (12)

However, this diagnostic is more difficult to apply with adjectives that are not derived from verbs. Another diagnostic that is used to identify adjectives with experiencers is whether they may be used felicitously with the *find* construction [2, 11, 17]. Such adjectives are routinely licensed in the comparative form in the complement of 'find':

We find this idea more shocking/disgusting/boring/offensive/astonishing than the previous one. (13)

She finds this pizza more delicious/saltier than that one. (14)

However, this diagnostic must be applied with care, as adjectives that do not come with an experiencer argument can appear in *find* constructions, as in "I find her tall." In line with the proposal defended in [8], I submit that for such uses to be felicitous, the attribution of tallness is made on the basis of the speaker's prior experience with different individuals' heights. Thus, notwithstanding appearance, the *find* construction introduces an experiential component, even if the latter need not be specifically tied to the property embedded under 'find'.

1.4 Evaluativity

Finally, a fourth criterion for classifying adjectives is whether they carry with their use an implication of a positive or negative attitude or evaluation on the part of the speaker. The paradigmatic examples of evaluative adjectives are *good* and *bad* (along with *mediocre, terrible, awesome*) and aesthetic adjectives: *beautiful, pretty, hideous, ugly*. One way to test for evaluativity is to look at conjunctions in which an evaluative adjective is used while the corresponding evaluative attitude is denied:

> ?This is a bad theory, but I don't value it negatively. (15)

> ?Paris is beautiful, but aesthetically, I don't value it positively. (16)

Conjunctions such as (15) and (16) sound bad, if not outright inconsistent. However, it remains an open issue whether the infelicity of such utterances stems from some semantic inconsistency, or is merely pragmatic.

Another way to test for evaluativity is to look once more at *find*-constructions – but, unlike what we saw in Sect. 1.3, it may be argued that genuinely evaluative adjectives are not entirely felicitous with 'find', and require using 'consider' instead. Compare:

> ?I find cheating on taxes to be morally wrong. (17)

> I consider cheating on taxes to be morally wrong. (18)

While (17) is not completely infelicitous, it differs from (18) in that it implicates that the value-judgement is based on a personal experience. In [8], we argued that the felicity in *find*-constructions may serve to distinguish genuinely evaluative adjectives from PPTs. An additional support to the argument comes from corpus data, which show that the adjectives that occur most frequently with *find* are: *difficult, hard/harder, easy/easier, useful, helpful, impossible, necessary, interesting, attractive*, and *strange*, none of which explicitly encodes a positive or negative attitude or evaluation. On the other hand, paradigmatic evaluative adjectives such as *good* seldom appear in *find*-constructions in corpora. The general failure to find evaluative adjectives with *find* strongly suggests that their evaluative component is not based directly on personal experience.

2 Expressing Value-Judgments in Context

2.1 Evaluative Predicates vs. Predicates used Evaluatively

While we customarily talk of evaluative predicates and evaluative adjectives, taking *good* and *bad* as their paradigms, and of aesthetic predicates/adjectives, taking *beautiful* and *ugly* as their paradigm, it remains an open question whether any set of linguistic criteria actually makes it possible to delineate the class of evaluative predicates (and then the subclass of aesthetic predicates). In philosophical literature in aesthetics, the following are considered as "aesthetic concepts": *unified, balanced, integrated, lifeless, serene, somber, dynamic, powerful, vivid, delicate, moving, trite, sentimental, tragic, graceful, delicate, dainty, handsome, comely, elegant, garish, dumpy*, and *beautiful*

([13]: 421). However, it takes little to see that many among these adjectives have primary meanings that are not at all aesthetic. For example, to describe a pastry mix as "unified", or the gender ratio at a conference program as "balanced", amounts to making purely descriptive claims. Similar observations may be made for *integrated, lifeless, dynamic, powerful*, and so on.

The crucial observation, then, is that many adjectives that are not by their very nature aesthetic may be *used* in order to make an aesthetic judgment. Similarly, many ordinary predicates may be used to express a value-judgment: thus describing e.g. a project as "ambitious" will, in a suitable context, express a positive stance towards the project; but in another context, it may express a negative stance. What distinguishes, then, genuine aesthetic adjectives (such as *beautiful* and *ugly*) from other adjectives such as *unified or lifeless*? My hypothesis is that the former, but not the latter, have it built into their lexical meaning that their role is to assign a certain aesthetic value to the object or individual to which they are attributed. What distinguishes genuine evaluative predicates (such as *good* and *bad*) from all sorts of predicates that can be used evaluatively? Again, I suggest that they have it built into their meaning that their primary role is to transmit a value-judgment.

2.2   The Context-Sensitivity of Valence

Many, if not most, value-judgments are expressed by means of vocabulary that is not primarily evaluative. In the moral domain, we morally value individuals and their actions by ascribing them properties that are not exclusively moral. In aesthetics, art critics hardly ever use adjectives like "beautiful" to express a positive evaluation of a work of art. To substantiate the latter claim, here is a quasi-random selection of excerpts from reviews of the movie *Mad Max: Fury Road*:

> Mad Max: Fury Road is totally insane … But in a good way.
> (source: Unhinged Reviews, 18/5/2015)   (19)

> It's an all − out rush of visceral, kinetic power and—yes—fury that mixes the old and the new in ways that are both dazzling and exhausting.   (20)
> (source: James Kendrick, Q Network Film Desk Reviews, 26/5/2015)

> [It is] one of the most harrowing, intense, thrilling action movies of all time. It is absolutely epic. (source: Anders Wright, the San Diego Union Tribune, 14/5/2015)   (21)

Natural though they are, these examples are also puzzling. As the reader will have figured out, all three reviews are *positive* reviews; and yet, many among the attributes ascribed to the movie are *normally* negative. In general, it is a bad thing to be insane, exhausting, or harrowing. Still, in this context, these are used in order to convey a positive evaluation of the movie.[2] Let us call this puzzle **Valence-switching**.

The second issue that the above examples raise is that adjectives which do not systematically carry either a positive or a negative valence, such as *intense*, and which I will call *evaluatively neutral*, may acquire a valence in the context. The question is how they acquire their valence? Let us call this puzzle **Valence-underspecification**.

Before we address the two puzzles in greater detail, let me stress that there are many situations in which the valence of the attitude expressed or conveyed crucially depends on the context. Though this may be said to hold more generally, here are examples in which the evaluative aspects may be traced to the use of an evaluatively neutral adjective:

What she did was audacious.   (22)

Their project is ambitious.   (23)

The plot of the movie is simple.   (24)

Proust's sentences are meticulous.   (25)

It is an easy exercise to imagine pairs of contexts such that each of the above conveys a positive vs. a negative value-judgement. Note that adjectives such as *audacious* are related to *thick concepts,* as discussed in moral philosophy, except that in discussions of thick concepts, the focus is on adjectives such as *courageous* and *cruel*, whose meanings are believed to encode respectively a positive and a negative valence. I shall return to the connection shortly.

2.3 Values in Context: Two Pragmatic Accounts

The question of how evaluatively neutral predicates, such as *intense* or *simple*, may acquire an evaluative use, and one whose valence depends on the context, has, to my best knowledge, been largely neglected both in the philosophical and in the linguistic literature. Two exceptions are Pekka Väyrynen's work on thick concepts ([18, 19]), and Eric McCready's work on expressive content ([7]). While coming from different traditions, both authors put forward accounts that are essentially *pragmatic*, in that the evaluation carried by a statement results from pragmatic mechanisms that appeal to the speaker's and hearer's beliefs, shared knowledge, intentions, expectations, and the like. Here, I will offer only a rough outline of their proposals, and will return briefly to them at the end of Sect. 3, after having outlined my own proposal.

Väyrynen's Proposal. Even though it originates in metaethics, Väyrynen's work constitutes an important step towards understanding the semantics and pragmatics of certain moral adjectives, called "thick terms", as he carefully takes into account their

---

[2] It is worth noting that it is not only *action* movies for which it may be a good thing that they should be harrowing. Interestingly, Haneke's drama *Amour* is also described as "harrowing, intense and thrilling" in a review by Blake Howard: http://www.graffitiwithpunctuation.net/2012/06/11/amour/.

linguistic behavior. Väyrynen's positive proposal (albeit simplifying somewhat) locates the evaluative import of thick moral terms entirely in pragmatics. Here is a quote that aptly summerizes his proposal: "The evaluations that thick terms and concepts may be used to convey are generalized but defeasible *conversational* implications of utterances involving such terms and concepts" ([18]: 267, my italics).

One important motivation for Väyrynen comes from the discussion of so called *objectionable* thick terms, such as *chaste* or *lewd*. Thus the use of *chaste* seems to carry a positive evaluation. However, the evaluation is based on the assumption that abstaining from sexual activity is good, and since this need not be the case, it seems possible to object to the use of the term without denying that the property effectively denoted by *chaste* (i.e. abstaining from sex) holds. By removing the evaluational aspects from the property actually denoted by the term, and placing them at the level of generalized conversational implicatures, Väyrynen aims to explain how in using such thick terms, evaluations are systematically triggered without being semantically entailed. Väyrynen generalizes his proposal to all thick terms, on the basis of the idea that in principle, *any* thick term may turn out to be objectionable.

As mentioned earlier, the adjectives that we are concerned with here, such as *intense*, are normally not included among thick terms, because there is no specific evaluation that is systematically associated with them. Nevertheless, Väyrynen's proposal can easily extend to them. What it would say is, simply, that the evaluation gets conveyed by the same mechanisms as any other conversational implicature. Towards the end of the paper, I will explain why I think that, in some cases at least, this is not a plausible analysis.

McCready's Proposal. McCready's interest, unlike Väyrynen's, is neither about thick terms nor adjectives more generally. Nevertheless, the puzzle that he discusses is about valence-underspecification. The expressions of interest to McCready are expressive terms such as *damn* or *fucking*. Here is a pair of examples from [7]:

Fucking Mike Thyson won another fight.   (26)

Fucking Mike Thyson got arrested again for domestic violence.   (27)

As McCready notes, while you can felicitously continue (26) by saying "He is great", you normally cannot do so in the case of (27). In other words, 'fucking' in (26) is used to express a positive evaluation, and in (27), a negative one.

Simplifying somewhat, McCready's proposal to account for this variability in valence goes as follows. An expected interpretation for an emotive expression (in a context) is computed on the basis of shared knowledge. For example, the conversational participants in (27) believe, and take each other to believe, that if someone is arrested for domestic violence, it must be because this person is violent, which is bad. This "normal" interpretation influences what the hearer expects to be a probable interpretation, and the speaker, being aware of this, decides whether to use an underspecified emotive term, anticipating the interpretation at which the hearer is likely to arrive.

There is an important difference between the sort of cases that McCready considers and the ones that interest us here. McCready's cases involve expressives, whose very function is to carry evaluative or, if you prefer, emotive content. In the cases such as (19)–(21) and (22)–(25), on the other hand, the expressions at stake ('insane', 'intense', 'ambitious' etc.) have a descriptive content of a perfectly familiar kind, and are not in need of being assigned any additional evaluative or emotive content. But setting this difference aside, McCready's proposal can be easily extended to the present cases. Whether it *should* be so extended is another question, to which I return in Sect. 3.3.

3   A Semantic Account

In this last section, I would like to outline a novel proposal that explains how certain evaluatively neutral predicates, such as *intense* or *audacious*, may acquire different valence in different contexts. In a nutshell, the proposal appeals to the fact that such predicates are multidimensional adjectives, and that the choice of the relevant dimensions, as well as their respective weights, may vary from one context to another. It further relies on the idea that some dimensions may be positive, or "good", and others negative, or "bad". If a positive dimension is salient, and if an object scores high on this dimension, then the overall evaluation conveyed about the object will be positive; and *mutatis mutandis* for the negative case. We will see, however, that there needs to be a further twist to this explanation, because whether a given dimension is positive or negative is, in turn, also a context-sensitive matter; or so I suggest. Nevertheless, this second sort of context-dependence is different, and in a way more basic, than that required for selecting and weighing the dimensions of a multidimensional adjective.

This section is structured as follows. In Sect. 3.1. I present the proposal regarding the way in which adjectives like *intense* acquire their valence from the context. In Sect. 3.2. I discuss the idea that whether a dimension is positive or negative is also context-dependent, but in a different way. I also sketch an explanation of how adjectives such as *harrowing*, while normally negative, sometimes get a positive valence. Finally, in Sect. 3.3, I will say something on why my proposal may be characterized as "semantic" and will briefly compare it to the two "pragmatic" proposals discussed in Sect. 2.3.

3.1   How Evaluatively Neutral Adjectives Inherit Their Valence from the Underlying Dimensions

Let me explain the proposal with the help of two examples, one from the aesthetic domain and another from the moral domain. Consider the adjective 'intense' in a statement like "Hardy's acting is intense". This sentence may be used to convey a positive value-judgment, but also a negative one, depending on the context. For example, in the context of a review of Mad Max: Fury Road as in example (21), it will be positive. But suppose that the context at stake is one in which we are talking about what was meant to be a light-hearted comedy but failed. Then it will likely be negative.[3] Now, 'intense' clearly passes the tests for multidimensionality (see Sect. 1.3):

$$\text{Hardy's acting is intense in every respect.} \qquad (28)$$

$$\text{Hardy's acting is intense, except for the way he speaks.} \qquad (29)$$

This being so, how is the truth value of the simple sentence "Hardy's acting is intense" to be determined? First, the context needs to determine the relevant dimensions. Second, it needs to determine the weight of each dimension, so that they may be combined into a single scale of intensity. Third, as with any gradable adjective (see Sect. 1.1), it needs to fix a threshold d on that single scale, and only to those entities that are above d will it be correct to apply the bare adjective 'intense'. With this fairly standard picture as our background, let us see where and how valence comes in. For simplicity, let us take it for granted that there are, in total, four dimensions relevant to assessing the application of 'intense': intensity in gesticulation (G), intensity in movement (M), intensity in speech production (SP), and intensity in display of emotion (DE). Let each of those correspond to a closed scale of 1 to 100, and let's assume that Hardy's acting, denoted by ha, figures on those scales objectively[4] as follows: $G(ha) = 83$, $M(ha) = 90$, $SP(ha) = 34$, $DE(ha) = 81$. Let us further assume that for an acting of the relevant sort, it is a *bad* thing that it be intense in movement, gesticulation, or speech production, and a *good* thing that it be intense in display of emotion. (As previously mentioned, we will come back to the question of what makes a given dimension a good one or a bad one.) Now consider two contexts, c_1 and c_2, such that in c_1, the weight is distributed equitably across the four dimensions, while in c_2, DE alone gets 60% of the weight and the other three, 13,33% each. Though the scales computed in the two contexts will be slightly different, the sentence "His acting is intense" will be true in both (assuming a reasonable threshold, say, so that the upper third of the scale licenses the application of 'intense'). However, in c_1 the sentence will likely convey a negative evaluation, because scoring high on two bad dimensions will trump scoring high on one good dimension, all of which have the same weight. On the other hand, in c_2, the dimension of intensity in display of emotion is so clearly dominant that it is its positive valence that the statement as a whole inherits and conveys.

Let me briefly give another example, and one in which (unlike above) there appears to be a more canonical way of determining which dimensions are good and which ones are bad. Consider the adjective 'audacious', which may convey something positive, but

---

[3] The negative use of 'intense' is illustrated by the following example, adapted from http:// www.culturedvultures.com/did-you-know-eric-stoltz-is-still-in-back-to-the-future/:   "Any self-proclaimed movie buff will be able to tell you of a time when Eric Stoltz was Marty McFly in *Back to the Future*. Coming across as intense and not really suited to the role, director Robert Zemeckis took the steps to replace him with Michael J. Fox." Thanks to Michael Murez for pointing this example out to me.

[4] Let me stress that what we are looking for are two contexts in which one and the same acting is correctly described as intense, yet in one the evaluation conveyed is positive and in the other it is negative. This is why the "objective" position of Hardy's acting on the different dimensions needs to be kept fixed.

also something negative.[5] Let's grant once more that there are several dimensions rele-vant to establishing the scale with respect to which the adjective is interpreted. Let's take one of those dimensions to be courage (hence positive), and another dimension to be recklessness or exposure to risk (hence negative). Then in a context in which the positive dimension is highly weighed, describing a person or an action as "audacious" will convey a positive evaluation; conversely, in a context in which a negative dimension is dominant, the overall evaluation will be negative.

3.2   What Is Good (or Bad) Is a Context-Sensitive Matter

I have outlined a solution to the puzzle of *Valence-underspecification*. However, the solution partly relies on the idea of "good" and "bad" dimensions, and one may legiti-mately ask: which dimensions are good, which ones are bad, and in general, how do we go about in deciding the question? Second, the solution only applies to adjectives that are both evaluatively neutral *and* multidimensional. But what about, on the one hand, adjectives that are not multidimensional, and on the other, adjectives that, at least *prima facie*, already come with a lexically marked valence, such as *harrowing*, yet whose valence may be overturned in an appropriate context (e.g. from negative to positive, in the context of a review such as (21))? In other words, how do we handle the puzzle of *Valence-switching*?

To fully answer the question of what makes certain dimensions good and others bad would be tantamount to addressing certain difficult questions from value-theory and metaethics. I have no such hopes here, but let me try to give a rough idea of the underlying picture. Whether something - say, a situation, or a course of events, or an action - is *good* is a question that only makes sense if it is asked in a specific context, with a specific background of considerations and often, while having some implicit beneficiary in mind. For example, is it a good thing that Osama Ben-Laden was killed? In answering yes, we typically mean that, given that he was the dangerous terrorist that he was, it was good - for the humanity - that he was killed. But this is of course compatible with the fact that for Osama himself, *qua* living organism, it was (probably) not a good thing to be killed. Here are a couple more examples that show that things are not good or bad *simpliciter*. Is drinking milk good? Well, milk has calcium, it is indispensible for babies' survival and growth, etc. But of course, if one is lactose-intolerant, then drinking milk is bad for such a person, rather than good. Or consider vomiting. If someone tells you that they have vomited all night, and you tell them "Bravo! I'm really happy for you!", that would normally be as an ironic reply, given that vomiting has lots of unpleasant consequences and is generally taken perceived as a bad thing. But if one has previously ingested poisonous food, then vomiting is one of the best things that might happen.

---

[5] Both uses are so systematic that many dictionaries posit two senses for 'audacious', one more positive and one more negative. E.g. The definition on Google: 1. showing a willingness to take surprisingly bold risks; 2. showing an impudent lack of respect. The definition in Webster: 1. having or exhibiting an unabashed or fearless spirit; 2. presumptuous; shameless, insolent. It is important to realize, however, that even if we restrict the interpretation to one sense only (say, the first), the valence may still vary with the context.

To bring the point home, there are usually many factors that need to be taken into account in order to decide whether something is, objectively speaking, good or bad. I suggest that the same goes for deciding whether scoring high on a given dimension is good or bad. To take up our example, being intense in gesticulation and movement may be a good thing for a protagonist of an action movie, but a bad thing for a protagonist of a light-hearted comedy. Similarly, being harrowing may be a bad thing for most events or situations, but a good thing, say, for a movie that a spectator precisely goes to watch with the expectation of that kind of experience. Just as it can be occasionally good to throw up, it can also be occasionally good to expose oneself to a harrowing experience.

The solution offered for *Valence-switching* is thus fairly simple. We think of harrowing as a negatively marked adjective because being harrowing is, in general, bad. However, there are exceptions to this generalization, and movies of a certain genre are precisely such. And because in such contexts being harrowing *is* a positive feature of a movie, to describe one as such conveys a positive evaluation.

The solution offered for *Valence-underspecification* will depend on the kind of adjective involved; there will be cases in which the same solution as for *Valence-switching* works out fine. However, the cases that I have focused on here call for a more elaborate solution. Consider again 'intense'. In determining which valence is conveyed, the context is required twice. First, it determines which dimensions are relevant to the application of the adjective, and how they combine into a single scale. Second, it determines which ones are positive and which ones are negative. These are two very different roles. The difference between the two roles is very similar to the way in which context is required to determine a truth value of a sentence. Consider "It is snowing". The context is needed to determine which place we are talking about. If the sentence is uttered in Tokyo, we understand that what it states is that it is snowing in Tokyo. Second, the context tells us which state of affairs we are in, i.e. what is the case and what isn't. If we are in a state of affairs in which it is snowing in Tokyo, the sentence is true; otherwise it is false.

3.3   Semantics or Pragmatics?

I have presented my account as "semantic." Yet, the account heavily relies on the notion of context and on the idea that the dimensions relevant to the application of a term such as 'intense' vary from one context to another and are salient to different degrees. In other words, my account crucially relies on features that are often considered to fall on the "pragmatic" rather than "semantic" side. So let me end the paper by clarifying certain aspects of the account and explaining how it differs from the two pragmatic accounts, Väyrynen's and McCready's, introduced in Sect. 2.3.

The sense in which the account is semantic is that the evaluation carried, in a given context, by a multidimensional adjective such as 'intense' derives from one or more dimensions that, in that context, combine into establishing a scale with respect to which it is to be determined whether the adjective applies or not. Even if the scale may vary from one context to another (because the dimensions may vary, and their respective weights may vary), this scale is what determines the application conditions of the adjective, and in this sense, it is an aspect of the semantic machinery.[6] There is a striking contrast with the two proposals discussed in Sect. 2.3, both of which appeal to a heavy pragmatic machinery. Thus Väyrynen's appeal to generalized conversational implicatures presupposes rationality principles and various reasoning mecanisms that are characteristic of Gricean pragmatics. McCready's appeal to the hearer's and speaker's mutual expectations and knowledge is similarly cast within a framework that requires reasoning about other's beliefs and expectations. My main worry with such pragmatic accounts is that they are led to posit costly pragmatic processes for a range of situations in which no pragmatic reasoning seems to be going on. For often (even if not always), a value-judgment can be conveyed spontaneously and directly, without calling for any pragmatic apparatus (Gricean or other). To be sure, in order to demonstrate this properly, what is needed is some evidence from psycholinguistics on how people process evaluative content. I must leave this task for future research.[7]

References

1. Bierwisch, M.: The semantics of gradation. In: Bierwisch, M., Lang, E. (eds.) Dimensional Adjectives, pp. 71–261. Springer, Berlin (1989)
2. Bylinina, L.: The grammar of standards: judge-dependence, purpose-relativity, and comparison classes in degree constructions. LOT Dissertation Series 347. LOT, Utrecht (2014)
3. Kennedy, C.: Vagueness and grammar: the semantics of relative and absolute gradable adjectives. Linguist. Philos. **30**, 1–45 (2007)
4. Kennedy, C., McNally, L.: Scale structure, degree modification and the semantics of gradable predicates. Language **81**, 345–381 (2005)
5. Klein, E.: A semantics for positive and comparative adjectives. Linguist. Philos. **4**, 1–45 (1980)
6. Lasersohn, P.: Context dependence, disagreement, and predicates of personal taste. Linguist. Philos. **28**, 643–686 (2005)
7. McCready, E.: Emotive equilibria. Linguist. Philos. **35**, 243–283 (2012)

---

[6] This may be disputed. More generally, one's stance regarding this issue will depend on how one thinks about the semantics-pragmatics interface. In [16], I urge to make room for the idea that there are aspects of meaning which are not semantic, but are not full-fledgedly pragmatic either. This could be a case at point.

[7] Earlier versions of this material have been presented at the 8th Latin Meeting in Milan, Italy (4 June 2015), at the 3rd Workshop on Semantic Content and Conversational Dynamics in Barcelona, Spain (30 June 2015), at the conference Context-Relativity in Semantics in Salzburg, Austria (3 July 2015), at Philosophy of Language and Linguistics in Dubrovnik, Croatia (9 September 2015), at the Mind and Language Seminar in Paris, France (23 September 2015), and especially, at the conference LENLS-12 in Tokyo, Japan (17 November 2015). I am grateful to the audiences at these events for helpful comments and feedback, and more particularly to Alain Pé-Curto, who commented on my paper at the Milan meeting. I would also like to acknowledge support from the following projects: MINECO FFI2016-80636-P, ANR-10-LABX-0087 IEC, ANR-10-IDEX-0001-02 PSL and Marie Sklodowska-Curie grant nb. 675415.

8. McNally, L., Stojanovic, I.: Aesthetic adjectives. In: Young, J. (ed.) The Semantics of Aesthetic Judgement. OUP, Oxford (forthcoming)
9. Pearson, H.: A judge-free semantics for predicates of personal taste. J. Semant. **30**, 103–154 (2013)
10. Rett, J.: Antonymity and evaluativity. In: Friedman, T., Gibson, M. (eds.) Proceedings of SALT XVII. Cornell University, Ithaca, NY, pp. 210–227 (2007)
11. Sæbø, K.J.: Judgment ascriptions. Linguist. Philos. **32**, 327–352 (2009)
12. Sassoon, G.W.: A typology of multidimensional adjectives. J. Semant. **30**, 335–380 (2013)
13. Sibley, F.: Aesthetic concepts. Philos. Rev. **68**, 421–450 (1959)
14. Stephenson, T.: Judge-dependence, epistemic modals, and predicates of personal taste. Linguist. Philos. **30**, 487–525 (2007)
15. Stojanovic, I.: Talking about taste: disagreement, implicit arguments, and relative truth. Linguist. Philos. **30**, 691–706 (2007)
16. Stojanovic, I.: Prepragmatics: widening the semantics-pragmatics boundary. In: Burgess, A., Sherman, B. (eds.) Metasemantics: New Essays on the Foundations of Meaning, pp. 311–326. OUP, Oxford (2014)
17. Umbach, C.: Evaluative propositions and subjective judgments. In: van Wijnbergen-Huitink, J., Meier, C. (eds.) Subjective Meaning. De Guyter, Berlin (2015)
18. Vayrynen, P.: Thick concepts: where's evaluation? In: Shafer-Landau, R. (ed.) Oxford Studies in Metaethics, vol. 7, pp. 235–270. Oxford University Press, Oxford (2012)
19. Vayrynen, P.: The Lewd, the Rude and the Nasty: A Study of Thick Concepts in Ethics. OUP, Oxford (2013)

Strong Permission in Prescriptive Causal Models

Linton Wang[✉]

Department of Philosophy, National Chung Cheng University, Chia-Yi, Taiwan
lintonwang@ccu.edu.tw

Abstract. This paper formulates strong permission in prescriptive causal models. The key features of this formulation are that (a) strong permission is encoded in causal models in a way suitable for interaction with functional equations, (b) the logic is simpler and more straightforward than other formulations of strong permission such as those utilizing defeasible reasoning or linear logic, (c) when it is applied to the free choice permission problem, it avoids paradox formation in a satisfactory manner, and (d) it also handles the embedding of strong permission, e.g. in conditionals, by exploiting interventionist counterfactuals in causal models.

Keywords: Strong permission · Free choice permission · Prescriptive causal model · Counterfactual · Intervention

1 Favoring Strong Permission

In a simplified version, the problem of free choice permission is a result of jointly holding the free choice inference (FCVE) and certain other intuitively attractive inferential principles such as (FCVI) and (FCCE); here, $May(\varphi)$ stands for that φ is permitted.

(FCVE)	(FCVI)	(FCCE)
$May(\varphi \vee \psi)$	$May(\varphi)$	$May(\varphi \wedge \psi)$
$\therefore May(\varphi)$	$\therefore May(\varphi \vee \psi)$	$\therefore May(\varphi)$

Though many find that the application of FCVE acceptable (e.g. 1a, b), allowing simultaneous application of both FCVE and FCVI leads to unacceptable results (e.g. 2a, b): φ is permitted only if any ψ is permitted.

(1) **FCVE**
   a. You may have tea or coffee.
   b. ∴You may have tea and you may have coffee.   ($\checkmark$)
(2) **FCVI+FCVE**
   a. You may do your homework.
   b. ∴ You may play video games.   ($\times$)

© Springer International Publishing AG 2017
M. Otake et al. (Eds.): JSAI-isAI 2015 Workshops, LNAI 10091, pp. 151–165, 2017.
DOI: 10.1007/978-3-319-50953-2_12

Among others, one of the main reasons to favor *strong* permission (rather than the *weak* permission defined as the dual of obligation) is that it aims to provide a logical framework which supports FCVE but rejects FCVI and FCCE, and so constitutes a semantic account for free choice permission without inducing paradox (cf. [2,3]).

In this paper, I will first briefly outline the general idea of strong permission, including how FCCE can itself also lead to undesirable results if strong permission is not properly designed (Sect. 2). Some further examples will be used to show that the frameworks of strong permission in [2,3], though they may properly handle the case of the free choice permission problem, are not yet equipped with the necessary tools to handle the practical aspect of strong permission (Sect. 3) and the embedding of strong permission in the context of conditionals (Sect. 4). It will also be indicated that, though counterfactual conditionals in causal modeling semantics (CMS) can be naturally extended to a framework for strong permission, the result does not actually do much better than previous approaches in handling the practical aspect and the embedding of strong permission (Sects. 5 and 6). Finally, a further extended causal modeling semantics, the *prescriptive* causal modeling semantics, will be introduced and shown to have various advantages over other approaches (Sect. 7).

2   General Constraints

The general idea to formalize strong permission is to make it one that differs from weak permission but also at the same time validates FCVE but rejects FCVI and FCCE. One strategy is to adopt some sort of *deontic reduction* strategy for deontic modalities (cf. [2,3]). For example, one may adopt a specific sentence letter S to represent something like "the sanction occurs" or "not immune to sanction", and then formulate standard obligation $O(\varphi)$ by $\neg A \to S$, weak permission $P_w(\varphi)$ by $\neg O(\neg\varphi)$, and strong permission $P(\varphi)$ by $\varphi \to OK$, in which $\to$ is some kind of conditional unspecified as of yet and OK is defined by $\neg S$ (cf. among others, [1–3]).

According to the above deontic reduction strategy, FCVE, FCVI, and FCCE have their correspondent formulations as follows, given the consequence relation $\models_?$ of a certain sort.

(FCVE*) $(\varphi \lor \psi) \to OK \models_? (\varphi \to OK) \land (\psi \to OK)$
(FCVI*) $\varphi \to OK \models_? (\varphi \lor \psi) \to OK$
(FCCE*) $(\varphi \land \psi) \to OK \models_? \varphi \to OK$

Nonetheless, as indicated in [2]: 308, one substantive challenge to a framework of strong permission is to validate FCVE by validating the simplification of disjunctive antecedents (SDA) but without at the same time validating the substitution of equivalent antecedents (SEA), for otherwise the undesirable antecedent strengthening (AS) follows.

(**SDA**) $(\varphi \lor \psi) \to \chi \models_? (\varphi \to \chi) \land (\psi \to \chi)$
(**SEA**) $\varphi \equiv \psi, \varphi \to \chi \models_? \psi \to \chi$
(**AS**) $\varphi \to \psi \models_? (\varphi \land \chi) \to \psi$

Proof.

1. $\varphi \,\square\!\!\rightarrow \psi$	Assum
2. $\varphi \equiv ((\varphi \wedge \chi) \vee (\varphi \wedge \neg\chi))$	Taut.
3. $(((\varphi \wedge \chi) \vee (\varphi \wedge \neg\chi)) \,\square\!\!\rightarrow \psi)$	2, SEA
4. $((\varphi \wedge \chi) \,\square\!\!\rightarrow \psi)$	3, SDA

With the seemingly acceptable FCCE, AS leads to the unacceptable consequence that once φ is permitted, any ψ is permitted. So the problem of free choice permission surfaces again.

(3) **AS+FCCE**
   a. You may invite John.
   b. $\therefore$ You may invite Bill.[1]   ($\times$)

It is then suggested that $\rightarrow$ should be weaker than conditionals such as the standard Stalnaker-Lewis counterfactual (cf. [2]: 308), since it allows (SEA).

In the literature, logical frameworks for strong permission do not simply adopt a deontic reduction strategy paired with a "weak" sense of conditional, but also adopts a "weak" sense of consequence relation corresponding to the conditional used in the deontic reduction strategy. For a conditional in a "weak" sense, in [2] Asher & Bonevac suggest to use a defeasible framework and in [3] Barker suggests to use a linear logic framework for the implementation of strong permission, so that "$\rightarrow$" is respectively understood as a defeasible conditional paired with a defeasible consequence relation, and the linear conditional paired with a corresponding linear consequence relation. These two frameworks are roughly presented as follows.

2.1 The Defeasible Framework

According to [2], Asher & Bonevac's deontic strategy implements strong permission $May(\varphi)$ by $\varphi > OK$, in which $>$ is a defeasible conditional, so that $\varphi > \psi$ roughly means *if φ then normally ψ*. To implement the logical feature of FCVE by what they call *C-disjunction*, a defeasible consequence $\vdash\!\!\sim$ is exploited.

   (C-Disjunction) $(\varphi \vee \psi) > \chi \vdash\!\!\sim (\varphi > \chi) \wedge (\psi > \chi)$

FCVE is implemented by $\vdash\!\!\sim$, for $\vdash\!\!\sim$ is defeasible, in that it is non-monotonic by nature because it allows for inferences to be "defeated" by "counterexamples", e.g. the fact that penguins do not fly defeats the conclusion that penguins fly follows from the defeasibility of the inference from the premises that birds fly and penguin are birds. [2] takes it that FCVE is only defeasible for the reason that the following is consistent.

(4) You may have tea or coffee, but you may not have tea or you may not have coffee.

The reason for them to find (4) consistent is that the mere fact that you may have tea or coffee is compatible with it being the case that you may not have both (cf. [2]: 311). Nonetheless, the "reason" they provide is more suitably represented by the consistency of the following example.

---

[1] By (AS), we infer $(p \wedge q) \rightarrow OK$ from $p \rightarrow OK$. By (FCCE), it then follows that $q \rightarrow OK$.

(5) You may have tea or coffee, but you may not have both tea and coffee.

Even if one finds (5) acceptable, it does not automatically follow that one will find (4) acceptable, unless one believes it to be the case that you may not have both tea and coffee implies (or is equivalent to) that you may not have tea or you may not have coffee.

(P-Dem) $\neg((\varphi \wedge \psi) > OK) \equiv \neg(\varphi > OK) \vee \neg(\psi > OK)$

It is unclear whether P-Dem is valid according to the semantics for $>$ given in [2]. However, if there is no independent reason to motivate something like P-Dem, then using defeasible inference to model inference in free choice permission becomes less attractive. Moreover, in Sect. 4, I will give an example to show that it is better for P-Dem not to hold in a framework for strong permission.

In [2], FCVI and FCCE do not hold even in the defeasible sense. This satisfies the requirement for their proposal to be a solution for the problem of free choice permission. Besides, it also has the advantage of making modus ponens for $>$ defeasible.

(Defeasible Modus Ponens)

$\varphi > \psi, \varphi \vdash \psi$, but

$\varphi > \psi, \varphi, \neg \psi \nvdash \psi$

If we only allow OK to defensibly follow from $\varphi > OK$ and φ, we can correctly avoid the inference from (6a) to (6b), which should not have followed:

(6) a.   The sanction occurs, and you pay compensation.

    b.   So, you may not pay compensation.   ([2]: 307)[2]

Any framework allow the inference from (6a) to (6b) is doomed to be inappropriate, for it can be inferred that once one violates some permission, he is no longer permitted to do anything else.

2.2   The Linear Logic Framework

In [3], Barker adopts a different sort of deontic reduction strategy. His strategy is differs in characters from [2] in two main aspects: strong permission is associated with linear-oriented operators and the linear consequence relation. Here, I will only illustrate the background idea in the linear logic metaphorically (and the metaphorical meaning can be read off from the sequent calculus for the linear logic) and refer readers to [3] for the formal details.

To begin with, consider the linear consequence $\varphi \vdash_L \psi$, which means that the resource provided by φ is enough to produce the result or product ψ. Notice that the resource in the linear consequence, once used, cannot be used again, meaning that a resource can be used only once, just as on a production line. The linear implication

---

[2] Assume that modus ponens for the deontic reduction based on conditional $\rightarrow$ is not defeasible, we show $\neg OK, p \nvDash_? \neg(p \rightarrow OK)$ to avoid inconsistency. Consider in some case that $\neg OK, p, p \rightarrow OK$ is consistent. If follows that $\neg OK \wedge OK$, which is inconsistent.

$\varphi \multimap \psi$ means that the resource φ can bring you the product ψ. When the linear implication is considered as a resource, meaning that it is located on the left hand side of the linear consequence, it means roughly that, continuing to use the production-line metaphor, the machine is adjusted in a manner so that $\varphi \multimap \psi$. On the other hand, when an linear implication is situated on the right hand side of the linear consequence, it means roughly that some resource makes it so that $\varphi \multimap \psi$.

According to [3], that it ought to be the case that φ is represented by $(\delta \multimap \varphi)$, where the sentence letter δ roughly means that *all required things are not violated*.[3] $\delta \multimap \varphi$ then means that if all required things are not violated, then φ is the case. Following this idea, a strong permission of φ is represented by $\varphi \multimap \delta$, meaning that if φ is executed, all required things are not violated.

Two other operators are relevant to the consideration of free choice permission in linear logic.

(Additive Conjunction) & is the additive conjunction.
(i) $\varphi \& \psi \vdash_L \chi$ means that at least one of φ and ψ is a sufficient resource for producing χ, and both of them are provided.
(ii) $\chi \vdash_L \varphi \& \psi$ means that χ is a sufficient resource for producing φ and χ is also a sufficient resource for producing ψ, but is not meant to produce both.

(Additive Disjunctive) $\oplus$ is the additive disjunction.
(i) $\varphi \oplus \psi \vdash_L \chi$ means that both φ and ψ are sufficient resources for producing χ, though only one of them is provided.
(ii) $\chi \vdash_L \varphi \oplus \psi$ means that the resource χ is sufficient for producing at least one of φ and ψ.

Given the above notions, strong permission is formulated as follows in linear logic.

(Free Choice Permission) $(\varphi \oplus \psi) \multimap \delta \vdash_L (\varphi \multimap \delta) \& (\psi \multimap \delta)$. This means that, if the machine or the situation is adjusted in the manner that both φ and ψ are sufficient resources for producing δ (given that only one of φ and ψ is provided), it follows that the machine or the situation is adjusted to one capable of making use of φ to produce δ and making use of ψ to produce δ, though both cases need not hold.

At this point, we obtain the inference pattern FCVE in linear logic. Other logical operators are needed in order to present the corresponding invalid inference patterns for FCVI and FCCE.[4]

---

[3] In [3]: 11, δ is understood as that *all things are as required*.
[4] [3] does not specifically address FCVI and FCCE. I suspect that, in linear logic, one might use the multiplicative operators (multiplicative conjunction $\otimes$ and multiplicative disjunction $\bindnasrepma$, which have meanings close to the classical operators) to represent them, i.e. $\varphi \multimap \delta \nvdash_L$ $(\varphi \bindnasrepma \psi) \multimap \delta$ for the invalidity of FCVI, and $(\varphi \otimes \psi) \multimap \delta \nvdash \varphi \multimap \delta$ for the invalidity of FCCE.

3 Permissions as Prescriptive Norms

Both of the two proposals for strong permission I have discussed rely on two things: (a) the implementation of strong permission by using a conditional in a certain weak sense, and (b) the orchestration of a non-classical consequence relation. These proposals may successfully capture the required logical-inferential features of strong permission. Nonetheless, more remains to be done.

No matter whether we are concerned with strong permission or obligation, we care about their practical aspects, in the sense that permissions and obligations are prescriptive norms suitable for practical guidance. In order to be suitable for practical guidance, it is required that by following the prescriptive norms, we are certain to avoid sanctions. For example, when we execute a strong permission, we want the execution to bring us to a situation in which we will not be in danger of stepping on somebody's toes, so that no sanction will result. However, the implementation of strong permission in both of the proposals does not capture this practical aspect appropriately. In [2], the defeasible conditional is excessively weak, so that the practice of strong permission does not guarantee that no sanction will occur (for it only promises that normally no sanction will occur). Similarly, since it is unclear how the linear implication can be understood in a way corresponding to a notion of prescriptive norm, it is also hard to see what is meant by a strong permission as formulated in Barker's style. For example, how is a strong permission represented by $(\varphi \multimap \delta)\&(\psi \multimap \delta)$ meant to guide us, in any practical sense? Are we actually allowed to do φ, and to do ψ?

Furthermore, the practical issue of strong permission in both proposals is even more vivid when we reconsider the orchestrated consequence relations designed for capturing the inference pattern FCVE. Since the consequence relations in both proposals are non-classical, it means that, given strong permission, what is non-classically implied is not guaranteed to be true, meaning that what is "implied" is not literally "true". If so, the agent making the inference faces the worry of being *in danger* of inferring the "wrong" strong permission and thus in turn faces the danger of stepping on somebody's toes (so that a sanction is activated). For example, when Asher & Bonevac claim in [2] that $(\varphi \vee \psi) > OK$ is compatible with $\neg(\varphi > OK) \vee \neg(\psi > OK)$, an agent cannot safely execute either φ or ψ without facing the possibility that his action will activate a sanction. Strong permission, as a prescriptive norm, should not allow this to happen.

There are other practical aspects of prescriptive norms that I intend to capture, but earlier proposals do not pay much attention to. One is *general* counterfactual applicability, to be further illustrated later in Sect. 4, that prescriptive norms can survive in a range (though not in all) of situations. A logical framework that can represent this feature should be desirable. What concerns me more is *specific* counterfactual applicability: one is permitted (or not permitted) to φ in a situation s just in case one is permitted (or not permitted) to φ in another situation which is exactly like s except that φ is executed in it. This practical aspect is important, for we need, as will further elaborated in Sect. 4, for example, when φ is permitted, φ to be kept permitted if one were to execute φ (so that one's φ-ing can be exempted from blaming); when φ is not permitted, φ should be kept not permitted even if one were to execute φ (so that one can still be blamed for his execution of φ).

Let us say that a strong permission is a prescriptive norm that promises us that when the permission is executed by an individual, the execution will bring the individual to a situation in which no sanction occurs or no obligations are violated. Let us also say that an inference to a strong permission reveals to us a different aspect of the permission encoded in the premises. The weak conditionals and non-classical consequence relations encoded in the previous two proposals, unfortunately, do not yield these aspects of permission, though the inference patterns they validate seem to capture the required inference patterns to solve the free choice permission problem.

4 Embedded Strong Permission

We say that a strong permission is one that when executed leads us to a situation in which no sanction occurs. We can go further. It is natural that some permission does not hold in the current situation, for the reason that even if the permission is executed, the execution may not bring us to a situation in which we are not susceptible to sanction. Nonetheless, it also happens that when the current situation has changed in a certain way, some previously non-applicable permission will be made applicable. I shall call this *strong permission embedding*, which is a case of general counterfactual applicability. Besides the suitable logical-inferential features to solve the free choice permission problem, I would also like to emphasize that a framework of strong permission should also properly capture the strong permission embedding. Let us look at some examples.

The sophisticated structure of prescriptive norms given in (Monty Hall) verifies various strong permission embeddings encoded by conditional permissions.

(Monty Hall) The guest is *permitted* to pick exactly one door from door 1, door 2, and door 3. Behind one door is a car and behind the other two are goats. After the guest chooses a door, the host Monty Hall is *permitted* to open a door which is not chosen by the guest and which has a goat behind it. After Monty Hall opens a door, the guest is *permitted* to change his choice.

There are some interesting features in this scenario of prescriptive norms. Consider some strong permissions to be derived from the norms in the scenario.

(7)   a. The guest may choose door 1, door 2, or door 3, but no more than one door.
       b. If the guest were to choose door 1 and the car were to be behind the door 1, then it would be the case that Monty Hall may open door 2 or door 3.
       c. If the guest were to choose door 1, the car were to be behind door 2, and Monty Hall were to open door 3, then it would be the case that the guest may change to door 2.

(7b, c) are examples of strong permission embeddings encoded under counterfactuals, which have the feature that the permissions in the consequents hold depending (or parasitic) on the actions and facts given in the antecedents. Not much about conditional permission is considered in the literature, since the conditional embedding gives rise to substantial complexity in a logic for strong permissions, concerning how strong permissions depend on other actions and facts.

A possible proposal is to embed defeasible permissions or linear strong permissions under Stalnaker-Lewis counterfactuals, e.g. $\psi \mathrel{\Box\!\!\rightarrow} (\varphi > OK)$. Nonetheless, from the perspective of specific counterfactual applicability, this move will not avoid the challenges to the deontic reduction strategy in the previous two proposals arising from the prescriptive norms in (Monty Hall). Assume that the guest actually chose to open the door 2 and Monty Hall also opened the door 2. Given this situation, it is still the case that the guest may open the door 2, so that the guest cannot be blamed, and Monty Hall may (and should) not open the door 2, so that sanction should apply to Monty Hall. However, this is neither a situation in which no sanction occurs (so that the guest may open the door 2) nor a situation in which no obligation is violated (so that the guest may open the door 2). This problem for both proposals, on the face of it, arises from the fact that they apply the deontic notions 'OK' and δ to the "whole" situation rather than specifically to the action to be executed.

Similar conditional permissions also arise from examples like the following.

(Pain Killers) There are two kinds of pain killers, A and B. They each are effective for relieving John's pain, so the doctor may use A and may use B. Though using both A and B is even more effective, some undesirable side effect will arise. A doctor's treatment should not lead to an undesirable side effect.

The above scenario *implicitly* verifies conditional permissions such as (8), even if this conditional permission is not written specifically in the hospital working manual.

(8) If the doctor were to use A, then it would be the case that he may not use B.

Furthermore, the following inference pattern seems to be reasonable.

(9)   a. If the doctor were to use both A and B, then the doctor's treatment would lead to an undesirable side effect.
    b. The doctor's treatment may not lead to an undesirable side effect.
    c. ∴ The doctor may not use both A and B.

It is unclear how a logic of strong permission validates the given inference from (9a, b) to (9c) based on the deontic reduction strategy, for the premise (9b) is formulated as a conditional rather than a plain non-conditional statement in the deontic reduction strategy. Moreover, the case and (9c) together constitute a straightforward case that P-Dem is unacceptable.

While previous attempts for a logic of strong permission focus on capturing a cluster of suitable inferential patterns, I would like to turn my attention to how we explicitly encode the structure of prescriptive norms in a semantics and make use of it for the truth conditions of strong permission. Given suitable logical-inferential features in a logical for strong permission, it does not follow directly that prescriptive norms are encoded properly in a given semantics. If we make a clear encoding of prescriptive norms and give the truth conditions of strong permission in a way based on that encoding, it will be more convincing that the given logic correctly captures the truth conditions of strong permission appropriately.

5   Causal Modeling Semantics

My plan is to use counterfactuals in the causal modeling semantics to formulate strong permission. To utilize causal modeling semantics (CMS), the formal framework is briefly introduced as follows (cf. [4,6]).

Definition 1 *(Causal Frame). A causal frame is a tuple $\mathbb{F} = <U, V, R, E>$, in which U is a set of exogenous variables, V is a set of endogenous variables, R is a function assigning a set of values to each variable in $U \cup V$, and E is a set of functional equations for every $V_i \in V$ in the form of $V_i = f_{V_i}(W_1, ..., W_n)$ such that $W_1, ..., W_n \in U \cup V$.*

A causal graph generated from a causal frame is a directed graph $G = <U \cup V, H>$ in which $H = \{<X, Y> \mid Y = f_Y (..., X, ...)\}$. I shall restrict the attention to the *recursive* frames which lack loops in their corresponding graphs (a loop in a graph $G = <U \cup V, H>$ is a sequence $X_1, ..., X_n$ such that $X_1 = X_n$ and $<X_i, X_{i+1}> \in H$). Formulas corresponding to causal frames are (a) atomic formulas in the form of $X_i = \kappa_i$ in which $X_i \in U \cup V$ and $\kappa_i \in R(X_i)$, (b) boolean combinations of formulas, and (c) counterfactual formulas in the form of $\varphi \,\square\!\!\rightarrow\, \psi$ such that there is no occurrence of '$\square\!\!\rightarrow$' in φ and ψ.

Definition 2 *(Causal Model). A causal model $\mathbb{M} = <\mathbb{F}, f>$, where $\mathbb{F}$ is a causal frame and f is a solution to the equations E in $\mathbb{F}$.*

If a frame $\mathbb{F}$ in a model is recursive, it follows that every different solution to $\mathbb{F}$ has at least some different value on their exogenous variables in the solutions (cf. [6]). To model counterfactuals in causal models, we need an extra piece of machinery called *interventions*.

Definition 3 *(Intervention and Submodel). An intervention Γ is a set of atomic formulas, and no atomic formulas $X_i = \kappa_i$ and $X_j = \kappa_j$ in Γ such that $X_i = X_j$ but $x_i \neq x_j$ (so that Γ is called intervenable). $\mathbb{M}^\Gamma = <U^\Gamma, V^\Gamma, R^\Gamma, E^\Gamma, f^\Gamma>$ is a submodel of (generated from intervention Γ on) $\mathbb{M} = <U, V, R, E, f>$, such that*

1. $U^\Gamma = U \cup \{X_i \mid X_i = \kappa_i \in \Gamma\}$,
2. $V^\Gamma = V - \{X_i \mid X_i = \kappa_i \in \Gamma\}$,
3. $R^\Gamma = R$,
4. $E^\Gamma = E - \{f_{X_i} \mid X_i = \kappa_i \in \Gamma\}$,
5. f^Γ *is a solution to $\mathbb{F}^\Gamma = <U^\Gamma, V^\Gamma, R^\Gamma, E^\Gamma>$ that satisfies the conditions (a) $f^\Gamma(X_i) = \kappa_i$, if $X_i = \kappa_i \in \Gamma$, and (b) $f^\Gamma(X_i) = f(X_i)$ if $X_i \in U^\Gamma$ but $X_i \notin \{X_i \mid X_i = \kappa_i \in \Gamma\}$.*

Simply speaking, intervention in a model by a formula $X_i = \kappa_i$ generates a submodel by setting X_i as an exogenous variable and its value to κ_i.

Definition 3 is designed for counterfactuals with antecedents only having the form of boolean conjunctions of atomic formulas. To extend to antecedents with any boolean antecedent but which still do not contain any counterfactual, we use the extension developed in [4].[5]

---

[5] See also the original proposal in [5]: 233–235.

Definition 4 *(State and Fusion, [4] : 154).*

1. *(Atomic State) For a variable X and $x_i \in R(X)$, $s_{X=x_i}$ is a state.*
2. *(Fusion) For distinct variables $X_1, ... X_n$ and $x_1 \in R(X_1), ..., x_n \in R(X_n)$, $s_{X_1=x_1} \sqcup ... \sqcup s_{X_n=x_n}$ is a state.*
3. *(Empty State) There is an empty state $\top$ such that for any state s, $s \sqcup \top = s$.*
4. *(Exclusion) For any $s_{X_i=x_i}$ and $s_{X_i=x_j}$ such that $x_i \neq x_j$, $... \sqcup s_{X_i=x_i} \sqcup ... \sqcup s_{X_i=x_j} \sqcup ...$ is not a state.*
5. *(Equivalence) For any state s, $... \sqcup s \sqcup ... \sqcup s \sqcup ... = s$.*

Definition 5 *(State Verification, [4]: 154–155).*

1. *$s \models X_i = \kappa_i$ if and only if $s = s_{X_i=\kappa_i}$.*
2. *$s \models \neg(X_i = \kappa_i)$ if and only if $s = s_{X_i=\kappa_j}$, where $\kappa_j \in R(X_i)$ and $\kappa_j \neq \kappa_i$.*
3. *$s \models \neg\neg\varphi$ if and only if $s \models \varphi$,*
4. *$s \models \top$ if and only if $s = \top$,*
5. *For any s, $s \not\models \bot$,*
6. *(a) $s \models \varphi \vee \psi$ if and only if $s \models \varphi$ or $s \models \psi$,[6]*
   (b) $s \models \neg(\varphi \vee \psi)$ if and only if $s = t \sqcup r$, in which $t \models \neg\varphi$ and $r \models \neg\psi$,
7. *(a) $s \models \varphi \wedge \psi$ if and only if $s = t \sqcup r$, in which $t \models \varphi$ and $r \models \psi$,*
   (b) $s \models \neg(\varphi \wedge \psi)$ if and only if $s \models \neg\varphi$ or $s \models \neg\psi$.

An intervention by any boolean formula can then be correspondingly defined as follows.

Definition 6 *(Extended Intervention). Consider a function f_s that maps every $s' = s_{X_1=\kappa_i} \sqcup ... \sqcup s_{X_n=\kappa_n}$ to a set $\Gamma_{s'} = \{X_1 = \kappa_1, ..., X_n = \kappa_n\}$. For a boolean formulas φ, we define a set of submodels $\mathbb{M}^\varphi = \{\mathbb{M}^{f_s(s')} | s' \models \varphi\}$ generated by intervention based on φ on $\mathbb{M}$.*

Based on the given definition of extended intervention, truth conditions for formulas are defined as follows.

Definition 7 *(Truth Conditions). For a model $\mathbb{M} = <\mathbb{F}, f>$, we define the truth conditions of formulas as follows.*

1. *$\mathbb{M} \models X_i = \kappa_i$ if and only if $f(X_i) = \kappa_i$,*
2. *$\mathbb{M} \models \neg\varphi$ if and only if $\mathbb{M} \not\models \varphi$,*
3. *$\mathbb{M} \models \varphi \wedge \psi$ if and only if $\mathbb{M} \models \varphi$ and $\mathbb{M} \models \psi$,*
4. *$\mathbb{M} \models \varphi \,\square\!\!\rightarrow \psi$ if and only if $\forall \mathbb{M}_i \in \mathbb{M}^\varphi$, $\mathbb{M}_i \models \psi$.*

Given the truth definition, Facts 1–4 follow directly, in which $\models$ is a classical consequence.

> **Fact 1** (SDA) $(\varphi \vee \psi) \,\square\!\!\rightarrow \chi \models (\varphi \,\square\!\!\rightarrow \chi) \wedge (\psi \,\square\!\!\rightarrow \chi)$
> **Fact 2** (DI Failure) $\varphi \,\square\!\!\rightarrow \chi \not\models (\varphi \vee \psi) \,\square\!\!\rightarrow \chi$
> **Fact 3** (CE Failure) $(\varphi \wedge \psi) \,\square\!\!\rightarrow \chi \not\models \varphi \,\square\!\!\rightarrow \chi$
> **Fact 4** (AS Failure) $\varphi \,\square\!\!\rightarrow \psi \not\models (\varphi \wedge \chi) \,\square\!\!\rightarrow \psi$

Though the causal modeling semantics for counterfactuals validates SDA, it does not follow that antecedent strengthening (AS) is valid, for the substitution of equivalent antecedent (SEA) is invalid, e.g., $p \equiv (p \wedge (q \vee \neg q))$, $p \,\square\!\!\rightarrow r \not\models (p \wedge (q \vee \neg q)) \,\square\!\!\rightarrow r$.

---

[6] In Briggs (2012), this principle is originally stated as $s \models \varphi \vee \psi$ if and only if $s \models \varphi$, $s \models \psi$, or $s \models \varphi \wedge \psi$. It will become clear in Sect. 6 that the original one does not suit our purpose for modeling strong permission.

6   Strategy One

To utilize CMS for implementing strong permission, we may consider a deontic reduction strategy that adopts a new variable OK with two values 0, 1 in a causal model, and take $OK = 1$ to stand for "sanction occurs" and $OK = 0$ to stand for "no sanction occurs". Based on the observed facts resulting from the truth definition in CMS, we have secured the following results suitable for a logic of free choice permission; here $\models$ is simply classical consequence.

Fact 5 (FCVE) $(\varphi \vee \psi) \,\square\!\!\rightarrow\, OK = 0 \models (\varphi \,\square\!\!\rightarrow\, OK = 0) \wedge (\psi \,\square\!\!\rightarrow\, OK = 0)$
Fact 6 (FCVI Failure) $\varphi \,\square\!\!\rightarrow\, OK = 0 \not\models (\varphi \vee \psi) \,\square\!\!\rightarrow\, OK = 0$
Fact 7 (FCCE Failure) $(\varphi \wedge \psi) \,\square\!\!\rightarrow\, OK = 0 \not\models \psi \,\square\!\!\rightarrow\, OK = 0$
Fact 8 (FCCI Failure) $\varphi \,\square\!\!\rightarrow\, OK = 0 \not\models (\varphi \wedge \psi) \,\square\!\!\rightarrow\, OK = 0$

The above facts fit the required inference patterns for free choice permission. There are also some other advantages. Since the counterfactual conditional $\square\!\!\rightarrow$ validates modus ponens if its consequent involves no counterfactuals, it satisfies the need for the practical aspect of strong permission. Since the consequence relation of current concern is classical, the inferred strong permission is suitable to represent the permission encoded in the premises without the need to worry about the danger of stepping on somebody's toes.

Though the above facts seem to fit the required logical and practical features for a logic of strong permission, it is unclear how to present the functional equation for variable OK so that a direct coding for prescriptive norms can be achieved. For example, the following two permissions are true in (Monty Hall), which $G = 1$ means that the guest opens the door 1, $C = 2$ means that the car is behind the door 2, and $M = 1$ means that Monty Hall opens the door 1.

(10) a. $(G = 1 \,\square\!\!\rightarrow\, OK = 0)$.
      b. $\neg((G = 1 \wedge C = 2 \wedge M = 1) \,\square\!\!\rightarrow\, OK = 0)$.

The question is, how can we design a causal model that makes true both (10a, b), under the conditions (a) that the variable OK is a variable whose value is dependent on variables G, M, and C, and also (b) that G, C, M are mutually independent variables with respect to each other? Suppose the factual situation is $G = 1 \wedge C = 2 \wedge M = 1$, meaning that the guest did something permissible but Monty Hall did something impermissible. Ought we consider the equation for OK to output the value 0 or 1? If it is the former, (10b) turns out to be false; if it is the latter, (10a) turns out to be false. This problem arises from the simple reason that *OK is a variable which evaluates the whole situation in a scenario, rather than evaluating whether a specific action is permitted.* I will come back to this point in the next section.

(10a, b) looks just like a case of the failure of antecedent strengthening, so one may suspect that it should be understood as harmless for a logic of counterfactuals, even in the causal modeling semantics. However, it is not so. Given that the variable OK has only two values and that it deterministically has one of the two values, it can be shown that a model which verifies both (10a, b) can only be one in which $\neg(G = 1 \wedge C = 2 \wedge M = 1)$. This result means that (10a, b) cannot both be true in a model in which $G = 1 \wedge C = 2 \wedge M = 1$. If this is the case, then we no longer have strong permissions as practically suitable prescriptive norms, for it is no longer suitable for Monty Hall to be condemned when he opened the door 1 under the situation that the guest opened the door 1. The specific counterfactual applicability that we desire is lost. One lesson to be learned is that a logical framework with correct logical features does not guarantee that framework being a suitable framework for strong permission, especially when we consider the practical aspect of strong permission. Alternative machinery may be worth considering.

7 Strategy Two

The situation described in examples (10a, b) fall into what I have called the specific counterfactual applicability of strong permission, in that (10a, b) is featured by their being applicable regardless of what the guest and Monty Hall have done. (10a, b) are also characterized by their general counterfactual applicability, in that they should be considered as true abstracting away from any action done (though not all permissions are true in this way), meaning that the truth of the permissions solely relies on how the causal frame is designed rather than what action is actually executed. This counterfactual applicability is important for the practical aspect, in that the strong permission does not just disappear when things are actually done. This is correct for any permissions true independent of any specific action actually executed, i.e. true by virtue of causal frames.

The view from counterfactual applicability also gives us a hint about how to provide a logic for strong permission but to escape from the deontic reduction strategy. If we take the deontic reduction strategy to be meant to capture the practical aspect of prescriptive norms, which in turn requires counterfactual applicability as an essential component, instead of considering whether an execution of an action leads to the situation OK = 1 or OK = 0, we can consider alternatively *whether an execution of an action leads specifically to whether the action's being executed is OK or not OK*. The former takes OK to be a variable talking about the whole situation, but the latter makes OK a property attributed to a specific action's being executed. Finally, the latter proposal means to capture OK as a property that directly observes the execution of actions.

This alternative proposal also implements strong permission by utilizing directly encoded prescriptive norms. To implement a direct encoding of prescriptive norms, we consider prescriptive causal models, a simple extension of causal models.

Definition 8 (*Prescriptive Causal Models*). *A prescriptive causal model* $\mathbb{M}^P$ =< *U, V, R, E, f, Ω* > *is one in which* $\mathbb{M}$ =< *U, V, R, E, f* > *is a causal model, and* Ω : *S F* → $\wp(\Pi)$ *is a prescriptive function that explicitly assigns situations in which an action is permitted (so that no sanction occurs), where S F is the set of formulas that are either atomic formulas or boolean conjunction of atomic formulas, and Π is the set of functions π such that π maps every variable X_i in $U \cup V$ to a value in $R(X_i)$.*

The distinctive feature of prescriptive causal models is the function Ω. Basically, Ω may be understood in this way: Ω maps a formula (an atomic formula or a conjunction of atomic formulas) to a set of situations (represented by a set of π) where the action (represented by the formula) is executed and is permissible. For example, in the case of

(Monty Hall), we map $G = 1$ to any possible value combination for G, C, M in which $G = 1$, since $G = 1$ is permitted no matter what the agents involved in the scenario actually do.

To utilize the direct encoding of prescriptive norms in prescriptive causal models, I take OK to be a unary operator whose truth conditions are to be defined rather than a variable in causal models. Instead of interpreting a sentence "φ is permitted" as $\varphi \Box\!\!\rightarrow OK = 1$, I interpret it as $OK(\varphi)$ (and accordingly, "φ is obligated" is interpreted as $O(\varphi)$). The truth conditions for boolean connectives and counterfactuals in prescriptive causal models are defined as in standard causal models (so that the prescriptive functions are inert in those definitions). Strong permission and obligation are defined as follows.

Definition 9 *(Strong Permission and Obligation). Let* $\mathbb{M}^P = <\, \mathbb{M}, \varOmega \,>$ *be a prescriptive causal model. For* $s' \models \varphi$ *(that the state* s' *verifies the boolean formula* φ*),* $\omega = \bigwedge f_s(s')$ *is a formula generated by the boolean conjunction on the set of atomic formulas* $f_s(s')$.

> $\mathbb{M}^P \models OK(\varphi)$ if and only if for any $s' \models \varphi$, $\pi \in \varOmega(\omega)$, where $\omega = \bigwedge f_s(s')$ and $\mathbb{M}^{f_s(s')} = <\, U, V, R, E, \pi >$.
>
> $\mathbb{M}^P \models O(\varphi)$ if and only if for any $s' \models \neg\varphi$, $\pi \notin \varOmega(\omega)$, where $\omega = \bigwedge f_s(s')$ and $\mathbb{M}^{f_s(s')} = <\, U, V, R, E, \pi >$.

Roughly, that $OK(\varphi)$ is true in a model means that every possible execution α of φ will lead to a α-permitted situation; $O(\varphi)$ being true in a model means that every possible execution α of $\neg\varphi$ will lead to a α-not-permitted situation. When weak permission is interpreted as the dual of obligation, saying that φ is weakly permitted is to say that not all possible executions of $\neg\varphi$ will lead to a $\neg\varphi$-not-permitted situation. Strong permission is distinguished from weak permission, for example, if we can find some execution of $\neg\varphi$ which is permissible, but nonetheless not all executions of φ are permissible. For example, we should find that, in (Monty Hall), when the guest actually opens the door 1, $OK(\neg(C = 1 \wedge M = 1))$ fails, but $\neg O(C = 1 \wedge M = 1)$ (so that $\neg(C = 1 \wedge M = 1)$ is weakly permitted) holds. Strictly speaking, weak permission may not tell you what you are really permitted to do.

Given the truth conditions for operator OK, the following facts hold.

Fact 9 (FCVE) $OK(\varphi \vee \psi) \models OK(\varphi) \wedge OK(\psi)$
Fact 10 (FCVI Failure) $OK(\varphi) \not\models OK(\varphi \vee \psi)$
Fact 11 (FCCE Failure) $OK(\varphi \wedge \psi) \not\models OK(\varphi)$
Fact 12 (FCCI Failure) $OK(\varphi) \not\models OK(\varphi \wedge \psi)$

Given facts 9–12, we have a logic for strong permission suitable for the analysis of free choice permission. Moreover, the logic does not make use of deontic reduction to implement strong permission.

To show that prescriptive causal models are useful for directly encoded prescriptive norms, we present the prescriptive causal model for (Monty Hall) as follows. A model for (Pain Killer) can be constructed similarly.

$\mathbb{M}_{MH}$ is a prescriptive causal model for (Monty Hall). $\mathbb{M}_{MH}$ includes a causal frame $\mathbb{F}_{MH} =< U, V, R, E >$ and a prescriptive function Ω.

1. $U =< G, C, M >$ where the variables G, C, M stands for the guest's choice of door 1, 2, or 3, C stands for the car's location at door 1, 2, or 3, and M stands for Monty Hall's action of choosing to open door 1, 2, or 3.
2. $V = \emptyset$.
3. R: each variable is assigned with a set of three possible values $\{1, 2, 3\}$.
4. $E = \emptyset$, meaning that each variable is causally independent from others.[7]
5. (a) $\Omega(G = 1) = \{$ any variable assignment function π that assigns 1 to variable $G \}$

    $\vdots$

    (b) $\Omega(M = 1) = \{$ any π that does not assign 1 to either G or $C \}$

    $\vdots$

    (c) $\Omega(C = 1 \wedge M = 2) = \{$ any π that does not assign $C = 2 \}$

    $\vdots$

It should be not difficult to see that strong permissions, embedded or not, such as (7a, b, c), are true in $\mathbb{M}_{MH}$.

In prescriptive causal models, we can further define the *range* of counterfactual applicability. For example, while the permission $G = 1$ is applicable in any variable-value combination of G, C, M, the permission of $M = 1$ is only applicable in those value assignments in which neither G nor C are assigned value 1. Formally, the counterfactually applicable situations of $OK(\varphi)$ are those solutions in the models in which $OK(\varphi)$ is true. Furthermore, we may state the following result: for any $\mathbb{M}^p$, $\mathbb{M}^p \models OK(\varphi)$ if and only if $(\mathbb{M}^p)^\varphi \models OK(\varphi)$. The specific counterfactual applicability follows.

Finally, I remark on some desirable logical features for utilizing prescriptive causal models. First, it can be shown that $\neg OK(\varphi \wedge \psi)$ is not equivalent to $\neg OK(\varphi) \vee \neg OK(\psi)$. This allows for (4) being inconsistent but (5) consistent. This makes the free choice permission more practically sensible. Second, we should see that the inference from (6a) to (6b) is invalid. This invalidity comes from the feature that OK in prescriptive causal models is an operator applied to a specific action rather than representing a feature in the situation brought about by an execution of an action.

8   Concluding Remarks

In this paper, I argued that a logical framework for strong permission may capture the inferential patterns required for free choice permission without leading to paradox, but some important aspects of strong permission may still be missing. Proposals for strong

---

[7] It is natural to think that which door the guest opens and which door the car locates have causal effect on which door Monty Hall opens. For example, given the guests choose the door 1 and the car is located at door 2, Monty Hall will open door 3. This causal concern is usefor for revealing the probabilistic dependency among variables under concern. When we construct prescriptive causal models, we assume that Monty Hall, one of the agents involved in the situation, are free to violated causal determination relations.

permission in the literature are shown to lack the necessary means to capture the practical aspect of prescriptive norms and strong permission embedding. Prescriptive causal models are further provided to achieve the goal without exploiting the usual deontic reduction strategy for strong permission.

References

1. Anderson, A.: The formal analysis of normative systems. In: Rescher, N. (ed.) The Logic of Decision and Action, pp. 147–213. University of Pittsburgh Press, Pittsburgh (1966)
2. Asher, N., Bonevac, D.: Free choice permission is strong permission. Synthese **145**, 303–323 (2005)
3. Barker, C.: Free choice permission as resource-sensitive reasoning. Semant. Pragmatics **3**, 1–38 (2010)
4. Briggs, R.: Interventionist counterfactuals. Philos. Stud. **160**, 139–166 (2012)
5. Fine, K.: Counterfactuals without possible worlds. J. Philos. **CIX**, 221–246 (2012)
6. Pearl, J.: Causality: Models, Reasoning and Inference. Cambridge University Press, Cambridge (2000)

Truth as a Logical Connective

Shunsuke Yatabe$^{(\boxtimes)}$

Graduate School of Letters, Center for Applied Philosophy and Ethics,
Kyoto University, Kyoto, Japan
shunsuke.yatabe@gmail.com

I swear, in the kingdom of generalities, you could be imperius rex.
(Haruki Murakami "A Wild Sheep Chase")

1   Introduction

Some truth theories allow to represent and prove *generalized* statements as "all that you said is true" or "all theorems of **PA** are true" in the sense of deflationism. But these theories are ω-inconsistent [HH05] caused by McGee's paradox [M85], Yablo paradox [Yb93] and so on. In this paper, we examine the relationship between generality and ω-inconsistency in terms of proof theoretic semantics. It is done by means of regarding the truth as a logical connective. Philosophically speaking, we do not try to assert that the truth conception should be represented by a logical connective instead of the predicate though some authors suggests the deflationistic truth is a logical expression [Gl16]: only we want to do here is to provide a new perspective to analyze the behavior of the truth predicate in ω-inconsistent truth theories.

The object of this analysis is Friedman-Sheared's truth theory **FS** [FS87] which is known to be ω-inconsistent. It consists of all axioms and schemata of **PA**, and the following special rules:

– Formal Commutativity (**FC**): for any logical connective $\circ$ and quantifier Q,

$$\mathbf{Tr}(x \dot{\circ} y) \equiv \mathbf{Tr}(x) \circ \mathbf{Tr}(y) \qquad \mathbf{Tr}(Qz(x(z))) \equiv Qz\mathbf{Tr}(x(z))$$

– two inference rules **NEC, CONEC**:

$$\frac{\varphi}{\mathbf{Tr}(\lceil \varphi \rceil)} \ \mathbf{NEC} \qquad \frac{\mathbf{Tr}(\lceil \varphi \rceil)}{\varphi} \ \mathbf{CONEC}$$

where $\lceil \varphi \rceil$ is a Godel code of φ. Since **NEC** looks like the introduction rule and **CONEC** looks like the elimination rule of **Tr**, it is natural to ask the following question from a viewpoint of a naive proof theoretic semantics [Hj12]:

Is it possible to think the truth predicate **Tr** of **FS** as a **logical connective?**

© Springer International Publishing AG 2017
M. Otake et al. (Eds.): JSAI-isAI 2015 Workshops, LNAI 10091, pp. 166–182, 2017.
DOI: 10.1007/978-3-319-50953-2_13

It is because — from the naive proof theoretic semantics viewpoint — to be a logical connective is to have its introduction and elimination rule.

However, there is a negative evidence that $\mathbf{Tr}(x)$ can be a logical connective because adding $\mathbf{NEC}, \mathbf{CONEC}$ could impact the **harmony** between the introduction and the elimination rule, that needs to be satisfied to be a logical connective. The harmony is a central principle of proof theoretic semantics. It requires the appropriate relationship between the introduction and the elimination rule and guarantees various preservation results — in particular the *normalizability* of proofs, that is, any proof which contains *roundabouts* (e.g. the application of the elimination rule just after the application of the introduction rule) as in the left-hand-side below can be reduced to a *normal* proof which does not contain the roundabouts as in the right-hand side.

$$
\frac{\dfrac{\vdots \quad \vdots}{A \quad B}}{\dfrac{A \wedge B}{A}} \begin{array}{l} \wedge + \\ \wedge - \end{array} \quad \Rightarrow \quad \begin{array}{c} \vdots \\ A \end{array}
$$

This reduction process is represented by a recursive function – whose input is (a Godel code of) a proof with roundabouts and whose output is (a Godel code of) a proof without them– which eventually terminates. Actually it is often regarded that the harmony of a logical connective ∘ between the introduction and the elimination rule is that the normalizability holds for the proof system with ∘ [D93]. The violation of the harmony is involved by an unexpected result, so called McGee's paradox, which says of the ω-inconsistency of **FS**. It is said that this is one of the most serious defects of **FS** and the similar theories [Fl08].

In this paper, we analyze this problem in terms of computer science. Contrary to what people believe about truth theories, the deflationists' truth expands not only their semantics but also their syntax (or, rather to say, the definability of formulae). According to deflationists, enough strong truth theories like **FS** is a sort of *the kingdom of generalities*, that is, the nature of truth (in **FS**) allows to say and prove a generalized statement which is formalized as follows. Let $\varphi_0, \varphi_1, \cdots$ be a recursive enumeration of sentences, and f be a recursive function s.t. $f(n) = \ulcorner \varphi_n \urcorner$, then the following represents the sentence "all $\varphi_0, \varphi_1, \cdots$ are true",

$$
\forall x \mathbf{Tr}(f(x))
$$

Intuitively, it is an *infinite conjunctions* $\varphi_0 \wedge \varphi_1 \wedge \cdots$. This sentence is *coinductive* (or *potentially infinite*), in the sense that it generates the infinite stream $\varphi_0 \wedge \varphi_1 \wedge \cdots$ though it is a finite sentence with a recursive machinery with the help of $\mathbf{Tr}$. In this way, coinductive formulae — *imperius rex* of the kingdom of generality — are definable of the form $\forall x \mathbf{Tr}(f(x))$ in **FS**. In McGee's paradox case, the paradoxical sentence γ is intuitively of the form $\neg\mathbf{Tr}(\mathbf{Tr}(\mathbf{Tr}(\cdots)))$.

Since γ is definable in **FS**, we should simulate to define γ by using $\mathbf{Tr}$ as a logical connective. It is done by thinking the system with the connective $\mathbf{Tr}$ allows to define the *streams*, formulae of the infinite length those are outputs of the recursive machinery in **FS**. The definability of such infinite streams seems to be a necessary consequence of the quest of the generality along the line of the deflationism.

Then the existence of these streams violates the concept of finiteness. It is easy to see **FC** violates the normalizability: **FC** proves γ, but we should apply **NEC** infinite times to prove the infinite stream $\neg\mathbf{Tr}(\mathbf{Tr}(\mathbf{Tr}(\cdots)))$ by the normal proof, therefore it is impossible to reduce such infinite proof to its corresponding normal proof by finite reduction steps.

This suggests traditional criteria of the harmony – such as normalizability – works well only for finite, inductively defined formulae effectively, but not for infinite streams with infinite proofs. From this viewpoint, **FC** is very problematic because it can be regarded as an infinitely operation though other inference rules are finitely: if both $\mathbf{Tr}(a)$ and $\mathbf{Tr}(b)$ are interpreted to be infinite streams, **FC** insists that the composition of two infinite streams, $\mathbf{Tr}(a) \rightarrow \mathbf{Tr}(b)$, can be reduced to the single infinite stream of the form $\mathbf{Tr}(a \rightarrow b)$ by just one step. **FC** makes it possible to process infinite objects as if they are purely finite: the resulting system **FS** becomes ω-inconsistent.

As a consequence, if we want to call **Tr** a logical connective, we should extend the concept of harmony appropriately. The way of such extension is already known in computer science: we introduce the solution of this problem by using the concept of *guarded corecursion* along the line of [S12], and we will show the positive answer. Let us explain the guarded corecursion: assume $\mathbf{Tr}(\mathbf{Tr}(\cdots))$ is an infinite stream, then the application of **NEC** to the stream seems somehow unproblematic; it is easy to think a non-terminate automaton whose program is just outputting **Tr** in each calculation step so its *log* of the behaviour is the infinite stream $\mathbf{Tr}(\mathbf{Tr}(\cdots))$ which is of infinite length and not computable as a whole. A guarded corecursive function is such finite operation iterated infinite times. **NEC** can be regarded a guarded corecursive function instead of recursive function for **CONEC** in ordinal proof theoretic semantics account for inductively defined formulae in this sense. If we extend the concept of the harmony in this way, and if we abandon **FC** because this violates the concept of finiteness, then we can regard that **Tr** with **NEC, CONEC** rule is a logical connective.

Many people believed that only sentences of finite length should be taken seriously in the context of truth theory. No philosophical theory of truth thoughts that such infinite sentences are the object of serious consideration. However, the consequence of this paper is opposite: we should take such coinductive nature of truth seriously.

2   Preliminaries

2.1   Friedman-Sheared's Axiomatic Truth Theory FS

It is well-known that the unrestricted form of T-schemata, of the form $\varphi \equiv \mathbf{Tr}(\lceil\varphi\rceil)$ for any formula φ, implies a contradiction by the liar paradox in classical logic. It is done by the diagonalization argument, so we have to weaken the T-schemata to prevent the truth theory from being inconsistent. Friedman-Sheared's Axiomatic truth theory **FS** is such a weakened truth theory. In **FS**,

T-sentence for φ is provable for any φ, but diagonalizing T-sentences is impossible, therefore the T-sentence of the liar sentence is not a theorem of **FS**.

Definition 1. *Friedman-Sheared's Axiomatic truth theory* **FS** *consists of the following axioms:*

– *Axioms and schemes of* **PA** *(mathematical induction for all formulae including* **Tr***)*
– *The formal commutability* (**FC**) *of the truth predicate:*
  • *logical connectives*
    ∗ *for any atomic formula* ψ, $\mathbf{Tr}(\lceil \psi \rceil) \equiv \psi$,
    ∗ $(\forall x \in \mathbf{Form})[\mathbf{Tr}(\dot{\neg}x) \equiv \neg\mathbf{Tr}(x)]$,
    ∗ $(\forall x, y \in \mathbf{Form})[\mathbf{Tr}(x\dot{\wedge}y) \equiv \mathbf{Tr}(x) \wedge \mathbf{Tr}(y)]$,
    ∗ $(\forall x, y \in \mathbf{Form})[\mathbf{Tr}(x\dot{\vee}y) \equiv \mathbf{Tr}(x) \vee \mathbf{Tr}(y)]$,
    ∗ $(\forall x, y \in \mathbf{Form})[\mathbf{Tr}(x\dot{\rightarrow}y) \equiv \mathbf{Tr}(x) \rightarrow \mathbf{Tr}(y)]$,
  • *quantifiers*
    ∗ $(\forall x \in \mathbf{Form})[\mathbf{Tr}(\dot{\forall}z\, x(z)) \equiv \forall z\mathbf{Tr}(x(z))]$,
    ∗ $(\forall x \in \mathbf{Form})[\mathbf{Tr}(\dot{\exists}z\, x(z)) \equiv \exists z\mathbf{Tr}(x(z))]$,
– *The* **introduction rule** *and the* **elimination rule** *of* $\mathbf{Tr}(x)$

$$\frac{\varphi}{\mathbf{Tr}(\lceil\varphi\rceil)} \ \mathbf{NEC} \qquad \frac{\mathbf{Tr}(\lceil\varphi\rceil)}{\varphi} \ \mathbf{CONEC}$$

where

– **Form** *is a definable predicate of Godel codes of formulae (and* $(\forall x \in \mathbf{Form})P(x)$ *is an abbreviation of the formula* $(\forall x)\mathbf{Form}(x) \rightarrow P(x))$,
– *for any connective* $\circ$, $\dot{\circ}$ *is an recursive function such that* $\lceil P \rceil \dot{\circ} \lceil Q \rceil = \lceil P \circ Q \rceil$.

It is easy to see that **FS** can prove many generalized statements as follows.

– "all formulae are true or false" $(\forall x \in \mathbf{Form})\mathbf{Tr}(x) \vee \mathbf{Tr}(\dot{\neg}x)$,
– "no contradictory formula is true" $(\forall x \in \mathbf{Form})\neg(\mathbf{Tr}(x) \wedge \mathbf{Tr}(\dot{\neg}x))$.

Intuitively, these statements can be regarded as infinite conjunctions as we explained in Sect. 1. In this sense, **FS** is a good candidate for deflationists of truth in the above sense:

> We believe that deflationist truth is a tool for formulating and *proving* generalizations (or, if you like, infinite conjunctions) [HH05, p. 208]

Next let us concentrate on the formal commutability **FC**. It is easy to prove the following lemma.

Lemma 1. **PA** + **NEC** + **CONEC** *proves the following commutability results for any concrete "real" formula as follows:*

- *for any atomic formula* ψ, $\mathbf{Tr}(\lceil\psi\rceil) \equiv \psi$,
- *for any formula* φ, $\mathbf{Tr}(\lceil\neg\varphi\rceil) \equiv \neg\mathbf{Tr}(\lceil\varphi\rceil)$,
- *for any formulae* φ, ψ, $\mathbf{Tr}(\lceil\varphi \wedge \psi\rceil) \equiv \mathbf{Tr}(\lceil\varphi\rceil) \wedge \mathbf{Tr}(\lceil\psi\rceil)$, $\mathbf{Tr}(\lceil\varphi \vee \psi\rceil) \equiv \mathbf{Tr}(\lceil\varphi \vee \psi\rceil)$, $\mathbf{Tr}(\lceil\varphi \to \psi\rceil) \equiv \mathbf{Tr}(\lceil\varphi\rceil) \to \mathbf{Tr}(\lceil\psi\rceil)$,
- *for any formula* φ, $\mathbf{Tr}(\lceil\forall z\,\varphi(z)\rceil) \equiv \forall z\mathbf{Tr}(\lceil\varphi(z)\rceil)$, $\mathbf{Tr}(\lceil\exists z\,\varphi(z)\rceil) \equiv \exists z\mathbf{Tr}(\lceil\varphi(z)\rceil)$.

The need of the formal commutability essentially depends on the case **Form** contains a nonstandard element.

2.2   A Proof Theoretic Semantics of T

In this section, we analyze the behavior of the truth predicate of **FS** from a naive proof theoretic semantics viewpoint. Proof theoretic semantics is a theory of meaning of logical connectives and quantifiers which is based on Wittgenstein's meaning as use theory *"the meaning of word is its use in the language"*. Since the use of a logical connective is determined by the introduction and the elimination rule, it involves the following (naive) proof theoretic semantics viewpoint: *the meaning of a logical connective is given by the introduction rule and the elimination rule.*

In the case of **FS**, as we saw, its two rules look like the introduction and the elimination rule. So let us define a connective **T** whose inference rules are the similar to **Tr**.

Definition 2.

- *Let* **T** *be a connective which has only two inference rules:*

$$\frac{\varphi}{\mathbf{T}\varphi}\ \textbf{NEC} \qquad \frac{\mathbf{T}\varphi}{\varphi}\ \textbf{CONEC}$$

- *A new system whose language is that of classical predicate logic* **CL∀** *plus new connective* **T**, *and whose system is* **CL∀** *plus* **NEC, CONEC** *is* **CL∀$^{\mathbf{T}}$**.

From the naive viewpoint, the above definition is enough: **T** seems to be a *logical* connective.

However, it is well-known that the naive proof theoretic semantics viewpoint has a serious defect [P60] as follows.

Example 1 (TONK). The following connective **tonk** seems to be a logical connective from the naive proof theoretic viewpoint.

$$\frac{A}{A\,\mathbf{tonk}\,B}\ \textbf{tonk+} \qquad \frac{A\,\mathbf{tonk}\,B}{B}\ \textbf{tonk−}$$

It is easy to see that **tonk** trivializes the system, i.e. for any theorem A and for any formula B, B is provable by using the **tonk**-introduction and the elimination rule.

$$\frac{\dfrac{\vdots}{\dfrac{A}{\dfrac{A\,\mathbf{tonk}\,B}{B}}}}{}\ \begin{array}{l}\\ \textbf{tonk+}\\ \textbf{tonk−}\end{array}$$

Therefore, we need to revise our criterion for connectives to be *logical* connectives. One of the most well-known candidates for that criterion is so-called the *harmony*: the connective concerned is a logical connective if its introduction rule and its elimination rule are in harmony, i.e. to have the special relationship. The origin of this idea is due to Gentzen [Gt34, p. 80]:

> The introductions represent, as it were, the *'definitions'* of the symbols concerned, and the eliminations are no more, in the final analysis, than *the consequences of these definitions.*

Here, they are in special relationship in the sense that the elimination rule is a derivation of the introduction rule. Until now, a few candidates of the harmony have been proposed as follows.

Definition 3 (harmony).

- **The conservativeness** [B63] *(so called "total harmony")*
  For any φ which does not include **Tr**, if **FS** proves φ, then **PA** can prove φ.
- **Normalizability of proofs** *(so called "intrinsic harmony" in* [D93]*)*
  Any proof can be rewritten, or **reduced**, to its normal form.

Two rules **NEC**, **CONEC** are the same to the necessitation and the co-necessitation rule of the normal modal logics. Then, it is easy to see the following:

Lemma 2.

(1) any connective and quantifier commutes **T**,
(2) $\mathbf{CL\forall^T}$ *is a conservative extension of* $\mathbf{CL\forall}$,
(3) any proof of $\mathbf{CL\forall^T}$ *is normalizable.*

Proof. (1) is just the same to Lemma 1.

As for (3), let us assume the proof lefthand-side which contains *indirect pair*, the roundabout that the application of **CONEC** just after that of **NEC**. It can be reduced to the normal proof as follows:

$$\frac{\dfrac{\vdots}{\dfrac{C}{TC}}}{C} \quad \Rightarrow \quad \begin{matrix}\vdots\\C\end{matrix}$$

Here the indirect pair disappears in the right-hand side.   □

Lastly we discuss the motivation and the involvement of regarding **Tr** as a logical connective. As for the motivation, this is encouraged from deflationists' viewpoint. Deflationists often think that the relationship between A and A *is true* is logical, therefore the expression *"is true"* is a logical expression. One of the most familiar logical expression is logical connective, therefore to examine what happens if we regard **Tr** as a logical connective [Gl16] must be valuable.

As for its involvement, let us assume ∘ is any connective. From the viewpoint of Dummett [D93], any proof is a demonstration of reducing the *justification of the consequence* to the *justification of the assumptions*. If the reduction is too complex to understand, for instance it contains many indirect pairs, it is epistemologically unreliable. The normalizability guarantees that any long and winding non-normal proof containing ∘ can be reduced to normal one which – might be longer but – is simpler therefore more epistemologically reliable. Thinking **Tr** is a logical connective means **Tr** is requested not to prevent such reduction process, and that's all. In this way, the role of **Tr** is very restricted – as much as deflationist requires - in proof theoretic semantics viewpoint.

3   A Problem: The Violation of Harmony

Like the Hilbert program, the interpreting **Tr** in **FS** as a logical connective project ends up in a, namely, failure. $\mathbf{Tr}(x)$ cannot be a logical connective because it violates the harmony between **NEC** and **CONEC**. That is mainly caused by the fact that **FS** is ω-inconsistent. Then, that affects to two criteria of Definition 3 as follows.

– There is a thread to the conservativeness condition (Sect. 3.1).
– This violates the normalizability (Sect. 3.2).

3.1   A Thread to Conservativeness

First, for **FS⁻** that is the truth theory **FS** minus **NEC, CONEC**, we can show the following.

Lemma 3. *Adding* **NEC, CONEC** *to* **FS⁻** *is not a conservative extension.*

Proof. It is because **FS** proves the consistency of **PA**: this is a contraposition of $(\exists x)\mathbf{provable}(x, \lceil 0 = 1 \rceil) \to \mathbf{Tr}(\lceil 0 = 1 \rceil)$ which is provable by the reflection principle [Hr12]. □

Since Godel's incompleteness theorem implies **PA** cannot capture its whole truth in a model, this is not a surprise. This just suggests **PA** is not a good framework to capture the concept of truth of arithmetic, therefore it is not a problem of **FS** but a problem of the base theory **PA**: the non-conservativeness of **FS** is act of God. Furthermore, we can also prove the following [Ha11].

Lemma 4 (Arithmetical soundness of FS). **FS** *is arithmetical sound, that is, if* **FS** *proves* φ *for any arithmetical formula* φ *then* φ *is true in the standard model* $\mathbb{N}$ *of* **PA**.

In other words, adding **NEC, CONEC** is a conservative extension over the *true arithmetic*. Therefore, we could still have hoped the *conservativeness of intended ontology*, that is, adding **NEC, CONEC** does not effect the intended model of the base theory.

However, this version of conservativeness negatively holds by McGee's unexpected theorem [M85].

Theorem 1 (McGee). *Any consistent truth theory T which include* **PA** *and* **NEC** *and satisfies the following:*

(1) $(\forall x, y)[x, y \in \mathbf{Form} \to (\mathbf{Tr}(x \dot{\to} y) \to (\mathbf{Tr}(x) \to \mathbf{Tr}(y)))]$,
(2) $\mathbf{Tr}(\bot) \to \bot$,
(3) $(\forall x)[x \in \mathbf{Form} \to \mathbf{Tr}(\dot{\forall} y x(y)) \to (\forall y \mathbf{Tr}(x(y)))]$

Then T is ω-inconsistent.

This theorem shows **FS** is ω-inconsistent. Since **PA** and *true arithmetic* has a standard model $\mathbb{N}$, adding **NEC, CONEC** violates the intended ontology. This is not a single result, for, the following similar result is well-known:

Theorem 2 (Yablo's paradox). *Truth theories in classical logic with enough-strong expressive power is either*

– *contradictory; it is provable without using self-referential sentences directly* [Yb93],
– *or it is ω-inconsistent* [Lt01].

Proof. of Theorem 1.
We define the following paradoxical sentence $\gamma \equiv \neg \forall x \mathbf{Tr}(g(x, \lceil \gamma \rceil))$ where g is a recursive function s.t.

$$g(0, \lceil \varphi \rceil) = \lceil \mathbf{Tr}(\lceil \varphi \rceil) \rceil$$
$$g(x + 1, \lceil \varphi \rceil) = \lceil \mathbf{Tr}(g(x, \lceil \varphi \rceil)) \rceil$$

Intuitively speaking, γ is the negation of the *truth teller*, "this sentence is true" $(\mathbf{Tr}(\lceil \mathbf{Tr}(\lceil \mathbf{Tr}(\lceil \cdots \rceil) \rceil) \rceil))$. Therefore the intuitive meaning of γ is

$$\gamma \equiv \neg \mathbf{Tr}(\lceil \mathbf{Tr}(\lceil \mathbf{Tr}(\lceil \mathbf{Tr}(\lceil \cdots \rceil) \rceil) \rceil) \rceil)$$

Then we can prove γ: roughly speaking, the proof is as follows.

$$\frac{\dfrac{\dfrac{\dfrac{\dfrac{\gamma \equiv \neg \mathbf{Tr}(\lceil \mathbf{Tr}(\lceil \mathbf{Tr}(\lceil \mathbf{Tr}(\lceil \cdots \rceil) \rceil) \rceil) \rceil)}{\gamma \to \neg \mathbf{Tr}(\lceil \mathbf{Tr}(\lceil \mathbf{Tr}(\lceil \mathbf{Tr}(\lceil \cdots \rceil) \rceil) \rceil) \rceil)}}{\mathbf{Tr}(\lceil \gamma \to \neg \mathbf{Tr}(\lceil \mathbf{Tr}(\lceil \mathbf{Tr}(\lceil \mathbf{Tr}(\lceil \cdots \rceil) \rceil) \rceil) \rceil) \rceil)}}{\mathbf{Tr}(\lceil \gamma \rceil) \to \mathbf{Tr}(\lceil \neg \mathbf{Tr}(\lceil \mathbf{Tr}(\lceil \mathbf{Tr}(\lceil \mathbf{Tr}(\lceil \cdots \rceil) \rceil) \rceil) \rceil) \rceil)} \, FC}{\mathbf{Tr}(\lceil \gamma \rceil) \to \neg \mathbf{Tr}(\lceil \mathbf{Tr}(\lceil \mathbf{Tr}(\lceil \mathbf{Tr}(\lceil \mathbf{Tr}(\lceil \cdots \rceil) \rceil) \rceil) \rceil) \rceil)} \, FC}{\mathbf{Tr}(\lceil \gamma \rceil) \to \gamma}$$

And

$$\frac{\dfrac{\neg \gamma \to \forall x \mathbf{Tr}(g(x, \lceil \gamma \rceil))}{\neg \gamma \to \mathbf{Tr}(g(\bar{0}, \lceil \gamma \rceil))}}{\neg \gamma \to \mathbf{Tr}(\lceil \gamma \rceil)}$$

Therefore $\mathbf{Tr}(\lceil \gamma \rceil) \to \gamma$ and $\neg \gamma \to \mathbf{Tr}(\lceil \gamma \rceil)$ implies, **FS** $\vdash \neg \neg \gamma$ though $\neg \mathbf{Tr}(g(\bar{n}, \lceil \gamma \rceil))$ is provable for any n.

This shows the ω-inconsistency: since γ is provable, all $\mathbf{Tr}(g(\bar{0}, \lceil \gamma \rceil))$, $\mathbf{Tr}(g(\bar{1}, \lceil \gamma \rceil))$, $\mathbf{Tr}(g(\bar{2}, \lceil \gamma \rceil))$, $\cdots$ are provable. Therefore, for any natural number n, $\mathbf{Tr}(g(\bar{n}, \lceil \gamma \rceil))$ is provable. However, γ, i.e. $\neg \forall x \mathbf{Tr}(g(x, \lceil \gamma \rceil))$ is provable. $\square$

3.2   The Failure of Normalizability

In this derivation of ω-inconsistency, the paradoxical formula γ plays a key role. The special feature of γ is its infinite iteration of **Tr**: this is possible by the combination of quantifier and the recursive function g. This is just the same mechanism which enables to represent *generalized sentences* (or *infinite* sentences) in the sense of deflationism. In this sense, deflationists must accept γ, and the ω-inconsistency, because the very mechanism which proves many generalized sentences provides γ simultaneously.

When we think **Tr** is a logical connective, we have to think that γ is not a formula defined by a recursive function (here **Tr** transforms Godel codes to real formulae), but a formula whose **Tr** expands the syntax.

The existence of such infinite sentences negatively impact to our claim. Normalizability does not hold for infinite formulae as γ. Let us assume that **T** is connective. Then, γ is of the form $\neg\mathbf{TT}\cdots$, therefore the normal proof of γ must be as follows.

$$\frac{\vdots \\ \mathbf{TT}\cdots}{\mathbf{TTT}\cdots}\ \mathbf{T}^{+}$$
$$\vdots$$
$$\frac{\perp}{\neg\mathbf{TTT}\cdots}\ \rightarrow^{+}$$

Since the formula of the form $\mathbf{TT}\cdots$ can only be introduced by iterating the **T**-introduction rule infinite times, the proof should be of the infinite length. Therefore, if the proof is of the infinite length, the reduction steps from a non-normal proof to the normal proof need not to be finite steps[1]. This certainly violates the normalizability.

4   Analysis: Induction, Coinduction and Harmony

In the previous section, we saw that two criteria of being a logical connective fail. The violation of the conservativeness of intended ontology itself is not a counter-evidence if we change the base theory, but it is still a negative one. It is the non-normalizability that has more serious consequences than the conservativeness. The ω-inconsistency is caused by that deflationists require generalized, infinite sentences which is impossible to rewrite them to their normal form (if any) in finite steps.

Let us concentrate on the non-normalizability. The existence of these infinite generalized sentences suggests traditional criteria of the harmony – such as normalizability – works well only for finite, inductively defined formulae effectively, but not for infinite sentences those must have proofs of the infinite length.

Responding to this, we should analyze the mechanism which causes ω-inconsistency in detail, and find a way (if any) to extend the criterion of the

---

[1] We note that this is the same reason why unrestricted form of the coinductive datatype implies a contradiction in Martin-Löf's intuitionistic type theory.

normalizability for finite sentences to one for infinite sentences naturally. If it is possible, then it must be a desirable result.

4.1   The Framework of the Analysis: A Truth Theory Without Arithmetization

From now on, we introduce a new framework which is used from now on to analyze this problem.

In many case, a truth theory is implemented over a formal arithmetic as **PA**. It is because the formal arithmetic is regarded as a standard *theory of syntax* [Hr12]. It seems to be a common sense that the use of arithmetic is not essential but practical for the study of conception of truth. It is purpose-neutral practical choice, so regarded. However, it turns out that using arithmetic is not so neutral. The problems of arithmetization, as the case of ω-inconsistency of **FS**, are now in the center of the investigation, but non-standard arithmetic is very difficult to study.

In this paper we try to formalize the conception of truth *without arithmetization*. The coding of formulae together with developing proof theory is possible in many theories: we choose the dependent type theory which is widely used in computer science. In computer science, the behavior of objects are focused. For example, in Type Theory, inductively defined formulae are members of a *inductive datatype* of formulae. The infinite stream as γ are defined as members of a *coinductive datatype* [C93], that is intended to represent behaviors, or the output log, of a non-terminate automaton with recursive machinery. This is more intuitive than the use of nonstandard arithmetic.

4.2   Proof Theoretic Semantics for Inductive Formulae

What Is Inductive Definition? To compare with the coinduction, first let us introduce a typical example of the *inductive definition*.

Definition 4 (Inductive datatype). *For any set A, the list of A can be constructed as $\langle A^{<\omega}, \eta : (1 + (A \times A^{<\omega})) \to A^{<\omega} \rangle$ by:*

– **the first step:** *empty sequence $\langle \rangle$*
– **the successor step:** *For all $a_0 \in A$ and sequence $\langle a_1, \cdots, a_n \rangle \in A^{<\omega}$*

$$\eta(a_0, \langle a_1, \cdots, a_n \rangle) = \langle a_0, a_1, \cdots, a_n \rangle \in A^{<\omega}$$

The inductive definition corresponds to the existence of *the least fixed point* of η.

Thinking Its Meaning from Proof Theoretic Semantics Viewpoint. Let us remember the explanation of the meaning of logical connective from the proof theoretic semantics viewpoint.

> The introductions represent, as it were, the 'definitions' of the symbols concerned, and the eliminations are no more, in the final analysis, than the consequences of these definitions. [Gt34, p. 80]

This shows that the meaning of induction is given by the introduction rule. Here, the constructor η represents *the introduction rule* For any $a_0 \in A$ and sequence $\langle a_1, \cdots, a_n \rangle \in A^{<\omega}$,

$$\eta(a_0, \langle a_1, \cdots, a_n \rangle) = \langle a_0, a_1, \cdots, a_n \rangle \in A^{<\omega}$$

Example 2. The introduction ruke $\gamma_\wedge$ of the logical connective γ is as follows. For any formula A, B, let p_A, p_B be their proofs. Then the proof p such that

$$\begin{array}{cc} \vdots\ p_A & \vdots\ p_B \\ A & B \\ \hline \multicolumn{2}{c}{A \wedge B} \end{array}$$

is represented by a recursive computation $\gamma_\wedge(p_A, p_B) = p$.

And *the elimination rule* γ is determined in relation to the introduction rule which satisfies the harmony.

The Elimination Rule and Recursive Computation. Contrary to the introduction rule, the meaning of the elimination rule is explained from the introduction rule. This is the *"inversion principle"*:

> Let α be an application of an elimination rule that has β as consequence. Then, deductions that satisfy the sufficient condition [...] for deriving the major premiss of α, when combined with deductions of the minor premisses of α (if any), already *"contain"* a deduction of β; the deduction of β is thus obtainable directly from the given deductions without the addition of α. [Pr65, p. 33]

Let us give an example.

Example 3. The inversion principle is about the reduction of the lefthand-side proof p_0 to the righthand-side proof p_1:

$$\cfrac{\cfrac{\begin{array}{cc}\vdots & \vdots \\ A & C\end{array}}{A \wedge C}\ \wedge+ \quad \cfrac{[v:A] \\ \vdots \\ \beta}{}}{\beta}\ \alpha \qquad \Rightarrow \qquad \begin{array}{c}\vdots \\ A \\ \vdots \\ \beta\end{array}$$

Here, the proof by the elimination rule should have the corresponding *recursive function*:

- whose **input:** the proof tree p_0 of β *with* α lefthand-side,
- whose **output:** the proof tree p_1 of β *without* α righthand-side.

Thinking its meaning, the *elimination rule* γ is represented as a *recursive function* which reduces p_0 to p_1 over the datatype of formulae inductively defined by the introduction rule, that is, $\gamma(p_0) = p_1$. The recursive function can be identified with the recursive procedure of the cut elimination or the normalization of proofs.

4.3   Proof Theoretic Semantics for Coinductive Formulae

What Is Coinduction?   In this section, we introduce what the coinduction is. As an example, we introduce the following computer scientific object: a vending machine. In many case, it takes a long time from booting to shutting down as compared to CPU clock cycle time. Therefore, there is practically no problem to regard its operating time as forever. In this sense, the behaviour of this vending machine can be described by using a cyclic automaton whose operating time is forever as follows:

(1) projecting "Welcome! Insert a coin please!" to the monitor,
(2) if someone put a coin, then turn the light of the selection button,
(3) if he (or she) push the button, he (or she) gets the can of a soft drink,
(4) if the sensor recognizes that he (or she) gets the can, turn off the monitor and the light of button,
(5) if the sensor finds a new customer, go to (1).

Every operating step of the machine is simple and finitely, but the whole operating time is infinite. The following directed graph shows the behaviour of the machine.

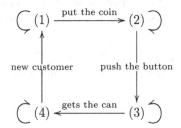

Since the operating time (and its behaviour log) is practically infinite, purely finite methods cannot describe the behaviour of this machine (this is conceptually infinite). However, every execution step is finite, in this sense it is *potentially infinite*. And the automaton itself has only 4 states, and the program itself consists of only finite number of cods. The *coinduction* makes it possible to describe the log of the behaviour of such finite size program which continues to operate forever.

   The situation is similar to the case of the liar paradox, for example. If we have the T-scheme $\Lambda \equiv \neg \mathbf{Tr}(\lceil \Lambda \rceil)$ for the liar sentence Λ, then it generates the 2-states automaton, the directed graph of the behaviour of the lair sentence, as follows:

$$\mathbf{Tr} \circlearrowleft \Lambda \overbrace{} \neg\mathbf{Tr}(\lceil\Lambda\rceil) \longleftrightarrow \mathbf{Tr}(\lceil\neg\Lambda\rceil) \overbrace{} \neg\Lambda$$

This generates the infinite stream $\neg\mathbf{Tr}(\lceil\neg\mathbf{Tr}(\lceil\neg\mathbf{Tr}(\lceil\cdots\rceil)\rceil)\rceil)$ of the log of the behaviour (if we describe the log in transfinite induction way, it is essentially the same to the *revision sequence* of Λ). Here, the original sentence are finitely, but it generates (or unfolds) the infinite sentence (in that sense this is *potentially infinite* or *coinductive*). We note that the liar paradox is the simplest case of coinductive paradox in truth theory: the McGee's paradox is a more complex example.

$$\mathbf{Tr} \circlearrowleft \neg\gamma \xrightarrow{\text{FC}} \gamma \circlearrowright \mathbf{Tr}$$

In this case, the cycle $\mathbf{Tr}(\lceil \neg\gamma \rceil) \equiv \gamma$ generates the infinite stream $\mathbf{Tr}(\mathbf{Tr}(\mathbf{Tr}(\cdots)))$; another cycle provides γ generates the stream $\mathbf{Tr}(\mathbf{Tr}(\mathbf{Tr}(\cdots\gamma\cdots)))$.

The keyword here is *coinductive*. Coinduction is a typical way to handle infinite streams generated by finite but non-terminate automata. Let us give an example of a typical coinductive object.

Definition 5 (Infinite streams of A). *Infinite streams of A is defined whose type is*

$$\langle A^\infty, \delta : A^\infty \to (A \times A^\infty) \rangle$$

Their intuitive meaning is infinite streams of the form $\langle a_0, a_1, \cdots \rangle \in A^\infty$

$$\delta(\langle a_0, a_1, \cdots \rangle) = (a_0, \langle a_1, \cdots \rangle) \in (A \times A^\infty)$$

δ is a function which pick up the first element a_0 of the stream. (but $\langle a_1, \cdots \rangle$ is still infinite stream) for $\langle a_0, a_1, \cdots \rangle \in A^\infty$.

The key concept of coinduction is the *productivity*.

Definition 6 (Productivity). *The productivity insists an intuition that finitely operations on the first element of the infinite sequence is possible to compute.*

In the context of truth theory, both **NEC** and **CONEC** only care about the productivity (or 1-step computation) if φ is coinductive.

$$\frac{\varphi}{\mathbf{Tr}(\varphi)} \ \mathbf{NEC} \qquad \frac{\mathbf{Tr}(\varphi)}{\varphi} \ \mathbf{CONEC}$$

Thinking the Meaning of Rules for Coinductive Formulae. The coinductive definition is a dual of the inductive definition. Therefore all roles are opposite. Contrary to the induction case, the meaning of coinduction is by *the elimination rule* as follows:

- de-constructor γ represents the elimination rule [S12], for any infinite stream $\langle a_0, a_1, \cdots \rangle \in A^\infty$,

$$\gamma(\langle a_0, a_1, \cdots \rangle) = (a_0, \langle a_1, \cdots \rangle) \in (A \times A^\infty)$$

  γ is a function which picks up the first element a_0.
- the introduction rule is known to have to satisfy the condition (*the guarded corecursion*) which corresponds to the failure of the formal commutativity of **Tr**.

The final algebra which has the same structure to streams is called weakly final coalgebra.

Next we introduce one of the most important concept the "*guardedness*". Let us remember that the productivity property of stream is regarding *the infinite as a whole but 1 step computation is possible*.

Example 4 (**Guarded corecursion**). Guarded corecursion guarantees the productivity of functions over coinductive datatypes. For any recursive function f, **map** : $(A \to B) \to A^\infty \to B^\infty$ is defined as

$$\mathbf{map}\, f \,\langle x, x_0, \cdots \rangle = \underbrace{\langle f(x) \rangle}_{\text{finite part}} \frown \underbrace{(\sharp\, \mathbf{map}\, f \langle x_0, x_1, \cdots \rangle)}_{\text{infinite part}}$$

where $\frown$ means the concatenation of two sequence.

Here recursive call of **map** only appears inside $\sharp$, the inside and outside of $\sharp$ are essentially different as if the those of modal operators are different, that is, $\square(\varphi \wedge \psi)$ and $(\square\varphi) \wedge \psi$ are different.

Let us show the meaning explanation of coinductive datatypes which is an analogy to the inductive case: the proof by the *introduction rule* should have the corresponding guarded corecursive function:

– whose **input:** a (coinductive) proof tree *with* the use of the introduction rule,
– whose **output:** a (coinductive) proof tree *without* the use of the introduction rule.

T-schemata only care about the productivity (or 1-step computation): it should satisfy the 1 step computability, and this involves the guarded corecursion.

The Harmony Extended. To sum up, to extend the concept of harmony between the introduction rule and the elimination rule for coinductive formulae, we define the following concept.

Definition 7 (The extended harmony for coinductive formulae). *The formulae and proofs are coinductive objects. Then, for any logical connective, the harmony between the introduction rule and the elimination rule is defined as follows:*

– *the elimination rule represents the de-constructor of the coinductive datatype of prooftrees,*
– *the introduction rule is defined as a guarded correcursive function.*

5   A Solution: How to be a Logical Connective and ω-consistent

In the previous section, we extended the concept of harmony, which was originally suitable only for finite formulae, to be suitable for streams, coinductive formulae. In this section, we analyze the McGee's paradox from the coinductive proof theoretic semantics viewpoint. Then we find what prevents **Tr** to be a logical connective (and what makes **FS** to be ω-consistent).

5.1 The Guarded Corecursion and the Failure of Formal Commutability

Let us see the **FC** violates the *guardedness*: let us remember the proof of McGee's theorem,

$$\frac{\mathbf{Tr}(\lceil \gamma \to \neg\mathbf{Tr}(\lceil \mathbf{Tr}(\lceil \cdots \rceil)\rceil)\rceil)}{\mathbf{Tr}(\lceil \gamma \rceil) \to \neg\mathbf{Tr}(\lceil \mathbf{Tr}(\lceil \mathbf{Tr}(\lceil \cdots \rceil)\rceil)\rceil)} \; FS$$

Here both γ and $\mathbf{Tr}(\lceil \mathbf{Tr}(\lceil \mathbf{Tr}(\lceil \cdots \rceil)\rceil)\rceil)$ are coinductive objects:

- To calculate the value of $\mathbf{Tr}(\lceil \gamma \to \neg\mathbf{Tr}(\lceil \mathbf{Tr}(\lceil \cdots \rceil)\rceil)\rceil)$, we should calculate the both the value of $\mathbf{Tr}(\lceil \gamma \rceil)$ and $\neg\mathbf{Tr}(\lceil \mathbf{Tr}(\lceil \mathbf{Tr}(\lceil \cdots \rceil)\rceil)\rceil)$.
- But $\mathbf{Tr}(\lceil \gamma \rceil)$ is *also coinductive object*, therefore the calculation of the first value $\mathbf{Tr}(\lceil \gamma \rceil)$ *never terminates*.

This is why **Tr** with the **FC** cannot be a logical connective[2].

However, if we abandon the **FC**, we can regard **Tr** as a logical connective in the extended sense. For example, the truth theory $\mathbf{G}^i$ over Heyting arithmetic within the intuitionistic logic with two rules **NEC, CONEC**, and various formal commutativity conditions except, in particular, $\mathbf{Tr}(A \to B) \to \mathbf{Tr}(A) \to \mathbf{Tr}(B)$, satisfies the following theorem [LgR12].

Theorem 3 (Leigh-Rathjen). $\mathbf{G}^i$ *is ω-consistent.*

$\mathbf{G}^i$ proves that it is consistent that **Tr** can be regarded as a logical connective and it is ω-consistent. This seems to suggest that their relationship between to be a logical connective and to be ω-consistent.

5.2 The Truth Connective as a De-Constructor of a Coinductive Datatype of Formulae

Some examples as $\mathbf{G}^i$, we can regard **Tr** as a logical connective in the extended sense, and we can develop coinductive proof theoretic semantics. It seems to say the following:

- the meaning of truth connective is given by the elimination rule (**CONEC**),
- the introduction rule (**NEC**) should be given in the form of the guarded corecursion function, this corresponds to the failure of the **FC**.

If we extend the concept of harmony for coinductive objects, then we can say "**Tr** in $\mathbf{G}^i$ is a logical connective". The nature of the truth concept here is to make it possible to define coinductive formula of infinite length. The truth connective is a de-constructor of such coinductive datatype. The truth conception here guarantees the *reducibility* of the argument, and the essence is that the truth connective never prevent the reducibility from the assumption to the consequence in the sense of proof theoretic semantics viewpoint.

---

[2] Contrary, $(\mathbf{Tr}(A) \to \mathbf{Tr}(B)) \to (\mathbf{Tr}(A \to B))$ is not problematic. Merging two streams is a typical guarded correcursive function.

We note that, deflationists might not accept this view: the nature of the truth conception makes it possible to define infinite streams. However this is the logical consequence that they admit to represent and prove generalized statements as "All sentences are equal, but some coinductive sentences are more equal than others".

References

[B63] Belnap, N.D.: Tonk, plonk and plink. Analysis **22**(6), 130–134 (1963)

[B08] Beal, J.: Spandrels of Truth. Oxford University Press, Oxford (2008)

[C93] Coquand, T.: Infinite objects in type theory. In: Barendregt, H., Nipkow, T. (eds.) TYPES 1993. LNCS, vol. 806, pp. 62–78. Springer, Heidelberg (1994). doi:10.1007/3-540-58085-9_72

[D73] Dummett, M.: Frege: Philosophy of Language. Duckworth, London (1973)

[D93] Dummett, M.: The Logical Basis of Metaphysics (The William James Lectures, 1976). Harvard University Press, Cambridge (1993)

[Fl08] Field, H.: Saving Truth From Paradox. Oxford University Press, Oxford (2008)

[FS87] Friedman, H., Sheared, M.: An axiomatic approach to self-referential truth. Ann. Pure Appl. Logic **33**, 1–21 (1987)

[Gl16] Galinon, H.: Deflationary truth: conservativity or logicality? Philos. Q. (to appear)

[GB] Gupta, A., Belnap, N.: The Revision Theory of Truth. MIT Press, Cambridge (1993)

[Gt34] Gentzen, G.: Untersuchungen über das logische Schließen. I. Math. Z. **39**(2), 176–210 (1934). Translated in: Investigations concerning logical deduction. In: Szabo, M. (ed.) The Collected Papers of Gerhard Gentzen, pp. 68–131. North-Holland, Amsterdam (1969)

[HPS00] Hájek, P., Paris, J.B., Shepherdson, J.C.: The liar paradox and fuzzy logic. J. Symbolic Logic **65**(1), 339–346 (2000)

[Ha11] Halbach, V.: Axiomatic Theories of Truth. Cambridge University Press, Cambridge (2011)

[HH05] Halbach, V., Horsten, L.: The deflationist's axioms for truth. In: Beal, J.C., Armour-Garb, B. (eds.) Deflationism and Paradox. Oxford University Press, Oxford (2005)

[Hj12] Hjortland, O.: HARMONY and the context of deducibility. In: Insolubles and Consequences. College Publications (2012)

[Hr12] Horsten, L.: Tarskian Turn. MIT Press, Cambridge (2012)

[LgR12] Leigh, G.E., Rathjen, M.: The Friedman-Sheard programme in intuitionistic logic. J. Symbolic Logic **77**(3), 777–806 (2012)

[Lt01] Leitgeb, H.: Theories of truth which have no standard models. Stud. Logica. **68**, 69–87 (2001)

[M85] McGee, V.: How truthlike can a predicate be? A negative result. J. Philos. Logic **17**, 399–410 (1985)

[Pr65] Prawitz, D.: Natural Deduction: A Proof-Theoretical Study. Dover Publications, New York (2006)

[P60] Prior, A.: The runabout inference ticket. Analysis **21**, 38–39 (1960–61)

[R93] Restall, G.: Arithmetic and Truth in Łukasiewicz's Infinitely Valued Logic. Logique et Analyse **36**, 25–38 (1993)

[S12] Setzer, A.: Coalgebras as types determined by their elimination rules. In: Dybjer, P., Lindström, S., Palmgren, E., Sundholm, G. (eds.) Epistemology versus Ontology, pp. 351–369. Springer, Heidelberg (2012)

[Yb93] Yablo, S.: Paradox without self-reference. Analysis **53**, 251–252 (1993)

JURISIN 9

Ninth International Workshop
on Juris-Informatics (JURISIN 2015)

Ken Satoh[1], and Takehiko Kasahara[2]

[1] National Institute of Informatics and Sokendai, Tokyo, Japan
[2] Toin University of Yokohama, Yokohama, Japan

Juris-informatics is a new research area which studies legal issues from the perspective of informatics. The purpose of this workshop is to discuss both the fundamental and practical issues among people from the various backgrounds such as law, social science, information and intelligent technology, logic and philosophy, including the conventional "AI and law" area. The Ninth International Workshop on Juris-Informatics (JURISIN 2015) was held on November 17 and 18, 2015, in association with the Seventh JSAI International Symposia on AI (JSAI-isAI 2015) with a support of the Japanese Society for Artificial Intelligence (JSAI).

This year, we have twenty one submissions and each paper was reviewed by at least three program committee members and as a result, nineteen papers was presented at the workshop.

Papers cover logical inference, ontology, natural language processing, information retrieval, and so on. As invited speakers, we have Giovanni Sartor from University of Bologna, Italy, Shiro Kawashima from Doshisha University, Japan, Do Kwan Jo from National Assembly Law Library, Korea and Miyoung Jin Kim, from Justice Law Firm, Korea. Moreover, Phan Minh Dung from AIT, Thailand, gave a joint-invited talk on his current research on an argumentation framework based on the strength of inference rules in cooperation with the 2nd International Workshop on Argument for Agreement and Assurance (AAA 2015).

Following the event started at JURISIN 2014, we held the second competition on legal information extraction/entailment (COLIEE-2015) on a legal data collection. The COLIEE competition is about performance on a legal question answering task; we structure the competition into three subtasks of retrieval, information extraction, and entailment.

Overall, the competition focuses on legal information processing related to answering yes/no questions from Japanese legal bar exams (the relevant data sets have been translated from Japanese to English):

- Phase 1 of the legal question answering task is to extract relevant civil codes given a legal bar exam question.
- Phase 2 of the legal question answering task is to decide whether relevant civil codes entail a legal bar exam question.
- Phase 3 task is combination of Phase 1 and Phase 2.

A variety of methods were used by competitors. For Phase 1, for example, ranking SVM, TFxIDF, n-gram features from the query and articles using lexical and morphological characteristics and their own relevance score are used. Furthermore, some extended information retrieval techniques have been used, e.g., Hiemstra, BM25 and PL2F, keyword weighting method and snippet scoring, and Hidden Markov Model. For Phase 2, the participants used AdaBoost with their own selected features, a convolutional neural network (CNN), and heuristic thresholds on snippet scores. Most of the systems do not depend on deep linguistic features, which require domain knowledge, but instead propose heuristic feature weighting and scoring methods.

After the workshop, fifteen papers were submitted for the post proceedings. They were reviewed by PC members again and eight papers were finally selected. Contributions of these papers are as follows. Gavanelli et al. extended their work on abducitive logic programming with constraint processing to process normative reasoning and ontological query answering. Arisaka proposed a belief revision framework to adapt new legal scenarios conflicting with the current legal statues. Robaldo et al. formalized real-world obligations by combining normative reasoning called "input/output logic" and natural language semantics called "reification-based approach". Bartolini et al. provided a bottom-up ontology describing the constituents of data protection domain and its relationships which is an urgent issue in European data protection policy. Sakamoto et al. enhanced a machine translation method using multi-word expressions which frequently appear in legal texts. Nishina et al. proposed an evaluation method of the acceptability of arguments called "reliability-based argumentation framework" and developed a tool for calculating acceptability of arguments. Kim et al. proposed solutions of COLIEE competition using a deep convolutional neural network for textual entailment which is the first to adapt deep learning for the task. Carvalho et al. also proposed solutions of the competition using combination of n-gram model for the civil code extraction and distributional semantic similarity for the textual entailment.

Finally, we wish to express our gratitude to all those who submitted papers, PC members, discussant and attentive audience.

JURISIN 2015 Workshop Co-chairs
Ken Satoh
Takehiko Kasahara

Abductive Logic Programming for Normative Reasoning and Ontologies

Marco Gavanelli[1]([✉]), Evelina Lamma[1], Fabrizio Riguzzi[2], Elena Bellodi[1], Zese Riccardo[1], and Giuseppe Cota[1]

[1] Dipartimento di Ingegneria, University of Ferrara,
via Saragat 1, 44122 Ferrara, Italy
{marco.gavanelli,evelina.lamma,elena.bellodi,
riccardo.zese,giuseppe.cota}@unife.it
[2] Dipartimento di Matematica e Informatica, University of Ferrara,
via Saragat 1, 44122 Ferrara, Italy
fabrizio.riguzzi@unife.it

Abstract. Abductive Logic Programming (ALP) has been exploited to formalize societies of agents, commitments and norms, taking advantage from ALP operational support as a (static or dynamic) verification tool. In [7], the most common deontic operators (obligation, prohibition, permission) are mapped into the abductive expectations of an ALP framework for agent societies. Building upon such correspondence, in [5], authors introduced $Deon^+$, a language where obligation and prohibition deontic operators are enriched with quantification over time, by means of ALP and Constraint Logic Programming (CLP).

In recent work [30,31], we have shown that the same ALP framework can be suitable to represent Datalog$^\pm$ ontologies. Ontologies are a fundamental component of both the Semantic Web and knowledge-based systems, even in the legal setting, since they provide a formal and machine manipulable model of a domain.

In this work, we show that ALP is a suitable framework for representing both norms and ontologies. Normative reasoning and ontological query answering are obtained by applying the same abductive proof procedure, smoothly achieving their integration. In particular, we consider the ALP framework named $\mathcal{S}$CIFF and derived from the IFF abductive framework, able to deal with existentially (and universally) quantified variables in rule heads and CLP constraints.

The main advantage is that this integration is achieved within a single language, grounded on abduction in computational logic.

1 Introduction

Norms represent desirable behaviors of members of a human or artificial society.

A normative system is a set of norms, together with mechanisms to reason about, apply, and modify them. Norms can be encoded by exploiting the notions of obligation, permission and prohibition, often modelled as modal operators, in the tradition of Deontic Logic [49].

M. Otake et al. (Eds.): JSAI-isAI 2015 Workshops, LNAI 10091, pp. 187–203, 2017.
DOI: 10.1007/978-3-319-50953-2_14

As for the structure of the formulas representing norms, a widely adopted approach is to encode norms as logical rules in the form of implications, since: (*i*) implications correspond intuitively to conditional norms, which state that some deontic consequence (such as the obligation for an agent to perform an action) follows from a state of affairs; (*ii*) rule-based systems also provide an operational support for reasoning, and draw conclusions (regarding, e.g., expected behavior, or norm violations and related sanctions). In the legal domain, the British Nationality Act was formalized using Logic Programming (LP) [47]; later, argument-based extended LP with defeasible priorities [43] and the use of defeasible logic was proposed [33]. Satoh et al.'s PROLEG is a Prolog implementation of the Presupposed Ultimate Fact Theory of the Japanese Civil Code [46]. The contract cancellation under the Japanese law was also formalized in computational logics [22].

Normative systems have been also advocated as a tool to model and reason upon a single agent, as in [13] where a normative system can be seen as a normative agent, equipped with mental attitudes, about which other agents can reason, choosing either to fulfill their obligations, or to face the possible sanctions.

More often, normative systems regulate interaction in multi-agent systems [12]. Among the organizational models, [24,25] exploit Deontic Logic to specify the society norms and rules. The whole research project ALFEBIITE [10] was focused on the formalization of an open society of agents using Deontic Logic.

The EU IST Project SOCS proposed a Computational Logic approach to multi-agent systems. The SOCS social model represents and verifies both social interaction protocols among members regulated via abductive expectations [2], and member specifications themselves [14] in Abductive Logic Programming (ALP). Both approaches have been later applied to model and reason about norms with deontic flavours [7,44].

In the EU project IMPACT [11,27], agent programs may be used to specify what an agent is obliged to do, what an agent may do, and what an agent cannot do on the basis of deontic operators of Permission, Obligation and Prohibition (whose semantics does not rely on a Deontic Logic semantics).

In the meantime, legal ontologies have proved crucial for representing, processing and retrieving legal information. A collective reflection on the theoretical foundations of legal ontology engineering is [45]. The ESTRELLA project [28] aimed at developing a standard based platform allowing public administrations to deploy comprehensive legal knowledge management solutions. To this purpose, the project developed a Legal Knowledge Interchange Format (LKIF), building upon OWL, and modeled European tax related legislation and national tax legislation of two European countries as case studies. Integration of rules and ontologies has been faced in [9], which proposes a normative language that combines expressivity of LP and Description Logic (DL) for hybrid knowledge bases modeling human laws, with examples from the Portuguese Penal Code. Another approach is the mapping of DL into computational logic; for example the $\mathcal{ALCN}$ Description Logics was also mapped into Open Logic Programming [48], an extension of ALP. Notable work has been done also in applying abductive reasoning to DL [26,38].

In Computational Logic, ALP was proved a powerful tool for knowledge representation and reasoning [35], taking advantage from ALP operational support as (static or dynamic) verification tool. ALP languages are usually equipped with a declarative (model-theoretic) semantics, and an operational semantics given in terms of a proof-procedure. Several abductive proof procedures have been defined, with many applications (diagnosis, monitoring, verification, etc.). Fung and Kowalski proposed the IFF abductive proof-procedure [29] to deal with forward rules, and with non-ground abducibles. It has been later extended [4], and the resulting proof procedure, named $\mathcal{S}$CIFF, can deal with both existentially and universally quantified variables in rule heads and Constraint Logic Programming (CLP) constraints [34]. The resulting system was used for modeling and implementing several knowledge representation frameworks, such as deontic logic [7], where the deontic notions of obligation and permission are mapped into special $\mathcal{S}$CIFF abducible predicates, normative systems [5], interaction protocols for multi-agent systems [8], Web services choreographies [1], etc.

In this work, we move a step forward, by showing that ALP, and $\mathcal{S}$CIFF in particular, is a suitable framework for representing and integrating both norms and Datalog$^{\pm}$ ontologies. Normative reasoning and ontological query answering are obtained by applying the $\mathcal{S}$CIFF abductive proof procedure. The main advantage is that this integration is achieved within a single language, grounded on abduction in Computational Logic.

We assume a basic familiarity with Logic Programming and Abductive Logic Programming, good introductions are, respectively [35, 40].

The paper is organized as follows. We first introduce ALP and the $\mathcal{S}$CIFF framework in Sect. 2, also mentioning its underlying proof procedure. Then, in Sect. 3, we introduce (a subset of) the deontic language $Deon^+$ and show its mapping into $\mathcal{S}$CIFF. Section 4 introduces Datalog$^{\pm}$ to formalize ontologies, and shows its correspondence to $\mathcal{S}$CIFF integrity constraints. This paves the way to the integration of norms and ontologies, discussed in Sect. 5, where we show a simple example, with norms and an ontology, and discuss the kind of inference supported by the $\mathcal{S}$CIFF proof procedure. More relevant related work is mentioned throughout the various sections. In Sect. 6 we conclude the paper, and outline future work.

2   ALP and the $\mathcal{S}$CIFF Language

Abductive Logic Programming (ALP) is a family of programming languages that integrate abductive reasoning into LP. An ALP program consists of a logic program, also called *knowledge base KB*, that is a set of clauses $h \leftarrow b$. As usual in LP, a set of clauses in which the h share a same functor symbol define a predicate. Differently from classical LP, in ALP there are also predicates that have no definition, that belong to a set $\mathcal{A}$ and are called *abducibles*. Abducible literals cannot occur in the h of a clause, but they can occur in the b. The aim in ALP is finding a set of abducibles $\Delta^{\mathcal{A}}$, built from symbols in $\mathcal{A}$, that, together with the KB, is an explanation for a given known effect (called *goal* $\mathcal{G}$) and satisfies a set of logic formulae, called *Integrity Constraints (ICs)*:

$$KB \cup \Delta^{\mathcal{A}} \models \mathcal{G} \qquad KB \cup \Delta^{\mathcal{A}} \models IC.$$

While in the early abductive proof-procedures [36] the set of abduced literals is ground, in later proof-procedures [23,29,37] it can also contain existentially-quantified atoms.

$\mathcal{S}$CIFF [4] is a language in the ALP class, extension of the IFF proof-procedure [29]. As in the IFF, integrity constraints are in the form *body* $\rightarrow$ *head* where *body* is a conjunction of literals and *head* is a disjunction of conjunctions of literals. While in the IFF the literals can be built only on defined or abducible predicates, in $\mathcal{S}$CIFF they can also be CLP constraints, occurring events (only in the body), or positive and negative expectations, as explained in the following.

Definition 1. *A $\mathcal{S}$CIFF Program is a pair $\langle KB, \mathcal{IC} \rangle$ where KB is a set of clauses and $\mathcal{IC}$ is a set of logic implications called* Integrity Constraints.

$\mathcal{S}$CIFF considers a (possibly dynamically growing) set of facts (called *history*) **HAP**, that contains ground atoms $\mathbf{H}(Event[, Time])$. This set can grow dynamically, during the computation, thus implementing a dynamic acquisition of events. Some distinguished abducibles are called *expectations*. A *positive expectation* $\mathbf{E}(Event[, Time])$ means that a corresponding event $\mathbf{H}(Event[, Time])$ is expected to happen, while $\mathbf{EN}(Event[, Time])$ is a *negative expectation*, and requires events $\mathbf{H}(Event[, Time])$ not to happen. To simplify the notation, we will omit the *Time* argument from events and expectations when not needed.

Variables occurring in positive expectations are existentially quantified (expressing the idea that a single event is enough to support them), while those occurring only in negative expectations are universally quantified, so that any event matching with a negative expectation leads to inconsistency with the current hypothesis. Nested existential quantifications are forbidden by the language syntax. CLP [34] constraints can be imposed on variables. The computed answer includes three elements: a substitution for the variables in the goal (as usual in Prolog), the constraint store (as in CLP), and the set $\Delta^{\mathcal{A}}$ of abduced literals.

The declarative semantics of $\mathcal{S}$CIFF includes the classic conditions of ALP:

$$KB \cup \mathbf{HAP} \cup \Delta^{\mathcal{A}} \models \mathcal{G} \qquad (1)$$
$$KB \cup \mathbf{HAP} \cup \Delta^{\mathcal{A}} \models \mathcal{IC} \qquad (2)$$

plus specific conditions to support the confirmation of expectations.

As the history can be dynamically growing, it makes sense to adopt either an open or closed world assumption on the history, depending on the envisaged application. $\mathcal{S}$CIFF can support both types of reasoning. In a *skeptical* reasoning attitude, all hypotheses that are not explicitly confirmed are rejected, assuming that no more events can happen (closed world assumption on the history): all the positive expectations that are not matched with an actual event are disconfirmed (symmetrically, the pending negative expectations are confirmed). In a *credulous* reasoning attitude, a set of hypotheses is acceptable even if some hypotheses are not explicitly confirmed, as long as the set is consistent. Declaratively, in skeptical reasoning positive expectations are confirmed if

$$KB \cup \mathbf{HAP} \cup \Delta^{\mathcal{A}} \models \mathbf{E}(X) \rightarrow \mathbf{H}(X), \tag{3}$$

which is not required in a credulous attitude. In this paper, we adopt a credulous reasoning attitude. In both cases, negative expectations should not be disconfirmed by actual events, and the same event cannot be expected both to happen and not to happen:

$$KB \cup \mathbf{HAP} \cup \Delta^{\mathcal{A}} \models \mathbf{EN}(X) \wedge \mathbf{H}(X) \rightarrow false. \tag{4}$$

$$KB \cup \mathbf{HAP} \cup \Delta^{\mathcal{A}} \models \mathbf{E}(X) \wedge \mathbf{EN}(X) \rightarrow false. \tag{5}$$

Note that in (3), (4), (5), additional object-level integrity constraints are introduced, to be accomplished by the declarative semantics.

Definition 2 ($\mathcal{S}$CIFF answer). *Given a $\mathcal{S}$CIFF program $\langle KB, \mathcal{IC} \rangle$ and a history $\mathbf{HAP}$, a goal $\mathcal{G}$ is a $\mathcal{S}$CIFF answer if there is a set $\Delta^{\mathcal{A}}$ such that (1), (2), (4) and (5) are satisfied. In this case, we write*

$$\langle KB, \mathcal{IC} \rangle \models_{\mathbf{HAP}} \mathcal{G}.$$

The operational counterpart to this declarative semantics is represented by $\mathcal{S}$CIFF proof procedure. $\mathcal{S}$CIFF is a rewriting system that searches a proof tree representing all abductive solutions and whose nodes represent states of the computation. A set of transitions rewrite a node into one or more children nodes. $\mathcal{S}$CIFF inherits the transitions of the IFF proof-procedure [29], and extends it in various directions. We recall the basics of $\mathcal{S}$CIFF; a complete description is in [4], with proofs of soundness, completeness, and termination. An efficient implementation of $\mathcal{S}$CIFF is described in [6].

Each node of the proof is a tuple $T \equiv \langle R, CS, PSIC, \Delta^{\mathcal{A}} \rangle$, where R is the resolvent, CS is the CLP constraint store, PSIC is a set of implications (called *Partially Solved Integrity Constraints*) derived from propagation of integrity constraints, and $\Delta^{\mathcal{A}}$ is the current set of abduced literals. The main transitions, inherited from the IFF are:

Unfolding replaces a (non abducible) atom with its definitions;
Propagation if an abduced atom $\mathbf{a}(X)$ occurs in the condition of an IC (e.g., $\mathbf{a}(Y) \rightarrow p$), the atom is removed from the condition (generating $X = Y \rightarrow p$);
Case Analysis given an implication containing an equality in the condition (e.g., $X = Y \rightarrow p$), generates two children in logical or (in the example, either $X = Y$ and p, or $X \neq Y$);
Equality rewriting rewrites equalities as in the Clark's equality theory [20];
Logical simplifications other simplifications like (true $\rightarrow A$) $\Leftrightarrow A$, etc.

The $\mathcal{S}$CIFF proof procedure includes also the transitions of CLP [34] for constraint solving.

In this paper we consider the *generative version* of $\mathcal{S}$CIFF, called g-$\mathcal{S}$CIFF [3,41], in which also the **H** events are considered as abducibles. Although from an ALP point of view the **H** events should be collected together with the other abducibles, we prefer to maintain a notation coherent with that used for the $\mathcal{S}$CIFF proof-procedure, and distinguish the set of abduced events from the set containing the remaining abducibles, Δ^A. A history **HAP** is provided as input, and further **H** atoms can be assumed like the other abducible predicates; they are then collected in a set $\mathbf{HAP'} \supseteq \mathbf{HAP}$.

Definition 3 (g-$\mathcal{S}$CIFF answer). *Given a $\mathcal{S}$CIFF program $\langle KB, \mathcal{IC} \rangle$ and a history* **HAP**, *we say that a goal $\mathcal{G}$ is a g-$\mathcal{S}$CIFF answer if there exist a set Δ^A and a set* $\mathbf{HAP'} \supseteq \mathbf{HAP}$ *such that*

$$KB \cup \mathbf{HAP'} \cup \Delta^A \models \mathcal{G} \tag{6}$$

$$KB \cup \mathbf{HAP'} \cup \Delta^A \models \mathcal{IC} \tag{7}$$

$$KB \cup \mathbf{HAP'} \cup \Delta^A \models \mathbf{EN}(X) \wedge \mathbf{H}(X) \rightarrow false. \tag{8}$$

$$KB \cup \mathbf{HAP'} \cup \Delta^A \models \mathbf{E}(X) \wedge \mathbf{EN}(X) \rightarrow false. \tag{9}$$

In this case, we write

$$\langle KB, \mathcal{IC} \rangle \models^g_{\mathbf{HAP} \rightsquigarrow \mathbf{HAP'}} \mathcal{G} \qquad \text{or simply} \qquad \langle KB, \mathcal{IC} \rangle \models^g_{\mathbf{HAP}} \mathcal{G}.$$

3   Norms in $\mathcal{S}$CIFF

In [5], $\mathcal{S}$CIFF was exploited to support legal regulations expressed in a deontic language, named $Deon^+$. In $Deon^+$, (positive) actions are represented by terms and, as usual in logic programming, terms can contain variables, constants, terms. Building upon this action language, obligations are represented as $\mathcal{S}$CIFF atoms $\mathbf{E}(A, T)$, where A is any action description, and T is a CLP variable, existentially quantified. For instance, the sentence "It is mandatory that John answers me", corresponds to:

$$\exists T \qquad \mathbf{E}(answer(john, me), T)$$

as any reply in any time complies to the obligation, and the obligation will no longer hold after John sends his answer. Note that $\mathbf{H}(answer(john, me), T)$ is different from $\mathbf{E}(answer(john, me), T)$: the first expresses that indeed John answers me (or, in g-$\mathcal{S}$CIFF, that we assume he answers), while the second expresses that he *should* answer, independently from the fact that he actually answers or not.

Prohibitions are represented as atoms $\mathbf{EN}(A, T)$, where again A is any action description, and T is a CLP variable, universally quantified (unless it occurs also in a **H** or **E** atom, in such a case it is existentially quantified). For instance, the sentence "It is forbidden that John smokes", corresponds to:

$$\forall T \qquad \mathbf{EN}(smoke(john), T)$$

because in any time John is not allowed to smoke, and the fact he did not smoke one minute ago does not allow him to smoke now and later on. In general, this is different from $not\mathbf{H}(smoke(john), T)$ (or, equivalently, $\mathbf{H}(smoke(john), T) \rightarrow false$) because the first expresses that he should not smoke, while the second expresses that he does not smoke. In this work, however, we adopt rule (4) that maps violations to failures, which makes the two equivalent in practice; in other works one can consider recovery actions to violations.

A notable advantage of adopting ALP is that the action language is not limited to the propositional case, as in the examples above, but it can contain variables in its turn, quantified (existentially or universally) as the T variable.

The adoption of CLP variables for representing time adds expressiveness to deontic operators and easily recovers deadlines by constraints over time variables. Constraints imposed on universally quantified variables are considered as quantifier restrictions [16]; a sentence like "It is forbidden that John leaves the meeting before 10" is therefore represented in $Deon^+$ as:

$$\forall T : T < 10 \quad \mathbf{EN}(leave(john, meeting), T),$$

and it is interpreted (coherently with the semantics of quantifier restrictions) as

$$\forall T, \quad T < 10 \rightarrow \mathbf{EN}(leave(john, meeting), T).$$

Integrity constraints can be also exploited to represent conditional obligatoriness and the deontic logic of deadlines, as shown in [7]. For instance, integrity constraints of the kind $\mathbf{H}(B) \rightarrow \mathbf{E}(A)$ are suitable to represent the obligatoriness of A given B, or Deontic logic with deadlines [15].

4   Datalog$^\pm$ Ontologies in $\mathcal{S}$CIFF

W3C has supported the development of a family of knowledge representation formalisms of increasing complexity for defining ontologies, called Web Ontology Language (OWL). DLs were chosen as the logic-based counterpart for the OWL family of languages.

In the Computational Logic realm, more recently, [18, 19] proposed Datalog$^\pm$, an extension of Datalog with existential rules for representing lightweight ontologies, encompassing the DL-Lite family, and achieving decidability and tractability [17] under appropriate syntactic conditions.

In short, Datalog$^\pm$ extends Datalog by allowing existential quantifiers, the equality predicate and the constant *false* in rule heads. Any Datalog$^\pm$ theory may, in fact, include three types of implication rules: Tuple-Generating Dependencies (TGDs), Negative Constraints (NCs) and Equality Generating Dependencies (EGDs), as shown in the following. In standard Datalog, we can represent a rule stating that the father X of any person Y is also a person:

$$\forall X \forall Y \, fatherOf(X, Y) \wedge person(Y) \rightarrow person(X)$$

In Datalog$^\pm$, we get higher expressiveness, and by a TGD rule we can state that any person X has a father Y who is also a person:

$$\forall X \, person(X) \rightarrow \exists Y \, fatherOf(Y, X) \wedge person(Y)$$

or that for any person X, and any couple of his/her fathers Y and Z, then Y and Z must be the same, by the following EGD:

$$\forall X \, \forall Y \, \forall Z \, fatherOf(Y, X) \wedge fatherOf(Z, X) \rightarrow Y = Z$$

Finally, we can also state, by the following NC, that for any X, his/her father and mother cannot be the same Y:

$$\forall X \, \forall Y \, fatherOf(Y, X) \wedge motherOf(Y, X) \rightarrow false$$

Declaratively, given a finite set of relation names $\mathcal{R}$, a Datalog$^\pm$ theory T (a set of TGDs, NCs and EGDs) on $\mathcal{R}$, and a database D (a set of ground atoms) for $\mathcal{R}$, the set of models of D given T, denoted $mods(D, T)$, is the set of all (possibly infinite) databases B such that $D \subseteq B$ and every $F \in T$ is satisfied in B.

In Datalog$^\pm$ the set of answers to a Conjunctive Query (CQ) q on D given T, denoted $ans(q, D, T)$, is the set of all tuples $\mathbf{t}$ such that $\mathbf{t} \in q(B)$ for all $B \in mods(D, T)$. With abuse of notation, we will write $q(\mathbf{t})$ to mean answer $\mathbf{t}$ for q on D given T.

Operationally, Datalog$^\pm$ query answering for CQs and Boolean Conjunctive Queries (BCQs) is achieved via the *chase*, a bottom-up algorithm for deriving atoms entailed by a given database D and a Datalog$^\pm$ theory. Informally, the chase works on the database D and extends it through the so-called TGD and EGD chase rules. When the body of a TGD is true in the database D, the TGD chase rule adds to D new atomic formulas corresponding to TGD's heads with new (null) variables for the existential ones, that are not already in D. The EGD chase rule, when the body of an EGD is true in the database D, tries to unify the two terms implied in the equality in the EGD's head. A *hard violation* is raised if the unification fails. A more formal description can be found in [17,18].

In [30,31] we showed that, by suitably extending the $\mathcal{S}$CIFF proof-procedure, we are able to represent in $\mathcal{S}$CIFF a Datalog$^\pm$ program, and to use it for ontological reasoning. $\mathcal{S}$CIFF abductive declarative semantics provides the model-theoretic counterpart to Datalog$^\pm$ semantics. Operationally, query answering is achieved bottom-up via the *chase* in Datalog$^\pm$, while in the ALP framework it is supported by the $\mathcal{S}$CIFF proof procedure, which uses both a top-down, backward reasoning from the goal, and a forward reasoning for integrity constraints.

In Datalog$^\pm$, tuples can be added to the database through the TGD chase rule, and unifications can be done via the EGD chase rule. As explained earlier, we mimic the chase through the propagation of $\mathcal{S}$CIFF integrity constraints. The $\mathcal{S}$CIFF syntax for integrity constraints is extended to allow for **H** literals in the head of integrity constraints. **H** atoms are now considered as abducible atoms, so that, through the propagation of integrity constraints, they are assumed as true if they occur in the head of a (transformed) TGD. Coherently with both the Datalog$^\pm$ and $\mathcal{S}$CIFF syntax, variables that occur only in the head of an IC and that occur in a **H** literal are implicitly existentially quantified.

The finite set of relation names of a Datalog$^\pm$ relational schema $\mathcal{R}$ is mapped into the set of terms occurring in the **H** predicates of the corresponding $\mathcal{S}$CIFF program. A Datalog$^\pm$ database D for $\mathcal{R}$ corresponds to the (possibly infinite) $\mathcal{S}$CIFF history **HAP**, since there is a one-to-one correspondence between each tuple in D and each (ground) fact in **HAP**. This mapping may seem unintuitive from an ALP viewpoint, since intensional predicate definitions are mapped into integrity constraints; on the other hand, as will be clear shortly, it lets one reuse the same implications used in a Datalog$^\pm$ theory.

In fact, a Datalog$^\pm$ theory T is mapped into a $\mathcal{S}$CIFF program with an empty KB, and $IC = \tau(T)$, where ICs have (conjunctions of) **H** atoms and CLP constraints (equalities in particular), or *false* in their heads, and are obtained by the τ mapping from the original TGDs, EGDs, and NCs.

The τ mapping is recursively defined as follows, where A is an atom, and F_1, F_2, ..., are Datalog$^\pm$ formulae:

$$\tau(Body \rightarrow Head) = \tau(Body) \rightarrow \tau(Head)$$
$$\tau(A) = \mathbf{H}(A)$$
$$\tau(F_1 \wedge F_2) = \tau(F_1) \wedge \tau(F_2)$$
$$\tau(false) = false$$
$$\tau(Y_i = Y_j) = Y_i = Y_j$$
$$\tau(\exists \mathbf{X}\ A) = \mathbf{H}(A)$$

where the last equation comes from the fact that the quantification for variables that occur only in **H** literals in the head of an IC is always existential, and it is implicit in the $\mathcal{S}$CIFF syntax.

A Datalog TGD $F = body \rightarrow head$ is mapped into the $\mathcal{S}$CIFF integrity constraint $IC = \tau(F)$, where the *body* is mapped into conjunctions of $\mathcal{S}$CIFF atoms, and *head* into conjunctions of $\mathcal{S}$CIFF abducible **H** atoms. Existential quantifications of variables occurring in the *head* of the TGD are maintained in the head of the $\mathcal{S}$CIFF IC, but they are left implicit in the $\mathcal{S}$CIFF syntax, while the rest of the variables are universally quantified with scope the entire IC.

Finally, Datalog$^\pm$ NCs are mapped into $\mathcal{S}$CIFF ICs with head *false*, and EGDs into $\mathcal{S}$CIFF ICs, each one with an equality CLP constraint in its head.

Other possible mappings could be considered; interesting research directions could be to map the database D to the KB, and/or atoms to normal abducible predicates, instead of **H** events.

Given a Datalog$^\pm$ theory T, let us denote the mapping of T into the corresponding set $\mathcal{IC}$ of $\mathcal{S}$CIFF integrity constraints, as $\mathcal{IC} = \tau(T)$.

Recall that, given a Datalog$^\pm$ theory T on $\mathcal{R}$, and a database D for $\mathcal{R}$, the set of models of D given T, denoted $mods(D, T)$, is the set of all (possibly infinite) databases B such that $D \subseteq B$ and every $F \in T$ is satisfied in B. For any such database B, in [30,31], we have proved that there exists an abductive explanation $\mathbf{HAP}' = \tau(B)$ such that:

$$\mathbf{HAP'} \models \mathcal{IC}$$

where $\mathbf{HAP'} \supseteq \mathbf{HAP} = \tau(D)$, and $\mathcal{IC} = \tau(T)$, in which Datalog$^\pm$ nulls are mapped to existentially quantified variables.

Therefore, informally speaking, the set of models of D given T, $mods(D, T)$, corresponds to the set of all the abductive explanations $\mathbf{HAP'}$ satisfying the set of $\mathcal{S}$CIFF integrity constraints $\mathcal{IC} = \tau(T)$.

In [30,31], we have stated and proved theorems for (model-theoretic) completeness of query answering. Informally, we recall here the completeness result for CQ-answering: for each answer $q(\mathbf{t})$ of a CQ $q(\mathbf{X}) = \exists\mathbf{Y}\Phi(\mathbf{X},\mathbf{Y})$ on D given T, in the corresponding $\mathcal{S}$CIFF program $\langle\emptyset, \mathcal{A}, \tau(T)\rangle$ there exists an answer substitution θ and an abductive explanation $\mathbf{HAP'}$ for goal $G = \tau(\Phi(\mathbf{X}, \_))$ (where the underscore stands for an unnamed variable) such that:

$$\langle\emptyset, \mathcal{IC}\rangle \models^g_{\mathbf{HAP}} G\theta$$

where $\mathbf{HAP} = \tau(D)$, $\mathcal{IC} = \tau(T)$, and $G\theta = \tau(\Phi(\mathbf{t}, \_))$,

The $\mathcal{S}$CIFF proof procedure was proved sound and complete w.r.t. $\mathcal{S}$CIFF declarative semantics in [4], thus for each abductive explanation δ for a given goal G in a $\mathcal{S}$CIFF program, there exists a $\mathcal{S}$CIFF-based computation producing a set of abducibles (positive expectations to our purposes) $\delta' \subseteq \delta$, and a computed answer substitution for goal G possibly more general than θ.

Example 1 (Real estate information extraction system in ALP). In [32], the authors present a simple ontology for a real estate information extraction system[1]:

$F_1 = ann(X, label), ann(X, price), visible(X) \rightarrow priceElem(X)$

If X is annotated as a label, as a price and is visible, then it is a price element.

$F_2 = ann(X, label), ann(X, priceRange), visible(X) \rightarrow priceElem(X)$

If X is annotated as a label, as a price range, and is visible, then it is a price element.

$F_3 = priceElem(E), group(E, X) \rightarrow forSale(X)$

If E is a price element and is grouped with X, then X is for sale.

$F_4 = forSale(X) \rightarrow \exists P\ price(X, P)$

If X is for sale, then there exists a price for X.

$F_5 = hasCode(X, C), codeLoc(C, L) \rightarrow loc(X, L)$

If X has postal code C, and C's location is L, then X's location is L.

$F_6 = hasCode(X, C) \rightarrow \exists L\ codeLoc(C, L), loc(X, L)$

If X has postal code C, then there exists L s.t. C has location L and so does X.

$F_7 = loc(X, L1), loc(X, L2) \rightarrow L1 = L2$

If X has the locations $L1$ and $L2$, then $L1$ and $L2$ are the same.

$F_8 = loc(X, L) \rightarrow advertised(X)$

If X has a location L then X is advertised.

The TGDs $F_1 - F_8$ from the Datalog$^\pm$ ontology above are one-to-one mapped into the following $\mathcal{S}$CIFF ICs:[2]

---

[1] The universal quantifiers are usually left implicit.

[2] We show for the sake of clarity the quantification of existentially quantified variables, although in the $\mathcal{S}$CIFF syntax the quantification is implicit.

$IC_1 : \mathbf{H}(ann(X, label)), \mathbf{H}(ann(X, price)), \mathbf{H}(visible(X)) \rightarrow \mathbf{H}(priceElem(X))$

$IC_2 : \mathbf{H}(ann(X, label)), \mathbf{H}(ann(X, priceRange)), \mathbf{H}(visible(X)) \rightarrow \mathbf{H}(price$
        $Elem(X))$

$IC_3 : \mathbf{H}(priceElem(E)), \mathbf{H}(group(E, X)) \rightarrow \mathbf{H}(forSale(X))$

$IC_4 : \mathbf{H}(forSale(X)) \rightarrow (\exists P)\ \mathbf{H}(price(X, P))$

$IC_5 : \mathbf{H}(hasCode(X, C)), \mathbf{H}(codeLoc(C, L)) \rightarrow \mathbf{H}(loc(X, L))$

$IC_6 : \mathbf{H}(hasCode(X, C)) \rightarrow (\exists L)\ \mathbf{H}(codeLoc(C, L)), \mathbf{H}(loc(X, L))$

$IC_7 : \mathbf{H}(loc(X, L1)), \mathbf{H}(loc(X, L2)) \rightarrow L1 = L2$

$IC_8 : \mathbf{H}(loc(X, L)) \rightarrow \mathbf{H}(advertised(X))$

The database in [32] is mapped into the following history **HAP**:

$\{\mathbf{H}(codeLoc(ox1, central)), \mathbf{H}(codeLoc(ox1, south)),$
  $\mathbf{H}(codeLoc(ox2, summertown)), \mathbf{H}(hasCode(prop1, ox2)), \mathbf{H}(ann(e1, price)),$
  $\mathbf{H}(ann(e1, label)), \mathbf{H}(visible(e1)), \mathbf{H}(group(e1, prop1))\}$

The $\mathcal{S}CIFF$ proof procedure applies ICs in a forward manner, and it infers the following set of abducibles from the program above:

$$\mathbf{HAP}' = \{\mathbf{H}(priceElem(e1)), \mathbf{H}(forSale(prop1)), \exists P\ \mathbf{H}(price(prop1, P)),$$
$$\mathbf{H}(loc(prop1, summertown)), \mathbf{H}(advertised(prop1))\}$$

Each of the (ground) atomic queries (BCQs) outlined in [32] is also entailed in the $\mathcal{S}CIFF$ program above. In particular, for the previous set **HAP**':

$$\mathbf{HAP}' \models \mathbf{H}(priceElem(e1)), \mathbf{H}(forSale(prop1)), \mathbf{H}(advertised(prop1))$$

Also, the CQ $\exists L\ \mathbf{H}(loc(prop1, L))$ is entailed as well (with unification $L = summertown$, as in [32]) since:

$$\mathbf{HAP}' \models \mathbf{H}(loc(prop1, summertown))$$

4.1   Related Work

A very related approach for mapping DL theories into ALP is contained in [48]. The authors consider $\mathcal{ALCN}$ Description Logics theories, i.e., ontologies where the terminological part consists of concept definitions of kind:

$$C \equiv F$$

where C is a concept symbol and F is a (non recursive) concept description. Given that R is a role, C a concept symbol and F, G concepts descriptions, valid concept descriptions are the terms:

$$C \mid \forall R.F \mid \exists R.F \mid F \cap G \mid F \cup G \mid \neg F \mid \leq nR \mid \geq nR$$

Any concept symbol C occurring in a concept definition of kind $C \equiv F$ is named a defined concept, otherwise it is named primitive.

Their mapping transforms concept definitions of such a kind into Open Logic Programming clauses [21], an extension of ALP, and they prove that this operation is equivalence preserving. They map concept definitions to program clauses, first translating the definitions into *general clauses*, where the head is an atom and the body any First Order Logic expression. Then, general clauses are transformed into a set of Horn clauses, by the Lloyd-Topor transformation [39].

Unluckily, in their mapping, they lose half of the definition. A concept definition of kind $C \equiv F$ is, in fact, transformed into a general clause of kind:

$$C(X) \leftarrow T'(F, X)$$

where T' is a mapping transformation inductively defined over the syntax of F (see above).

For instance, given the ontological definition stating that a father is any male person having at least one child:

$$Father \equiv Male \cap Person \cap \exists child.Person$$

their mapping produces only the clause:

$$father(X) \leftarrow male(X), person(X), child(X, Y), person(Y)$$

thus losing the second part of the definition, which corresponds to:

$$father(X) \rightarrow male(X), person(X), \exists Y(child(X, Y), person(Y)).$$

In this sense, even if [48] captures the intuition that DL theories, with the Open World Assumption approach, have much in common with Abductive or Open Logic Programming, the mapping and representation they provide do not support the notion of *concept definition* which is peculiar of terminological systems, and which lets one reason both ways when considering a definition $C \equiv F$.

The major issue for mapping DL theories into Logic Programming languages is exactly the need to represent implications having in their heads existentially quantified variables (as the example above outlines). Datalog$^\pm$ is, instead, a logic programming language enriched with implications having also existentially quantified variables in their heads. This feature is fundamental to fully represent even the simplest $\mathcal{ALC}$ Description Logic.

Moreover, Datalog$^\pm$ conflates logic programming clauses, forward rules and integrity constraints, thus proving an integration of ontological reasoning with rule-based programing, in a single language.

As well, $\mathcal{S}$CIFF is a language naturally providing the same syntax extension, and smoothly integrating into ALP both ontological representation and rule-based programming.

5   Integrating Norms and Ontologies in $\mathcal{S}$CIFF

After mapping a Datalog$^\pm$ ontology into $\mathcal{S}$CIFF, we are now ready to show how normative reasoning is smoothly integrated with ontological reasoning within the

$\mathcal{S}$CIFF abductive framework. We will show this via a simple example, starting from the real estate ontology, and enriching it with normative rules for some interacting agents in a real estate scenario.

As discussed in Sect. 3, the set of regulations holding in the given society of agents can be expressed again through integrity constraints.

In the previous example, suppose that a real estate agent (rea) is in charge of selling a property, and it uses the data from the real estate information extracted from the ontological knowledge. If another agent has seen an announcement that a property (e.g., a flat) is for sale, it can request to buy it. In such a case, for some fixed time ΔT, the real estate agent is obliged to reserve the flat for that client. For ΔT time units, the real estate agent cannot sell the flat to other agents and, if the buyer issues a payment, the flat must be declared sold (within some deadline D). The expected behavior of the real estate agent can be defined by the following rules:

$$\mathbf{H}(tell(X, rea, buy(E), T), \mathbf{H}(forSale(E))$$
$$\rightarrow \mathbf{EN}(sell(rea, Y, E), T_s), Y \neq X, T_s < T + \Delta T$$

$$\mathbf{H}(tell(X, rea, buy(E), T), \mathbf{H}(forSale(E)), \mathbf{H}(price(E, P)),$$
$$\mathbf{H}(pay(X, rea, P), T_p), T_p < T + \Delta T \rightarrow \mathbf{E}(sell(rea, X, E), T_s), T_s < T_p + D.$$

If some agent e asks to buy property $prop1$ at price $e1$ at time 1, i.e., $\mathbf{H}(tell(e, rea, buy(prop1), 1)$, the proof procedure is able to infer the following information about the expected behavior of the rea agent:

$$\forall T_s \text{ s.t. } T_s < 1 + \Delta T, \ \forall Y \neq e \quad \mathbf{EN}(sell(rea, Y, prop1), T_s)$$

i.e., agent rea is not allowed to sell property $prop1$ to any other agent until ΔT time units have passed. Let us suppose that $\Delta T = 5$ and $D = 3$; if agent e actually executes the payment, e.g. $\mathbf{H}(pay(e, rea, e1), 3)$, agent rea is now expected to sell $prop1$, and the following expectation is raised:

$$\exists T_s \text{ s.t. } T_s < 6 \quad \mathbf{E}(sell(rea, e, prop1), T_s).$$

In this way, not only the $\mathcal{S}$CIFF proof procedure is able to infer the knowledge from the ontological database, but also to provide the expected behavior of the agents (in our example, the rea agent) including obligations and prohibitions.

6 Conclusions and Future Work

We have shown that Abductive Logic Programming (ALP) is a powerful tool for knowledge representation and reasoning about norms and ontologies. We have focused in detail about the $\mathcal{S}$CIFF ALP framework, used in the past to model and verify agent societies, interaction protocols for multi-agent systems [8], Web services choreographies [1], but also powerful enough to represent deontic operators [7] and normative systems [5]. Its underlying $\mathcal{S}$CIFF proof procedure [4], derived from the IFF one, considers, in rule heads, atoms that can contain existentially and universally quantified variables, as well as CLP constraints. Recently [30,31], $\mathcal{S}$CIFF has been also proved useful for representing ontologies expressed in Datalog$^{\pm}$.

In this work, we have exploited the $\mathcal{S}$CIFF framework for representing and integrating both norms and Datalog$^{\pm}$ ontologies. Normative, rule-based reasoning and ontological query answering are obtained by applying the $\mathcal{S}$CIFF abductive proof procedure. Both norms and a Datalog$^{\pm}$ theory can be encoded as $\mathcal{S}$CIFF integrity constraints. The main advantage is that this integration is achieved within a single language, grounded on ALP. Nonetheless, different ALP approaches might be possible.

A number of issues are subject of future work. First of all, we have not focused here on complexity results. Identifying syntactic conditions that guarantee tractable ontologies in $\mathcal{S}$CIFF, in the style of what has been done for Datalog$^{\pm}$, is crucial for achieving nice computational performance.

A second issue for future work concerns experimentation with real cases, in the normative and legal domain, and on real-size ontologies.

Finally, different mappings of Datalog$^{\pm}$ to ALP might exist, possibly enjoying different properties: another research direction is oriented toward identifying the mapping that suits best for legal reasoning.

Acknowledgements. We would like to thank the anonymous referees for their helpful comments, and JURISIN2015 participants for questions and discussions during the workshop. The second author would like to thank NII, Tokyo (J), for supporting her travel to Japan and for making possible to attend JURISIN2015.

References

1. Alberti, M., Chesani, F., Gavanelli, M., Lamma, E., Mello, P., Montali, M.: An abductive framework for a-priori verification of web services. In: Maher, M. (ed.) Proceedings of the Eighth Symposium on Principles and Practice of Declarative Programming, pp. 39–50. ACM Press, New York (2006)
2. Alberti, M., Chesani, F., Gavanelli, M., Lamma, E., Mello, P., Torroni, P.: Compliance verification of agent interaction: a logic-based software tool. Appl. Artif. Intell. **20**(2–4), 133–157 (2006)
3. Alberti, M., Chesani, F., Gavanelli, M., Lamma, E., Mello, P., Torroni, P.: Security protocols verification in abductive logic programming: a case study. In: Dikenelli, O., Gleizes, M.-P., Ricci, A. (eds.) ESAW 2005. LNCS (LNAI), vol. 3963, pp. 106–124. Springer, Heidelberg (2006). doi:10.1007/11759683_7
4. Alberti, M., Chesani, F., Gavanelli, M., Lamma, E., Mello, P., Torroni, P.: Verifiable agent interaction in abductive logic programming: the SCIFF framework. ACM Trans. Comput. Logic **9**(4), 29:1–29:43 (2008)
5. Alberti, M., Gavanelli, M., Lamma, E.: $Deon^{+}$: abduction and constraints for normative reasoning. In: Artikis, A., Craven, R., Kesim Çiçekli, N., Sadighi, B., Stathis, K. (eds.) Logic Programs, Norms and Action. LNCS (LNAI), vol. 7360, pp. 308–328. Springer, Heidelberg (2012). doi:10.1007/978-3-642-29414-3_17
6. Alberti, M., Gavanelli, M., Lamma, E.: The CHR-based implementation of the SCIFF abductive system. Fundamenta Informaticae **124**(4), 365–381 (2013)

7. Alberti, M., Gavanelli, M., Lamma, E., Mello, P., Sartor, G., Torroni, P.: Mapping deontic operators to abductive expectations. Comput. Math. Organ. Theory **12**(2–3), 205–225 (2006)
8. Alberti, M., Gavanelli, M., Lamma, E., Mello, P., Torroni, P.: Specification and verification of agent interactions using social integrity constraints. Electr. Notes Theor. Comput. Sci. **85**(2), 94–116 (2003)
9. Alberti, M., Gomes, A.S., Gonçalves, R., Leite, J., Slota, M.: Normative systems represented as hybrid knowledge bases. In: Leite, J., Torroni, P., Ågotnes, T., Boella, G., Torre, L. (eds.) CLIMA 2011. LNCS (LNAI), vol. 6814, pp. 330–346. Springer, Heidelberg (2011). doi:10.1007/978-3-642-22359-4_23
10. ALFEBIITE: a logical framework for ethical behaviour between infohabitants in the information trading economy of the universal information ecosystem. IST-1999-10298 (1999)
11. Arisha, K.A., Ozcan, F., Ross, R., Subrahmanian, V.S., Eiter, T., Kraus, S.: IMPACT: a platform for collaborating agents. IEEE Intell. Syst. **14**(2), 64–72 (1999)
12. Boella, G., van der Torre, L., Verhagen, H.: Introduction to normative multiagent systems. Comput. Math. Organ. Theory **12**, 71–79 (2006)
13. Boella, G., van der Torre, L.: Attributing mental attitudes to normative systems. In: Rosenschein, J.S., Sandholm, T., Wooldridge, M., Yokoo, M. (eds.) Proceedings of AAMAS-2003, 14–18 July 2003, pp. 942–943. ACM Press (2003)
14. Bracciali, A., Demetriou, N., Endriss, U., Kakas, A., Lu, W., Mancarella, P., Sadri, F., Stathis, K., Terreni, G., Toni, F.: The KGP model of agency for global computing: computational model and prototype implementation. In: Priami, C., Quaglia, P. (eds.) GC 2004. LNCS, vol. 3267, pp. 340–367. Springer, Heidelberg (2005). doi:10.1007/978-3-540-31794-4_18
15. Broersen, J., Dignum, F., Dignum, V., Meyer, J.-J.C.: Designing a deontic logic of deadlines. In: Lomuscio, A., Nute, D. (eds.) DEON 2004. LNCS (LNAI), vol. 3065, pp. 43–56. Springer, Heidelberg (2004). doi:10.1007/978-3-540-25927-5_5
16. Bürckert, H.: A resolution principle for constrained logics. Artif. Intell. **66**, 235–271 (1994)
17. Calì, A., Gottlob, G., Kifer, M.: Taming the infinite chase: query answering under expressive relational constraints. In: International Conference on Principles of Knowledge Representation and Reasoning, pp. 70–80. AAAI Press (2008)
18. Calì, A., Gottlob, G., Lukasiewicz, T.: A general datalog-based framework for tractable query answering over ontologies. In: Symposium on Principles of Database Systems, pp. 77–86. ACM (2009)
19. Calì, A., Gottlob, G., Lukasiewicz, T.: Tractable query answering over ontologies with Datalog$^{\pm}$. In: International Workshop on Description Logics, CEUR Workshop Proceedings, vol. 477. CEUR-WS.org (2009)
20. Clark, K.L.: Negation as failure. In: Gallaire, H., Minker, J. (eds.) Logic and Data Bases, pp. 293–322. Plenum Press, New York (1978)
21. Console, L., Theseider Dupré, D., Torasso, P.: On the relationship between abduction and deduction. J. Logic Comput. **1**(5), 661–690 (1991)
22. De Vos, M., Padget, J., Satoh, K.: Legal modelling and reasoning using institutions. In: Onada et al. [42], pp. 129–140 (2010)
23. Denecker, M., De Schreye, D.: SLDNFA: an abductive procedure for abductive logic programs. J. Logic Program. **34**(2), 111–167 (1998)
24. Dignum, V., Meyer, J.J., Weigand, H.: Towards an organizational model for agent societies using contracts. In: Castelfranchi, C., Lewis Johnson, W. (eds.) AAMAS-2002, 15–19 July 2002, pp. 694–695. ACM Press, Bologna, Italy (2002)

25. Dignum, V., Meyer, J.J., Weigand, H., Dignum, F.: An organizational-oriented model for agent societies. In: Proceedings of International Workshop on Regulated Agent-Based Social Systems: Theories and Applications, AAMAS 2002, Bologna (2002)

26. Du, J., Wang, K., Shen, Y.D.: A tractable approach to ABox abduction over description logic ontologies. In: Brodley, C., Stone, P. (eds.) Proceedings of AAAI 2014 (2014)

27. Eiter, T., Subrahmanian, V., Pick, G.: Heterogeneous active agents, I: semantics. Artif. Intell. **108**(1–2), 179–255 (1999)

28. ESTRELLA: European project for standardized transparent representations in order to extend legal accessibility. IST-2004-027655 (2004). http://www.estrellaproject.org

29. Fung, T.H., Kowalski, R.A.: The IFF proof procedure for abductive logic programming. J. Logic Program. **33**(2), 151–165 (1997)

30. Gavanelli, M., Lamma, E., Riguzzi, F., Bellodi, E., Zese, R., Cota, G.: An abductive framework for Datalog$^\pm$ ontologies. In: Eiter, T., Toni, F. (eds.) Technical Communications of the 31st International Conference on Logic Programming (ICLP 2015). CEUR Workshop Proceedings, Sun SITE Central Europe, Aachen, Germany (2015)

31. Gavanelli, M., Lamma, E., Riguzzi, F., Bellodi, E., Zese, R., Cota, G.: Abductive logic programming for Datalog$^\pm$ ontologies. In: Ancona, D., Maratea, M., Mascardi, V. (eds.) Proceedings of the 30th Italian Conference on Computational Logic (CILC2015), Genova, Italy, 1–3 July 2015. CEUR Workshop Proceedings, vol. 1459, pp. 128–143. Sun SITE Central Europe, Aachen (2015)

32. Gottlob, G., Lukasiewicz, T., Simari, G.I.: Conjunctive query answering in probabilistic datalog+/− ontologies. In: Rudolph, S., Gutierrez, C. (eds.) RR 2011. LNCS, vol. 6902, pp. 77–92. Springer, Heidelberg (2011). doi:10.1007/978-3-642-23580-1_7

33. Governatori, G., Rotolo, A.: BIO logical agents: norms, beliefs, intentions in defeasible logic. Auton. Agents Multi-agent Syst. **17**(1), 36–69 (2008)

34. Jaffar, J., Maher, M.: Constraint logic programming: a survey. J. Logic Program. **19–20**, 503–582 (1994)

35. Kakas, A.C., Kowalski, R.A., Toni, F.: Abductive logic programming. J. Logic Comput. **2**(6), 719–770 (1993)

36. Kakas, A.C., Mancarella, P.: On the relation between truth maintenance and abduction. In: Fukumura, T. (ed.) Proceedings of PRICAI-90. Ohmsha Ltd. (1990)

37. Kakas, A.C., van Nuffelen, B., Denecker, M.: $\mathcal{A}$-System: problem solving through abduction. In: Nebel, B. (ed.) Proceedings of IJCAI-01 (2001)

38. Klarman, S., Endriss, U., Schlobach, S.: ABox abduction in the description logic $\mathcal{ALC}$. J. Autom. Reason. **46**(1), 43–80 (2011)

39. Lloyd, J., Topor, R.: Making prolog more expressive. J. Logic Program. **1**(3), 225–240 (1984)

40. Lloyd, J.W.: Foundations of Logic Programming. Springer, Heidelberg (1987)

41. Montali, M., Torroni, P., Alberti, M., Chesani, F., Gavanelli, M., Lamma, E., Mello, P.: Verification from declarative specifications using logic programming. In: Garcia de la Banda, M., Pontelli, E. (eds.) ICLP 2008. LNCS, vol. 5366, pp. 440–454. Springer, Heidelberg (2008). doi:10.1007/978-3-540-89982-2_39

42. Onada, T., Bekki, D., McCready, E. (eds.): JSAI-isAI 2010. LNCS (LNAI), vol. 6797. Springer, Heidelberg (2011)

43. Prakken, H., Sartor, G.: Argument-based extended logic programming with defeasible priorities. J. Appl. Non-classical Logics **7**(1), 25–75 (1997)

44. Sadri, F., Stathis, K., Toni, F.: Normative KGP agents. Comput. Math. Organ. Theory **12**(2–3), 101–126 (2006)
45. Sartor, G., Casanovas, P., Biasiotti, M.A., Fernández-Barrera, M. (eds.): Approaches to Legal Ontologies. Law, Governance and Technology, Springer, Netherlands (2011)
46. Satoh, K., Asai, K., Kogawa, T., Kubota, M., Nakamura, M., Nishigai, Y., Shirakawa, K., Takano, C.: PROLEG: an implementation of the presupposed ultimate fact theory of Japanese civil code by PROLOG technology. In: Onada et al. [42], pp. 153–164 (2010)
47. Sergot, M.J., Sadri, F., Kowalski, R.A., Kriwaczek, F., Hammond, P., Cory, H.T.: The British nationality act as a logic program. Commun. ACM **29**, 370–386 (1986)
48. Van Belleghem, K., Denecker, M., De Schreye, D.: A strong correspondence between description logics and open logic programming. In: Naish, L. (ed.) Proceedings of Fourteenth International Conference on Logic Programming. MIT Press, Cambridge (1997)
49. Wright, G.: Deontic logic. Mind **60**, 1–15 (1951)

A Belief Revision Technique to Model Civil Code Updates

Ryuta Arisaka[✉]

National Institute of Informatics, Chiyoda, Japan
ryutaarisaka@gmail.com

Abstract. A scenario that was not considered at the time of enforcing a civil code article may be discovered later. In case application of the civil code article in the discovered scenario is not consistent with the intention of the article, it is necessary that the article be appropriately updated. We show that this kind of civil code update that is induced upon reaction to augmentation of knowledge can be modelled in a belief revision theory. We develop our formal framework, and show one instantiation of the framework with case application of a civil code article.

1 Introduction

A scenario that was not considered at the time of enforcing a civil code article may be discovered later. In the event that application of the civil code article in the discovered scenario is in conflict with the intention of the article, it is necessary that the article be appropriately updated.

A previous Japanese civil case over lease contract termination is an exemplar. According to the Japanese Civil Code Article 612 which states:

- A lessee may not assign the lessee's rights or sublease a leased thing without obtaining the approval of the lessor [Paragraph 1]; and
- If the lessee allows any third party to make use of or take profits from a leased thing in violation of the provisions of the preceding paragraph, the lessor may cancel the contract [Paragraph 2],

when there is a contract between A as a lessor and B as a lessee, A is granted the right to terminate the contract with B if B has allowed a third party to make use of the leased property or has subleased it without obtaining approval of A for doing so. The Supreme Court ruled (Supreme Court Case:1966.1.27, 20-1 Minsyu 136), however, that the article should really be interpreted as: a lessor A may terminate a contract with a lessee B if B has either transferred B's rights as a lessee or has subleased the leased property to a third party without A's approval, and *if such action as taken by B has undermined A's trust in B so irreparably that it would be no longer possible for A to continue the contract.* In other words, where there is no evidence of the abuse of A's confidence in B, even if B transfers the rights as a lessee or subleases the leased property to a

© Springer International Publishing AG 2017
M. Otake et al. (Eds.): JSAI-isAI 2015 Workshops, LNAI 10091, pp. 204–216, 2017.
DOI: 10.1007/978-3-319-50953-2_15

third party, the Article 612 is not applicable, and A will not be granted the legal right to terminate the contract.

An interesting aspect about this civil case is that, while it very much appears to highlight the need for default reasoning, it is not correct to say that the Article 612 normally applies unless there is evidence of no abuse of confidence. That would not justify the point that the Article 612 ceases to apply after the Supreme Court Case wherever there is shown no evidence of abuse of confidence. Rather, we should view it as a particular kind of update on reaction to knowledge augmentation. To expound, let us suppose that enforcement of a civil code article is done at time t_1, and that we stand at some future time t_2. At both t_1 and t_2, some knowledge concerning the civil code article is available, say K_1 and, respectively, K_2. It is reasonable to assume that the intention of the article is adequately expressed in what the article states at time t_1. The adequacy, however, is clearly relative to K_1. Suppose some knowledge, say K_3, then. If for instance K_2 is the summation of K_1 and K_3, it could be that K_3 consists of some scenarios relevant to the article which were not considered at t_1. In that case there is clearly no guarantee that what the article states, which sufficiently characterises its intention under K_1, also does under K_2. And if there should occur any mismatches between the intention of the article and what the article states, updating the article by taking K_3 into account is an appropriate action to resolving them. It is this need that better portrays the above-mentioned civil case example.

Detailed elaboration. Let t_1 be the time at which the Civil Code Article 612 was enforced. There is certain knowledge K_1 at t_1 under which the Article was written. The Article should be codification of certain intention **I**, and should preserve one-to-one correspondence to it.

If it were possible at all to codify **I** directly, that is, not through any medium, then it would be trivially possible to attain the one-to-one correspondence, and the Article that could be so derived would never need to be altered: the fictitious Article at t_1 is then the right codification of **I** not just at t_1 but also at any future time t_x to come after it. However, this can never materialise, for the codification of **I** requires, on the part of anybody who wants to codify it, knowledge of a language upon which the very act of codification depends, as well as knowledge to know or see the entirety of **I**.

Given this consideration, the Civil Code Article 612 that is codified at t_1 is actually $\mathbf{I}_1$ which is some observation of **I** under K_1. Insofar as $\mathbf{I}_1$ is indistinguishable from **I** under K_1, i.e. $\mathbf{I}_1 =_{K_1} \mathbf{I}$, the Civil Code Article 612 is an appropriate codification of **I** at t_1. To be able to see that it is actually not adequate, that is, to see the difference between $\mathbf{I}_1$ and **I**, the knowledge with which **I** is observed should be larger than K_1. The exceptional case in the given civil case example is just such addition to it. Let t_2 be the moment of discovery of the exception, let K_2 be the new knowledge at t_2, and let $\mathbf{I}_2$ be some observation of **I** under it. At t_2 it holds true that $\mathbf{I}_1 \neq_{K_2} \mathbf{I} =_{K_2} \mathbf{I}_2$. Hence the Civil Code Article 612 which is supposed to be codification of **I** should cover $\mathbf{I}_2$ instead of $\mathbf{I}_1$. The Supreme Court Ruling facilitated the alignment.

In this work, we show that we can model this kind of update on reaction to the discovery of unconsidered scenarios by adapting the principle of the belief revision theory with latent beliefs [2]. Roughly, a belief revision theory [1,4,11] is a theory that minimally updates a consistent database upon addition of new data or removal of existing data into another consistent database. Each datum is a formal expression, and the minimality is measured (ensured) by a set of conditions. As in [2] we will have two types of data. One of them will comprise civil code articles and presented facts, both represented as sentences in first-order logic, whereas the other will comprise code article update instructions which are triggered only if some trigger condition (which is a fact) logically follows from combination of the civil code articles and the presented facts. Being a belief revision theory, our framework can also deal with other types of updates such as deletion of a code article. We will: formalise our belief revision technique (in Sect. 2); show an instantiation of our framework with the example of the Article 612 we touched upon informally in this section (in Sect. 3); and compare our approach with relevant background in the literature (in Sect. 4), before drawing conclusions.

2 A Belief Revision Theory on Code Articles, Facts and Exceptions

We make use of first-order logic with equality but without function symbols of arity greater than or equal to 1, specifically the following languages $\mathbf{K}$ with or without a subscript ($\mathbf{K}, \mathbf{K_1}, \mathbf{K_2}, \ldots$) consisting of: (1) a fixed set of logical symbols, which we denote by **Log**; and (2) a finite number of non-logical symbols, which we denote by **NonLog** with or without a subscript.

The fixed **Log** contains the following items.

- $\forall, \exists, \wedge, \vee, \supset, \neg, \top$. These are the usual first-order logic symbols. $\supset$ is material implication.
- An equality symbol $=$.
- Parentheses, brackets, and punctuation symbols.
- An infinite number of variables, each of which is referred to by x with or without a subscript.

A **NonLog** contains the following sets:

- **Const**: consists of a finite number of finite sequences of English letters in Italic font beginning with a small letter; e.g. *lessor*, *lessee*. Each such sequence of letters is called a constant.
- **Pred**: consists of a finite number of finite sequences of English letters in sans-serif font beginning with a small letter; e.g. allowUse, sublease with some number of arguments. Each such sequence of letters with n arguments is called a n-ary predicate symbol.

The formulas are defined in the usual way, and the binding precedence, the free/bound variables, the semantics, and the satisfiability follow the standard convention. A formula without a free variable is a sentence. We denote a sentence by F with or without a subscript. We say that F in some language $\mathbf{K}$ is a logical consequence of a set of sentences $F_1, F_2, \ldots$ in the same language iff any model of the set of sentences conjunctively understood is also a model of F. Hereafter, we presume a consequence operator Cn that satisfies the following two conditions.

1. $Cn(\{F_1, F_2, \ldots\})$ contains all the sentences each of which is a logical consequence of $\{F_1, F_2, \ldots\}$, but contains no other formulas (Logical closure).
2. If F is in $Cn(\{F_1, F_2, \ldots\})$, then there exists a finite subset X of $\{F_1, F_2, \ldots\}$ such that F is in $Cn(X)$. (Compactness).

2.1   Belief Sets and New Information

The belief revision theories [1,2,4,11] define postulates about belief changes, stating how a particular set of beliefs should rationally change as new information is added to it or existing information is removed from it. Let us define a belief set in our context. Let B with or without a subscript denote a pair: $(Cn(\mathbf{Articles} \cup \mathbf{Facts}), \mathbf{Exceptions})$ where

- **Articles** is a finite set of code articles formalised as sentences.
- **Facts** is a finite set of (ultimate) facts [9,12,13] again as sentences.
- Both **Articles** and **Facts** are sentences in some language $\mathbf{K}$. However, the sentences that belong to **Articles** are kept disjoint from the sentences that belong to **Facts**.
- **Exceptions** is a finite set of triples $F_1[F_a, F_2]$ for some sentence F_1 in **Articles**, and some sentences F_a and F_2 in some possibly different language(s) than the above language $\mathbf{K}$.

Example 1. For an example of B, consider: (1) The Japanese Civil Code Article 612 for **Articles**; (2) the fact that a lessee has subleased a leased house to his friend for **Facts**; and (3) no exceptions for **Exceptions**. Then we can for instance formalise them in the pair: $(Cn(\{F_1\} \cup \{F_a\}), \emptyset)$, where

- F_1 is $[\exists x_1 \, \exists x_2.x_1 \neq x_2 \wedge (\mathsf{allowUseTo}(lessee, x_1, x_2) \vee \mathsf{subleaseTo}(lessee, x_1, x_2)) \wedge \neg\mathsf{gainPermissionFrom}(lessee, lessor)] \supset \mathsf{cancelContract}(lessor, lessee)$.
- F_a is $\mathsf{subleaseTo}(lessee, lesseeFriend, leasedHouse)$.

Now, let `mapBtoK` be a function such that `mapBtoK`$(B) = (\mathbf{Log}, \mathbf{NonLog})$ and that **NonLog** is the set of all predicates and constants that occur in **Articles** or **Facts** of B. We say that $\mathbf{K}$ is the language of B iff $\mathbf{K} = $ `mapBtoK`(B). When we say 'the first component of' or 'the second component of' B, we mean $Cn(\mathbf{Articles} \cup \mathbf{Facts})$, or respectively **Exceptions** of B. The exact manner in which **Exceptions** interacts with $Cn(\mathbf{Articles} \cup \mathbf{Facts})$ will be formally defined later. Nonetheless, the informal meaning of a triple $F_1[F_a, F_2]$ for some F_1, F_a and F_2 is:

1. It is an instruction to update F_1 in **Articles** to F_2 in case F_a is discovered in $Cn(\textbf{Articles} \cup \textbf{Facts})$. An informal example: let F_1 be a formalisation of the Japanese Civil Code Article 612; let F_b be the formalisation of the fact that confidence is abused; let F_a be $\neg F_b$; and let F_2 be $F_b \supset F_1$. Then $F_1[F_a, F_2]$ says that, if F_1 is in **Articles**, i.e. the civil code includes the Article 612, then if F_a is in $Cn(\textbf{Articles} \cup \textbf{Facts})$, i.e. abuse of confidence is discovered, F_1 updates to F_2 on reaction to the discovery, the applicability of the civil code article getting narrowed down. A triple is disposed of once the update that it is suggesting is conducted.
2. F_a and F_2 do not have to be sentences in the language of B. The rationale behind it is that, while $Cn(\textbf{Articles} \cup \textbf{Facts})$ is supposed to be the currently available knowledge, which is hence in the language of B, F_a may be some exceptional fact that could be revealed only at some point in future, if revealed at all. Naturally, the future potential knowledge is not already visible in the current knowledge, and does not have to be expressible in the language of B. Likewise, the emergence of F_2 depends on the visibility of F_a. In case F_a is currently not in $Cn(\textbf{Articles} \cup \textbf{Facts})$, it does not have to be expressible within the language of B.

Incidentally, in this paper we do not consider how such exceptional facts are constructed by means of for instance abduction. Now, let us list four desirable conditions. We assume $\Gamma(B) = \bigcup_{F_1[F_a, F_2] \in \textbf{Exceptions of } B} \{F_a\}$.

1. If $F_1[F_a, F_2]$ is in **Exceptions** of B, then F_1 is in **Articles** of B (Safety).
2. If $F_1[F_a, F_2]$ is in **Exceptions** of B, then F_a is not in $Cn(\textbf{Articles} \cup \textbf{Facts})$ of B (No-unused-triggers).
3. $Cn(\textbf{Articles} \cup \textbf{Facts})$ is consistent, i.e. if $F \in Cn(\textbf{Articles} \cup \textbf{Facts})$, then $\neg F \notin Cn(\textbf{Articles} \cup \textbf{Facts})$ (Consistency).
4. If $F_1[F_a, F_2]$ is in **Exceptions** of B, then for all $F_1[F_b, F_3]$ also in **Exceptions** of B, if $Cn^*(\{F_a\}) = Cn^*(\{F_b\})$ then $F_2 = F_3$ (Uniqueness). Assume that Cn^* is for the language which extends $\texttt{mapBtoK}(B)$ with any predicate or constant that occurs in some $F \in \Gamma(B)$.

Definition 1 (Belief sets). *We say that a given B is a belief set iff B satisfies (Safety), (No-unused-triggers), (Consistency), and (Uniqueness).*

Let us describe the intuition behind the conditions above. The (Safety) says that in order that we be able to update F_1 to F_2 when a fact F_a is presented/discovered, we must have the code article F_1 that is supposed to be updated already in **Articles**. The intent of the (No-unused-triggers) is to do any update whenever possible. For instance, B may contain just one triple $F_1[F_a, F_2]$ in the second component, and F_1 in **Articles** of the first component. Then the triple is suggesting an update of F_1 into F_2 for B. By (No-unused-triggers), we prevent such an intermediate B that can be still updated from being counted as a belief set. By (Uniqueness) it is guaranteed that every code article can have at most one update trigger per fact at any given moment.[1]

---

[1] It having at most one update trigger per fact does not mean it having at most one update trigger.

Information to be added to or removed from a belief set B can be a code article, a fact, or an exception. It satisfies the following conditions.

1. If it is either a code article or a fact, say F, then
   (a) F is consistent (Consistency of a code article and a fact).
   (b) F is not a tautology (Non-triviality of a code article and a fact).
2. If it is an exception, say $F_1[F_a, F_2]$, then
   (a) F_2 is consistent (Consistency of the updated article).
   (b) F_2 is not a tautology (Non-triviality of the updated article).

2.2 Belief Change Postulates

In this subsection we define belief change operations for our belief sets: **ContractBy** for contracting it by some information `Info`; and **ReviseBy** for revising it.[2] Contraction is informally an operation that removes some information off a belief set to derive a new belief set; and revision is an operation that adds new information to a belief set, ensuring that the result be a belief set which is as consistent as possible. As some preparation, let $+$ be an operator between some $B = (Cn(\textbf{Articles} \cup \textbf{Facts}), \textbf{Exceptions})$ not necessarily a belief set and some `Info` such that

- $B + \texttt{Info} = (Cn(\textbf{Articles}_1 \cup \textbf{Facts}_1), \textbf{Exceptions}_1)$ where
  - If `Info` is a code article, then $\textbf{Articles}_1 = \textbf{Articles} \cup \{\texttt{Info}\}, \textbf{Facts}_1 = \textbf{Facts}$ and $\textbf{Exceptions}_1 = \textbf{Exceptions}$.
  - Else if `Info` is a fact, then $\textbf{Articles}_1 = \textbf{Articles}, \textbf{Facts}_1 = \textbf{Facts} \cup \{\texttt{Info}\}$ and $\textbf{Exceptions}_1 = \textbf{Exceptions}$.
  - Else if `Info` is some $F_1[F_a, F_2]$, then,
    * If F_1 is in $Cn(\textbf{Articles} \cup \textbf{Facts})$, then $\textbf{Articles}_1 = \textbf{Articles}, \textbf{Facts}_1 = \textbf{Facts}$ and $\textbf{Exceptions}_1 = \textbf{Exceptions} \cup \{\texttt{Info}\}$.
    * Else if F_1 is not in $Cn(\textbf{Articles} \cup \textbf{Facts})$, $\textbf{Articles}_1 = \textbf{Articles}, \textbf{Facts}_1 = \textbf{Facts}$ and $\textbf{Exceptions}_1 = \textbf{Exceptions}$ (Adequacy).

The (Adequacy) is compatible with (Safety).

Also, let $\subseteq$, the subset relation, extend to a pair of sets, so that, for any B_1 and B_2 not necessarily belief sets, $B_1 \subseteq B_2$ iff the first and respectively the second component of B_1 are subsumed by the first and respectively the second components of B_2.

Belief Contraction is the simpler belief change operation.
ContractBy$(B, \texttt{Info})$ satisfies all the following axioms.

1. B is a belief set $(Cn(\textbf{Articles} \cup \textbf{Facts}), \textbf{Exceptions})$ for some **Articles**, **Facts** and **Exceptions** (Pre-condition).

---

[2] Usually a paper defines either belief revision only, or belief contraction, revision and expansion. The last, expansion, is not touched upon in this work, since there is hardly any point in distinguishing revision from it in our setting.

2. **ContractBy**$(B, \texttt{Info})$ is a belief set (Closure).
3. **ContractBy**$(B, \texttt{Info}) \subseteq B$ (Inclusion).
4. If $\texttt{Info}$ is not in B,[3] then **ContractBy**$(B, \texttt{Info}) = B$ (Vacuity).
5. If $\texttt{Info}$ is a triple $F_1[F_a, F_2]$, then
   (a) No $F_1[F_b, F_2]$ such that $Cn^{**}(\{F_a\}) = Cn^{**}(\{F_b\})$ is in
       ContractBy$(B, \texttt{Info})$ (Success 1). We assume that Cn^{**} for the language which extends $\texttt{mapBtoK}(B)$ with any predicate or constant that occurs either in $F \in \Gamma(B)$ or in F_a. We also assume that F_b is expressible in the extended language.
   (b) $B \subseteq$ **ContractBy**$(B, \texttt{Info}) + \texttt{Info}_1$ where $\texttt{Info}_1$ is $F_1[F_b, F_2]$ for some F_b such that $Cn^{**}(\{F_a\}) = Cn^{**}(\{F_b\})$ (Reinstatement).
6. If $\texttt{Info}$ is a sentence, then
   (a) $\texttt{Info}$ is not in **ContractBy**$(B, \texttt{Info})$ (Success 2).[4]
   (b) The first component of B is a subset of the first component of **ContractBy**$(B, \texttt{Info}) + \texttt{Info}$ (Partial recovery).
   (c) If $F_x[F_y, F_z]$ for some F_x, F_y and F_z is in the second component of B and if F_x is in the first component of **ContractBy**$(B, \texttt{Info})$, $F_x[F_y, F_z]$ is in the second component of **ContractBy**$(B, \texttt{Info})$ (Preservation).
   (d) **ContractBy**$(B, \texttt{Info}) = $ **ContractBy**$(B, \texttt{Info}_1)$ for any $\texttt{Info}_1$ such that $Cn^{***}(\{\texttt{Info}\}) = Cn^{***}(\{\texttt{Info}_1\})$ (Extensionality). We assume that Cn^{***} is for the language that extends $\texttt{mapBtoK}(B)$ with any predicate or constant that occurs in $\texttt{Info}$ or $\texttt{Info}_1$.

Let us provide informal explanations to **ContractBy**. The (Vacuity) says that nothing that is not in B is removed. The (Success 1) and the (Success 2) say that the contraction operation ensures that what is equivalent to $F_1[F_a, F_2]$ up to the logical consequence is not in the result. The (Reinstatement) says that the removal of $\texttt{Info}$ is done minimally, so that there exists a single $F_1[F_b, F_2]$ for $Cn^{**}(\{F_a\}) = Cn^{**}(\{F_b\})$ which, if added to the result, will completely reinstate all the elements in B. The (Preservation) says that, for any sentence in the result, it has the same set of triples as had in B.

The result can be represented also set-theoretically. Let Δ be a mapping from belief sets and new information into belief sets. For any belief set B and $\texttt{Info}$, we say that a belief set B_1 satisfying $B_1 \subseteq B$ is a maximal subset of B iff

1. $\texttt{Info}$ is not in B_1.
2. If $\texttt{Info}$ is a triple, then for any belief set B_2, if $B_1 \subset B_2 \subseteq B$, then $B_2 = B$.
3. Else if $\texttt{Info}$ is a sentence, then
   (a) For any $F_1[F_a, F_2]$, if it is in the second component of B and if F_1 is in the first component of B_1, then $F_1[F_a, F_2]$ is in the second component of B_1.

---

[3] In the sense that if $\texttt{Info}$ is a sentence, it is not in $Cn(\textbf{Articles} \cup \textbf{Facts})$, and if it is a triple, it is not in **Exceptions**.

[4] A belief revision theory usually puts an additional condition that $\texttt{Info}$ is not a tautology in order for this condition to apply, which, in our setting, is ensured by (Non-triviality of a code article and a fact).

(b) For any belief set B_2, if $B_1 \subset B_2 \subseteq B$, then Info is in B_2.

We define $\Delta(B, \text{Info})$ to be the set of all the maximal subsets, provided that B is a belief set. We also define a function γ, so that, if $\Delta(B, \text{Info})$ is not empty, then $\emptyset \subset \gamma(\Delta(B, \text{Info})) \subseteq \Delta(B, \text{Info})$ holds true; or if $\Delta(B, \text{Info})$ is empty, $\gamma(\Delta(B, \text{Info}))$ is simply B.

Lemma 1 (Alchourrón et al. [1]). *Suppose B is a belief set, and B_1 is a maximal subset of B for Info. If Info is a sentence, then the first component of B equals the first component of $B_1 + \text{Info}$.*

Theorem 1 (Representation theorem). *Let B be a belief set. Then it follows that for each outcome of $\textbf{ContractBy}(B, \text{Info})$ there exists an outcome of $\bigcap(\gamma(\Delta(B, \text{Info})))$ such that they are equal, and for each outcome of $\bigcap(\gamma(\Delta(B, \text{Info})))$, and of $\textbf{ContractBy}(B, \text{Info})$, there exists an outcome of $\textbf{ContractBy}(B, \text{Info})$, and $\bigcap(\gamma(\Delta(B, \text{Info})))$, such that they are equal.*

Proof. Trivial if Info is a triple. Consider when Info is a sentence. The proof is similar to the one in [1]; however, we show cases of one direction of the proof with details for not very straightforward ones. The other direction is proved similarly. We show that for any particular belief set as results from $\textbf{ContractBy}(B, \text{Info})$, say B', there exists a set of specific choices of γ such that $B' \subseteq \bigcap(\gamma(\Delta(B, \text{Info})))$. Suppose, by way of showing contradiction, that there exists α which is either some F or some $F_1[F_a, F_b]$ such that $\alpha \in B'$ and $\alpha \notin \bigcap(\gamma(\Delta(B, \text{Info})))$. Suppose that $\alpha = F$. By (Closure) and (Inclusion), we have that α is in the first component of B. There are two cases: either α is a tautology, or otherwise. In the latter case, there are two possibilities: either (1) Info is not in the first component of B or (2) otherwise. For the latter case, define Conflict to be $\{F_x$ in the first component of $B \mid \text{Info} \in Cn(\{F_x\})\} \cup \{F_u$ in the first component of $B \mid Cn(\{F_u\}) = Cn(\{F_v \vee F_w\})$ and $\text{Info} \in Cn(\{F_v\})$ and there is no F_z in the first component of B such that $F_z \in Cn(\{F_w\})\}$. Then by (Success 2), $\alpha \notin \text{Conflict}$. There are two cases here. If $\alpha \in Cn(\{\text{Info}\})$ such that $Cn(\{\text{Info}\}) \neq Cn(\{\alpha\})$, then by the first condition satisfied by a maximal subset, we can choose γ appropriately so that any selected maximal subset(s) includes α, contradiction to the supposition. Otherwise, we have that α is in (the first component of B)\Conflict)\$\{F_x \mid F_x \in Cn(\{\text{Info}\})\}$. By Lemma 1, contradiction to the supposition is drawn. Hence there indeed exists some choice of γ with which the first component of B' is a subset of the first component of $\bigcap(\gamma(\Delta(B, \text{Info})))$. However, then by (Preservation) and the third condition of a maximal subset, we also have that the second component of B' is a subset of the second component of $\bigcap(\gamma(\Delta(B, \text{Info})))$. So far we have supposed $\alpha = F$. It is trivial if $\alpha = F_1[F_a, F_b]$ instead. This completes the first part of the one direction of the proof. We then also show that there exists some choice γ_1 among those γ with which $B' \subseteq \bigcap(\gamma(\Delta(B, \text{Info})))$ such that $B' \supseteq \bigcap(\gamma_1(\Delta(B, \text{Info})))$. But this second part of the same direction of the proof is similar. $\square$

Belief Revision is an iterating process, similar to the one in [3], which is best described as a procedure or a state transition system.

   Description of ReviseBy(B, Info)

   Inputs: a belief set B and some information Info (each satisfying the conditions stated earlier).

   Output: B_{out}.

L$_0$ Assign B to B_{out}.

L$_1$ If Info is a sentence, then assign **ContractBy**(B_{out}, ¬Info) + Info to B_{out}.

L$_2$ If Info is a triple $F_1[F_a, F_2]$, then assign **ContractBy**($B, F_1[F_a, F_x]$) + Info to B_{out}, where F_x is: F_3 if $F_1[F_a, F_3]$ occurs in B; F_2, otherwise.

L$_{3_0}$ While there exists some $F_1[F_a, F_2]$ in B_{out} such that F_a is in the first component of B_{out}, do:

    L$_{3_{1_0}}$ If $\{F_2\} \cup X$ for X being the first component of **ContractBy**(B_{out}, F_1) is consistent, then;

        L$_{3_{1_1}}$ Assign **ContractBy**(B_{out}, F_1) to B_{out}.

        L$_{3_{1_2}}$ Assign (**ContractBy**($B_{out}, ¬F_2$)) + F_2 to B_{out}.

    L$_{3_2}$ Else assign **ContractBy**($B_{out}, F_1[F_a, F_2]$) to B_{out}.

L$_4$ End of While loop

Let us suppose some belief set B, and the new information Info to revise B by. The first point of focus (up to Line 2) is to see whether insertion of Info into B would: (1) contradict the first component of B, i.e. the first component of B includes ¬Info, in which case we must minimally change B to B_x in order that B_x + Info be consistent in its first component, which is done by **ContractBy**(B, ¬Info); (2) or contradicts (Uniqueness) condition, i.e. Info is some triple $F_1[F_a, F_2]$ and the second component of B contains some $F_1[F_b, F_3]$ such that $Cn^*(\{F_a\}) = Cn^*(\{F_b\})$, in which case we must remove the existing $F_1[F_b, F_3]$ off B into B_x in order that B_x + Info still satisfy (Uniqueness). Now, for the part from Line 3 to Line 4, it describes the following iterative process. We have B_y. If it is a belief set, our revision is already done, and we simply end this iteration process; otherwise, the set B_y we have does not satisfy (No-unused-triggers), since it satisfies (Safety), (Consistency) and (Uniqueness).[5] This means that there is some $F_1[F_a, F_2]$ in the second component of B_y such that F_a is in the first component of B_y, which means that F_1 in **Articles** may update to F_2. 'May update' because the updating can make the first component of the resulting set inconsistent, which we do not permit and discard $F_1[F_a, F_2]$ off B_y by **ContractBy**($B_y, F_1[F_a, F_2]$) in that case,[6] but otherwise the update takes place by **ContractBy**(B_y, F_1), to derive B_z, in the first component of which F_1 does not appear, followed by **ContractBy**($B_z, ¬F_2$) + F_2. We regard the resulting set as B_y and repeat the iteration.

---

[5] See the proof of the theorem right below for details.

[6] We apply this principle on the supposition that when judges see an update necessary, they should have already checked that the update would not contradict the present civil code.

Theorem 2. *Let B be a belief set, and* `Info` *new information. Then* ***ReviseBy***$(B,$ `Info` $)$ *is a belief set.*

Proof. We show that $B_{out} = (Cn(\mathbf{Articles}_x \cup \mathbf{Facts}_y), \mathbf{Exceptions}_z)$ satisfies (Safety), (No-unused-triggers), (Consistency) and (Uniqueness).

(Safety) ReviseBy is characterised by $+$ and **ContractBy**, both of which guarantee the condition.

(No-unused-triggers) From the condition of the While loop.

(Consistency) Since **ContractBy** ensures consistency and since B is assumed consistent, B_{out} on Line 0 is consistent. B_{out} on Line 1 and Line 2 are consistent, since **ContractBy** ensures consistency. B_{out} on Line 3_{1_1} is consistent for the same reason. B_{out} on Line 3_{1_2} is consistent by the condition of the If on Line 3_{1_0}. Finally, B_{out} on Line 3_2 is consistent. Hence the output of **ReviseBy**$(B,$ `Info` $)$ satisfies Consistency.

(Uniqueness) Be `Info` a sentence or a triple, on Line 3_0 B_{out} satisfies Uniqueness, due to Line 1 or Line 2. Then the output straightforwardly satisfies Uniqueness, since F_2 on Line 3_{1_2} is a sentence, and, in that case, the second component of $B_{out} + F_2$ is the same as that of B_{out} by the definition of $+$. $\square$

3   The Example Revisited

We instantiate our framework with the example in Sect. 1. To model the civil code before the civil litigation, we set our language **K** to be (**Log**, (**Const**, **Pred**)) where

- **Const**: comprises *lessee, lessor, thirdParty,* and *leasedLand.*
- **Pred**: comprises ternary predicates allowUseTo and subleaseTo; binary predicates gainApprovalOf and cancelContract; and unary predicates isLessee, isLessor, isThirdParty and isLeasedThing.

We then set the initial B (which is $(Cn(\mathbf{Articles} \cup \mathbf{Facts}), \mathbf{Exceptions}))$, as follows. DISTINCT$(x_1, x_2, x_3, x_4)$ is the abbreviation of $\bigwedge_{i,j\in\{1,2,3,4\} \text{ such that } i\neq j} x_i \neq x_j$.

- **Articles**: comprises, for brevity, only the Article 612: $\exists x_1 \ \exists x_2 \ \exists x_3 \ \exists x_4.$ isLessee(x_1) $\wedge$ isLessor(x_2) $\wedge$ isThirdparty(x_3) $\wedge$ isLeasedThing(x_4) $\wedge$ DISTINCT (x_1, x_2, x_3, x_4) $\wedge$ (allowUseTo(x_1, x_3, x_4) $\vee$ subleaseTo(x_1, x_3, x_4)) $\wedge$ ¬gainApprovalOf (x_1, x_2) $\supset$ cancelContract(x_2, x_1).
  Let this sentence be F.
- **Facts**: is empty.
- **Exceptions**: has $F[\mathsf{p}_8, F_x]$ where
  - p_8 is noConfidenceAbuseBy$(lessor, lessee)$.
  - F_x models the updated Article 612:
    $\exists x_1 \ \exists x_2 \ \exists x_3 \exists x_4.$¬noConfidenceAbuseBy$(x_2, x_1)$ $\wedge$ isLessee(x_1) $\wedge$ isLessor(x_2) $\wedge$ isThirdparty(x_3) $\wedge$ isLeasedThing(x_4) $\wedge$ DISTINCT(x_1, x_2, x_3, x_4) $\wedge$ (allowUseTo(x_1, x_3, x_4) $\vee$ subleaseTo(x_1, x_3, x_4)) $\wedge$ ¬gainApprovalOf(x_1, x_2) $\supset$ cancelContract(x_2, x_1).

B hence represents the civil code that has just one article which may update on reaction to the discovery of p_8.

At this point a civil litigation over cancellation of a lease contract begins between two parties. The contract states that 'lessee' is the lessee and that 'lessor' is the lessor who has let a land 'leasedLand' to 'lessee'. There is an evidence that 'lessee' has let a third party, 'thirdParty', use the land.

Let us reflect these. We add isLeasedThing(*leasedLand*) (denote it by p_1), isLessee(*lessee*) (by p_2), isLessor(*lessor*) (by p_3), isThirdParty(*thirdParty*) (by p_4), and allowUseTo(*lessee, thirdParty, leasedLand*) (by p_5) as well as DISTINCT(*lessee, lessor, thirdParty, leasedLand*) (by p_6) into **Facts** of B, deriving a new belief set B_1: $\mathbf{ReviseBy}^{\times 6}((((((B, p_1), p_2), p_3), p_4), p_5), p_6)$, which in this example is the same as $(((((B + p_1) + p_2) + p_3) + p_4) + p_5) + p_6$.

The lessor now proves that the lessee let the third party use the land without his approval.

We have $B_2 = \mathbf{ReviseBy}(B_1, \neg\text{gainApprovalOf}(lessee, lessor))$. At this point, it is true that cancelContract(*lessor, lessee*) is in the first component of B_2.

However, judges at the Supreme Court notice an exceptional circumstance in this litigation, which is that there is no evidence that confidence of the lessor in the lessee has been abused by the lessee. They conclude that the civil code article is applicable only if abuse of confidence is evidenced.

Here B_2 is revised by new information noConfidenceAbuseBy(*lessor, lessee*), which incidentally is p_8. We have: $\mathbf{ReviseBy}(B_2, p_8)$. Notice that the language of B_2, which is still $\mathbf{K}$, does not involve p_8. Once the belief revision takes place, however, we see to it that it becomes $\mathbf{K_1} : (\mathbf{Log}, (\mathbf{Const}, \mathbf{Pred} \cup \{\text{noConfidenceAbuseBy}\}))$, noConfidenceAbuseBy the p_8 being the newly discovered fact that triggers the update of F to F_x. And because of this, the cancelContract can no longer be in the first part of the resulting belief set: let B_a be the belief set that is revised by the triple; by $\mathbf{L3_{1_1}}$ of $\mathbf{ReviseBy}$, contraction of B_a by F takes place, at which point cancelContract(*lessor, lessee*) must be removed by (Closure) and (Success 2) of $\mathbf{ContractBy}$, and the addition of the updated code article F_x on $\mathbf{L3_{1_2}}$ does not recover cancelContract(*lessor, lessee*) because it is not a logical consequence of F_x in $\mathbf{K_2}$.

4   Related Works and Discussion

Traditional belief revision approaches revise a belief set (in the AGM sense [1]) with new information by possibly removing existing information in contention first, and by then adding the new information. There are viewpoints that revision/contraction by modification rather than by deletion of existing information reflect norm changes more faithfully [5,10]. In [10] Maranhão proposes a weak belief acceptance model. When new information F_1 is in conflict with existing information, his model accepts $F_2 \supset F_1$ instead of F_1, that way precluding inconsistency in the resulting belief set. He also considers the cases where existing information is instead weakened by such a conditional proposition. These refinement operations can be used to accommodate new information as much as possible without contradicting an existing belief set. In [5], Giusto and Governatori consider what they call the minimax revision for belief bases not necessarily closed under the logical consequence relation. They, too, show how propositions can be weakened to avoid inconsistencies. Compared to their approaches, our approach is in a way more traditional. To update F_1 into F_2 in a belief set in our framework, the belief set is contracted (deleted) by F_1, which is then revised by F_2. However, because we make use of an extended belief set which has not only propositions but also proposition triples as latent update instructions, the two belief change operations: contraction first and then revision, are coupled very tightly as if they were one single belief change operation. In this sense our approach can also be used to model norm changes. Flexibility is ensured by not predestining in what way an existing belief as a civil code article must be changed. Also, that the update instructions do not actively interfere with the propositions in the same belief set means that the AGM postulates are left untouched, which is a nice property from a compositional perspective. Also, we believe that our approach can be adjusted to capture different ways of changing norms such as annulments and abrogations [6,7] as well as various ways of deleting unwanted legal effects [8], and further research to extend our study should be interesting.

5 Conclusion

We developed a formal framework to handle civil code updates on reaction to the discovery of exceptional circumstances, which we went through with one civil case example. Based firmly on a belief revision theory, our framework guarantees that a civil code update be modelled in a logically reasonable manner.

Acknowledgements. This work would not have existed without insight and helpful comments given to us by Ken Satoh. Proofreading was kindly done by Thomas Given-Wilson. Reviewers helped us improve this paper.

References

1. Alchourrón, C.E., Gärdenfors, P., Makinson, D.: On the logic of theory change: partial meet contraction and revision functions. J. Symb. Log. **50**, 510–530 (1985)
2. Arisaka, R.: How do you revise your belief set with %$;@*? arXiv e-prints:1504.05381 (2015)
3. Arisaka, R.: Latent belief theory and belief dependencies: a solution to the recovery problem in the belief set theories. arXiv e-prints: 1507.01425 (2015)

4. Darwiche, A., Pearl, J.: On the logic of iterated belief revision. Artif. Intell. **89**, 1–29 (1997)
5. Giusto, P., Governatori, G.: A new approach to base revision. In: Barahona, P., Alferes, J.J. (eds.) EPIA 1999. LNCS (LNAI), vol. 1695, pp. 327–341. Springer, Heidelberg (1999). doi:10.1007/3-540-48159-1_23
6. Governatori, G., Palmirani, M., Riveret, R., Rotolo, A., Sartor, G.: Norm modifications in defeasible logic. In: JURIX, pp. 13–22 (2005)
7. Governatori, G., Rotolo, A.: Changing legal systems: legal abrogations and annulments in defeasible logic. Log. J. IGPL **18**(1), 157–194 (2010)
8. Governatori, G., Rotolo, A., Olivieri, F., Scannapieco, S.: Legal contractions: a logical analysis. In: ICAIL, pp. 63–72 (2013)
9. Ito, S.: Lecture Series on Ultimate Facts, Shojihomu (2008). (In Japanese)
10. Maranhão, J.: Refinement. In: ICAIL, pp. 52–60 (2001)
11. Katsuno, H., Mendelzon, A.O.: Propositional knowledge base revision and minimal change. Artif. Intell. **52**(3), 263–294 (1991)
12. Satoh, K., Asai, K., Kogawa, T., Kubota, M., Nakamura, M., Nishigai, Y., Shirakawa, K., Takano, C.: PROLEG: an implementation of the presupposed ultimate fact theory of Japanese civil code by PROLOG technology. In: Onada, T., Bekki, D., McCready, E. (eds.) JSAI-isAI 2010. LNCS (LNAI), vol. 6797, pp. 153–164. Springer, Heidelberg (2011). doi:10.1007/978-3-642-25655-4_14
13. Satoh, K., Kubota, M., Nishigai, Y., Takano, C.: Translating the Japanese presupposed ultimate fact theory into logic programming. In: JURIX, pp. 162–171. IOS Press (2009)

Combining Input/Output Logic and Reification for Representing Real-World Obligations

Livio Robaldo[1(✉)], Llio Humphreys[1], Xin Sun[1], Loredana Cupi[2],
Cristiana Santos[3], and Robert Muthuri[2]

[1] University of Luxembourg, Luxembourg City, Luxembourg
{livio.robaldo,llio.humphreys,xin.sun}@uni.lu
[2] University of Turin, Turin, Italy
loredana.cupi@unito.it, muthuri.r@gmail.com
[3] University of Barcelona, Barcelona, Spain
cristiana.teixeirasantos@gmail.com

Abstract. In this paper, we propose a new approach to formalize *real-world* obligations that may be found in existing legislation. Specifically, we propose to formalize real-world obligations by combining insights of two logical frameworks: Input/Output logic, belonging to the literature in deontic logic and normative reasoning, and the Reification-based approach of Jerry R. Hobbs, belonging to the literature in Natural Language Semantics. The present paper represents the first step of the ProLeMAS project, whose main goal is the one of filling the gap between the current logical formalizations of legal text, mostly propositional, and the richness of Natural Language Semantics.

1 Introduction

Legal scholars and practitioners are feeling increasingly overwhelmed with the expanding set of legislation and case law available these days, which is assuming more and more of an international character. Consider, for example, European legislation, which is estimated to be 170,000 pages long, of which over 100,000 pages have been produced in the last ten years.

Legal informatics is an under-researched area in IE, and there is a lack of suitable annotated data. The idiosyncratic nature of legal text poses new challenges for the task of extracting such information using NLP, in order to associate norms with semantic representations on which to perform reasoning [4].

Livio Robaldo has received funding from the European Unions Horizon 2020 research and innovation programme under the Marie Sklodowska-Curie grant agreement No 661007 for the project "ProLeMAS: PROcessing LEgal language in normative Multi-Agent Systems". Llio Humphreys is supported by the National Research Fund, Luxembourg. Loredana Cupi has received funding from the European Union's Horizon 2020 research and innovation programme under the Marie Skodowska-Curie grant agreement No 690974 for the project "MIREL: MIning and REasoning with Legal texts". We would like to thank prof. Leon van der Torre for fruitful discussions.

© Springer International Publishing AG 2017
M. Otake et al. (Eds.): JSAI-isAI 2015 Workshops, LNAI 10091, pp. 217–232, 2017.
DOI: 10.1007/978-3-319-50953-2_16

The ProLeMAS project[1] has been specifically proposed to address these challenges. The main goal of the project is to overcome two main limitations of current approaches in normative reasoning and deontic logic:

(1) a. Several proposals in deontic logic are typically propositional, i.e. their basic components are whole propositions. A proposition basically refers to a whole sentence. On the other hand, natural language (NL) semantics includes a wide range of fine-grained intra-sentence linguistic phenomena: named entities, anaphora, quantifiers, etc. It is then necessary to move beyond the propositional level, i.e. to enhance the expressivity to formalize the meaning of the *phrases* constituting the sentences.
   b. Few proposals in deontic logic have been implemented and tested on real legal text. Most of them are only promising methodologies, which overcome the limits of other approaches on the theoretical side. In order to make the logical framework really useful and worth being implemented, its design has to be guided by the analysis of *real* norms.

We started by studying a corpus of EU legislation in English. The corpus includes twenty EU directives from 1998 to 2011, covering a range of subjects, e.g., the profession of lawyer, passenger ships, biotechnological inventions, etc.

Our initial experiments of norm representation in ProLeMAS is conducted on the English version of Directive 98/5/EC of the EU Parliament to facilitate practice of the profession of lawyer on a permanent basis in a Member State other than that in which the qualification was obtained. There are 36 obligations, 13 powers, 10 legal effects, 8 definitions, 6 permissions, 6 applicability types, 6 rationales, 2 rights, 1 exception, and 1 hierarchy. In this paper, we use the following example of obligation for explanatory purposes:

(2) *A lawyer who wishes to practise in a Member State other than that in which he obtained his professional qualification shall register with the competent authority in that State.*

The approach proposed in this paper merges two specific logical frameworks into a new one: (1) Input/Output (I/O) logic, belonging to the literature in deontic logic and normative reasoning, and (2) the Reification-based approach of Jerry R. Hobbs, belonging to the literature in Natural Language Semantics.

I/O logic appears as one of the new achievements in deontic logic in recent years [9]. The key feature of I/O logic is that it adopts *operational* semantics and not truth-conditional ones. Thus, it allows to determine which obligations are operative in a situation that already violates some of them. It is not possible to achieve such a characterization of norms in terms of a truth-conditional semantics: a violation would correspond to an inconsistency, which will make the whole knowledge base inconsistent.

---

[1] http://www.liviorobaldo.com/ProLeMAS.htm.

On the other hand, Hobbs's logic is a wide-coverage logic for Natural Language Semantics centered on the notion of Reification. Reification is a concept originally introduced by Donald Davidson in [6]. Modern logical frameworks based on Reification are known in the literature as "neo-Davidsonian" approaches. Reification allows a wide variety of complex natural language (NL) statements to be expressed in First Order Logic (FOL). NL statements are formalized such that events, states, etc., correspond to constants or quantifiable variables of the logic. Following [2], we use in this paper the term 'eventuality' to denote both the reification of a state and the one of an event.

Reification allows us to move from standard FOL notations such as "$(give\ a\ b\ c)$", asserting that "a" gives "b" to "c", to another notation "$(give'\ e\ a\ b\ c)$", which is again in FOL, where e is the *reification* of the giving action. In other words, the expression "$(give'\ e\ a\ b\ c)$" says that "e" is a giving event by "a" of "b" to "c". "e" is a FOL term exactly as "a", "b", and "c".

Many neo-Davidsonian logical frameworks have been proposed in Natural Language Semantics and also in Legal Informatics (cf. Sect. 2 below). The peculiarity of Hobbs's with respect to all other neo-Davidsonian approach is the *total avoidance* of subformulae within the scope of other operators. In other words, the formulae are mere conjunctions of atomic predications. It has been argued in [14,25] that many interpretations available in NL require the *parallel* evaluation of two or more logical operators (e.g., modal operators or quantifiers). Section 2 presents some example. Hobbs's logic, by avoiding embeddings of operators within the scope of other operators, straightforwardly and uniformly handles these readings, that are intrinsically prevented in many traditional logical frameworks for Natural Language Semantics.

2 Related Work

Some previous approaches try to model, in some deontic settings, sentences coming from *existing norms*. The most representative work is perhaps [30]. Examples of real norms formalized in deontic logic may be also found in [1,11].

Many current state-of-the-art approaches try to formalize legal knowledge via Event Calculus [16,22]. Event Calculus is a neo-Davidsonian logical language that extends the original account of Reification (see [10] for a discussion).

A recent approach in the line is [12]. In [12], it is argued that Event Calculus predicates for handling time cannot be directly used for handling also deontic meaning. Therefore, a new version is proposed to incorporate the deontic effect of norms, so that they can be used for compliance checking. Similar proposals are [7,8,24]. However, [12] appears to be superior in that it identify and formalize much more fine-grained and complex obligation modalities.

The mentioned approaches in Event Calculus focus on business process compliance. In other words, they do not specifically focus on formalizing norms coming from existing legislation.

Thus, they cannot be directly compared with the present proposal, where Natural Language Semantics has a prominent role.

To our knowledge, the approach that appears closest to the one we are going to propose below is perhaps McCarty's Language for Legal Discourse (LLD) [20,21]. LLD is strongly drawn on previous studies on Natural Language Semantics, it uses Reification, and it has been developed specifically to model real legal text.

An example of McCarthy's Language for Legal Discourse (LLD) is shown in (3). The sentence in (3) is represented via the formula below it.

(3) "The petitioner contends that the regulatory takings claim should not have been decided by the jury"

```
sterm(contends, A,
      [nterm(petitioner, B, [])
       /det(The, nn),
       sterm(decided, C,
             [D,
              aterm(regulatory,E,[F]) &
              nterm(takings,G,[]) &
              nterm(claim,F,[])
              /det(the, nn)])
      && H^pterm(by, H,
                 [C,
                  nterm(jury,I,[])
                  /det(the, nn)])
/[modal(should),negative,perfect,passive]
```

sterm, nterm, aterm, and pterm are reified terms of different kind. For instance, sterms denote reified relations. Thus, the sterm on the first line refers to the eventuality denoted by the main verb "contends".

Space constraints forbid us to illustrate all technical details of LLD. We focus only on the two architectural choices most relevant for the present work:

(4) a. Each *term is associated with a lexical entry, e.g. "contends", "petitioner", "decided", etc. This is the first argument of the *term.
     b. *terms may outscope by other *terms. E.g., the sterm associated with the main verb "contends" outscopes the nterm associated with "petitioner" which in turn outscopes the sterm associated with "decided".

(4.a-b) make McCarthy's logic very reminiscent of standard representation formalisms used in Natural Language Semantics such as Discourse Representation Theory (DRT) [15] and Minimal Recursion Semantics (MRS) [5].

Nevertheless, [14,25] argues that (4.a-b) intrinsically prevents the proper representation of several readings that are actually available in NL utterances. For instance, consider sentences in (5), drawn from the large range of examples considered by Hobbs and Robaldo in their past research in NL semantics.

(5) a. Permission may be obtained, but *it* could take more than one month.
     b. The city does not have a train station, *but* it has a bus station.
     c. If the parents of a student earn *less than 20k* euros per year, then the student is eligible.

Sentence (5.a) highlights that reification can easily give rise to a practical formalism for NL semantics. The pronoun "it" refers to the permission to be obtained. The referent of the pronoun could be then directly identified by the FOL term reifying the eventuality denoted by the main verb of the first clause (see [14] and several other earlier publications by the same author[2]).

(5.b) is an example of concessive relation, one of the trickiest semantic relations occurring in NL: the first clause creates the expectation that the city is unreachable by public transportation. The second clause denies that expectation. A practical way to properly model concessive relations is to reify the eventuality corresponding to the expectation, as proposed in [28]. Note that the expectation is a "hidden" eventuality, i.e., it is not denoted by any lexical item. Thus, (5.b) cannot be represented in LLD via its basic constructs, due to (4.a).

Finally, (5.c) is an example of cumulative reading. The meaning of (5.c) is that if the money *cumulatively* earned by either one of the parents, or by both together[3] is less than 20k, the student is eligible. Cumulative readings have been extensively studied in a reification setting in [26,27,29].

It is quite hard to represent (5.a-c) by embedding operators within the scope of other operators, as it is done in DRT or MRS. For instance, in (5.c) we have two operators/quantifiers: "Two" and "Less than 20k". By embedding the latter within the scope of the former, we get a reading where the student is eligible if *either* parent independently earns less than 20k euros, but the sum of the two earnings is superior to 20k. On the other hand, in order to get the meaning of cumulative readings, the two quantifiers must be evaluated *in parallel*, i.e., none of the two must outscope the other.

This paper defines a reified deontic logic characterized by the *total avoidance* of embeddings in the instantiated formulae, in line with [14,29].

The next two sections introduce the formal instruments at the base of our logical formalization: Hobbs's logic and Input/Output logic. The subsequent sections illustrates how the former can be integrated into the latter, while retaining the advantages of both formalisms.

3   Hobbs' Logical Framework

Jerry R. Hobbs defines a wide-coverage logic for Natural Language Semantics centered on the notion of Reification. Hobbs's logic uses two related kinds of predicates: primed and unprimed. For instance, the predication "(*give a b c*)" seen above is associated with "(*give' e a b c*)", where "*e*" is the reification of the giving action. Hobbs' implements a fairly large set of linguistic and semantic concepts including sets, composite entities, scales, change, causality, time, event structure, etc., into an integrated first order logical formalism.

Eventualities may be *possible* or *actual*. In Hobbs', this distinction is represented via a unary predicate *Rexist* that holds for eventualities really existing

---

[2] See http://www.isi.edu/~hobbs/csknowledge-references/csknowledge-references.html and http://www.isi.edu/~hobbs/csk.html.

[3] For instance, suppose they rent an apartment that they co-own.

in the world. To give an example cited in Hobbs, if I want to fly, my wanting exists in reality, but my flying does not. This is represented as:

$$\exists_e[\ (Rexist\ e) \wedge (want'\ e\ \mathtt{I}\ e_1) \wedge (fly'\ e_1\ \mathtt{I})\ ]$$

Eventualities can be treated as the objects of human thoughts. Reified eventualities are inserted as parameters of such predicates as *believe, think, want,* etc. Reification can be applied recursively. The fact that *John believes that Jack wants to eat an ice cream* is represented as an eventuality e such that it holds:

$$\exists_e\exists_{e_1}\exists_{e_2}\exists_{e_3}[\ (Rexist\ e) \wedge (believe'\ e\ \mathtt{John}\ e_1) \wedge (want'\ e_1\ \mathtt{Jack}\ e_2) \wedge$$
$$(eat'\ e_2\ \mathtt{Jack}\ \mathtt{Ic}) \wedge (iceCream'\ e_3\ \mathtt{Ic})\ ]$$

Every relation on eventualities, including logical operators, causal and temporal relations, and even tense and aspect, may be reified into another eventuality. For instance, by asserting $(imply'\ e\ e_1\ e_2)$, we reify the implication from e_1 to e_2 into an eventuality e and e is, then, thought of as "the state holding between e_1 and e_2 such that whenever e_1 really exists, e_2 really exists too". Negation is represented as $(not'\ e_1\ e_2)$: e_1 is the eventuality of e_2's not existing.

The predicates *imply'* and *not'* are defined to model the concept of 'inconsistency'. Two eventualities e_1 and e_2 are said to be inconsistent if and only if they (respectively) imply two other eventualities e_3 and e_4 such that e_3 is the negation of e_4. The definition is as follows:

(6) $(forall\ (e_1\ e_2)$
            $(iff\ (inconsistent\ e_1\ e_2)$
                $(and\ (eventuality\ e_1)\ (eventuality\ e_2)$
                    $(exists\ (e_3\ e_4)\ (and\ (imply\ e_1\ e_3)$
                        $(imply\ e_2\ e_4)(not'\ e_3\ e_4))))))$

(6) is an example of an 'axiom schema'. In this logic, an 'axiom *schema*' provides one or more different axioms for each predicate p. The axiom schemas of the predicates used in formulae are generally stored into a separate ontology.

Higher order operators, such as modal and temporal operators, are modelled by introducing new predicates and by defining axiom schemas to restrict their meaning. However, it is important to note that thanks to reification, the formulae representing natural language utterances *never* feature any kind of embedding of predicates within other operators. In other words, formulae are *always* conjunctions of atomic FOL predicates applied to FOL terms. See for instance [28], which propose a formalization in Hobbs' of concessive relations.

As [14], pp. 5, states: "There has been an attempt to make the notation as 'flat' as possible. All knowledge is knowledge of predications[4]. FOL terms are only handles. The intuition is that in natural language we cannot communicate entities directly. We can only communicate properties and hope that the listener can determine the entity we are attempting to refer to."

---

[4] And, it may be separately asserted in axiom schema.

It should be clear that the main peculiarity of Reification-based logical frameworks is their formal simplicity. This eases the handling of several natural language phenomena. Hobbs' past research particularly addresses the proper treatment of anaphora (cf. in particular [14]). For instance, (7.a) may be represented via formula (7.b). For simplicity, in (7.b) the semantic relation between the two clauses is simply represented as a material implication (predicate $imply'$).

(7) a. If John goes to Mary's house, he tells her before.

  b. $(Rexist\ e) \land (imply'\ e\ e_1\ e_2) \land (goTo'\ e_1\ \text{J M}) \land$
     $(tell'\ e_2\ \text{J}\ e_1\ \text{M}) \land (happenBefore\ e_2\ e_1)$

e_1 is the event of John's going to Mary, while e_2 is the event of John's telling to Mary *the fact that he will come to her*. The predicate $(happenBefore\ e_2\ e_1)$ states that e_2 must occur before e_1. Note that e_1 and e_2 are only hypothetical eventualies: (7.b) does not assert that they exist in the real world. In (7.a), the (hidden) pronoun "it", which is the object of the verb "tell", is straightforwardly represented: the eventuality e_1 is directly inserted as the second parameter of the $tell'$ predicate in (7.b). Without Reification, it would be necessary to introduce some 2-order operators in the logic in order to get the intended meaning of (7.a).

Hobbs' formula are formulae in first order logic with a very restricted syntax. They are basically *conjunctions* of atomic predicates instantiated on FOL terms. From a formal point of view, eventualities are FOL terms exactly as "Jack" in example (7.b), the only difference being that they refers to facts and actions occurring in the world. Facts and actions are taken to be individuals of the domain like persons, dogs, etc.

The logic we are going to use as the object logic of I/O systems - which we will call "ProLeMAS object logic" - is a further simplification of Hobbs' logic. The formulae will be more verbose, but, in our view, the simpler syntax will enhance readability and it will facilitate the definition of a reference ontology storing the available predicates and the axiom schemas modelling their meaning.

In fact, it is easy to see that the structure of the formulae strictly resemble the technique of rewriting relations of arbitrary arity as binary relations, used in AI as entity-attribute-value (EAV) triples in the last decades, which is at the basis of the subject-predicate-object of the RDF/OWL data model[5].

In ProLeMAS object logic, there is a single type of predicate, i.e. there is no distinction between primed and unprimed predicates. Predicates will be always unary or binary predicates. Thus, for instance, "$(give\ a\ b\ c)$" is not an acceptable predicate in our logic. N-ary relations are modelled by making thematic roles explicit. This is done by introducing other FOL predicates referring each to an available thematic role. For example, "$(give\ a\ b\ c)$" is translated into:

(8) $(give\ e_1) \land (agent\ e_1\ a) \land (patient\ e_1\ b) \land (recipient\ e_1\ c)$

The meaning of (8) is obvious: e_1 is a giving event whose agent is a, whose patient is b, and whose recipient is c. "Agent", "patient", and "recipient" are

---

[5] http://www.w3.org/TR/owl-ref.

thematic roles. The ontology specifies, for each kind of eventuality, the available thematic roles and - via further axioms - the restrictions on these thematic roles.

For instance, if we want to impose that agents of giving eventualities can be only human beings, we add the following axiom to the ontology:

(9)   *(forall (e a) (if (and (give e) (agent e a)) (humanBeing a)))*

Of course, the computational ontology has to be designed and developed with respect to the application and the domain where we want to concretely use the formulae. In our future research, we aim at specifically designing and implementing a legal ontology for modelling the meaning of norms expressed in natural language [3].

We now formally defines[6] the syntax of ProLeMAS object logic. For reasons that will be clear below, ProLeMAS object logic includes existential quantifiers but not universal ones. And, free variables are allowed. Those will be bound by an (external) universal quantifier.

Definition 1 (Syntax of ProLeMAS object logic). *ProLeMAS object logic is a fragment of First Order Logic (FOL). Where:*

- *The vocabulary includes FOL terms (constants, variables, and functions), FOL unary or binary predicates, the boolean connective "∧" and the existential quantifier "∃".*
- *If "p" and "q" are, respectively, a unary and a binary predicate, while "a" and "a" are terms, "p(a)" and "q(a,b)" are atomic formulae.*
- *If $\Phi_1 \ldots \Phi_n$ are atomic formulas, "$\Phi_1 \wedge \ldots \wedge \Phi_n$" and "$\exists_{x_1} \ldots \exists_{x_m} [\Phi_1 \wedge \ldots \wedge \Phi_n]$", where $x_1 \ldots x_m$ occurring in $\Phi_1 \ldots \Phi_n$, are non-atomic formulas, possibly containing free variables.*

4   Input/Output Logic

Input/Output logic (I/O logic) was originally introduced by Makinson and van der Torre in [19]. For a comprehensive survey and a techinqual introduction of I/O logic, see [23,31] respectively. Strictly speaking, I/O logic is not a single logic but a family of logics, just like modal logic is a family of logics containing systems K, KD, S4, S5, etc. In the first volume of the handbook of deontic logic and normative systems [9], I/O logic appears as one of the new achievements in deontic logic in recent years.

I/O logic originated from the study of conditional norms. Unlike modal logic, which usually uses possible world semantics, I/O logic mainly adopts operational semantics: an I/O system is conceived as a deductive machine, like a black box which produces deontic statements as output, when we feed it factual statements as input. In the original paper of Makinson and van der Torre, i.e. [19], four I/O

---

[6] Definition 1 only includes the boolean connective "∧". Other boolean connectives are modelled by introducing special predicates as *imply'* and *not'* in (6).

logics are defined: out_1, out_2, out_3, and out_4. They vary on the axioms used to constrain the deductive machine that produces the output against a valid input.

Let $\mathbb{P} = \{p_0, p_1, \ldots\}$ be a countable set of propositional letters and L be the propositional language built upon $\mathbb{P}$. Let $G \subseteq L \times L$ be a set of ordered pairs of formulas of L. G represents the deduction machine of the I/O logic: whenever one of the heads is given in input, the corresponding tails are given in output. Each pair (a, b) in G is called a "generator" and it is read as "given a, it ought to be x". In this paper, "a" and "b" are respectively termed as the *head* and the *tail* of the generator (a, b). Formally, G is a function from 2^L to 2^L such that for a set A of formulas, $G(A) = \{x \in L : (a, x) \in G$ for some $a \in A\}$. Makison and van der Torre [19] define the semantics of I/O logics from out_1 to out_4 as follows:

(10)   – $out_1(G,A)=Cn(G(Cn(A)))$.
       – $out_2(G,A)=\bigcap\{Cn(G(V)) : A \subseteq V, V$ is complete$\}$.
       – $out_3(G,A)=\bigcap\{Cn(G(B)) : A \subseteq B = Cn(B) \supseteq G(B)\}$.
       – $out_4(G,A)=\bigcap\{Cn(G(V) : A \subseteq V \supseteq G(V)), V$ is complete$\}$.

Here Cn is the classical consequence operator of propositional logic, and a set of formulas is *complete* if it is either *maximal consistent* or equal to L. These four logics are called *simple-minded output*, *basic output*, *simple-minded reusable output* and *basic reusable output* respectively. For each of these four logics, a *throughput* version that allows inputs to reappear as outputs, defined as $out_i^+(G, A) = out_i(G_{id}, A)$, where $G_{id} = G \cup \{(a, a) \mid a \in L\}$. When A is a singleton, we write $out_i(G, a)$ for $out_i(G, \{a\})$.

I/O logics are given a proof theoretic characterization. We say that an ordered pair of formulas is derivable from a set G iff (a, x) is in the least set that extends $G \cup \{(\top, \top)\}$ and is closed under a number of derivation rules. The following[7] are the rules we need to define out_1 to $out_4{}^+$:

(11)   – SI (strengthening the input): from (a, x) to (b, x) whenever $b \vdash a$.
       – OR (disjunction of input): from (a, x) and (b, x) to $(a \vee b, x)$.
       – WO (weakening the output): from (a, x) to (a, y) whenever $x \vdash y$.
       – AND (conjunction of output): from (a, x) and (a, y) to $(a, x \wedge y)$.
       – CT (cumulative transitivity): from (a, x) and $(a \wedge x, y)$ to (a, y).
       – ID (identity): from nothing to (a, a).

The derivation system based on the rules SI, WO and AND is called $deriv_1$. Adding OR to $deriv_1$ gives $deriv_2$. Adding CT to $deriv_1$ gives $deriv_3$. The five rules together give $deriv_4$. Adding ID to $deriv_i$ gives $deriv_i^+$ for $i \in \{1, 2, 3, 4\}$. $(a, x) \in deriv_i(G)$ is used to denote the norms (a, x) derivable from G using rules of derivation system $deriv_i$. In [19], it is proven that each $deriv_i^{(+)}$ is sound and complete with respect to $out_i^{(+)}$.

I/O logic is a general framework for normative reasoning, used to formalize and reason about the detachment of obligations, permissions and institutional facts from conditional norms. I/O logic is not defined in terms of a *truth-conditional* semantics. Rather, as pointed out above, I/O logic adopts *operational*

---

[7] In (11), $\vdash$ is the classical entailment relation of propositional logic.

semantics. As explained in [17,18], "directives *do not* carry truth-values. Only declarative statements may bear truth-values, but norms are items of another kind. They may be respected (or not), and may also be assessed from the standpoint of other norms, for example when a legal norm is judged from a moral point of view (or viceversa). But it makes no sense to describe norms as true or as false."

Thus, I/O logic allows to determine which obligations are operative in a situation that already violates some of them. To achieve this, we must look at the family of all maximal subsets G' such that G'⊆G and out_i(G', A) is consistent with A. The family of such out_i(G', A) is called the outfamily of (G, A).

To understand this concept, consider the following example. Suppose we have the following two norms: "The cottage should not have a fence or a dog" and "if it has a dog it must have both a fence and a warning sign." that we may formalize as the I/O logic generators "$(\top, \neg(f \lor d))$" and "$(d, f \land w)$" respectively.

Suppose further that we are in the situation that the cottage has a dog. In this context, the first norm is violated. And, the outfamily of (G, A) determines[8] that, still, we are obliged to build a fence with a warning sign around the cottage:

$$G \equiv \{(\top, \neg(f \lor d)), (d, f \land w)\}, A \equiv \{d\}, \text{outfamily(G, A)} \equiv \{Cn(f \land w)\}$$

Although Input/Output logic is an adequate framework for representing and reasoning on norms, only propositional logics have been used for asserting the generators and the input so far. This is because of issues related to the complexity of the framework. The complexity of input/output logic is at least as difficult as the objective logic. By the complexity of input/output logic, we mean the complexity of the following fulfillment problem:

$$\text{Given finite } G, A \text{ and } x, \text{ is } x \in out_i(G, A)?$$

[32] shows that the complexity of the fulfillment problem for out_1, out_2, out_4 is coNP complete, for out_3 the lower bound is coNP while the upper bound is P^{NP}.

There have been no efforts to represent norms coming from *existing* legal texts as those from the corpus described above in Sect. 2. For representing concrete existing norms, the expressivity of propositional logic is not sufficient.

First order (object) logics are needed in order to fill the gap between the Input/Output logic and the richness of NL semantics. To this end, in the next section, we propose a merger of the ProLeMAS Object Logic as defined in Sect. 3 with Input/Output logic. There is no precedent in the literature of a first-order Input/Output logic. However, it is worth noticing that this proposal is not the first deontic logic employing first-order relational variables.

5   The ProLeMAS Logic

The present section merges together Input/Output logic and the ProLeMAS object logic, whose syntax has been defined above in Definition 1. The resulting logic will be termed as "the ProLeMAS logic".

---

[8] In this example, outfamily(G, A)≡$\{Cn(f \land w)\}$ for all Input/Output logic $out_i^{(+)}$, with i=1,2,3,4.

We are not interested here in proposing first order versions of *all* definitions shown in the previous section, but we will restrict our attention to only those needed for the aims of the ProLeMAS project. In particular, the axiom OR in (11) does not appear to be suitable for legal reasoning. To see why consider the following obligations: "If someone kills a dog, s/he has to spend two years in prison" and "If someone robs a bank s/he has to spend two years in prison". And, suppose John did one of the two, but there is no way to understand *which* one, i.e. if either he killed a dog or he robbed a bank. Logically, John must spend two years in prison. But from a legal reasoning perspective, he must not: only if concrete evidence of what he did is found, obligations apply. Thus, in the rest of the paper we will no longer consider the OR axiom and, consequently, the Input/Output logic out_2 and out_4. On the other hand, we will focus in particular on the CT (cumulative transitivity) axiom, used to define the *simple-minded reusable output* logic out_3. It will also be easy to apply the considerations below to the remaining axioms SI, WO, AND, and ID, so that we will skip formal definitions about them.

From a formal point of view, recalling that the syntax of the ProLeMAS object logic admits free variables, the last ingredient needed to merge ProLeMAS object logic and out_3 are quantifiers bounding each a free variable. We impose these free variables to occur both in the head and the tail of a generator, and we bound them via universal quantifiers. This establishes a "bridge" between the head and the tail, needed to "carry" individuals from the input to the output. Consider these simple (toy) examples:

(12)   a. Each lawyer *must* run.
      b. A lawyer who runs *must* wear a pair of shoes.
      c. If John goes to Mary's house, he'll *have to* tell her before.

We propose to represent (12.a–c) via the following generators:

(13)   a. $\forall_x (lawyer(x), \exists_{e_r}[(Rexist\ e_r) \wedge run(e_r) \wedge agent(e_r, x)])$
      b. $\forall_x \forall_{e_r} (lawyer(x) \wedge (Rexist\ e_r) \wedge run(e_r) \wedge agent(e_r, x),$
$$\exists_{e_w} \exists_y [(Rexist\ e_w) \wedge wear(e_w) \wedge agent(e_w, x) \wedge$$
$$patient(e_w, y) \wedge shoes(y)])$$
      c. $\forall_{e_g} ((Rexist\ e_g) \wedge go(e_g) \wedge agent(e_g, \text{J}) \wedge to(e_g, \text{M}),$
$$\exists_{e_t} [(Rexist\ e_t) \wedge tell(e_t) \wedge agent(e_t, \text{J}) \wedge receiver(e_t, \text{M}) \wedge$$
$$theme(e_t, e_g) \wedge (happenBefore\ e_t\ e_g)])$$

Note that, in (13.b), x occurs in *both* the head *and* the tail of the generator, while e_r *only* occurs in the head. On the other hand, y and e_w are existentially quantified variables that occur *only* in the tail: every time a lawyer runs, there is a different "wearing" eventuality and (possibly) a different pair of shoes.

On the other hand, in (13.c), the eventuality e_g occurs in both the head and the tail. That's because the sentence means: "If John goes to Mary's house, he'll have to tell *Mary* before *that he goes to Mary*".

In our solution, free variables occurring in the heads (and possibly *also* in the tails) are outscoped by universal quantifiers. Free variables occurring *only* in the tails are outscoped by the existential quantifiers of the ProLeMAS object logic syntax (cf. Definition 1). Formally:

Definition 2 (ProLeMAS logic generators). *A generator in ProLeMAS logic is a construct in the form:*

$$\forall_{x_1,\ldots,x_n,y_1,\ldots,y_m} (\Phi(x_1,\ldots,x_n, y_1,\ldots,y_m), \Psi(y_1,\ldots,y_m)) \in \mathsf{G}$$

where $x_1,\ldots,x_n$ are free variables occurring only in Φ while $y_1,\ldots,y_n$ occur both in Φ and in Ψ. Φ and Ψ do not contain any other free variable. Furthermore, Φ does not contain existential quantifiers.

Sentence (2), copied in (14) for reader's convenience, which come from an EU directive in our corpus, can be represented in ProLeMAS logic in a straightforward manner. We simply increase the *size* of the formula, but not its *complexity*.

(14) A lawyer who wishes to practise in a Member State other than that in which he obtained his professional qualification shall register with the competent authority in that State.

The formula is:

$$\forall_x \forall_y \forall_{e_w} \forall_{e_p} (cond(x, y, e_w, e_p), action(x, y))$$

where:

$$cond(x, y, e_w, e_p) \Leftrightarrow lawyer(x) \wedge memberState(y) \wedge different(y, f_w(x)) \wedge$$
$$Rexist(e_w) \wedge want(e_w) \wedge practise(e_p) \wedge agent(e_w, x) \wedge$$
$$patient(e_w, e_p) \wedge agent(e_p, x) \wedge at(e_p, y)$$

$$action(x, y) \Leftrightarrow \exists_{e_r}[\, Rexist(e_r) \wedge register(e_r) \wedge agent(e_r, x) \wedge$$
$$patient(e_r, x) \wedge with(e_r, f_c(y)) \,]$$

where x, y, e_w, e_p, and e_r are variables denoting a lawyer, a Member State, and the eventualities of "wanting", "practising" and "registering". *agent*, *patient*, *at*, and *with* are thematic roles of the two eventualities. $f_w(x)$ is a function referring to the Member State where x obtained his professional qualification while $f_c(y)$ is a function that given a member state y returns the competent authority of y. The meaning of the formula is obvious: for every tuple of lawyer, Member State, and "wanting", and "practising" actions that satisfy together the predicates in *cond*, the predicates in *action* are instantiated on x and y.

6   Working with ProLeMAS Formulae

The main peculiarity of Reification-based logical frameworks is their formal simplicity. By instantiating FOL predicates on non-variable FOL terms, we obtain again propositional formulae. Thus, it is easy to see that ProLeMAS's generators do not increase the complexity of the Input/Output logic originally defined in [19], provided that two requirements are met: (1) the domain is finite; and (2) the input formulae are only atomic formulae in ProLeMAS object logic, i.e., FOL predicates instantiated on non-variable FOL terms, i.e., propositional formulae.

The aim of the ProLeMAS project is to build concrete NLP-based applications to be used in practical applications, where both requirements (1) and (2) are met. The domains of individuals will always be finite (e.g., the set of all lawyers in the EU). And, we are interested in performing normative reasoning on specific[9] individuals only, e.g., deriving all obligations a specific lawyer must obey, according to a specific normative code.

Universal quantifiers are just a compact way to refer to all individuals in the domain. We obtain equivalent formulae by simply substituing the universally quantified variables with all constants referring each to an individual of the domain. For instance, assume G contains generator (13.a) only:

$$G = \{\forall_x(lawyer(x), \exists_{e_r}[(Rexist\ e_r) \land run(e_r) \land agent(e_r, x)])\}$$

And, suppose the domain is made up of the individuals John and Jack, the former being a lawyer, the latter not. G is equivalent to the following G':

$$G'=\{\ (lawyer(\text{John}), \exists_{e_r}[(Rexist\ e_r) \land run(e_r) \land agent(e_r, \text{John})])$$
$$(lawyer(\text{Jack}), \exists_{e_r}[(Rexist\ e_r) \land run(e_r) \land agent(e_r, \text{Jack})])\ \}$$

Since John is a laywer while Jack is not, the *propositional* symbol "*lawyer*(John)" belongs to our initial facts while "*lawyer*(Jack)" does not, i.e. $lawyer(\text{John}) \in A$.

In out_1, we infer that John is obliged to run, i.e.

$$out_1(G', A)=\{\ run(f_1(\text{John})) \land agent(f_1(\text{John}), \text{John})\ \}$$

Where we substituted[10] the existential quantifier on e_r with a Skolem function f_1, so that $out_1(G', A)$ again contains propositional symbols only. Thus, it satisfies requirement (2) and, in out_3, it could be reused to trigger other obligations, e.g., being applied to an obligation such as "every runner must wear a pair of shoes".

We stress again that, thanks to Reification, we are not increasing the complexity of the original I/O logic. On finite domains, the formulae we are going to use turns out to be propositional. Original I/O logic definitions and proofs of soundness and completeness still hold, modulo generalizations via universal and existential quantifiers. For instance, CT is modified as follows:

---

[9] It could be the case that such applications will have to reason on *sets*, e.g., a sets of ten laywers. To properly deal with sets, Hobbs introduces in his framework the notion of *typical element* (cf. [13,14]).

[10] Skolemization is merely a formality to meet requirement (2). Alternatively, we could allow existential quantifiers on inputs and define a different pattern-matching rule between the input and the heads of the generators.

CT (cumulative transitivity):

from

$$\forall_{x_1..x_n}(\Phi(x_1, \ldots, x_n), \exists_{z_1..z_k}[\Psi(y_1, \ldots, y_m, z_1, \ldots, z_k)]),$$
$$\text{with } \{y_1, \ldots, y_m\} \subseteq \{x_1, \ldots, x_n\}$$

and

$$\forall_{x_1..x_n w_1..w_r}(\Phi(x_1, \ldots, x_n) \wedge \Psi(w_1, \ldots, w_r), \exists_{k_1..k_i}[\Upsilon(t_1, \ldots, t_l, k_1, \ldots, k_i)]),$$
$$\text{with } \{t_1, \ldots, t_l\} \subseteq \{x_1, \ldots, x_n, w_1, \ldots, w_r\}$$

to

$$\forall_{x_1..x_n}(\Phi(x_1, \ldots, x_n), \exists_{k_1..k_i m_1..m_s}[\Upsilon(p_1, \ldots, p_c, k_1, \ldots, k_i, m_1, \ldots, m_s)]),$$
$$\text{with } \{m_1, \ldots, m_s\} \subseteq \{w_1, \ldots, w_r\} \text{ and } \{m_1, \ldots, m_s\} \cup \{p_1, \ldots, p_c\} \equiv \{t_1, \ldots, t_l\}$$

7   Future Work and Conclusions

This paper is part of the ProLeMAS research project, which aims at (1) filling the gap between the current logical formalizations of legal text, mostly propositional, and the richness of Natural Language Semantics, and (2) formalizing norms extracted from existing legal documents.

The first step is to move beyond the propositional level towards first-order logical frameworks, in order to enhance the expressivity fit to formalize the meaning of the phrases constituting the sentences. ProLeMAS proposes to achieve such a result by using the Reification-based constructs from Hobbs. The key feature of Hobbs's approach, which distinguishes it from other neo-Davidsonian approaches, is the total avoidance of embeddings of logical operators within the scope of other logical operators.

This paper shows how it is possible to integrate Hobbs's account within I/O logic by adding universal and existential quantifiers to the latter. It also discusses how, provided the domain is finite, the complexity of the resulting framework does not increase with respect to that of propositional I/O logic. We consider this a great result, due to the complexity issues related to I/O logic.

Our next steps will involve the following future work:

(15)   a. Studying how the ProLeMAS logic could deal with other kinds of norms, such as permissions, powers, etc. And, eventually, extending the account to provide a proper representation of their meaning.

b. Designing suitable legal ontologies to represent and restrict the meaning of relevant predicates, in order to trigger automatic reasoning on the individuals in the Abox. We are particularly interested in developing legal ontologies in the data protection domain. Under the pressure from technological developments during the last few years, the EU legislation on data protection has shown its weaknesses, and is currently undergoing a long and complex reform that is finally approaching completion.

c. Building a concrete pipeline to populate the Abox of the ontology. In ProLeMAS, the pipeline will firstly process the documents via dependency parsing, then it will define a syntax-semantic interface from the dependency trees to the final formulae.

References

1. Araszkiewicz, M., Pleszka, K. (eds.): Logic in the Theory and Practice of Lawmaking. Springer, Heidelberg (2015)
2. Bach, E.: On time, tense, and aspect: An essay in english metaphysics. In: Cole, P. (ed.) Radical Pragmatics, pp. 63–81. Academic Press, New York (1981)
3. Bartolini, C., Muthuri, R.: Reconciling data protection rights and obligations: An ontology of the forthcoming EU regulation. In: Proceedings of the Workshop on Language and Semantic Technology for Legal Domain (LST4LD) (2015)
4. Boella, G., Di Caro, L., Ruggeri, A., Robaldo, L.: Learning from syntax generalizations for automatic semantic annotation. J. Intell. Inf. Syst. **43**(2), 231–246 (2014)
5. Copestake, A., Flickinger, D., Sag, I.A.: Minimal recursion semantics: An introduction. J. Res. Lang. Comput. **3**(2), 281–332 (2005)
6. Davidson, D.: The logical form of action sentences. In: Rescher, N. (ed.) The Logic of Decision and Action. University of Pittsburgh Press (1967)
7. Evans, D., Eyers, D.: Deontic logic for modelling data flow and use compliance. In: Proceedings of the 6th International Workshop on Middleware for Pervasive and Ad-hoc Computing, pp. 19–24. ACM, New York (2008)
8. Fornara, N., Colombetti, M.: Specifying artificial institutions in the event calculus. In: Dignum, V. (ed.) Handbook of Research on Multi-Agent Systems: Semantics and Dynamics of Organizational Models, pp. 335–366. IGI Global (2009)
9. Gabbay, D., Horty, J., Parent, X., van der Meyden, R., van der Torre, L. (eds.): Handbook of Deontic Logic and Normative Systems. College Publications, London (2013)
10. Galton, A.: Operators vs. arguments: The ins and outs of reification. Synthese **150**(3), 415–441 (2006)
11. Governatori, G., Olivieri, F., Rotolo, A., Scannapieco, S.: Computing strong and weak permissions in defeasible logic. J. Philos. Logic **42**(6), 799–829 (2013)
12. Hashmi, M., Governatori, G., Wynn, M.T.: Modeling obligations with event-calculus. In: Bikakis, A., Fodor, P., Roman, D. (eds.) RuleML 2014. LNCS, vol. 8620, pp. 296–310. Springer, Cham (2014). doi:10.1007/978-3-319-09870-8_22
13. Hobbs, J.R.: Monotone decreasing quantifiers in a scope-free logical form. In: Semantic Ambiguity and Underspecification, pp. 55–76 (1995)
14. Hobbs, J.R.: The logical notation: Ontological promiscuity. In: Discourse and Inference (1998). Chapter 2, http://www.isi.edu/~hobbs/disinf-tc.html
15. Kamp, H., Reyle, U.: From Discourse to Logic: An Introduction to Modeltheoretic Semantics, Formal Logic and Discourse Representation Theory. Kluwer Academic Publishers, Dordrecht (1993)
16. Kowalski, R., Sergot, M.: A logic-based calculus of events. New Gener. Comput. **4**(1), 67–95 (1986)
17. Makinson, D., van der Torre, L.: Permission from an input/output perspective. J. Philos. Logic **32**(4), 391–416 (2003)

18. Makinson, D., van der Torre, L.: What is input/output logic? In: Lwe, B., Malzkom, W., Rsch, T. (eds.) Foundations of the Formal Sciences II. Trends in Logic, vol. 17, pp. 163–174. Springer, Netherlands (2003)

19. Makinson, D., van der Torre, L.: Input/output logics. J. Philos. Logic **29**(4), 383–408 (2000)

20. McCarty, L.T.: A language for legal discourse I. basic features. In: Proceedings of the 2nd International Conference on Artificial Intelligence and Law (ICAIL 1989). ACM Press (1989)

21. McCarty, L.T.: Deep semantic interpretations of legal texts. In: 2007 Proceedings of the Eleventh International Conference on Artificial Intelligence and Law, 4–8 June 2007, Stanford Law School, Stanford, California, USA, pp. 217–224 (2007)

22. Miller, R., Shanahan, M.: The event calculus in classical logic alternative axiomatizations. Electron. Trans. Artif. Intell. **16**(4), 77–105 (1999)

23. Parent, X., van der Torre, L.: Input/output logic. In: Horty, J., Gabbay, D., Parent, X., van der Meyden, R., van der Torre, L. (eds.) Handbook of Deontic Logic and Normative Systems. College Publications, London (2013)

24. Paschke, A., Bichler, M.: SLA representation, management and enforcement. In: 2005 IEEE International Conference on e-Technology, e-Commerce, and e-Services (EEE 2005), 29 March–1 April 2005, Hong Kong, China, pp. 158–163 (2005)

25. Robaldo, L.: Interpretation and inference with maximal referential terms. J. Comput. Syst. Sci. **76**(5), 373–388 (2010)

26. Robaldo, L.: Distributivity, collectivity, and cumulativity in terms of (in)dependence and maximality. J. Logic Lang. Inf. **20**(2), 233–271 (2011)

27. Robaldo, L.: Conservativity: a necessary property for the maximization of witness sets. Logic J. IGPL **21**(5), 853–878 (2013)

28. Robaldo, L., Miltsakaki, E.: Corpus-driven semantics of concession: Where do expectations come from. Dialogue Discourse **5**(1), 1–36 (2014)

29. Robaldo, L., Szymanik, J., Meijering, B.: On the identification of quantifiers' witness sets a study of multi-quantifier sentences. J. Logic Lang. Inf. **23**(1), 53–81 (2014)

30. Sergot, M.J., Sadri, F., Kowalski, R.A., Kriwaczek, F., Hammond, P., Cory, H.T.: The british nationality act as a logic program. Commun. ACM **29**(5), 370–386 (1986)

31. Sun, X.: How to build input/output logic. In: 15th International Workshop on Computational Logic in Multi-Agent Systems, pp. 123–137 (2014)

32. Sun, X., Ambrossio, D.A.: On the complexity of input/output logic. In: Hoek, W., Holliday, W.H., Wang, W. (eds.) LORI 2015. LNCS, vol. 9394, pp. 429–434. Springer, Heidelberg (2015). doi:10.1007/978-3-662-48561-3_38

Using Ontologies to Model Data Protection Requirements in Workflows

Cesare Bartolini[1]([⊠]), Robert Muthuri[2], and Cristiana Santos[3]

[1] University of Luxembourg, Luxembourg, Luxembourg
cesare.bartolini@uni.lu
[2] University of Turin, Turin, Italy
robert.kiriinya@unito.it
[3] Institute of Law and Technology,
University of Barcelona (IDT-UAB), Barcelona, Spain
cristiana.teixeirasantos@gmail.com

Abstract. Data protection, currently under the limelight at the European level, is undergoing a long and complex reform that is finally approaching its completion. Consequently, there is an urgent need to customize semantic standards towards the prospective legal framework. The aim of this paper is to provide a bottom-up ontology describing the constituents of data protection domain and its relationships. Our contribution envisions a methodology to highlight the (new) duties of data controllers and foster the transition of IT-based systems, services, tools and businesses to comply with the new General Data Protection Regulation. This structure may serve as the foundation for the design of data protection compliant information systems.

Keywords: Legal ontology · Data protection · General data protection regulation · Compliance · Business process · BPMN

1 Introduction

The goal of the privacy and data protection domains of law is to protect the personal information of individuals (normally referred to as personal data) in a given jurisdiction. With the advent of social media and the uptake of digital technology, the availability of digital services and the soon-to-be Internet of Things have dramatically increased the amount of information collected and processed by governments and companies. Accordingly, businesses are continually developing techniques such as machine learning, big data analytics, natural language processing and applications to exploit data assets, to the detriment of new concerns of profiling, identification and re-identification risks.

The European Union (EU) is in the process of upgrading the current data protection law, which is based on the so-called Data Protection Directive (DPD), to a more modern and uniform legislation [36], in accordance with the recent technological progresses. The objective is to enhance individuals' rights, give

© Springer International Publishing AG 2017
M. Otake et al. (Eds.): JSAI-isAI 2015 Workshops, LNAI 10091, pp. 233–248, 2017.
DOI: 10.1007/978-3-319-50953-2_17

them more control over their own data, simplify the regulatory environment for businesses, and set the foundation for the Digital Single Market [15]. The main legislative document of the reform is the General Data Protection Regulation (GDPR), which constitutes the basis for the general protection of personal data. Although the new legislation is in its final stages, it will not be in force before 2018. The text of the GDPR is not finalized yet, and the latest official version released by the Commission dates back to early 2012[1].

A data subject is the individual to whom the personal data relate. On the other hand, a data controller is the natural or legal person who determines the purposes and means processing. The controller may delegate the actual processing to another entity called a data processor. Data Protection Authorities (DPAs) are mandated with regulating the controller and the processor while helping subjects to enforce their rights. In the light of the importance that the processing of personal data has attained over the last decade, the reform is trying to clarify and strengthen the rights of the data subjects. Correspondingly, the duties of the controller and of the processor become more burdensome and require new technical measures. As per the latest version of the GDPR, DPAs will have inquisitory powers with the possibility to levy fines as high as 5% of the annual global turnover [28]. Enterprises will thus be pressed to avoid infringements. However, most of the duties of the data controller are expressed in evaluative terms, making it difficult for the controller to know the exact extent of its obligations. For instance, the draft Regulation requires "appropriate technical and organizational measures" to ensure secure processing of personal data albeit, without further elaboration[2].

The foundations of European data protection have been laid out and evolved over several decades. Data protection involves a large number of stakeholders, including the controller, processor, data subject, recipient of transfer, national authority, legislator, auditor, and the data protection officer - a new role introduced in the draft Regulation. Additional roles which do not exist at the European level have been introduced in the legislation of some Member States. Such a context, along with the importance of the interests involved, entails a complex set of rules where each stakeholder has different powers, rights, and obligations. The technical evolution of the last decades has also significantly changed the environment in which the rules operate, blurring the distinction between the controller and the data subject [41]. Consequently, data protection in the legal domain nowadays represents a major challenge for any business or public administration involved in the processing of personal data, and a potential source of liability if its rules are not complied with correctly.

Achieving compliance is no easy task. The transition of a firm's organizational and technical measures could be eased if appropriate standards existed for it to adopt. However, no significant standards currently exist for data protection, much less in the light of the upcoming reform. Within computer science,

---

[1] However, versions amended by the Parliament and the Council have either been published or leaked to the general public.

[2] Article 30 of the draft Regulation.

data protection is often referred to as *privacy* and considered a subset of the security domain [27,32]. Significant differences exist between the two terms from a legal perspective, although some overlapping does exist. For example, some provisions in data protection legislation require that the data processing be performed under appropriate security measures. An early-stage research [6] aims at evaluating the overlapping between the GDPR and security standards, such as the ISO 27000 family, and in particular ISO 27001:2013 [24], to measure the degree of coverage of the data protection rules a security standard would cover. This facilitates controllers to understand what is required of them when they adopt a widespread security standard relying on many years of expertise and consolidated audit firm methodologies.

Our previous work [7] defined the specific research problem, the context within which it arose, the rationale behind a potential solution, and an ontology of the data protection domain in the context of the GDPR. Its objectives were focused on the scope and extent of the duties and obligations of the data controller to facilitate compliance with the GDPR.

In this paper we illustrate the design and development of the ontology following the initial stage described therein. As a proof of concept, we introduce an approach that uses the ontology to enrich a workflow model such as a business process, with annotations that express data protection requirements. In other words, the ontology will constitute the knowledge base from which the concepts to annotate the workflow model are extracted. Such an approach can provide benefits for a number of stakeholders:

– data controllers would have a clearer view of their duties with respect to data protection in the context of their business;
– the auditors would have a first-look model to assess the GDPR compliance;
– DPAs would have a structured approach to detect potential violations.

The paper is organized as follows. In Sect. 2, we describe the related work concerning domain legal ontologies within data protection and privacy, and business processes. Section 3 presents the ontology definition, explaining how to describe data protection concepts by means of ontologies and describing the ontology requirements and construction; finally, it summarizes some preliminary evaluation of the ontology. Section 4 portrays a sample extension of business processes using the envisioned legal ontology. Finally, in Sect. 5, we give a set of conclusions and future work.

2 Related Work

"Domain ontologies" in the legal field focus on a particular area of law, but their relevance is constrained by their subject-matter modeling [10] and only some have been applied beyond the prototype stage. Some of the pertinent domain ontologies are briefly mentioned in terms of their purpose, subject-matter, reusability, and availability. Despite efforts in modeling data protection domain, according to the best of our knowledge, there is no ontological representation that specifically addresses the data protection legislation in the light of the reform, the duties of data controllers and the corresponding rights of data subjects.

The LegLOPD ontology [29] was applied for the preservation of privacy in location-based services. It modeled concepts from the Spanish data protection law. The essential structure to be protected in LegLOPD is the concept of *private data*, derived from an LRI Core [8] abstract concept.

The OntoPrivacy [9] ontology modeled a glossary of keywords from the Italian Personal Data Protection Code. A bottom-up approach was used as the lexicon was the basis to build the ontology. It consisted of a domain ontology reusing top level ontologies. OntoPrivacy has been created to support a tool that allows to query the functional profile of legislative data.

The Neurona Ontologies [11] are application-oriented, and modeled the knowledge for the development of data protection compliance to offer reports regarding the correct application of security measures to data files containing personal data. Their design is based on a Data Protection Knowledge Ontology, which contains the core concepts of the system, and a Data Protection Reasoning Ontology, to assess data protection compliance. These ontologies provide legal professionals and citizens with better access to legal information, but could also support data protection and privacy compliance in organizations and administrations. However, there are several problems that make them unsuited for the purposes of the current research: the surveyed ontologies are proprietary, and their point of view is not focused on the duties of the data controller.

The Privacy by Design (PbD) approach requires that data protection measures be implemented prior to the means of processing being determined[3]. An ontology framework based on the PbD approach [26] consists of nine base ontologies, eight domain ontologies and four application specific ontologies. Another interesting approach is presented in [33]. However, that work is not focused on the obligations of the data controller, but rather on expressing the legal norms using an ontology to enforce access control policies.

The idea of using ontologies to extend notations is not novel [31,34]. It has been acknowledged in [22] that ontologies can be integrated in the Software Development Life Cycle (SDLC) in any situation where requirements in a domain are frequently used, e.g., the data protection requirements in our case. However, the proposal of this paper addresses the use of the ontology in software design not for the purposes of detailing the application domain of the software, but to specify legal constraints with which the software, or more generally the business process, must comply with. This approach will allow a more consistent interaction between the data controller, the auditors, and the DPAs to ease the transition to the GDPR.

---

[3] Article 23 of the GDPR, addressing the design and the implementation of a system.

3   An Ontology for Data Protection Rules

3.1   Ontology Engineering

Ontology Engineering refers to the set of activities that concern the ontology development process, the ontology life cycle, the methodologies for building ontologies, the tool suited and languages that support them [20]. For legal knowledge formalization we use the legacy guiding methodologies: METHON-TOLOGY and Neon specification tasks [39] to ensure a sustainable modeling. METHONTOLOGY [17,18] is a structured method to build ontologies, also applied to legal knowledge formalization [12], carrying out the whole *ontology development* process (through the specification, conceptualization, formalization, implementation, and maintenance tasks of the ontology), and its *support activities* (knowledge acquisition, integration, evaluation and documentation), tasks that we describe below.

The ontology *specification* phase expressed in the Ontology Requirement Specification [38] facilitates the ontology development and refers to the activity of collecting the requirements that the ontology should fulfill: (a) the purpose, intended scenarios of use, end-users, etc.; (b) level of formality of the implemented ontology; c) scope. In particular, the Ontology Requirement Specification Document (ORSD) (1) allows the identification of the particular knowledge that should be represented in the ontology; (2) facilitates the reuse of knowledge resources by means of focusing the resource search towards the particular knowledge to be represented; and (3) permits the verification of the ontology with respect to the requirements that the ontology should fulfill.

Accordingly, our ontological commitment [14] provides a foundational structure in relation to the new data protection reform. In particular it identifies the scope and extent of the obligations of the data controller, especially in relation to the rights of the data subject.

Pursuing the context of use (users and use), this work anticipates the impact that the GDPR is likely to have on firms once it enters into force. While businesses have a legitimate interest in collecting personal data as assets to achieve their business goals, they should also comply with regulatory requirements. The chosen context envisions integration/interoperation within a business process.

Functional requirements are represented in the form of informal Competency Questions (CQs) that the ontology must be able to answer. A CQ [40] is a natural language sentence that expresses a pattern for a type of questions the domain experts expect an ontology to answer. The ability to answer the CQs hence becomes a functional requirement of the ontology. We extracted the CQs from external expert generated content sources declared below. For our data protection ontology, the following are CQs: 1. What are the obligations of a data controller? 2. What are the functions of a data processor? 3. What are the rights of the data subject? 4. How do the rights of the data subject relate to the obligations of the data controller and the functions of the processor? 5. How can a data subject interact and/or enforce their rights against a data controller? 6. What are the possible fines and sanctions issued in response to violations by data controllers? 7. Who supervises a data controller?

As for the knowledge acquisition phase, we elicited domain expert conceptual knowledge to support our modeling decisions. We manually harvested from normative frameworks, particularly the DPD, the GDPR[4], and the Handbook on European data protection law [16].

Concerning non-functional requirements, this ontology is expressed in Web Ontology Language (OWL) [5] and uses Protégé [30] as the ontology development environment. A graphical depiction of the ontology is shown in Fig. 1. The framework presented in this paper relies on previous efforts of the community in the field of legal knowledge representation, therefore we reuse concepts from LKIF Core and SKOS.

3.2 Describing Data Protection Concepts

The conceptualization activity implies the organization and conversion of the informally perceived image of our domain into a semi-formal specification. Therefore, ontology components (concepts, attributes, relations, formal axioms and instances) were compiled using the sources described, and are here articulated through a task-oriented approach, to restate the informal competency questions.

A glossary of data protection terms was built and is provided together with the ontology[5].

The ontology's architecture follows the high-level partitioning structure of European data protection rules outlined by [16], and therefore it is made up of the following blocks:

1. the basic data protection principles;
2. the rules of data processing (constituting most of the duties of the data controller);
3. the data subject's rights.

An ontology entails a given level of consensus in a particular community. Within the data protection domain this includes basic data protection principles, as they have been established over the years by the Council of Europe (CoE), the EU, and the national DPAs. These serve as the foundation for our ontology. It is from these concepts that we derive and define the conceptual obligations of the data controller while contrasting them to the rights of the data subject. The result of the principles analysis is a set of ontology classes, their attributes and the relations between them.

The following is an enumeration of some of the principles, as classified under the European Data Protection Handbook [16]: lawfulness principle; purpose limitation principle (personal data must be processed for specified and lawful purposes); data quality principles (data must be adequate, relevant and not in excess

---

[4] Subject to changes in the final text - we used the official Commission text, COM (2012) 11 final. To better sharpen the scope, the ontology does not refer to decisions of courts or DPAs. The purpose is not to define a model of the legal text, but to model the requirements that the controller must meet to be compliant with the legislation.

[5] See footnote 12 *infra*.

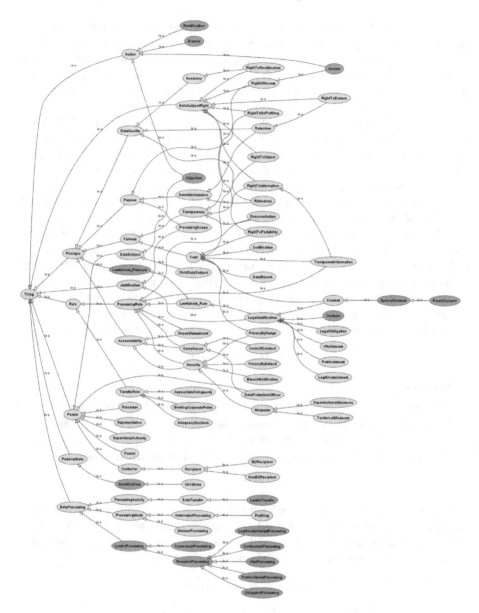

Fig. 1. Schema of the data protection ontology.

in relation to the purpose of the processing, accurate, up to date); principle of data minimization, among others. A more detailed description of the principles underlying the ontology is given in [7].

The data protection principles constitute the unifying harmony underlying a controller's obligations (called *Rules* in the ontology, to ensure consistency with the knowledge sources) and data subject's rights. Since they are reifications of the general principles, in the ontology every data processing rule or data subject's right is a subclass of some principles. For example, the LawfulnessPrinciple entails a LawfulnessRule, which *is a* processing rule, and can consist of the data subject's Consent, a LegalObligation of the controller, a VitalInterest, a Contract, and so on.

To relate the data subject's rights with the corresponding rules of the controller, we define the deontic concepts in terms of correlative relations between right and rule (obligation), assuming symmetric roles. For example, the data subject's right to access corresponds to the obligation of the controller to provide means to request access to the data. To exercise the right, the data subject must perform a single access, which, by means of an object property, is defined in terms of the right to access, and is bound by a relationship with the data for which access is requested; similarly, the data subject can exercise the right to object to the processing of personal data. The objection, connected to the right to object, is related to a specific processing by a functional property called isObjected, defined in the domain of Processing. This property is also used to define the lawfulness of the processing, because personal data cannot be lawfully processed if the data subject has exercised the right to object.

Table 1 shows the hierarchy of the main concepts of the ontology.

Table 1. Top-level hierarchy.

Root classes	Subclasses
Data processing	Processing activity, processing mode, lawful processing
Data subject right	Right to rectification, right to object, right to no profiling, right to portability, right to erasure
Processing rule	Compliance, impact assessment, transparent information, security, lawfulness rule

Relations bind two resources (normally classes), and for each relation a *domain* and a *range* can be defined. A domain is the set of possible classes where the relation can be applied, and a range is the set of possible values of a relation. Table 2 shows the main relations in the ontology.

Table 2. Main relations.

Relation	Domain	Range
hasObligation	Controller	Legal obligation
notifyBreach	Controller	Data breach
consentGrantedBy	Consent	Data subject
AccessData	Right of access	Personal data

To formalize the ontology, a useful subset of classes were reused from LKIF Core in order to offer a solid support for the acquisition, sharing and reuse of legal knowledge. LKIF Core [23] is an established legal ontology. Our most generic concepts were linked with LKIF-Core concepts (such as the right, rule, legal person and natural person) using the SKOS data model[6]. Our alignment is compliant to it, but axiomatizes domain concepts of data protection, which is our priority and ontological commitment. There was therefore no need to extend the core ontology.

The main ontology metrics are summarized in Table 3.

Table 3. Ontological components.

Axiom	822
Logical axiom count	279
Class count	88
Object property count	42
Data property count	3
Individual count	16
DL expressivity	ALCHOIQ(D)
SubClassOf axioms count	114
EquivalentClasses axioms count	25
DisjointClasses axioms count	7

3.3 Evaluation

To evaluate the technical quality and consistency checking of the ontology, we used OntOlogy Pitfall Scanner! (OOPS!)[7] as pitfall detector. The results of the analysis were evaluated, and we assert no problems or inconsistencies were found in the ontology. The ontology documentation, containing the classes, properties and individuals, is available online[8], built using the Live OWL Documentation Environment (LODE) tool.

The usage of informal CQs for ontology requirements' description and its further evaluation has already been accounted [21] in ontology design methodologies. In fact, the ability to answer a CQ meaningfully can be regarded as a functional requirement that must be satisfied by the ontologies. The CQs presented in Subsect. 3.1 were built into a set of SPARQL Protocol and RDF Query Language (SPARQL) queries. The execution of the evaluation environment[9] showed that the ontology is able to answer those CQs (except #6,

---

[6] http://www.w3.org/2009/08/skos-reference/skos.html.

[7] http://oops.linkeddata.es/.

[8] http://www.essepuntato.it/lode/owlapi/https://raw.githubusercontent.com/guerret/lu.uni.eclipse.bpmn2/master/resources/dataprotection.owl.

[9] The environment is available together with the Eclipse plugin described in Subsect. 4.1. See footnote 12 *infra*.

since the fines are not modeled in the ontology yet). For example, a SPARQL query requesting the rights of the data subject returns the following result: RightToPortability, RightToInformation, RightToObject, RightToRectification, RightOfAccess, RightToErasure, RightToNoProfiling, TransparentInformation.

4 Extending Business Process Notation

The ontology described in Sect. 3 can be used to aid a data controller in being compliant with the GDPR. When developing a software system, the PbD approach mentioned in Sect. 2 means that the development cycle should address data protection. The development cycle is a workflow which can be expressed by means of formal notations such as Unified Modeling Language (UML) [25]. However, UML is domain-neutral. To express data protection, the data protection ontology would be useful: by exploiting UML's extensibility features [3], such as profiles, the expressiveness of (for example) activity or sequence diagrams can be enhanced, to specify data protection activities or requirements that a certain software routine, component, class should address.

But this is not sufficient for GDPR compliance. Many of the obligations of the GDPR involve organizational requirements as a risk assessment, and sometimes manual processing is required. Some of these activities have nothing to share with software development, but are still subject to the GDPR.

In this perspective, business processes [13] are more suited to embrace all the activities that can be subject to the GDPR, whether they are performed manually or software-based, or have a technical or organizational nature. Business processes are used to provide a description of the relationships between the various activities performed within a business, at various degrees of detail [37].

Various notations exist for specifying business processes, the most popular of which are Web Services Business Process Execution Language (WS-BPEL) [4] and Business Process Model and Notation (BPMN) [2], which have some similarities but still differ in scopes and domains [35]. They are based on an eXtensible Markup Language (XML) grammar and have extensibility features, including the possibility of using tags from different XML languages such as the OWL/XML serialization of the ontology. We chose to implement an extension of BPMN to demonstrate the possibilities offered by the present work, but the methodology is a general one that can be applied to any extensible notation.

Introducing data protection requirements by means of an ontology is a methodology that can be used in conjunction with different technologies, and it also provides a means to make heterogeneous models interoperable. In other words, the description of a workflow process might use different models at different levels e.g., UML and BPMN: if both are extended using the same ontology, the data protection requirements would be consistent, thus easing the integration and auditing of the overall workflow.

It would be easy to extend the methodology to use different ontologies. By using an ontology expressing the legal requirements in a specific domain (e.g., regulations for financial or healthcare services), this can be an effective method to model a clear and immediate view of the requirements in a workflow.

4.1   BPMN Implementation

The proposed approach has been integrated, although at a basic level, in a BPMN 2.0 modeling tool. BPMN does not have a uniform implementation. Although it is defined as a standard, it is designed so that its implementation is platform-specific. For the purposes of the present paper, we have selected the Eclipse BPMN2 Modeler[10]. It is an Eclipse plugin which implements BPMN features using Model-Driven Engineering (MDE) techniques and the Ecore metamodel. The Eclipse version used is 4.5 (Mars).

The BPMN standard defines several different diagrams (PROCESS, COLLABORATION, CHOREOGRAPHY and CONVERSATION) which serve different purposes. For this example, we only focused on the PROCESS diagram, although the same methodology can be extended to all diagram types. We created an Eclipse extension plugin for BPMN, defining a new type of TASK called DATA PROTECTION TASK. The new task type has a distinctive graphical appearance (marked with a red icon) and supports annotations extracted from the ontology. The properties of the new DATA PROTECTION TASK include a new tab which allows to introduce the annotations for data protection.

The implementation of the form to add the annotations parses through the data protection ontology using the OWL Application Programming Interface (API)[11]. Since our purpose is to offer a way to specify the activities that a data controller must perform for GDPR compliance, the reasoner selects the OWL classes that are descendants of the Rule class. This is a rough implementation, but it can be refined at the desired level, using the ontology structure or its instances, adding extra parameters and so on.

Figure 2 shows the interface of the extension plugin in operation and a sample application of the extended notation[12]. The example, which is built upon the official BPMN example from [1, p. 170], is not a real business process, but only aims at showing the possibilities of our approach. Some TASKs have been replaced with DATA PROTECTION TASKs. So, for example, the Handle Order activity has been annotated with the following three ontology classes:

Consent because the data subject must consent to the processing;
Security to ensure the protection of security measures;
AppropriateSafeguards because the customer's data might have to be transmitted to a vendor which might be located in a non-EU country.

---

[10] https://www.eclipse.org/bpmn2-modeler/.
[11] http://owlapi.sourceforge.net/.
[12] The sources are available at https://github.com/guerret/lu.uni.eclipse.bpmn2. The "resources" folder contains the OWL file with the ontology, the SPARQL queries and the glossary.

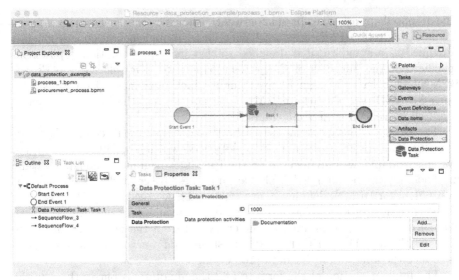

(a) The BPMN extension plugin.

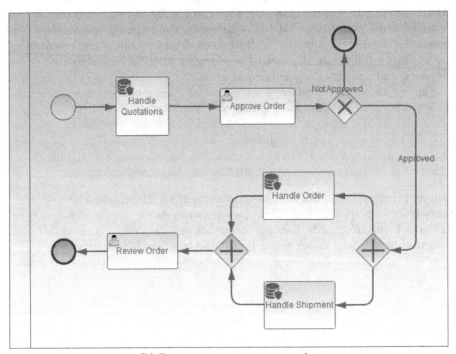

(b) Procurement process example.

Fig. 2. The data protection ontology extension plugin.

5   Conclusions

In this work, the authors have presented two artifacts: an ontology to model data protection requirements, and an approach for integrating it into a workflow to express the GDPR requirements within a business process by means the ontology. Our main objective is that the ontology will to assist data controllers in achieving compliance with the upcoming data protection reform. We aim to achieve this with a rigorous evaluation of the ontology and its extension to business process modelling within the data protection domain in the next phase of this research.

The ontology was modeled by a legal expert. The provisions that contain duties for the data controller and rights for the data subject have been selectively identified and built into the ontology. The granularity of the ontology is still coarse. High detail would be required in a judicial perspective, but not in the scope of the current research. However, some of the concepts expressed in the ontology appear to be generic or evaluative because they are expressed as such in the law, and not fit for direct usage. These concepts must be coordinated with knowledge from other domains. Computer security standards can partly fill these gaps, so understanding the relationship between them and the GDPR would be key to a fast transition to the new legislation.

This ontology is by definition a work in progress. It will have to be adapted to the changes in the legal text when a final version of the GDPR is released. However, in the final text, the core concepts expressed in the ontology won't drift significantly from the current ones. This structure is the basis for further refinements. It will act as a starting point which was necessary to pursue the long-term goal of verifying compliance with the GDPR. An improved version of the ontology is currently under development. It will feature a much broader and complete perspective on the GDPR, and will be designed to address many provisions not covered by the current version.

The workflow integration is an intuitive and simple way of expressing the GDPR requirements within the workflow. While not as rich and complex as some of the languages and models used in requirements engineering (such as SysML [19]), it clearly expresses the relationship between specific duties of the data controller and the workflow activities where the duties apply.

The approach presented in this work may ease the transition from the DPD to the GDPR and provide a basis for the PbD model. It can provide benefits to all end-users. Data controllers and processors will be able to determine what their duties are, on the basis of the rights of the data subject. Auditors will have a structured knowledge that can dissipate the mists of terminological uncertainties. DPAs can speed up their procedures thanks to a clearer notation. The formalization of the meaning of legal terms in an ontology could help compare the impact of the new legislation on the existing national regimes, as well as overcome linguistic differences in data protection across the EU. Also, expressing the controllers' requirements through an ontology will allow them to easily adapt designs to changes in the law and its interpretation, in a dynamic perspective.

The ontology may encompass automated classification to facilitate finding documents. Querying performance is foreseen as a future development in our ontology, using SPARQL-DL to ascertain the corresponding rights and duties. For example, a database structured according to the ontology could be queried by data subjects to retrieve the rights and remedies in case of breaches and violations; by data controllers, to understand their obligations; by data processors, to clarify their functions.

The long-term aims of the current research focus on assessing compliance to the GDPR by means of security standards. This purpose will require development a similarly-structured ontology for security standards and a methodology to compare the degree of overlapping between the two normative bodies.

From a technical perspective, there are a number of improvements that can be investigated as well. The sample plugin introduced in Sect. 4 could benefit from a more formal implementation using MDE, for example by defining the meta-model of the extension, integrating it with the meta-models of BPMN and OWL, and using it to generate the supporting classes.

Regardless of the underlying technologies used, the integration of the SDLC or business process notation with the data protection annotations from the ontology could also be enhanced with metrics to analyze the degree of coverage of the GDPR. Finally, when the ontology reaches a sufficient degree of maturity, a full-fledged real-world scenario will be modeled using the proposed notation.

References

1. BPMN 2.0 by example. Technical report. dtc/2010-06-02, Object Management Group, June 2010
2. Business process model and notation (BPMN). Technical report. formal/2011-01-03, Object Management Group, January 2011
3. Alhir, S.S.: Guide to Applying the UML. Springer Professional Computing, New York (2002)
4. Alves, A., Arkin, A., Askary, S., Barreto, C., Bloch, B., Curbera, F., Ford, M., Goland, Y., Guízar, A., Kartha, N., Liu, C.K., Khalaf, R., König, D., Marin, M., Mehta, V., Thatte, S., van der Rijn, D., Yendluri, P., Yiu, A.: Web services business process execution language version 2.0. Technical report, OASIS, April 2007. http://docs.oasis-open.org/wsbpel/2.0/OS/wsbpel-v2.0-OS.html
5. Antoniou, G., van Harmelen, F.: Web ontology language: OWL. In: Staab, S., Studer, R. (eds.) Handbook on Ontologies. International Handbooks on Information Systems, 2nd edn., pp. 67–92. Springer, Heidelberg (2004). Chapter 4
6. Bartolini, C., Gheorghe, G., Giurgiu, A., Sabetzadeh, M., Sannier, N.: Assessing IT security standards against the upcoming GDPR for cloud systems. In: Proceedings of the Grande Region Security and Reliability Day (GRSRD 2015), pp. 40–42, March 2015
7. Bartolini, C., Muthuri, R.: Reconciling data protection rights and obligations: an ontology of the forthcoming eu regulation. In: Proceedings of the Workshop on Language and Semantic Technology for Legal Domain (LST4LD), Recent Advances in Natural Language Processing (RANLP), September 2015
8. Breuker, J., Hoekstra, R.: Epistemology and ontology in core ontologies: FOLaw and LRI-Core, two core ontologies for law. In: Proceedings of the Workshop on Core Ontologies in Ontology Engineering (EKAW), October 2004

9. Cappelli, A., Lenzi, V.B., Sprugnoli, R., Biagioli, C.: Modelization of domain concepts extracted from the Italian privacy legislation. In: Proceedings of the 7th International Workshop on Computational Semantics (IWCS-7), January 2007
10. Casellas, N.: Legal Ontology Engineering Methodologies, Modelling Trends, and the Ontology of Professional Judicial Knowledge. Law, Governance and Technology Series, vol. 3. Springer, Netherlands (2011)
11. Casellas, N., Nieto, J.E., Roig, A., Meroño, A., Torralba, S., Reyes, M., Casanovas, P.: Ontological semantics for data privacy compliance: the Neurona project. In: Proceedings of the Intelligent Privacy Management Symposium, pp. 34–38, March 2010
12. Corcho, O., Fernández-López, M., Gómez-Pérez, A., López-Cima, A.: Building legal ontologies with METHONTOLOGY and WebODE. In: Benjamins, V.R., Casanovas, P., Breuker, J., Gangemi, A. (eds.) Law and the Semantic Web. Lecture Notes in Computer Science, vol. 3369, pp. 142–157. Springer, Berlin Heidelberg (2005)
13. Davenport, T.H., Short, J.E.: The new industrial engineering: information technology and business process redesign. Sloan Manag. Rev. 31(4), 11–27 (1990). Summer
14. Davis, R., Shrobe, H., Szolovits, P.: What is a knowledge representation? AI Mag. 14(1), 17–33 (1993). Spring
15. European Commission: A digital single market strategy for Europe, May 2015. http://eur-lex.europa.eu/legal-content/EN/TXT/PDF/?uri=CELEX: 52015DC0192&from=EN
16. European Union Agency for Fundamental Rights: Handbook on European data protection law, April 2014
17. Fernández, M., Gómez-Pérez, A., Juristo, N.: METHONTOLOGY: from ontological art towards ontological engineering. In: Proceedings of the Ontological Engineering AAAI-1997 Spring Symposium Series, pp. 33–40, March 1997
18. Fernández López, M., Gómez-Pérez, A., Pazos Sierra, J., Pazos Sierra, A.: Building a chemical ontology using methontology and the ontology design environment. IEEE Intell. Syst. 14(1), 37–46 (1999)
19. Friedenthal, S., Moore, A., Steiner, R.: A Practical Guide to SysML: The Systems Modeling Language, 3rd edn. Morgan Kaufmann, San Francisco (2014)
20. Gómez-Pérez, A., Fernández-López, M., Corcho, O.: Ontological Engineering: With Examples from the Areas of Knowledge Management, e-Commerce and the Semantic Web. Advanced Information and Knowledge Processing. Springer, London (2004)
21. Grüninger, M., Fox, M.S.: The role of competency questions in enterprise engineering. In: Rolstadås, A. (ed.) Benchmarking — Theory and Practice. IFIP, pp. 22–31. Springer, Boston, MA (1995). doi:10.1007/978-0-387-34847-6_3
22. Hesse, W.: Ontologies in the software engineering process. In: Lenz, R., Hasenkamp, U., Hasselbring, W., Reichert, M. (eds.) Proceedings of the 2nd GI-Workshop on Enterprise Application Integration (EAI), pp. 3–15, June 2005
23. Hoekstra, R., Breuker, J., Di Bello, M., Boer, A.: LKIF core: principled ontology development for the legal domain. In: Breuker, J., Casanovas, P., Klein, M.C., Francesconi, E. (eds.) Law, Ontologies and the Semantic Web: Channelling the Legal Information Flood, Frontiers in Artificial Intelligence and Applications, vol. 188, pp. 21–52. IOS Press, January 2009
24. International Organization for Standardization: ISO/IEC 27001 - Information technology - Security techniques - Information security management systems - Requirements, 2nd edn., October 2013

25. Jacobson, I., Booch, G., Rumbaugh, J.: The Unified Software Development Process. Addison-Wesley, Reading (1999)

26. Kost, M., Freytag, J.C., Kargl, F., Kung, A.: Privacy verification using ontologies. In: Proceedings of the Sixth International Conference on Availability, Reliability and Security (ARES), pp. 627–632, August 2011

27. Massacci, F., Prest, M., Zannone, N.: Using a security requirements engineering methodology in practice: the compliance with the Italian data protection legislation. Technical report. University of Trento, November 2003

28. Mikkonen, T.: Perceptions of controllers on EU data protection reform: a finnish perspective. Comput. Law Secur. Rev. **30**(2), 190–195 (2014)

29. Mitre, H.A., González-Tablas, A.I., Ramos, B., Ribagorda, A.: A legal ontology to support privacy preservation in location-based services. In: Meersman, R., Tari, Z., Herrero, P. (eds.) On the Move to Meaningful Internet Systems 2006: OTM 2006 Workshops. LNCS, vol. 4278, pp. 1755–1764. Springer, Heidelberg (2006)

30. Noy, N.F., Sintek, M., Decker, S., Crubézy, M., Fergerson, R.W., Musen, M.A.: Creating semantic web contents with Protégé-2000. IEEE Intell. Syst. **16**(2), 60–71 (2001)

31. Paulheim, H., Probst, F.: Ontology-enhanced user interfaces: a survey. Int. J. Semant. Web Inf. Syst. **6**(2), 36–59 (2010)

32. Pfleeger, C.P., Pfleeger, S.L.: Security in Computing, 4th edn. Prentice Hall, Upper Saddle River (2006)

33. Rahmouni, H.B., Solomonides, T., Casassa Mont, M., Shiu, S.: Privacy compliance and enforcement on European healthgrids: an approach through ontology. Phil. Trans. R. Soc. A **368**(1926), 4057–4072 (2010)

34. Rebstock, M., Fengel, J., Paulheim, H.: Ontologies-Based Business Integration. Business Information Systems. Springer, Heidelberg (2008)

35. Recker, J.C., Mendling, J.: On the translation between BPMN and BPEL: conceptual mismatch between process modeling languages. In: Latour, T., Petit, M. (eds.) The 18th International Conference on Advanced Information Systems Engineering. Proceedings of Workshops and Doctoral Consortium, pp. 521–532. Namur University Press, June 2006

36. Reding, V.: The upcoming data protection reform for the European Union. Int. Data Priv. Law **1**(1), 3–5 (2011). https://academic.oup.com/idpl/article/1/1/3/759666/The-upcoming-data-protection-reform-for-the

37. Reijers, H.A.: Design and Control of Workflow Processes: Business Process Management for the Service Industry. Lecture Notes in Computer Science, vol. 2617. Springer, Heidelberg (2003)

38. Suárez-Figueroa, M.C., Gómez-Pérez, A., Villazón-Terrazas, B.: How to write and use the ontology requirements specification document. In: Meersman, R., Dillon, T., Herrero, P. (eds.) On the Move to Meaningful Internet Systems: OTM 2009. Lecture Notes in Computer Science, vol. 5871, pp. 966–982. Springer, Heidelberg (2009)

39. Suárez-Figueroa, M.C., Gómez-Pérez, A., Motta, E., Gangemi, A. (eds.): Ontology Engineering in a Networked World. Springer, Heidelberg (2012)

40. Uschold, M., Gruninger, M.: Ontologies: principles, methods and applications. Knowl. Eng. Rev. **11**(2), 93–136 (1996)

41. Van Alsenoy, B., Ballet, J., Kuczerawy, A., Dumortier, J.: Social networks and web 2.0: are users also bound by data protection regulations? Identity Inf. Soc. **2**(1), 65–79 (2009)

Utilization of Multi-word Expressions to Improve Statistical Machine Translation of Statutory Sentences

Satomi Sakamoto[1], Yasuhiro Ogawa[1,2(✉)], Makoto Nakamura[3],
Tomohiro Ohno[1,2], and Katsuhiko Toyama[1,2]

[1] Graduate School of Information Science, Nagoya University,
Furo-cho, Chikusa-ku, Nagoya 464-8601, Japan
satomi@kl.i.is.nagoya-u.ac.jp, yasuhiro@is.nagoya-u.ac.jp
[2] Information Technology Center, Nagoya University,
Furo-cho, Chikusa-ku, Nagoya 464-8601, Japan
[3] Graduate School of Law, Japan Legal Information Institute, Nagoya University,
Furo-cho, Chikusa-ku, Nagoya 464-8601, Japan

Abstract. Statutory sentences are generally difficult to read because of their complicated expressions and length. Such difficulty is one reason for the low quality of statistical machine translation (SMT). Multi-word expressions (MWEs) also complicate statutory sentences and extend their length. Therefore, we proposed a method that utilizes MWEs to improve the SMT system of statutory sentences. In our method, we extracted the monolingual MWEs from a parallel corpus, automatically acquired these translations based on the Dice coefficient, and integrated the extracted bilingual MWEs into an SMT system by the single-tokenization strategy. The experiment results with our SMT system using the proposed method significantly improved the translation quality. Although automatic translation equivalent acquisition using the Dice coefficient is not perfect, the best system's score was close to a system that used bilingual MWEs whose equivalents are translated by hand.

Keywords: Multi-word expressions · Statistical machine translation · Legal information sharing

1 Introduction

As human globalization continues, the translation of Japanese statues into foreign languages is required to meet such demands as promoting investment and facilitating international transactions. In 2009, the Japanese Ministry of Justice released the Japanese Law Translation Database System (JLT) [7], which translates Japanese statues into English. However, JLT's translation speed and volume are inadequate because of the difficulty of legal translations.

Many reasons have been offered for this difficulty: excessive technical terms, complicated dependencies in the sentence structure, long sentences, etc.

© Springer International Publishing AG 2017
M. Otake et al. (Eds.): JSAI-isAI 2015 Workshops, LNAI 10091, pp. 249–264, 2017.
DOI: 10.1007/978-3-319-50953-2_18

These reasons are not independent. A sentence that includes complicated expressions with many modifiers becomes too long. This difficulty degrades the quality of statistical machine translation (SMT). SMT systems learn how likely it is that the words in the target language are the translations of words in the source language in a parallel corpus, which is a collection of bilingual aligned sentences, by automatic word alignment. The quality of the word alignment affects the translation. For long sentences, word alignment tends to fail due to excessive alignment candidates, which makes the translation quality degraded.

Bui et al. [6] split long statutory sentences into sub-sentences based on their logical structure and improved the translation of a tree-based SMT system. In contrast, we focus on multi-word expressions (MWEs), which are defined as "idiosyncratic interpretations that cross word boundaries (or spaces)" [17]. Statutory sentences include many MWEs, such as technical terms and boilerplate phrases. In this paper, we use MWEs in the broader sense, where they include not only compound nouns and idiomatic phrases but also functional expressions and other meaningless segments. Since some of them tend to be translated into a foreign language in a non-compositional way, their automatic word alignments suffer from noise. Tsvetkov et al. [19] focused on the fact that MWEs cause incorrect word alignment in a parallel corpus and proposed a general methodology to extract bilingual MWEs with such misalignments. They used GIZA++ [11], which is a popular bilingual word aligner, and acquired the translations of extracted MWEs from its phrase table. However, since their candidates are based on misalignment, the translation quality is unreliable. Thus, we introduce another method based on the Dice similarity coefficient to acquire the translations of MWEs and introduce a single-tokenization technique to integrate bilingual MWEs with SMT.

The purpose of this paper is to improve the translation quality of statutory SMT with MWEs. We propose a utilization method that extracts MWEs by Tsvetkov's method and acquire their translations by word alignment based on the Dice coefficient, and integrate the extract bilingual MWEs into an SMT system by Pal's single-tokenization strategy [12]. We evaluate the contribution of our method on SMT with translation experiments.

This paper is organized as follows: in the next section, we describe the related works of MWEs. We propose our MWE extraction method in Sect. 3. In Sect. 4, we describe and evaluate four experiments. The first experiment is an acquisition bilingual MWE dictionary using our proposed method. The other three are translations using an SMT with the acquired bilingual MWEs. Finally, we summarize and conclude our paper in Sect. 5.

2   Related Works

2.1   Multi-word Expressions for Natural Language Processing

MWEs are the expressions of compound words. Examples include conjunctions ("as well as"), idioms ("keep one's fingers crossed" that means to hope for a positive result), phrasal verbs ("find out"), compound nouns ("bus stop"), phrasal prepositions ("according to"), etc. [17]. MWEs appear in all text genres and pose significant problems for every kind of natural language processing (NLP) [17]. Since some have a meaning that cannot be derived from the component words, identifying and understanding MWEs is essential to language understanding and is important for any NLP applications that address robust language meaning and use. For example, MWEs play a role in tasks of word sense disambiguation since they are less polysemous than mono-words on average. Finlayson and Kulkarni [4] argued that the word *world* has nine different senses and *record* has fourteen, but the MWE *world record* has only one. In addition, many NLP tasks and applications like optical character recognition, morphological and syntactic analysis, information retrieval, computer-aided lexicography, and foreign language learning can be supported by MWEs [15].

2.2   Statistical Machine Translation with MWEs

Bilingual MWEs can contribute to SMT improvement. Improving the performance of an SMT system needs a good quality word and phrase alignment that acquires translation knowledge from a parallel corpus. MWEs can solve the major lexical ambiguity problems for any language and improve the quality of word alignment.

Ren et al. [16] proposed three methods to improve the translation model of SMT system using the bilingual MWEs extracted by an automatic process. Their first method is model retraining in which they take the automatically extracted bilingual MWEs as parallel sentence pairs, add them to the training corpus, and retrain the model using GIZA++. In the second method, the additional feature, they append one feature to a bilingual phrase table to indicate whether a bilingual phrase contains bilingual MWEs. The third method is the additional phrase table of bilingual MWEs in which they construct an additional phrase table that contains automatically extracted bilingual MWEs and combine the original phrase table and the newly constructed bilingual MWE table. These methods improved the SMT system and achieved the highest improvement with the additional features. SMT systems must train MWEs as special expressions that are different from common phrases.

Following the above study, Pal et al. [12] proposed a method of handling MWEs by tokenizing them into a single word: *single-tokenization*. The single-tokenization of MWEs is a process where the spaces in MWEs are replaced by a special symbol, "_", and an MWE is regarded as a single word. They successfully improved the SMT quality using this method to integrate bilingual MWEs that were automatically extracted.

2.3   MWE Extraction Methods

Since MWEs resemble collocations, early approaches to identifying them focused on their collocational behavior. However, in fact, collocation measures are inadequate for identifying MWEs. Hybrid methods, which combine word statistics with such linguistic information as morphological, static, and semantic idiosyncrasies, need to extract idiomatic MWEs [14].

Some works have focused on the semantic properties of MWEs. One work proposed a method that used Latent Semantic Analysis [8] and measured semantic properties to replace semantically related terms [2].

In recent years, other works have exploited the translational correspondences of MWEs from a parallel corpus. Caseli et al. [1] identified MWEs by an alignment-based approach. They extracted source sequences whose lengths exceeded two words and aligned them with one or more words in the target text. Then they filtered this set to determine whether it complies with the predefined POS patterns or whether they are sufficiently frequent in the parallel corpus. They marked precision below 40% and recall around 5%. Zarrieß and Kuhen [20] also used aligned parallel corpora and focused on one-to-many word alignments.

The previous works focused on specific syntactic patterns and assumed that MWEs are distinguished as phrases in the processing of alignment systems. While these methods rely on the quality of automatic word alignment, there are cases when their quality is insufficient due to the shortage of language resources. Tsvetkov et al. [19] proposed a method for identifying MWEs in bilingual corpora for their Hebrew-English domain that suffers from a dearth of parallel corpora, semantic dictionaries, and syntactic parsers. They extracted MWEs with automatic word alignment and focused on where the word alignment system failed. We adopt part of this scheme in our proposed method and explain it in detail in the next section.

3   Proposed Method of Bilingual MWE Utilization

In this section, we propose a method of bilingual MWE utilization to improve the SMT system of statutory sentences. Our methodology consists of three steps. First, we extract monolingual MWEs from a parallel corpus by Tsvetkov's method [19], which is language independent. Although Tsvetkov's method can extract bilingual MWEs, the translation quality is inadequate for our scheme. Thus, secondly, we acquire their translations by word alignment based on the Dice coefficient. Finally, we integrate the bilingual MWEs into an SMT system by Pal's single-tokenization strategy [12]. Figure 1 shows our proposed method in the form of flow diagram.

3.1   Monolingual MWE Extraction from Misalignments

The simple way of translation is the replacement each word of the source language with the equivalent word of the target language. If all of the expression of the source language has the compositional translation, that means the word by word translation, we can get the perfect translation by this simple way. Regrettably, the way is disturbed by the existence of MWEs. Because the MWEs are often translated into the expression of other language by non-compositional

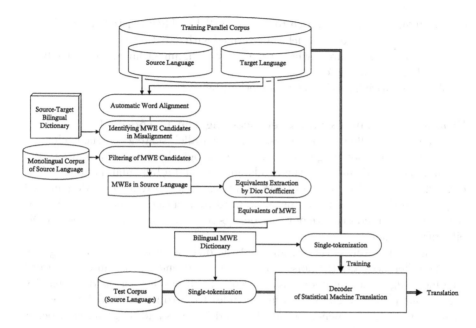

Fig. 1. Flowchart of proposed method

way. For example, an English expression "go Dutch" is translated the non-compositional way. This is the idiom and means 'split the bill'. Although "go" and "Dutch" are the constituents of "go Dutch", they cannot compose the meaning because each word has no meaning related to 'split the bill'. Therefore "go Dutch" is an MWE and cannot be translated by the compositional way. This is the example of MWEs in narrow sense, but we expand the definition of MWE in translating. For example, a Japanese expression '*fufuku moushi tate*' consists of three words: '*fufuku*' (means 'compliant'), '*moushi*' (means 'speak') and '*tate*' (means 'up'), and if it is translated by compositional way, its English translation is "speak up compliant". However, it is translated into "appeal" in the Legal Terms Dictionary of JLT. We regard such a translation with unbalanced number of word as non-compositional. Such a non-compositional translation cannot be made by simple replacement and leads an asymmetry alignment of words. Therefore we consider '*fufuku moushi tate*' as an MWE and can find such MWEs from asymmetry alignments.

Tsvetkov et al. [19] concluded that MWEs have a non-compositional meaning and may be translated into a single word in a foreign language. They assumed that there may be three sources to misalignments (anything without a one-to-one word alignment) in parallel texts: MWEs that trigger one-to-many or many-to-many alignments, language-specific differences (e.g., the source language lexically captures notions that are realized morphologically, syntactically or in some other way in the target), or noise (e.g., poor translations, low-quality sentence alignment, and the inherent limitations of word alignment algorithms). They focused on misalignments as a resource to extract MWEs. They trusted the quality of the one-to-one alignments, which they verified with a dictionary and searched for MWEs exactly in the areas where proper word alignment failed without relying on the alignment in these cases.

We summarize Tsvetkov et al.'s method below.

Process I: Resources and Preprocessing the Corpora. Their methodology needs the following resources: a small parallel, sentence-aligned bilingual corpus; two large monolingual corpora; morphological processors (analyzers and disambiguation modules) for the two languages; and a bilingual dictionary. The parallel corpus is the source of the MWE candidate extraction. Monolingual corpora are the sources to compute the word frequencies for measuring the statistics of the candidates. We only use one-to-one word alignment entries in their bilingual dictionary.

First, we preprocess the corpora and the dictionary to reduce the language specific differences and the noise of the word alignment because automatic word alignment algorithms are noisy, and given a small parallel corpus data sparsity is a serious problem. The processing includes tokenizing, lemmatizing, etc. We then compute the frequencies of all the word bigrams and unigrams that occur in each of the monolingual corpora.

Process II: Word Alignment and Identifying MWE Candidates. Next, we compute the word alignment on the parallel corpus. We use GIZA++ [11] to word-align the text and capture the alignments merged in both directions. We look up all the one-to-one alignments in the merged alignments in the dictionary. If a pair exists in their bilingual dictionary, we remove it from the sentence and replace it with a special symbol: "*". By the replacement, we can remove a word that can be translated into a word in the other language from the MWE candidates.

Figure 2(a) shows a Japanese sentence and its English translation in the parallel corpus. Here, '*fufuku moushi tate*' (means 'appeal') and '*to no*' (means 'between') are an MWE that cannot be literally translated. Figure 2(b) shows after the preprocessing and Fig. 2(c) shows after the word alignment. Since there are three pairs ('*to*' (means 'and') and "and", '*sosho*' (means 'lawsuit') and "lawsuit", '*kankei*' (means 'relation') and "relation") in the dictionary, we replace them with "*" (Fig. 2(d) shows).

Process III: Filtering MWE Candidates. Now the object sentences are word sequences separated by "*"s, and we can address each sequence that contains MWE candidates. But these candidates also include noise caused by the misalignment. In this stage, we do not rely on the alignments; we concentrate on distinguishing whether the bigram of each word is a compound expression.

We use association measure PMI^k to prune these candidates. PMI^k is a PMI-based score that measures the co-occurrence among two things. Its formula is defined as follows:

	fufuku moushi tate to sosho to no kankei
(a)	不服申立てと訴訟との関係
	compliant speak up and lawsuit – – relation

Relations Between Appeal and Lawsuit

	fufuku	moushi	tate	to	sosho	to	no	kankei
(b)	不服	申	立て	と	訴訟	と	の	関係
	compliant	speak	up	and	lawsuit	–	–	relation

relations between appeal and lawsuit

	fufuku	moushi	tate	to	sosho	to	no	kankei
(c)	不服	申	立て	と	訴訟	と	の	関係
	compliant	speak	up	and	lawsuit	–	–	relation
			$\mid$	$\mid$				$\mid$
		appeal		and	lawsuit	between	relation

	fufuku	moushi	tate	*	*	to	no	*
(d)	不服	申	立て	*	*	と	の	*
	compliant	speak	up	*	*	–	–	*
		appeal		*	*	between		*

Fig. 2. Example of word alignment and replacement

# of unigram	4,399	11,742	7,467		519,883	3,092,086
	fufuku	moushi	tate		to	no
	不服	申	立て		と	の

# of bigram	1,092 7,443	12,696
PMI^k	3.091 324.2	0.074

Fig. 3. Examples of PMI^k-score of the candidate

$$\mathrm{PMI}^k(x,y) = \frac{P(x,y)^k}{P(x)P(y)}, \tag{1}$$

where $P(x)$ is the number of occurrences of unigram "x" in the monolingual corpus and $P(x,y)$ is that of bigram "$x\ y$". We set weight k based on heuristics. A word sequence of any length is considered an MWE if all of its adjacent bigrams contain a score above the threshold. We computed PMI^k-scores based on the statistics of a large monolingual corpus. Finally, we restored the original forms of the words in the candidates and extracted them as MWEs.

In Fig. 2(d), which shows the replacement by "*", we consider each substring ('fufuku moushi tate' and 'to no') an MWE candidate and verify it by the PMI^k-score in the monolingual corpus. Figure 3 shows the result of the PMI^k-score of these candidates. If the PMI^k threshold is set to 1.0, term 'fufuku moushi tate' is an MWE, and term 'to no' is not, so we extract only 'fufuku moushi tate' as an MWE.

3.2   Translation of Extracted MWEs Based on the Dice Coefficient

Even though the above method of Tsvetkov et al. is reasonable, it has a problem extracting MWE translations. They extract the translations of candidate MWEs from these alignments. For each MWE in the source-language sentence, they consider the translations of all the words in the target-language sentence that are aligned to the word constituents of the MWE as long as they form a contiguous string. Since their candidates are based on misalignment, the translation quality is unreliable. They removed translations that are longer than four words because they are often wrong. Moreover, although MWEs can be translated into different expressions in different target sentences, they did not show a strategy for ranking translation candidates. Since our statutory corpus has many long terms, some of which may be translated into different expressions depending on the context, we want to adequately acquire their translations.

Thus, based on the Dice coefficient, we introduce another word alignment method to acquire the translations of MWEs. To measure the similarity between two terms, the Dice coefficient is described as follows:

$$Dice(x, y) = \frac{2 \cdot freq(x, y)}{freq(x) + freq(y)} \quad (0 \leq Dice(x, y) \leq 1), \tag{2}$$

where $freq(x)$ and $freq(y)$ denote the numbers of occurrences of term x in the source sentences and term y in the target ones, and $freq(x, y)$ denotes the number of co-occurrences of x and y in the aligned sentences.

In our proposed method, we give x as a candidate of source-language MWE and calculate its Dice coefficient to determine the highest y that consists of one or more words in the target-language sentences as a translation equivalent of x. Next we gather the remaining sentences that do not contain the equivalent and repeat the calculation of the Dice coefficient if its value exceeds a threshold to get multiple translation equivalents for one MWE.

3.3   Integration Strategy by Single-Tokenization Using Bilingual MWEs

We use the single-tokenization technique proposed by Pal et al. [12] to integrate our bilingual MWEs with the SMT system. The single-tokenization method embraces active processing to replace the spaces in MWEs as "_". It helps the SMT system treat MWEs as single words for training and decoding. We single-tokenize the bilingual MWEs if they appear in both the source and target sentences as preprocessing in the training corpus.

4   Experiments and Discussions

In these experiments, we extracted bilingual MWEs from a statutory corpus and evaluated them based on their contributions to improving the SMT system. First, we extracted bilingual MWEs using our proposed method. Second, we investigated an adequate threshold of the Dice coefficient. Third, we evaluated the single-tokenization of MWEs to improve the translation model. Finally, we evaluated the extracted translations using Dice coefficient by comparing it with Tsvetkov's extraction method.

4.1   Extraction Experiment

In the first experiment, we extracted bilingual MWEs using our proposed method.

Resources. We prepared three resources: a Japanese-English statutory parallel corpus, a Japanese statutory monolingual corpus, and a general bilingual dictionary. The parallel corpus, which contains 151,951 parallel sentences, is part of the corpus of the Japanese Law Translation Database System[1], where we kept the rest of the corpus (randomly identified 15,026 sentences) for the next experiment's test. We used the Japanese subset of the parallel corpus as the monolingual corpus. The bilingual dictionary is Eijiro [3] to which we added 1,004 of our own translations from Chinese numerals to Roman numerals. The 5,618 entries of the bilingual dictionary occur in our parallel corpus with one-to-one word alignment.

Preprocess. We lowercased and tokenized both corpora by morphological analyzers: MeCab [10] for Japanese and tokenizer.perl [9] for English. We also lemmatized them by scripts: MeCab for Japanese and Ruby Lemmatizer[2] for English. Since GIZA++ cannot handle long sentences, we cleaned up sentences longer than 80 words in Japanese or in English from the parallel corpus.

Bilingual MWE Extraction. We extracted statutory bilingual MWEs using the method proposed in Sect. 3. We set weight k to 2.7 and the threshold to 1 for PMI^k. Since the extracted monolingual MWEs remained noisy, we cleaned up some of them, including their punctuation, and acquired English translations of the Japanese MWEs based on the Dice coefficient.

Results. After removing the punctuation, the number of Japanese MWE candidates was 2,829. Figure 4 shows the numbers of acquired MWE equivalents of each threshold in line charts. Since we allowed one Japanese MWE to have one or more English equivalents, the number of equivalents exceeds the number of Japanese MWEs. 544 candidates could not acquire any English equivalent even when the threshold value was 0.1.

---

[1] http://www.japaneselawtranslation.go.jp/.
[2] https://github.com/yohasebe/lemmatizer/.

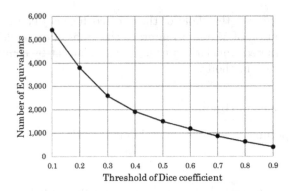

Fig. 4. Equivalents of each threshold of Dice coefficient

4.2   Translation Experiment of Dice Coefficient

To evaluate our bilingual MWEs that we extracted above, we compared the translation models trained on the corpus with MWEs. In this experiment, bilingual MWEs include some improper translation equivalents since automatic translation equivalent acquisition using Dice coefficient is not perfect. We evaluated the influence of such mistranslations by comparing the translations using the bilingual MWEs by the Dice coefficient with those that used the bilingual MWEs whose equivalents were translated by hand.

Training Data. As training data, we used the same parallel corpus that consist of 151,951 parallel sentences from previous section and prepared three translation model types: baseline, Dice, and manual. We lowercased and tokenized the corpus, where the baseline model was trained on the normally tokenized corpus, and the others were trained on the corpus whose MWEs were single-tokenized. For the Dice models, we acquired translations of the MWEs by the Dice coefficient and prepared nine models by changing the threshold from 0.1 to 0.9. For the manual models, we modified the MWE translations by hand after acquiring them by the Dice coefficient, and prepared nine models.

Since GIZA++ cannot handle long sentences, we cleaned up sentences longer than 80 words from the corpus, as in the previous section. Since the single-tokenization technique reduces the sentence length, the Dice and manual models used a slightly larger corpus than the baseline model.

Test Data. We prepared 15,026 Japanese statutory sentences as test data, none of which overlapped with the training data. We lowercased and tokenized them, like the training data; we single-tokenized the MWEs in the case of translating with the Dice and manual models.

Translation. We used the following freely available translation tools: GIZA++, SRILM [18], and Moses [9]. GIZA++ trains a translation model from the parallel

Table 1. Scores of models using bilingual MWEs by the single-tokenization

	BLEU score	
	Dice	Manual
0.1	**31.19**	30.98
0.2	30.89	**31.14**
0.3	30.75	**31.29**
0.4	**31.08**	**31.22**
0.5	**30.98**	**31.31**
0.6	**31.24**	**31.42**
0.7	**31.26**	**31.37**
0.8	**31.15**	**31.08**
0.9	30.51	30.57
Baseline		30.32

corpus described above. In this experiment, we used '-grow-diag-final-and -msd-bidirectional-fe' as the command options. SRILM trains a language model in English from the parallel corpus. We used '-ukndiscount -interpolate' as the smoothing command option. The other parameters were set to default.

Evaluation. To evaluate the outputs of the translation systems, we used automatic evaluation metric BLEU [13], which scores the system's outputs by referring to human translations. We removed the '‿'s of the single-tokenized MWEs in the translated sentences in the evaluations.

Results and Discussion. Table 1 shows the evaluation results. The scores that are significantly higher ($p < 0.05$) than the baseline model are written in bold. Figure 5 shows the line charts of the scores.

Both the Dice and manual models significantly improved the BLEU scores, where the peaks of the Dice threshold were 0.7 in the Dice model and 0.6 in the Manual model. The Dice coefficient's threshold is related to the number of bilingual MWEs (Fig. 4). Table 1 and Fig. 4 imply that the scores are reduced by too many MWEs, which are extracted by the low threshold. Because of the incorrect translations acquired by the Dice coefficient, these noisy equivalents may disturb the training and decoding in the Dice model. In the Manual model, on the other hand, the scores are also not improved by adding many more bilingual MWEs, even though their translation equivalents are correct. One reason is because of the variant of the translations. The Dice coefficient scores tend to be low when the MWEs have plural translations, which include variants representing tense, voice, singular, plural, etc. In our method, we single-tokenized the bilingual MWEs when they co-occur in the parallel text in the training corpus. If the MWEs have plural translations, our bilingual MWE dictionary is not easy

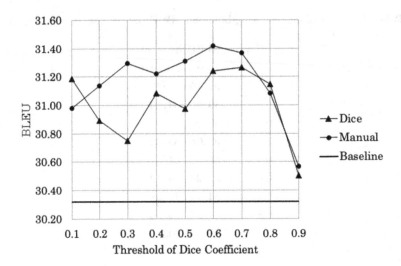

Fig. 5. Scores of models of using bilingual MWEs by the single-tokenization

to cover all the translation. Therefore, we have the problem that whether the MWEs are single-tokenized or not depends on the context. Because these variations negatively affect the reduction of the search space, the scores were not so good in the lower thresholds of the Manual model.

Although some of the translation equivalents acquired by the Dice coefficient are incorrect, as mentioned above, the scores of the Dice and manual models are close. We examined the MWEs acquired at the 0.7 threshold that marked the highest score in the Dice model, and found some of their equivalents are partially correct. Table 2 shows their examples. As this table shows, although the translations used in Dice model are incorrect, they are partially correct translation of that used in manual model. Single-tokenization integrates MWEs into a single word and reduces the search space of automatic word alignment. These partially correct translation co-occur with the MWEs with a high probability. If the translation is incorrect, they can help reduce the search spaces. As a result, even if we acquire the partial fragments of the correct equivalents, they can improved the translation quality. This shows that the Dice model suffices our purpose without requiring human hands and setting the thresholds to 0.7 is suitable for utilizing the bilingual MWEs in our method.

The number of acquired MWE translations at the 0.7 threshold was 863. While most of them are noun phrases, some are verb phrases, and the others are stereotyped expressions. We acquired translations over four words that cannot be extracted by the method of Tsvetkov et al. [19]. For example, the long noun MWE, '*chiiki mittchaku gata kaigo yobou sa-bisu hi*', got a proper translation that consisted of six words: "allowance for community-based long-term preventative care".

Table 2. Bilingual MWEs extracted at 0.7 threshold

Japanese MWE	Acquired translation	Correct translation
銀行持株会社 *ginko mochikabu gaisya*	Bank holding	**Bank holding** company **Bank holding** companies
産前産後 *sanzen sango*	After childbirth	Before or **after childbirth** Before and **after childbirth**
行為に対する罰則 *koui nitaisuru bassoku*	Penal provisions to	**Penal provisions to** actions **Penal provisions to** actions **Penal provisions to** an act **Penal provisions to** any acts
政令で定める *seirei de sadameru*	Cabinet order	Prescribed by **cabinet order** Designated by **Cabinet order** Set forth in **Cabinet order** Stipulated by **cabinet order** ...

Table 3. Scores of models using bilingual MWEs by each integration method

Model	BLEU
Baseline	30.32
Train	30.58
Table	30.48
Single-tokenization	**31.26**

4.3 Translation Experiment of Single-Tokenization

In our third experiment, we evaluated MWE single-tokenization by comparing it with other methods that utilize MWEs. We used the bilingual MWEs extracted by the Dice threshold of 0.7.

Integration Methods. We tested two other naïve methods to utilize MWEs, which are *train* and *table* methods, because Ren et al. [16] used them for evaluation. For the train method, we added bilingual MWEs to the training corpus and obtained, as results, new alignments and a phrase table. For the table method, bilingual units are incorporated as a new phrase table in addition to the baseline model's phrase table, where the Moses system uses multiple phrase tables. We gave the probability weighting by the Dice coefficient scores to the new phrase table of the MWEs.

Results and Discussion. Table 3 shows the evaluation results by the BLEU metric. The scores, which are significantly higher ($p < 0.05$) than the baseline, are written in bold.

Compared to the baseline, the train and table methods improved the BLEU scores, but not significantly. Our methodology using single-tokenization is more effective than these naïve methods.

4.4 Translation Experiment with Extraction Method Using Word Alignment by GIZA++

In our fourth experiment, we examined the model with the extraction method using the word alignment by GIZA++. This naïve method, adopted by Tsvetkov et al. to acquire MWE translations, extracts the translations by the word alignment calculated by GIZA++. Since the MWEs are extracted based on misalignment, we assumed that the quality of the translations extracted by the word alignment is unreliable and introduced the method using the Dice coefficient. Therefore, we evaluated this assumption by comparing the models using each method.

Extraction Method Using Word Alignment by GIZA++. We tested the naïve method that Tsvetkov et al. adopted to extract MWE translations described in Sect. 3.2. We made the bilingual MWEs dictionary by using GIZA++ instead of the Dice coefficient and integrated this dictionary into the SMT system by single-tokenization.

Results and Discussion. The number of acquired MWE translations was 23,350. This is over twenty times as many as the number of translations extracted by the Dice coefficient at the 0.7 threshold. This method extracted more different translations for one MWE than the method using the Dice coefficient. Although not all the translations are correct, we cannot choose good translations. The reason is that Tsvetkov et al. did not use the measure to evaluate translations, while the method using the Dice coefficient can use a threshold value. Therefore, we integrated all translations into the SMT system by single-tokenization and evaluated its translation quality by comparing it with other systems.

Table 4 shows the evaluation result, in which the score in bold is significantly higher ($p < 0.05$) than the baseline. Compared to the baseline, the model using GIZA++ did not improve the BLEU score and its score was significantly lower than that of the model using the Dice coefficient. This is due to the same reason as described in the Sect. 4.2: too many MWE translations reduce the translation quality. In contrast, we solved this problem by setting a threshold value of the Dice coefficient score. Therefore, our method is more effective than the naïve extraction method using GIZA++.

Table 4. Scores of models by each extraction method

Model	BLEU
Baseline	30.32
GIZA++	29.69
Dice coefficient	**31.26**

5   Conclusion and Future Work

We described a methodology for utilizing multi-word expressions to improve the statistical machine translation of statutory sentences. We extracted the monolingual MWEs from a parallel corpus, automatically acquired their translations based on the Dice coefficient, and integrated the extracted bilingual MWEs into the SMT system by a single-tokenization strategy. The SMT system with our proposed method significantly improved the translation quality more than both the baseline system and one using the other naïve methods by MWEs. The system worked best when we set its Dice coefficient threshold to 0.7. Although automatic translation equivalent acquisition using the Dice coefficient is not perfect, the system's best score was close to one that used bilingual MWEs whose equivalents were translated by hand.

Our future works have four directions. The first is the further analyze of our method's performance. We wonder why the score of the Dice model at the 0.1 threshold is close to that of the model at the 0.7 and whether the performance depends on kinds of statute. We should evaluate the result in more detail to answer these questions. The second is to apply our method to other languages. Since our method is the language independent approach and MWEs appear in all text genres, we expect it to perform as well as it does in this paper. The third is further improvement of the SMT system's translation quality. Since our method is useful to reduce the search space where the word alignment system works, we believe that it is compatible with word reordering approach [5]. The last is the construction of multilingual MWE dictionaries for human translators working on Japanese statues. The acquired MWE dictionary includes both noun and verb phrases, which benefit human translators. We also plan to introduce a screening technique and select proper entries from it.

Acknowledgements. This research was partly supported by the Japan Society for the Promotion of Science KAKENHI Grant-in-Aid for Scientific Research (S) No. 23220005, (A) No. 26240050 and (C) No. 15K00201.

References

1. Caseli, H.M., Villavicencio, A., Machado, A., Finatto, M.J.: Statistically-driven alignment-based multiword expression identification for technical domains. In: Proceedings of the Workshop on Multiword Expressions: Identification, Interpretation, Disambiguation and Applications, pp. 1–8 (2009)
2. Van de Cruys, T., Moirón, B.V.: Semantics-based multiword expression extraction. In: Proceedings of the Workshop on A Broader Perspective on Multiword Expressions, pp. 25–32 (2007)
3. EDP, ALC Press Inc.: Eijiro, 8 edn. (2014)
4. Finlayson, M.A., Kulkarni, N.: Detecting multi-word expressions improves word sense disambiguation. In: Proceedings of the Workshop on Multiword Expressions: From Parsing and Generation to the Real World, pp. 20–24 (2011)

5. Isozaki, H., Sudoh, K., Tsukada, H., Duh, K.: Head finalization: a simple reordering rule for SOV languages. In: Proceedings of the Joint 5th Workshop on Statistical Machine Translation and Metrics MATR, pp. 244–251 (2010)

6. Bui, T.H., Nguyen, L.M., Shimazu, A.: Translating legal sentence by segmentation and rule selection. Int. J. Nat. Lang. Comput. **2**(4), 35–54 (2013)

7. Toyama, K., Saito, D., Sekine, Y., Ogawa, Y., Kakuta, T., Kimura, T., Matsuura, Y.: Design and development of Japanese law translation memory database system. In: Law via the Internet 2011, 12 p. (2011)

8. Katz, G., Giesbrecht, E.: Automatic identification of non-compositional multi-word expressions using latent semantic analysis. In: Proceedings of the Workshop on Multiword Expressions: Identifying and Exploiting Underlying Properties, pp. 12–19 (2006)

9. Koehn, P., Hoang, H., Birch, A., Callison-Burch, C., Federico, M., Bertoldi, N., Cowan, B., Shen, W., Moran, C., Zens, R., et al.: Moses: open source toolkit for statistical machine translation. In: Proceedings of the 45th Annual Meeting of the ACL on Interactive Poster and Demonstration Sessions, pp. 177–180 (2007)

10. Kudo, T., Yamamoto, K., Matsumoto, Y.: Applying conditional random fields to Japanese morphological analysis. In: Proceedings of the 2004 Conference on Empirical Methods on Natural Language Processing, pp. 230–237 (2004)

11. Och, F.J., Ney, H.: A systematic comparison of various statistical alignment models. Comput. Linguist. **29**(1), 19–51 (2003)

12. Pal, S., Naskar, S.K., Bandyopadhyay, S.: MWE alignment in phrase based statistical machine translation. In: Proceedings of the XIV Machine Translation Summit, pp. 61–68 (2013)

13. Papineni, K., Roukos, S., Ward, T., Zhu, W.J.: BLEU: a method for automatic evaluation of machine translation. In: Proceedings of the 40th Annual Meeting of the Association for Computational Linguistics, pp. 311–318 (2002)

14. Piao, S.S., Rayson, P., Archer, D., McEnery, T.: Comparing and combining a semantic tagger and a statistical tool for MWE extraction. Comput. Speech Lang. **19**(4), 378–397 (2005)

15. Ramisch, C.: Multiword Expressions Acquisition: A Generic and Open Framework. Springer, Cham (2014)

16. Ren, Z., Lü, Y., Cao, J., Liu, Q., Huang, Y.: Improving statistical machine translation using domain bilingual multiword expressions. In: Proceedings of the Workshop on Multiword Expressions: Identification, Interpretation, Disambiguation and Applications, pp. 47–54 (2009)

17. Sag, I.A., Baldwin, T., Bond, F., Copestake, A., Flickinger, D.: Multiword expressions: a pain in the neck for NLP. In: Gelbukh, A. (ed.) CICLing 2002. LNCS, vol. 2276, pp. 1–15. Springer, Heidelberg (2002). doi:10.1007/3-540-45715-1_1

18. Stolcke, A.: SRILM - an extensible language modeling toolkit. In: Proceedings of the 7th International Conference on Spoken Language Processing, vol. 2, pp. 901–904 (2002)

19. Tsvetkov, Y., Wintner, S.: Extraction of multi-word expressions from small parallel corpora. In: Proceedings of the 23rd International Conference on Computational Linguistics, pp. 1256–1264 (2010)

20. Zarrieß, S., Kuhn, J.: Exploiting translational correspondences for pattern-independent MWE identification. In: Proceedings of the Workshop on Multiword Expressions: Identification, Interpretation, Disambiguation and Applications, pp. 23–30 (2009)

Argumentation Support Tool with Reliability-Based Argumentation Framework

Kei Nishina[✉], Yuki Katsura, Shogo Okada, and Katsumi Nitta

Department of Computational Intelligence and Systems Science, Tokyo Institute of Technology, 4259 Nagatsuta-cho, Midori-ku, Yokohama, Kanagawa, Japan
{nishina,katsura_y,okada,nitta}@ntt.dis.titech.ac.jp

Abstract. In legal debates, it is a matter of importance whether one's own argument is accepted or not. For this, we propose evaluation method for calculating the acceptability of arguments, and a tool developed based on the measures. This method is called reliability-based argumentation framework (RAFs), extended from argumentation framework, seeking for multivalued dialectical validities of arguments reliable to some extent. The modular reliability-based argumentation framework (MRAF) based on RAFs is able to integrate the RAF semantics in every module. This leads to an over-all valuation of the acceptability of argumentations including several local arguments. The argumentation-support tool can represent the utterance logs of those who join an debate, the argumentation diagram its users made, and the argumentation framework converted from this, contributing to the intuitive comprehension of the logical structures of arguments and their acceptability. This tool also enables represented argumentation framework to be converted into modular structures of local AFs, leading to an overall valuation of the acceptability of arguments.

Keywords: Argumentation theory · Argumentation framework · Argumentation support tool · Logical analysis of real complicated discussions · Reliability of argument · Modular structure

1 Introduction

A legal debate is a discussion between lawyers, legal academics and others. Arguments are series of statements typically used to present reasons to let him or her accept a conclusion. It is an important problem for participants to know which ones of the all the arguments come to be finally accepted. To solve this problem, argument diagrams, visual representation of the structure of arguments made in the discussion, can be seen being often used. There have been several researches done about such argument diagrams, such as the Araucaria system [1] which provides an interface which supports the Toulmin's diagramming [2] process and then saves the result, and the theory of Argumentation Framework

© Springer International Publishing AG 2017
M. Otake et al. (Eds.): JSAI-isAI 2015 Workshops, LNAI 10091, pp. 265–281, 2017.
DOI: 10.1007/978-3-319-50953-2_19

(AF) [3] which is one of the fundamental theories of the field of argumentation theory, and has been extended in various ways to accommodate different kinds of abstraction about argumentation [4–6] and etc.

On the other hand, several educational argumentation support tools have been developed to teach legal debate skills [7–11]. Some of them have functions not only to visualize but also to evaluate some aspects of argumentation skills to give advice for arguing, and etc. For example, efficiency of using the argumentation support tool in the law schools is studied by Pinkwart by using LARGO [9,10], which can provide detailed educational support and various function for legal students. These diagram based argumentation support tools can't provide the function to judge the dialectical superiority or inferiority of each participant, which is closely related to the important problem as mentioned above in real time. As these informations are useful to select the next move during the debate, developing a real time argumentation support tool based on the theory of AF is a promising way for legal debate education.

The AF diagram tends to become bigger as the discussion gets longer or more complicated, which loses visibility of the diagram and causes much cost to calculate semantics. To cope with this problem, we need additional functions of the argumentation support tool. The additional function is to build an AF displayed on the screen into the module structures consisting of multiple local AFs, and to calculate the overall semantics by integrating semantics of local AFs to address these issues. To extend the argumentation support tool, we also have to extend the AF theory by structuring it into distinct modules in hierarchical structure. There has already been several theoretical studies of AFs whose structure is of a modular structure which resembles our view about the support tool, such as Modular Assumption-Based Argumentation (MABA) [12], Argumentation Context System (ACS) [13], and hierarchical Extended Argumentation Framework (hEAF) [14].

However, their objectives of employing modular structure is different from ours. Their modularization theories are mainly based on PAF [15] or VAF [16] which are used to introduce relative priorities or relative ordering among values of arguments to AF theory. In the meta level module, they argue the relative priority, and in the object module, the semantics of the meta level module result is used to calculate the semantics of PAF or VAF. On the contrary, our objective is to keep visibility of AF graph and to reduce the calculation time by dividing the overall AF graph into sub AF graphs. The semantics of an AF of the object module is calculated by integrating semantics of AFs of sub modules. However, the original theory of AF doesn't support the way to integrate semantics of sub AFs. Like hEAF is based on VAF theory, our modularization method needs a new AF theory.

Therefore, one of our research objectives is to propose Reliability-based Argumentation Framework (RAF) whose semantics consider the reliability of each argument. It is important to note that the reliability of an argument is different from the "trust" of trust-extended argumentation graph [17], which comes from the relationship of trust between the participants of discussion, and the appearance frequency of the argument of PrAF [18]. In a sense, the concept of the semantics of RAF is similar to the one of Authority Degree-Based Evaluation Strategy's concept [19] to the effect that the argument made by the expert well-acquainted with relative fields of its content has more significance than the one made by the outsiders. But, in RAF semantics, the arguments made by the outsiders which agrees with the content of opinions of experts in related fields is regarded as more significant than the gratuitous ones, and the number of the attackers and defenders of each argument in discussion is not taken into account because we follow the part of the concept of the acceptability of AF semantics.

The second objective is the analytical approach developing a dynamic argumentation support tool for real time analysis of real discussions whose function is visualization of argument diagram, to derive the evaluation results of the acceptance of each argument, and to provide strategic direction. The paper is structured as follows: In Sect. 2, we review the background of support tool and argumentation theory. In Sects. 3 and 4, we define RAF, Modular RAF(MRAF), and their semantics to evaluate the dialectical validity of each argument in a discussion. Furthermore, we outline the architecture, the functions of the support tool, and the support provided by it in Sect. 5, and conclude this paper in Sect. 6.

2   Background

2.1   Diagram Based Argumentation Support Tools

There are several argumentation support tools for law school students as described in the above. For example, LARGO Intelligent Tutoring System [9, 10] is a support system for legal student to understand or reflect upon the transcripts of complex real-world discussion such as legal debates in the U.S. Supreme Court. While using the system, students read through the transcript and produce a graphical markup of it, identifying the key tests, hypotheticals, responses, facts, and the relationships between them. LARGO can help students by capitalizing on the pedagogical value of argument diagrams and giving feedback in the form of self-explanation prompts.

2.2   Argumentation Framework

An argumentation framework [3] is a tuple $AF = (Ar, attacks)$, where Ar is a set of arguments, and attacks $\subseteq Ar \times Ar$ is a binary attack relation on Ar. An attack from an argument $a \in Ar$ to an argument $b \in Ar$ is expressed as $(a, b) \in attacks$.

- The set $S \subseteq Ar$ is *conflict-free* iff $^\forall x, y \in S, (x, y) \notin attacks$.
- For any $x \in Ar$, x is *acceptable* with respect to some $S \subseteq Ar$ iff $^\forall y \in Ar$ s.t. $(y, x) \in attacks$ implies $^\exists z \in S$ s.t. $(z, y) \in attacks$.
- $F_{AF} : 2^{Ar} \to 2^{Ar}$, and $F_{AF}(S) = \{a \in Ar \mid a$ is acceptable w.r.t. $S\}$.

Let $S \subseteq Ar$ be *conflict-free*. Then, each extension which is the family of the sets is defined as follows:

- S is a set of arguments which is a complete extension of AF iff $F_{AF}(S) = S$.
- S is a set of arguments which is a stable extension of AF iff it is a maximal set in complete extension (w.r.t. set inclusion) which satisfies the following: $\forall y \notin S, \exists x \in S : (x, y) \in attacks$.
- S is a set of arguments which is a preferred extension of AF iff it is a maximal set in complete extension (w.r.t. set inclusion).
- S is a set of arguments which is the grounded extension of AF iff it is the only minimal set in complete extension (w.r.t. set inclusion).

Furthermore, the contraposition of Proposition 1 is valid, too.

Caminada's theory of reinstatement labelling [20] is more expressive than the extensions defined in Dung's theory. AF-labelling is a total function $L :$ $Ar \rightarrow \{$in, out, undec$\}$. The label **in** indicates that the argument is explicitly accepted, the label **out** indicates that the argument is explicitly rejected, and the label **undec** indicates that the status of the argument is undecided. This theory can convert every extension of Dung's AF into the set of in-labelled arguments by each reinstatement labelling.

Caminada's theory of reinstatement labelling [20] is more expressive than the extensions defined in Dung's theory. AF-labelling is a total function $L :$ $Ar \rightarrow \{$in, out, undec$\}$. The label **in** indicates that the argument is explicitly accepted, the label **out** indicates that the argument is explicitly rejected, and the label **undec** indicates that the status of the argument is undecided. This theory can convert every extension of Dung's AF into the set of in-labelled arguments by each reinstatement labelling.

3   Reliability-Based Argumentation Framework

In this section, we formally define Reliability-Based Argumentation Framework (RAF) and their semantics by extending the AF defined by Dung. An RAF is a directed graph consisting of nodes which have one status of reliability and links between nodes. The nodes represent argument and the links represent attack relation on arguments, like the AF defined by Dung. Furthermore, each argument has one status of the reliability.

Definition 1. *A reliability-based argumentation framework (RAF) is a tuple* $(Ar, attacks, ST)$, *where* Ar *is a set of argument, attacks* $\subseteq Ar \times Ar$, *and* ST *is a function:* $Ar \rightarrow \{$sk, cr, def, unc$\}$.

Here, *sk*, *cr*, *def*, and *unc* stand for skeptical, credulous, defeated and uncertain, respectively, and each of them represents the status of the reliability. In the function ST, any one of them is assigned to an argument $A \in Ar$ as follows.

- A is assigned to *sk* if A has sufficient evidences to be true. The evidence may be a book, a TV program, a research report, or the conclusion of other discussion.

- *A* is assigned to *cr* if *A* is true or not is controversial.
- *A* is assigned to *def* if there are sufficient evidences not to be true.
- *A* is assigned to *unc* if there is no information to judge *A*'s reliability.

We will show a simple example of these assignments of the status of the reliability to argument.

A1 : "Since the lady saw the boy killed his father through the window, he is the murderer."
A2 : "Since the lady is short-sighted, she may mistake the murderer."
A3 : "Since the lady may tell a lie, he is not the murderer."
A4 : "Since the lady wasn't there, it is impossible for her to see him."

In this example, *A*2, *A*3 and *A*4 attack *A*1. The original AF theory doesn't discriminate these three counter argument because there is no difference from the view of the graph structure. However, if, by the inspection, the facts that she was short-sighted and that she was there become clear, status of *A*2, *A*3 and *A*4 become *sk*, *cr* and *def*, respectively. This inspection result may be represented by AF of sub module, or may be given by other mechanism beyond the argumentation theory. We must note that even if an argument is assigned to *sk*, it doesn't guarantee that the argument holds because it may be attacked by another counter argument.

Extensional semantics for RAFs: Extensional semantics for RAF define the detailed acceptability of arguments in real discussions. We propose three kinds of semantics based on three kinds of acceptability of arguments in *Ar*. They come from different intuitive perspectives on the reliabilities of arguments. Firstly, we adopt an intuitive perspective that unfounded arguments and the attack relations concerning them should be judged to be invalid in persuasiveness of arguments. Then, we define RAF-Non-Def semantics in which no argument whose reliability is *def* in *Ar*, (*defeated reliable argument*) is included in any RAF-Non-Def complete extension of a RAF, and every attack relation concerning any *defeated reliable argument* is invalid in a RAF. Secondly, we adopt the optimistic perspective that the argument which isn't rejected in the discussion dialectically and has the status of reliability is *cr* or *unc* should be included in the result of this discussion considering the content of the meta-information because there is no evidence in the meta- information which shows it is false. Then, we define RAF-Optimistic semantics which regards a set of argument which is a RAF-Non-Def complete extension of a RAF is a RAF-Optimistic complete extension of it, and every argument which isn't included in any set of argument which is RAF-Non-Def complete extension and is attacked only from the arguments whose status of reliability is *cr*, *unc*, or *def* satisfies the following: It is included in at least one RAF Optimistic complete extension of a RAF. Finally, we adopt the pessimistic perspective that the argument which isn't rejected in the discussion dialectically and has the status of reliability is *cr* or *unc*, shouldn't be included in this result of the discussion considering the content of the meta-information because there is no evidence in the meta- information which shows it is absolutely true. Then, we define RAF-Pessimistic semantics in which any set which is any extension of RAF consists of the arguments whose reliability is sk.

Hence, the three semantics are formally defined based on the labelling extending Caminada's reinstatement labelling [20]: We first define RAF-labelling and RAF-*conflict-free* labelling in Definitions 2 and 3, following AF-labelling and AF-*conflict-free* labelling:

Definition 2. *A RAF-labelling in Δ is a total function $L : Ar \rightarrow \{in, out, undec\}$. The set of in-labelled arguments by L, that of out-labelled arguments by L, and that of undec-labelled arguments by L are expressed as $in(L)$, $out(L)$, and $undec(L)$.*

RAF labellings assign any one status of $\{in, out, undec\}$. to each argument of RAF. These statuses which indicates the acceptability of each argument in RAF.

Definition 3. *A RAF-labelling L is a RAF-conflict-free labelling iff no in-labelled argument by L attacks any other in-labelled argument by L and itself. The set $in(L)$ is a RAF-conflict-free set iff L is a RAF-conflict-free labelling.*

Note that there is no attack relation between the arguments included in the set $in(L)$ if L is a RAF-conflict-free labelling. Secondly, we define RAF-σ ε extension($\varepsilon \in \{$complete, stable, preferred, grounded$\}$) of RAF-σ semantics ($\sigma \in \{$Non-Def, Optimistic, Pessimistic$\}$) as follows:

Definition 4. *Let Δ be a RAF, and let L is a RAF-σ ε labelling of Δ. Then, $in(L)$ is a RAF-σ ε extension of Δ.*

Thirdly, we define each complete labelling in RAF-Non-Def semantics, RAF-Optimistic semantics, and RAF-Pessimistic semantics in Definitions 5, 6, and 7.

Definition 5. *Let Δ be a RAF, and let L be a RAF-conflict-free labelling in Δ. Then, L is a RAF-Non-Def complete labelling iff it satisfies the following:*

$$^\forall a \in Ar : (L(a) = in \equiv \{ST(a) \neq def\} \wedge \{^\forall b \in Ar : [(b,a) \in attacks \supset L(b) = out]\}) \, and$$
$$^\forall a \in Ar : (L(a) = out \equiv \{ST(a) \neq def\} \vee \{^\exists b \in Ar : [(b,a) \in attacks \wedge L(b) = in]\}).$$

Let L is a RAF-Non-Def complete labelling in Δ and S is a set of argument that is the same as $in(L)$. Then, S is a RAF-Non-Def complete extension and Definition 5 indicates that any element of S satisfies the following condition:

(a) It is not attacked from any argument in Δ or
(b) It is attacked only from the arguments which are *defeated reliable argument* in Δ or the arguments which are attacked only from any one element of S.

Definition 6. *Let Δ be a RAF, and let L be a RAF-conflict-free labelling of Δ. Then, L is a RAF-Optimistic complete labelling iff it satisfies the following:*

$$^\forall a \in Ar : (L(a) = in \equiv \{ST(a) \neq def\} \wedge \{^\forall b \in Ar : [(b,a) \in attacks \supset L(b) = out]\}) \, and$$
$$^\forall a \in Ar : (L(a) = out \equiv \{ST(a) = def\} \vee \{^\exists b \in Ar : [(b,a) \in attacks \wedge L(b) = in]\}$$
$$\vee \{(ST(a) = cr \, or \, unc) \wedge (^\forall b \in Ar : [(b,a) \in attacks \supset L(b) = out])\}).$$

Note that any RAF-Non-Def complete labelling of Δ is a RAF-Optimistic complete labelling of Δ, too. Therefore, any RAF-Non-Def complete extension of Δ is a RAF-Optimistic complete extension of Δ. In addition, the arguments which are *credulous reliable* or *uncertain reliable* and are attacked only from the out-labelled arguments can be labelled **in** or **out** in RAF-Optimistic complete labelling even though they can be labelled only **in** in RAF-Non-Def complete labelling. Furthermore, let L is a RAF-Optimistic complete labelling in Δ and S is a set of argument that is the same as $in(L)$. Then, S is a RAF-Optimistic complete extension and Definition 6 indicates that any element of S satisfies the following condition:

(c) It is included in any one of Non-Def complete extension or
(d) It is the argument which attacked from the argument satisfying the following condition: the arguments which are *credulous reliable* or *uncertain reliable* and are attacked only from the arguments attacked from any one of the element of S.

Definition 7. *Let Δ be a RAF, and let L be a RAF-conflict-free labelling of Δ. Then, L is a RAF-Pessimistic complete labelling iff it satisfies the following:*

$$^{\forall}a \in Ar : (L(a) = \textbf{in} \equiv \{ST(a) = sk\} \wedge \{^{\forall}b \in Ar : [(b,a) \in attacks \supset L(b) = \textbf{out}]\}) \text{ and}$$

$$^{\forall}a \in Ar : (L(a) = \textbf{out} \equiv \{ST(a) = def\} \vee \{^{\exists}b \in Ar : [(b,a) \in attacks \wedge L(b) = \textbf{in}]\}).$$

Let L is a RAF-Pessimistic complete labelling in Δ and S is a set of argument that is the same as $in(L)$. Then, S is a RAF-Pessimistic complete extension and Definition 7 indicates that any element of S satisfies the following condition:

(e) It is not attacked from any argument which is *skeptical reliable* in Δ or
(f) It is the argument which is *skeptical reliable* and attacked only from the arguments attacked from any one of the element of S.

Definition 8. *Let Δ be a RAF, and let L be a RAF-σ-complete labelling of Δ, where $\sigma \in \{Non\text{-}Def, Optimistic, Pessimistic\}$).*

– *L is a RAF-σ-stable labelling of Δ iff it is a RAF-Non-Def complete labelling of Δ and $undec(L) = \emptyset$.*
– *L is a RAF-σ-preferred labelling of Δ iff $in(L)$ is maximal (w.r.t. set inclusion) among all RAF-σ-complete labelling of Δ.*
– *L is a RAF-σ-grounded labelling of Δ iff it is a RAF-Non-Def complete labelling of Δ and $in(L)$ is minimal (w.r.t. set inclusion) among all RAF-σ-complete labelling of Δ.*

4 Modular Reliability-Based Argumentation Framework

Each module in a modular reliability-based argumentation framework (MRAF) is a RAF representing the information of a local discussion. Two modules are in the hierarchical relationship when the argument expressed in one module by referring to the other one in the other module. We call this hierarchical relationship between two modules *reference relation*. One of the two modules in the *reference relation* is a higher module which refers to the argument in the other module, which is a lower module. Note that the meta-information of the discussion which is represented in the higher module is the content which is represented in the lower module by the form of RAF. Hence, a MRAF represents the discussions which are end targets of analysis as the highest modules and the meta-information which is relative to them as lower modules.

Definition 9. *A Modular Reliability-Based Argumentation Framework is a tuple (MO, refers), where MO is a set of RAFs. Refers in the tuple is defined as* refers $\subseteq MO \times MO$, *which is a set of reference relations between modules in MO, and whose power set contains any element constructing the circular relationship of the elements of MO. The module $M \in MO$ is a higher module of the module $M' \in MO$, where M' is a lower module of M when $(M, M') \in$ refers holds.*

We introduce the way of defining the reliability of the referred argument in the higher module derived by the function ST in the following, Definition 10. Note that the reliability of an argument in each RAF is defined uniquely in MRAF. Then, we express as an RAF taken as the module M as $\Delta_M = (Ar_M, attacks_M, ST_M)$.

Definition 10. *In every module, each argument's acceptability is calculated by only RAF-σ-semantics. The ranked reliability of the argument, a in module M, $ST_M(a)$ is defined as the following 1, 2, 3, and 4:*

1. *$ST_M(a) = sk$ holds iff the argument, a satisfies either of the (i) and (ii) in each lower module of M.*
   (i) a is an element of RAF-σ-grounded extension.
   (ii) a doesn't exist.
2. *$ST_M(a) = def$ holds iff the argument, a satisfies the following (i) and (ii) in at least one lower module of M.*
   (i) a is not an element of RAF-σ-complete extension.
   *(ii) a is labelled **out** in at least one RAF-σ-complete labelling.*
3. *$ST_M(a) = unc$ holds iff the argument, a satisfies the following (i) and (ii) in at each lower module of M.*
   (i) a is not an element of RAF-σ-complete extension.
   *(ii) a is labelled **undec** in every RAF-σ-complete labelling.*
4. *$ST_M(a) = cr$ holds the argument a satisfies none of (1) or (2) or (3).*

Furthermore, $ST_M(x) = sk$ for $^\forall x \in Ar_M$ holds iff there is no lower module of M.

– *The Interrelation between MRAF semantics and AF semantics*

Let a MRAF be $\Gamma = (MO, refers)$, where $MO = \{M_i, M_j\}$ and $refers = \{(M_i, M_j)\}$. Let M_i be $\Delta_i = (Ar_i, attacks_i, ST_i)$, and let M_j be $\Delta_j = (Ar_j, attacks_j, ST_j)$. We express $(Ar_i \cap Ar_j)$ as Ar_{ij}. Let an AF be $A_0 = (Ar_0, attacks_0)$, where $Ar_0 = (Ar_i \cup Ar_j)$, $attacks_i \cap attacks_j = \emptyset$, and $attacks_0 = (attacks_i \cup attacks_j)$ holds. Then, we can detect one interrelation between the MRAF semantics in M_i and the AF semantics in A_0 as described in the following Proposition:

Proposition 1. *If the set S which satisfying $Ar_{ij} \subseteq S \subseteq Ar_0$ is complete extension of A_0, then, the set $(S \cap Ar_i)$ is RAF-Non-Def complete extension of Δ_i.*

Proof: *Let any argument which attacks the elements of S in M_0 be x and let any argument which is in the attack relation of M_0 be y. Then, if $x \in (Ar_i \backslash Ar_j)$, $y \in Ar_i$, and if $x \in (Ar_j \backslash Ar_i)$, $y \in Ar_j$. Based on these points, the following (i) and (ii) hold because S is complete extension of A_0:*

(i) For each argument such that attacks an element of $(S \cap Ar_i)$ in M_i, there is always at least one argument such that attacks it and included in $(S \cap Ar_i)$.
(ii) For each argument such that attacks an element of $(S \cap Ar_j)$ in M_j, there is always at least one argument such that attacks it and included in $(S \cap Ar_j)$.

Furthermore, the following thing holds because there is no lower module of M_j in Γ: Each argument in Ar_j is skeptical reliable. Because this point and (ii) holds, then the following thing holds: $(S \cap Ar_j)$ is RAF-Non-Def complete extension in M_j. Based on this point, every element of $(S \cap Ar_i)$ is sk reliable or cr reliable in M_i because every element of Ar_{ij} is included by $(S \cap Ar_i)$, which is RAF-Non-Def complete extension in M_j and every element of $(Ar_i \backslash Ar_j)$ doesn't exist in M_j. Because this point and (i) holds, therefore, $(S \cap Ar_i)$ is RAF-Non-Def complete extension in M_i. □

When a set of argument where every argument is ranked as sk ($ST_0(x) = sk$ holds for $\forall x \in Ar_0$) is RAF-Non-Def complete extension of Δ_0 ($\Delta_0 = (Ar_0, attacks_0, ST_0)$), it is complete extension of A_0, too. Then, Proposition 1 and the following (a) are equivalent. Furthermore, the contraposition of Proposition 1, (b) is valid, too.

(a) If the set S satisfying $Ar_{ij} \subseteq S \subseteq Ar_0$ is RAF-Non-Def complete extension of Δ_0, then, the set $(S \cap Ar_i)$ is RAF-Non-Def complete extension of Δ_i.
(b) Let $S' \in Ar_i$ be a set which isn't RAF-Non-Def complete extension of Δ_i, and let $S'' \in Ar_i$ be a superset of S'. Then, S' and S'' are not complete extension of A_0.

Note that the above interrelation (b) is useful, when A_0 is too huge to calculate the sets which are complete extension of it. We can convert it into a MRAF consisting of two modules. Hence, the interrelation (b) can give us the perspective which argument in A_0 isn't included by any complete extension of it by calculating RAF-Non-Def complete extension of Δ_i which is partially alternative to the semantics of A_0. In short, we can decrease the number of the possible candidates of the elements of the sets which are complete extension of A_0. Furthermore, in Proposition 1, (a), and (b), the words of "RAF-Non-Def complete extension" can be interpreted as the words of "RAF-Optimistic complete extension" because a set which is RAF-Non-Def complete extensions of a RAF is always RAF-Optimistic complete extension of it.

5   Argumentation Support Tool Based on RAF

5.1   Overview of the Argumentation Support Tool

Our argumentation support tool is assumed to be used by participants of a discussion who have tablet PCs and it is already installed to each of them. Each of their tablet PCs are connected to the argumentation server, and the users of the tool can exchange utterances and argumentation diagrams consisting of two kinds of nodes (Data and Claim) and two kinds of links (support and attack), which are similar to Toulmin diagram [2] through it. When users input utterances and the diagrams representing the contents of the utterances and the attack relations, the information of its construction is send to the argumentation server. In the server, it is converted into logical formulas which are illustrated in it, and they are displayed to users as arguments by the converter. Furthermore, when users input the diagram including the attack relation between their utterances and its information is send to the argumentation server, then, in the argumentation server, the converter converts the diagram including the attack relation into the graph structure of AF and the calculator calculates all extensions of the AF, which are displayed to users. Displaying a graph structure and helps users to grasp logical structures of whole argumentation and attack relation between arguments. Furthermore, this tool can produce strategic planning for each user to persuade other people in the discussion dialectically by giving them hints indicating how each user makes his or her arguments to be elements of any extension in the AF. We have showed that these functions are useful for participants of actual discussions [21]. However, when the content of the discussion is complicated or there are many conflicts of views in it, the graph structure of AF will become complicated and huge, and users of this tool cannot understand whole structure of discussion well. To solve this problem, we extended our tool by attending the function to convert MRAF to make it easier to grasp the content of the discussion. To integrate the semantics of each module which is described as one RAF, we employed the semantics of MRAF. The new functions of our tool are as follows.

- Status assignment to each argument and visualization of the graph structure of RAF This tool can assign the status of reliability to each argument in the graph structure of AF by users' manual input and is able to visualize the graph structure of RAF which is converted from it.
- Calculating all extensions of each RAF semantics. This tool can calculate all extensions of RAF-Non-Def semantics as long as the graph structure of RAF is not too huge or too complicated to calculate.
- Conversion of the graph structures of an AF which is displayed by our tool into the ones of a MRAF. Furthermore, this tool can focus on particular nodes

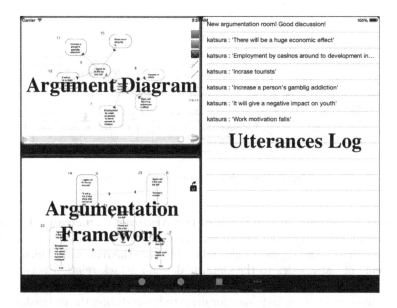

Fig. 1. Snapshot of the screen of the argumentation support tool

on these graphs. When the graph structure of an AF representing the content of the discussion becomes too complicated to see, users can convert it into a MRAF, whose structure is determined by users. We will explain the details of this conversion in Sect. 5.3.

Figure 1 shows the snapshot of the main screen of our tool. The display window of argument diagram is at the upper left, one of the argumentation frameworks (AF and MRAF) is at the lower left, and the utterances log is at the right side of the screen.

5.2 System Operational Procedures

Figure 2 shows the system operating procedures. When users input utterances in the form of argument diagram, then attack relations between them in the frame of Argument Diagram at the upper left, and the information about them is sent to the server. In the server, firstly, they are converted into logical expressions as described in (B). Secondly, by generating the nodes which indicate the logical expressions and the links between nodes which indicate attack relation between utterances, the graph structure of the AF converted from the inputted information is generated as described in (C). Finally, all sets which are each extensions of the AF are calculated as described. When these steps are finished, the server returns the information about the graph structure of AF and all sets of arguments which are any extensions of it to the tablet PC and the information is displayed in the frame of AF at the lower left of the main screen.

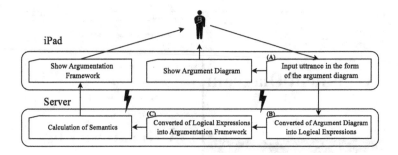

Fig. 2. Operating diagram of this system

(A) Input Utterance in the Form of the Argument Diagram. The argument diagram represents the logical structure of utterances consisting of data nodes, claim nodes and support links, which indicate the reasoning which is from the contents of data nodes to the claim node, and attack relation between the components of them. This style of the representation by the diagram is very simple because we want to make it easy for users to input during the discussion. This diagram saves time-series information of the argument, while accepting utterances input of user. To input the user's utterance in the form of the argument diagram, the user creates a node on the screen, makes a link from the node to the other node, and inputs his utterance in the node.

(B) Conversion of Argument Diagram into Logical Expressions. The information of the structure of argument diagram is sent to the argumentation server and logical formulas are constructed from it. The formula is classified to R_s (A set of strict inference rules), R_d (A set of defeasible inference rules), K_p (A set of the ordinary premises) and K_n (A set of the axioms). Let an argument diagram be $(A, attacks, supports)$. For any member of attacks and supports, R_d and K_p are obtained as follows:

$$\forall a, b \in A, \ (b, a) \in attacks \implies \{\neg a \Leftarrow b\} \in R_d, \ \{b\} \in K_p$$
$$\forall a, b \in A, \ (b, a) \in supports \implies \{a \Leftarrow b\} \in R_d, \ \{b\} \in K_p$$

Furthermore, the conversion of logical formulas into AF is carried out according to the conversion process of Aspic+ [22].

5.3 Modularization of Argumentation Framework

In this section, we explain how to construct an MRAF from an AF in our tool. Let (a) of Fig. 3 be the original AF. To convert this AF into the MRAF consisting of two modules, at first, users generates the "Modularization frame" at the right of the original frame which. In this frame, there are two frames, the "lower frame" and the "higher frame" to represent the graph structure of higher module and lower module of the MRAF. In these frame, user can build the graphs of each

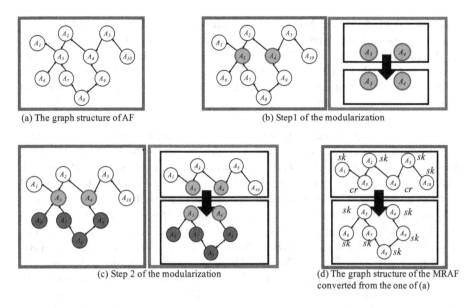

(a) The graph structure of AF

(b) Step1 of the modularization

(c) Step 2 of the modularization

(d) The graph structure of the MRAF converted from the one of (a)

Fig. 3. Modularization of argumentation framework

module of the MRAF which they want to construct by copying a part of the graph of the original AF.

Then, users select some nodes on the graph of the original AF which they want to be shared among the graphs of two modules in MRAF (In Fig. 3(b)). We call these nodes the "shared nodes" (In Fig. 3(b), A_3 and A_4 are shared nodes).

Next, users select some nodes which they want to be contained in the graph of the lower module of the MRAF from all nodes in the graph of the original AF other than shared nodes. We call these nodes the "targeted nodes." And then, one new graphs of AF are constructed in each frame in the "Modularization frame" (Fig. 3(c)). Note that the graph of an AF in the lower frame is build by copying the shared nodes, the targeted nodes, and the attack relations between arguments which are shared nodes or targeted nodes on the original graph structure of AF. Furthermore, the graph of an AF in the higher frame is build by copying the shared arguments and the all components of the original graph other than the one of the graph of an AF in lower module. After these two graphs are constructed, the status of reliability, sk is assigned to each argument in lower module automatically. Then, a RAF of each module is constructed and its semantics is calculated as follows.

Firstly, the server calculate each extension of the lower module M_l in each semantics of RAF. Let $\Delta_l = (Ar_l, attacks_l, ST_l)$, then, $Ar_l = \{A_3, A_4, A_6, A_7, A_8, A_9\}$, $attacks_l = \{(A_3, A_6), (A_6, A_3), (A_7, A_3), (A_8, A_7), (A_8, A_9), (A_9, A_4), (A_9, A_8)\}$, and $ST_l(x) = sk$ for $\forall x \in Ar_l$ because there is no lower module of this module. Here are the lists of the sets of arguments which is RAF-σ ϵ extension($\epsilon \in \{$complete, grounded$\}$, $\sigma \in \{$Non-Def, Optimistic,

Pessimistic}) of Δ_l in the MRAF whose component is not yet completed and, later, becomes Γ: RAF-σ complete extension: $\{\}$, $\{A_3, A_4, A_8\}$, $\{A_4, A_8\}$, $\{A_4, A_6, A_8\}$, $\{A_6\}$, $\{A_6, A_7, A_9\}$, RAF-σ grounded extension: $\{\}$. Note that the set of argument which is RAF-Non-Def complete extension of Δ_l, the one which is RAF-Optimistic complete extension Δ_l, and the one which is RAF-Non-Def complete extension Δ_l are the same because $ST_l(x) = sk$ for $\forall x \in Ar$, and similar relation is satisfied also in grounded extension of each RAF semantics. Furthermore, the set of argument which is RAF-σ complete extension of Δ_l and the one which is complete extension of AF_l ($AF_l = (Ar_l, attacks_l)$) are the same. Moreover, the server calculate each extension of the higher module M_h in each semantics of RAF. Based on the result of the calculation According to the way of defining the reliability in Definition 10, the status of reliability of A_3 and the one of A_4 in the higher module are cr. Let $\Delta_h = (Ar_h, attacks_h, ST_h)$, then, $Ar_h = \{A_1, A_2, A_3, A_4, A_5, A_{10}\}$, $attacks_h = \{(A_2, A_3), (A_3, A_1), (A_4, A_2), (A_5, A_4), (A_{10}, A_5)\}$, and ST_h is the following function: $ST_h(x) = sk$ for $\forall x \in Ar \backslash \{A_3, A_4\}$, $ST_h(A_3) = ST_h(A_4) = cr$ because there is no lower module of this module. Therefore, the graph structure of the MRAF showed in (d) is constructed in the server. Let the MRAF be Γ, then $\Gamma = (MO, refers)$, $MO = \{M_h, M_l\}$, and $refers = (M_h, M_l)$". Here are the lists of the sets of arguments which is RAF-σ ϵ extension($\epsilon \in \{$complete, grounded$\}$, $\sigma \in \{$Non-Def, Optimistic, Pessimistic$\}$) of Δ_h in Γ: RAF-Non-Def complete extension: $\{A_3, A_4, A_{10}\}$, RAF-Non-Def grounded extension: $\{A_3, A_4, A_{10}\}$. RAF-Optimistic complete extension: $\{A_1, A_2, A_{10}\}$, $\{A_3, A_4, A_{10}\}$, $\{A_{10}\}$, RAF-Optimistic grounded extension: $\{A_{10}\}$. RAF-Pessimistic complete extension: $\{A_{10}\}$, RAF-Pessimistic grounded extension: $\{A_{10}\}$.

5.4   Example of Modularization

We show the advantage of the modularization of AF by citing an experimental data done by us. The arguments in an AF illustrated in Fig. 4 represents a part of a discussion for the restart of nuclear power stations in Japan where, two examinees who are inexpert in any relative domain of nuclear power, and they discussed whether they should be restarted or not for over 40 min. Such an AF tends to become larger and more complicated steadily and in the end, becomes such whose graph structure is too complicated to grasp the content of the discussion, similar to that illustrated in Fig. 4, which is a screen shot of one part of the display window of AF. In such a case, we can convert it as one modularized AF including all components of it as described in Sect. 5.3. Figure 5 is a screen shot of another part of the display window of AF, and illustrates one modularized AF based on factors. Each module is represented as one smaller AF representing the dialectical content of each local discussion. Furthermore, we regard module 0 as the main module in the modularized AF and module 1 is the sub module in Fig. 5. The topics of the discussions represented by the main module and sub module is as follows: The main topic is "Restart of nuclear power stations in Japan" and The subtopic is "Alternative energy." The only shared argument of Module 0 and Module 1 is the argument surrounded with

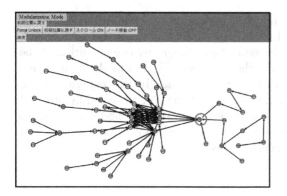

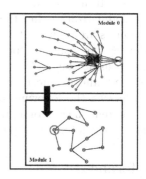

Fig. 4. Entire structure of the AF about reopen of nuclear facility problem

Fig. 5. Module structure of the AF about reopen of nuclear facility problem

a circle in Figs. 4 and 5, which represents the argument "Since thermal power stations in Japan can be operated, we shouldn't restart nuclear power stations in Japan." We can enumerate three kinds of advantages of modularization in Fig. 5. Firstly, users can easily understand that the content of each argument in module 0 is relative to alternative energy. In fact, we can easily see that the content of the shared argument surrounded with a circle is relative to the main topic and the subtopic by looking at the diagram in Fig. 5. Secondly, the users can grasp the structure of the diagram by looking at two smaller AF, instead of looking at a huge AF. In fact, while there are 64 arguments in the original AF illustrated in Fig. 4, 52 arguments in module 0 and 13 arguments in module 1 are illustrated in Fig. 5. Thirdly, note that further modularizations can be applied to the graph structure of the AF illustrated by module 0, which is still complicated, to simplify the entire structure of the diagram to get a simpler graph than the graph in Fig. 5. These modularizations can be based not only topics but also simplification of graph structure, types of discussion, and etc. Therefore, modularization of AF can contribute to the users of our tool getting a more understandable visualization of the dialectical content of discussions.

6   Conclusion

We introduced an argumentation support tool based on the theory of MRAF proposed this paper. MRAF is an extended theory of AF defined by Dung, and has three kinds of semantics which integrates more than one RAF semantics which depends on the reliability of each argument and gives theoretical foundation of MRAF. Based on MRAF theory, our tool has a function to modularize the original AF, which helps users to grasp the logical structure of a huge discussion and the acceptability of each argument and evaluates the dialectical superiority in an AF or a MRAF.

As a future work, we will conduct experiments to evaluate how effective the modularization function is. In this experiment, as explained in Sect. 5.4, examinees are required to argue the same topic under any one condition of the followings: by using no tool, by using the tool disabled from modularizing AFs, and by using the tool equipped with its all functions. In this experiment, we will ascertain the effects of the modularization of the AFs provided by our tool by checking the following points: how often examinees use modularization, to what extent the size of a graph is kept un-increased when it is modularized, and whether the interface of the tool is easy to use.

References

1. Reed, C., Rowe, G.: Araucaria: software for argument analysis, diagramming and representation. Int. J. Artif. Intell. Tools **13**(04), 961–979 (2004)
2. Stephen, T.: The Uses of Argument. Cambridge University Press, Cambridge (1958)
3. Dung, P.M.: On the acceptability of arguments and its fundamental role in non-monotonic reasoning, logic programming and n-person games. Artif. Intell. **77**(2), 321–357 (1995)
4. Atkinson, K., Bench-Capon, T., Dunne, P.E.: Uniform argumentation frameworks. In: Computational Models of Argument: Proceedings of COMMA 2012, vol. 245, p. 165 (2012)
5. Dunne, P.E., Hunter, A., McBurney, P., Parsons, S., Wooldridge, M.: Weighted argument systems: basic definitions, algorithms, and complexity results. Artif. Intell. **175**(2), 457–486 (2011)
6. Torre, L., Villata, S.: An aspic-based legal argumentation framework for deontic reasoning. In: Computational Models of Argument: Proceedings of COMMA 2014, pp. 266–421 (2014)
7. Muntjewerff, A.J., Breuker, J.A.: Evaluating PROSA, a system to train solving legal cases. In: Artificial Intelligence in Education: Ai-Ed in the Wired and Wireless Future, pp. 278–285 (2001)
8. Aleven, V.: An intelligent learning environment for case-based argumentation. Technol. Instr. Cognit. Learn. **4**(2), 191–241 (2006)
9. Pinkwart, N., Aleven, V., Ashley, K., Lynch, C.: Evaluating legal argument instruction with graphical representations using LARGO. Front. Artif. Intell. Appl. **158**, 101 (2007)
10. Pinkwart, N., Lynch, C., Ashley, K., Aleven, V.: Re-evaluating LARGO in the classroom: are diagrams better than text for teaching argumentation skills? In: Woolf, B.P., Aïmeur, E., Nkambou, R., Lajoie, S. (eds.) ITS 2008. LNCS, vol. 5091, pp. 90–100. Springer, Heidelberg (2008). doi:10.1007/978-3-540-69132-7_14
11. Ravenscroft, A., Pilkington, R.M.: Investigation by design: developing dialogue models to support reasoning and conceptual change. Int. J. Artif. Intell. Educ. **11**(1), 273–298 (2000)
12. Dung, P.M., Thang, P.M., Hung, N.D.: Modular argumentation for modelling legal doctrines of performance relief. Argument Comput. **1**(1), 47–69 (2010)
13. Brewka, G., Eiter, T.: Argumentation context systems: a framework for abstract group argumentation. In: Erdem, E., Lin, F., Schaub, T. (eds.) LPNMR 2009. LNCS (LNAI), vol. 5753, pp. 44–57. Springer, Heidelberg (2009). doi:10.1007/978-3-642-04238-6_7

14. Modgil, S.: Reasoning about preferences in argumentation frameworks. Artif. Intell. **173**(9), 901–934 (2009)
15. Amgoud, L., Cayrol, C.: On the acceptability of arguments in preference-based argumentation. In: Proceedings of the Fourteenth conference on Uncertainty in Artificial Intelligence, pp. 1–7. Morgan Kaufmann Publishers Inc. (1998)
16. Bench-Capon, T.J.: Persuasion in practical argument using value-based argumentation frameworks. J. Logic Comput. **13**(3), 429–448 (2003)
17. Tang, Y., Cai, K., McBurney, P., Sklar, E., Parsons, S.: Using argumentation to reason about trust and belief. J. Logic Comput., exr038 (2011)
18. Li, H., Oren, N., Norman, T.J.: Probabilistic argumentation frameworks. In: Modgil, S., Oren, N., Toni, F. (eds.) TAFA 2011. LNCS (LNAI), vol. 7132, pp. 1–16. Springer, Heidelberg (2012). doi:10.1007/978-3-642-29184-5_1
19. Pazienza, A., Esposito, F., Ferilli, S.: An authority degree-based evaluation strategy for abstract argumentation frameworks (2013)
20. Caminada, M.: On the issue of reinstatement in argumentation. In: Fisher, M., Hoek, W., Konev, B., Lisitsa, A. (eds.) JELIA 2006. LNCS (LNAI), vol. 4160, pp. 111–123. Springer, Heidelberg (2006). doi:10.1007/11853886_11
21. Katsura, Y., Okada, S., Nitta, K.: Dynamic argumentaion support tool using argument diagram. Ann. Conf. J. Soc. Artif. Intell. **29**, 1–4 (2015). (in Japanese)
22. Prakken, H.: An overview of formal models of argumentation and their application in philosophy. Stud. Logic **4**(1), 65–86 (2011)

Applying a Convolutional Neural Network to Legal Question Answering

Mi-Young Kim[✉], Ying Xu, and Randy Goebel

Alberta Innovates Centre for Machine Learning, Department of Computing Science,
University of Alberta, Edmonton, Canada
{miyoung2,yx2,rgoebel}@ualberta.ca

Abstract. Our legal question answering system combines legal information retrieval and textual entailment, and we describe a legal question answering system that exploits a deep convolutional neural network. We have evaluated our system using the training/test data from the competition on legal information extraction/entailment (COLIEE). The competition focuses on the legal information processing related to answering yes/no questions from Japanese legal bar exams, and it consists of three phases: ad-hoc legal information retrieval, textual entailment, and a learning model-driven combination of the two phases. Phase 1 requires the identification of Japan civil law articles relevant to a legal bar exam query. For that phase, we have implemented a combined TF-IDF and Ranking SVM information retrieval component. Phase 2 requires the system to answer "Yes" or "No" to previously unseen queries, by comparing extracted meanings of queries with relevant articles. Our training of an entailment model focuses on features based on word embeddings, syntactic similarities and identification of negation/antonym relations. We augment our textual entailment component with a convolutional neural network with dropout regularization and Rectified Linear Units. To our knowledge, our study is the first to adapt deep learning for textual entailment. Experimental evaluation demonstrates the effectiveness of the convolutional neural network and dropout regularization. The results show that our deep learning-based method outperforms our baseline SVM-based supervised model and K-means clustering.

Keywords: Legal question answering · Recognizing textual entailment · Information retrieval · Convolutional neural network

1 Task Description

Legal question answering can be considered as a number of intermediate steps. For instance, consider a question such as "Is it true that a special provision that releases warranty can be made, but in that situation, when there are rights that the seller establishes on his/her own for a third party, the seller is not released of warranty?" In this example, a system must first identify and retrieve relevant documents, typically legal statutes, and subsequently, identify a most relevant sentence. Finally, it must compare the semantic connections between question and the relevant sentence, and determine whether an entailment relation holds.

© Springer International Publishing AG 2017
M. Otake et al. (Eds.): JSAI-isAI 2015 Workshops, LNAI 10091, pp. 282–294, 2017.
DOI: 10.1007/978-3-319-50953-2_20

Deep Neural Networks (DNNs) are an emerging technology that has recently demonstrated dramatic success in several areas, including speech feature extraction and recognition. Incorporation of convolution and subsequent pooling into a neural network has provided the basis for a technique called Convolutional Neural Networks (CNNs) [22]. CNNs have shown good performance in image and speech recognition [18], and many studies have proposed applying CNNs to natural language processing [11, 12]. Here we adapt a CNN for legal question answering, especially focused on textual entailment. One primary motivation for using deeper models such as neural networks with many layers is that they have the potential to be much more representationally efficient compared with shallower neural network models. In textual entailment, we will extract linguistic features between two sentences, and determine textual entailment by comparing the features. In this task, not all linguistic features are directly related to each other, so we intend to capture related features, then connect them locally. One major motivation for CNNs is to restrict the network architecture through the use of local connections known as receptive fields.

The Competition on Legal Information Extraction/Entailment (COLIEE) 2015 focuses on two aspects of legal information processing related to answering yes/no questions from legal bar exams: legal document retrieval (Phase 1), and textual entailment for Yes/No question answering of legal queries (Phase 2).

Phase 1 is an ad-hoc information retrieval (IR) task. The goal is to retrieve relevant Japan civil law articles that are most relevant to a legal bar exam. We approach this problem with two models based on statistical information. One is the TF-IDF model [1], i.e., term frequency-inverse document frequency. The relevance between a query and a document depends on their intersection word set. The importance of words is measured with a function of term frequency and document frequency as parameters. Our terms are lemmatized words, e.g., the verbs "attending," "attends," and "attended" are lemmatized as the same base form "attend."

Another popular model for text retrieval is a Ranking SVM model [2]. That model is used to re-rank documents that are retrieved by the TF-IDF model. The features used to train this model are lexical words, dependency path bigrams and TF-IDF scores. The intuition is that the supervised model can learn weights or priority of words based on training data, in addition to or as an alternative to TF-IDF.

The goal of Phase 2 is to construct Yes/No question answering systems for legal queries, by confirming the entailment of questions from the relevant articles. The answer to a question is typically determined by measuring some kind of heuristically-determined semantic similarity between question and answer. While there are many possible approaches, we note that neural network-based distributional sentence models have achieved success in many natural language processing tasks such as sentiment analysis [12], paraphrase detection [13], document classification [14], and question answering [11]. As a consequence of this success, it appears natural to approach textual entailment using similar techniques. Here we show that a neural network-based sentence model can be applied to the task of textual entailment. After constructing a set of pre-trained semantic word embeddings using the *word2vec* [20], we used a supervised method to learn a heuristic semantic-matching model between question and corresponding articles.

In addition to semantic word embeddings, our system uses features that depend on some components of the syntactic structure, and the presence of negation. We employ a convolutional neural network algorithm with dropout regularization and Rectified Linear Units, and compare its performance with baseline system based on support vector machines.

2 Phase 1: Legal Information Retrieval

2.1 IR Models

2.1.1 TF-IDF Model

Here we introduce our TF-IDF and Ranking SVM models. Queries and articles are all tokenized and parsed by the Stanford NLP tool. For the IR task, the similarity of a query and an article is based on the terms within them. Our terms for TF-IDF are lemmatized words.

For TF-IDF, we use the simplified version of Lucene's similarity score of an article to a query as suggested in [15]:

$$tf-idf(Q,A) = \sum_{t} [\sqrt{tf(t,A)} \times \{1 + \log{(idf(t))}\}^2]$$

The score tf-$idf(Q,A)$ is a measure which estimates the relevance between a query Q and an article A. First, for every term t in the query A, we compute $tf(t,A)$, and $idf(t)$. The score $tf(t,A)$ is the term frequency of t in the article A, and $idf(t)$ is the inverse document frequency of the term t, which is the number of articles that contain t. After some normalization computed within the Lucene package, we multiply $tf(t,A)$ and $idf(t)$, and then we compute the sum of these multiplication scores for all terms t in the query A. This summation result is tf-$idf(Q,A)$. The bigger tf-$idf(Q,A)$ is, the more relevant between the query Q and the article A. The real version has some normalized parameters in terms of an article's length to alleviate the functions biased towards long documents. The parameters are set as the default of the Lucene's TF-IDF model.

2.1.2 Ranking SVM

The Ranking SVM model was proposed by Joachims [2]. That model ranks a set of retrieved documents based on a selection of attributes from user's data. Given the feature vector of a training instance, i.e., a retrieved article set given a query, denoted by $\Phi(Q, Ai)$, the model tries to find a ranking that satisfies constraints:

$$\emptyset(Q,A_i) > \emptyset(Q,A_j)$$

where A_i is a relevant article for the query Q, while A_j is less relevant.

We adopt the same model and features suggested in [15]. The three types of features are as follows:

- Lexical words: the lemmatized normal form of surface structure of words in both the retrieved article and the query. In the conversion to the SVM's instance representation, this feature is converted into binary features whose values are one or zero, i.e., depending on if a word exists in the intersection word set or not.
- Dependency pairs: word pairs that are linked by a dependency link. The intuition is that, compared with the bag of words information, syntactic information should improve the capture of salient semantic content. Dependency parse features have been used in many NLP tasks, and improved IR performance [3]. This feature type is also converted into binary values.
- TF-IDF score (Sect. 2.1.1).

We set the Ranking SVM model's parameter c according to the cross validation on the training set.

2.2 Experiments

The legal IR task that we use to test our system has several sets of queries paired with the Japan civil law articles as documents (724 articles in total). Here follows one example of the query and a corresponding relevant article.

Question: A person who made a manifestation of intention which was induced by duress emanated from a third party may rescind such manifestation of intention on the basis of duress, only if the other party knew or was negligent of such fact.

Related Article: (Fraud or Duress) Article 96 (1)Manifestation of intention which is induced by any fraud or duress may be rescinded. (2)In cases any third party commits any fraud inducing any person to make a manifestation of intention to the other party, such manifestation of intention may be rescinded only if the other party knew such fact. (3)The rescission of the manifestation of intention induced by the fraud pursuant to the provision of the preceding two paragraphs may not be asserted against a third party without knowledge.

Before the final test set was released, we received 6 sets of queries for a "dry run" in COLIEE 2015. The 6 sets of data include 267 queries, and 326 relevant articles (average 1.22 articles per query). We used a corresponding 6-fold leave-one-out cross validation evaluation. The metrics for measuring our IR model performance is Mean Average Precision (MAP):

$$MAP(Q) = \frac{1}{|Q|} \sum_{q \in Q} \frac{1}{m} \sum_{k \in (1,m)} precision(R_k)$$

where Q is the set of queries, and m is the number of retrieved articles. R_k is the set of ranked retrieval results from the top until the k-th article. In the following experiments, we set m as 5 for all queries, corresponding to the column $MAP@5$ in Table 1.

Table 1. IR results on dry run data with different models.

Id	Models	MAP@5
1	TF-IDF with lemma	0.294
2	SVM-ranking	0.302

Table 1 shows the results of experiments with our two IR models on the legal IR task on the training set. The ensemble SVM-Ranking model is slightly better than the TF-IDF model. Table 2 shows the results of our SVM-ranking model on the final test set. The test data size is 79 queries for Phase 1. The performance of our system was ranked first among the submitted systems in the Competition on Legal Information Extraction/Entailment (COLIEE) 2015 [23].

Table 2. IR results on test data using the SVM-ranking model

Participant ID	Performance on Phase 1	
UA (University of Alberta)	* The number of submitted articles: 79	
	* The number of correctly submitted articles: 50	
	Precision	0.6329
	Recall	0.4902
	F-measure	**0.5525**

3   Phase 2: Answering 'Yes'/'No' Questions Using a Convolutional Neural Network

Our system uses syntactic information in addition to word embedding to predict textual entailment. We exploit syntactic similarity features, negation and antonyms in Kim et al. [15]. Details are provided in the next subsections.

3.1   Our System

3.1.1   Model Description

The problem of answering a legal yes/no question can be viewed as a binary classification problem. Assume a set of questions Q, where each question $q_i \in Q$ is associated with a list of corresponding article sentences $\{a_{i1}, a_{i2}, ..., a_{im}\}$, where $y_i = 1$ if the answer is 'yes' and $y_i = 0$ otherwise. We choose the most relevant sentence a_{ij} using the algorithm of Kim et al. [15], and we simply treat each data point as a triple (q_i, a_{ij}, y_i). Therefore, our task is to learn a classifier over these triples so that it can predict the answers of any additional question-article pairs.

Our solution assumes that correct answers have high semantic similarity to questions. We model questions and answers as vectors using word embedding and linguistic information, and evaluate the relatedness of each question-article pair in a shared vector space. Following Yu et al. [11], given the vector representations of a question q and a most relevant article sentence a, the probability of the answer being correct is

$$p(y = 1|q, a) = rectifier(q^T Ma + b)$$

Where the bias term b and the transformation matrix M in $R^{d \times d}$ are model parameters. This formulation can be understood as follows: we first generate a question through the transformation $q' = Ma$, and then measure the similarity of the generated question q' and the given question q by their dot product. The rectifier function is used as an activation function.

As mentioned above, a Convolutional Neural Network (CNN) is a biologically-inspired variant of a multi-layer perceptron. CNN employs two component techniques: (1) restricting the network architecture through the use of local connections known as receptive fields; and (2) constraining the choice of synaptic weights through the use of weight-sharing. Most of the applications of CNNs include a max-pooling layer, which reduces and integrates the neighboring neurons' outputs. CNNs also exploit spatially-local correlation by enforcing a local connectivity pattern between neurons of adjacent layers.

CNN-based models have been proved to be effective in applications such as twitter sentiment prediction [12] and semantic parsing [16]. Figure 1 illustrates the architecture of the CNN-based sentence model in one dimension. We use word embedding and linguistic features with one convolutional layer and one pooling layer. A word embedding is a parameterized function mapping words in some language to high-dimensional vectors. We use the Bag-of-words model of [11] for word embedding.

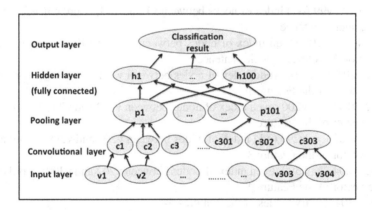

Fig. 1. Our CNN architecture

Given word embeddings, the bag-of-words model generates the vector representation of a sentence by summing over the embeddings of all words in the sentence, after removing stopwords. The vector is then normalized by the length of the sentence.

$$s = \frac{1}{n} \sum_{i=1}^{n} s_i$$

In the formula above, s and s_i are d-dimensional vectors. s_i corresponds to the vector of the i-th word in the sentence, n is the number of words in the sentence, and s is vector representation of the sentence. We used *word2vec* [20] word embedding technique ($d = 50$, which is typically used), and used the training data of COLIEE 2014 for training *word2vec*. *Word2vec* [20] used a neural network consisting of input layer, projection layer, and output layer and they removed a hidden layer to improve learning speed.

In addition to word embeddings, the types of features we use are as follows:

(a) Word Lemma
(b) Tree structure features (considering only roots)

Feature 1: $w_{root}(condition_{query\_n})$
Feature 2: $w_{root}(condition_{article\_n})$
Feature 3: $w_{root}(conclusion_{query\_n})$
Feature 4: $w_{root}(conclusion_{article\_n})$
Feature 5: $neg\_level(condition_{query\_n})$
Feature 6: $neg\_level(condition_{article\_n})$
Feature 7: $neg\_level(conclusion_{query\_n})$
Feature 8: $neg\_level(conclusion_{article\_n})$

In Fig. 1, the input layer consists of the following word embedding vectors and binary values:

(1) v1, v3, ..., v99 (odd index of nodes between v1 and v99): word embedding vector of the query sentence
(2) v2, v4, ..., v100 (even index of nodes between v2 and v100): word embedding vector of the relevant article sentence
(3) v101, v103, ..., v199 (odd index of nodes between v101 and v199): word embedding vector of the Feature 1
(4) v102, v104, ..., v200 (even index of nodes between v102 and v200): word embedding vector of the Feature 2
(5) v201, v203, ..., v299 (odd index of nodes between v201 and v299): word embedding vector of the Feature 3
(6) v202, v204, ..., v300 (even index of nodes between v202 and v300): word embedding vector of the Feature 4
(7) v301-v304: binary values of the Features 5-8

In the features above, $article_n$ is the most relevant article of the query $query_n$. First we detect condition part and conclusion part in the question and corresponding article, and also compute negation value ($neg\_level()$) of each part according to Kim et al. [15]. The following is an example of condition and conclusion detection:

<Civil Law Article 177> *Acquisitions of, losses of and changes in real rights concerning immovable properties may not be asserted against third parties, unless the same are registered pursuant to the applicable provisions of the Real Estate Registration Act (Law No. 123 of 2004) and other laws regarding registration.*

(1) *Conclusion* =>*Acquisitions of, losses of and changes in real rights concerning immovable properties may* **not** *be asserted against third parties,*

(2) *Condition* => **unless** *the same are registered pursuant to the applicable provisions of the Real Estate Registration Act (Law No. 123 of 2004) and other laws regarding registration.*

In Features 1-4, $w_{root}(s)$ means the root word in the syntactic tree of the sentence s. Features 1-4 consider both lexical and syntactic information, and Features 5-8 incorporate negation and antonym information. We use some morphological and syntactic analysis to extract lemma and dependency information. Details of the morphological and syntactic analyzer are given in Sect. 3.2.

As shown in Fig. 1, the nodes of the input layer of even indices between v1 and v100 (such as v2, v4, v6.... and v100) indicate the word embedding vector for a relevant article sentence, and the nodes of odd indices between v1 and v100 (such as v1, v3, ... and v99) show the word embedding vector for a query sentence. The nodes between v101 and v304 are linguistic features. Because the adjacent two nodes indicate the same feature type (e.g., v1 and v2 indicate the first index value of each word embedding vector), we make a convolutional layer constructed from the adjusting two input nodes.

The convolutional vector t in R^2 (the 2-dimensional real number space) projects adjacent two nodes into a feature value c_i, computed as follows:

$$C_i = rectifier(t * v_{i:i+1} + b),$$

where $rectifier(x) = max(0,x)$. We explain the rectifier function in the Subsect. 3.1.2.

In the pooling layer, we just do summation of 3 adjacent c_i values, to reduce the features. The number 3 was just chosen for this experiment, and we can find an optimal number in future work. The p_i values in the pooling layer as follows:

$$p_i = \sum_{k=3i-2}^{3i} c_i \quad i = 1, 2, \dots, 101$$

The training is done with multithreaded mini-batch gradient descent in the weka[1] tool.

3.1.2 Dropout Regularization and Rectified Linear Units

When a neural network is trained on a small training set, it typically performs poorly on test data. This "overfitting" is greatly reduced by randomly omitting some of the feature detectors on each training case. It is called 'dropout'. The dropout prevents complex co-adaptations in which a feature detector is only helpful in the context of several other specific feature detectors. Random dropout gives big improvements on many benchmark tasks and sets new records for speech and object recognition [21]. We found that the

---

[1] https://weka.wikispaces.com/Unofficial+packages+for+WEKA+3.7.

dropout rate needs to be between 0.6 and 0.7 for a hidden layer and 0.1 for an input layer in order to make it effective in achieving low errors.

We also employ the Rectified Linear Unit for CNN. Neural networks with rectified linear unit (ReLU) non-linearities have been highly successful for computer vision tasks and have been shown to be faster to train than standard sigmoid units [17].

ReLU is a neuron which uses a rectifier instead of a hyperbolic tangent or logistic function as an activation function. *Rectifier f(x) = max(0,x)* allows a network to easily obtain sparse representations.

3.1.3   Supervised Learning with SVM

We have compared our method with SVM, as a kind of supervised learning model. Using the SVM tool included in the Weka [4] software, we performed cross-validation for the 179 questions of the dry run data in COLIEE 2014 using word embedding vector and linguistic features in Sect. 3.1.1. We used a linear kernel SVM because it is popular for real-time applications as they enjoy both faster training and classification speeds, with significantly reduced memory requirements than non-linear kernels, because of the compact representation of the decision function.

3.2   Experimental Setup for Phase 2

In the general formulation of the textual entailment problem, given an input text sentence and a hypothesis sentence, the task is to make predictions about whether or not the hypothesis is entailed by the input sentence. We report the accuracy of our method in answering yes/no questions of legal bar exams by predicting whether questions are entailed by the corresponding civil law articles.

There is a balanced positive-negative sample distribution in the dataset (55.87% yes, and 44.13% no) for a dry run of COLIEE 2014 dataset, so we consider the baseline for true/false evaluation is the accuracy when returning always "yes," which is 55.87%. Our data for our dry run has 179 questions.

The original examinations are provided in Japanese and English, and our initial implementation used a Korean translation, provided by the Excite translation tool (http://excite.translation.jp/world/). The reason that we chose Korean is that we have a team member whose native language is Korean, and the characteristics of Korean and Japanese language are similar. In addition, the translation quality between two languages ensures relatively stable performance. Because our study team includes a Korean researcher, we can easily analyze the errors and intermediate rules in Korean. We used a Korean morphological analyzer and dependency parser [5].

3.3   Experimental Results

To compare our performance with Kim et al. [15], we measured our system's performance on the dry run data of COLIEE 2014. Table 3 shows the experimental results. An SVM-based model showed accuracy of 60.12%, and a convolutional neural network with pre-trained semantic word embeddings and dropout showed best performance of

63.87% with the setting of input layer dropout rate of 0.1, hidden layer dropout rate of 0.6, and 100 hidden layer nodes. When we did not use the dropout regularization, the accuracy was lower by 1.22%. Without dropout and word embedding, the accuracy was 56.30%, which showed no significant difference with the baseline accuracy. We also compare our performance with the Kim et al. [15] performance using the same dataset. In [15], they used their linguistic features for SVM learning, and also proposed a model combining rule-based method and k-means clustering. Our CNN performance outperformed both of their SVM model and combined model.

Table 3. Experimental results on dry run data for Phase 2

Our method	Accuracy (%)
Baseline	55.87
Cross-validation with supervised learning (SVM) [15]	59.43
Rule-based model + K-means clustering [15]	61.96
Cross-validation with supervised learning (SVM) using our features	60.12
Convolutional neural network with word embedding + linguistic features + dropout	63.87
Convolutional neural network with word embedding +linguistic features	62.65
Convolutional neural network with only linguistic features	56.30

Table 4 shows the experimental results using formal run data of COLIEE 2015. The formal run data size of COLIEE 2015 is 66 queries for Phase 2 from the bar exam of 2013. For Phase 3, we use the same test data of Phase 1, which consists of 79 queries extracted from the bar exam of 2012. Our performance of textual entailment (phase 2) is 66.67%, and the performance of combined phase (phase 3) is 65.82%. This result is ranked first among the systems in COLIEE 2015 competition [23].

Table 4. Experimental Results on the formal run data of COLIEE 2015

Our method	Accuracy (%)
Entailment results (Phase 2)	66.67
Combined results (Phase 3)	65.82

From unsuccessful instances, we classified the error types as shown in Table 5. We could not identify the errors arising from the Neural Network architecture or embedding vectors, so we just classified the errors into 7 cases which are shown in Table 5. The biggest error arises, of course, from the paraphrasing problem. The second biggest error is because of complex constraints in conditions. As with the other error types, there are cases where a question is an example case of the corresponding article, and the corresponding article embeds another article. There has been also errors in the case that a question is an exceptional case of the corresponding legal law article. In further work, we will need to complement our knowledge base with some kind of paraphrasing dictionary employing a paraphrase detection method.

Table 5. Error types

Error type	Accuracy (%)	Error type	Accuracy (%)
Specific example case	6.28	Paraphrasing	38.85
Exceptional case	9.14	Complex constraints in condition	28.09
Incorrect detection of condition, conclusion	3.84	Reference to another article	4.10
Etc.	9.7		

4   Related Work

Only very recently have researchers started to apply deep learning to question answering [11, 16, 19]. Relevant work includes Yih et al. [16] who constructed models for single-relation question answering with a knowledge base of triples. In the same direction, Bordes et al. [19] used a type of siamese network for learning to project question and answer pairs into a joint space. Finally, Yu et al. [11] selected answer sentences, which includes the answer of a question. They modelled semantic composition with a recursive neural network. However these tasks differ from the work presented here in that our purpose is not to make a choice of answer selection in a document, but to answer "yes" or "no."

A textual entailment method from W. Bdour et al. [6] provided the basis for a Yes/No Arabic Question Answering System. They used a kind of logical representation, which bridges the distinct representations of the functional structure obtained for questions and passages. This method is also not appropriate for our task. If a false question sentence is constructed by replacing named entities with terms of different meaning in the legal article, a logic representation can be helpful. However, false questions are not simply constructed by substituting specific named entities, and any logical representation can make the problem more complex. Nielsen et al. [7] extracted features from dependency paths, and combined them with word-alignment features in a mixture of an expert-based classifier. Zanzotto et al. [8] proposed a syntactic cross-pair similarity measure for RTE. Harmeling [9] took a similar classification-based approach with transformation sequence features. Marsi et al. [10] described a system using dependency-based paraphrasing techniques. All these previous systems uniformly conclude that syntactic information is helpful in RTE: we also use syntactic information.

As further research, we will try unsupervised pre-training with a CNN to solve the problem of small training datasets. We are also considering adopting more convolutional layers and pooling layers in the CNN architecture, and investigating the effect of more layers in the textual entailment problem. The challenge is the management of the tradeoff between encoding attribute dependencies and learning effective models.

5   Conclusion

We have described our implementation for the Competition on Legal Information Extraction/Entailment (COLIEE) Task.

For Phase 1, legal information retrieval, we implemented a Ranking-SVM model for the legal information retrieval task. By incorporating features such as lexical words, dependency links, and TF-IDF score, our model shows better mean average precision than TF-IDF.

For Phase 2, we have proposed a method to answer yes/no questions from legal bar exams related to civil law. We used a convolutional neural network model using dropout regularization and Rectified Linear Units with pre-trained semantic word embeddings. We also extract deep linguistic features with lexical, syntactic information based on morphological analysis and dependency trees. We show the improved performance over previous systems, using a convolutional neural network.

Acknowledgements. This research was supported by the Alberta Innovates Centre for Machine Learning (AICML) and the Natural Sciences and Engineering Research Council (NSERC). We are indebted to Ken Satoh of the National Institute for Informatics, who has had the vision to create the COLIEE competition.

References

1. Jones, K.S.: A statistical interpretation of term specificity and its application in retrieval. In: Willett, P. (ed.) Document Retrieval Systems, pp. 132–142. Taylor Graham Publishing, London (1988)
2. Joachims, T.: Optimizing search engines using clickthrough data. In: Proceedings of the Eighth ACM SIGKDD International Conference on Knowledge Discovery and Data Mining, KDD 2002, pp. 133–142. ACM, New York (2002)
3. Maxwell, K.T., Oberlander, J., Croft, W.B.: Feature-based selection of dependency paths in ad hoc information retrieval. In: Proceedings of the 51st Annual Meeting of the Association for Computational Linguistics, vol. 1. Long Papers, pp. 507–516. Association for Computational Linguistics, Sofia, Bulgaria, August 2013
4. Hall, M., Frank, E., Holmes, G., Pfahringer, B., Reutemann, P., Witten, I.H.: The WEKA data mining software: an update. SIGKDD Explor. **11**(1), 10–18 (2009)
5. Kim, M-Y., Kang, S-J., Lee, J-H.: Resolving ambiguity in inter-chunk dependency parsing. In: Proceedings of 6th Natural Language Processing Pacific Rim Symposium, pp. 263–270 (2001)
6. Bdour, W.N., Gharaibeh, N.K.: Development of yes/no arabic question answering system. Int. J. Artif. Intell. Appl. **4**(1), 51–63 (2013)
7. Nielsen, R.D., Ward, W., Martin, J.H.: Toward dependency path based entailment. In: Proceedings of the Second PASCAL Challenges Workshop on RTE (2006)
8. Zanzotto, F.M., Moschitti, A., Pennacchiotti, M., Pazienza, M.T.: Learning textual entailment from examples. In: Proceedings of the Second PASCAL Challenges Workshop on RTE (2006)
9. Harmeling, S.: An extensible probabilistic transformation-based approach to the third recognizing textual entailment challenge. In: Proceedings of ACL PASCAL Workshop on Textual Entailment and Paraphrasing (2007)

10. Marsi, E., Krahmer, E., Bosma, W.: Dependency-based paraphrasing for recognizing textual entailment. In: Proceedings of ACL PASCAL Workshop on Textual Entailment and Paraphrasing (2007)
11. Yu, L., Hermann, K.M., Blunsom, P., Pulman, S.: Deep learning for answer sentence selection. arXiv preprint arXiv:1412.1632 (2014)
12. Kalchbrenner, N., Grefenstette, E., Blunsom, P.: A convolutional neural network for modelling sentences. In: Proceedings of ACL (2014)
13. Socher, R., Huval, B., Manning, C.D., Ng, A.Y.: Semantic compositionality through recursive matrix-vector spaces. In: Proceedings of EMNLP-CoNLL (2012)
14. Hermann, K.M., Blunsom, P.: Multilingual models for compositional distributional semantics. In: Proceedings of ACL (2014)
15. Kim, M-Y., Xu, Y., Goebel, R.: Alberta-KXG: legal question answering using ranking SVM and syntactic/semantic similarity. In: JURISIN Workshop (2014)
16. Yih, W.-T., He, X., Meek, C.: Semantic parsing for single-relation question answering. In: Proceedings of ACL (2014)
17. Dahl, G.E., Sainath, T.N., Hinton, G.E.: Improving deep neural networks for LVCSR using rectified linear units and dropout. In: Proceedings of Acoustics, Speech and Signal Processing (ICASSP), pp. 8609–8613 (2013)
18. Deng, L., Abdel-Hamid, O., Yu, D.: A deep convolutional neural network using heterogeneous pooling for trading acoustic invariance with phonetic confusion. In: Proceedings of Acoustics, Speech and Signal Processing (ICASSP), pp. 6669–6673. IEEE (2013)
19. Bordes, A., Chopra, S., Weston, J.: Question answering with subgraph embeddings. In: Proceedings of EMNLP (2014)
20. Mikolov, T., Sutskever, I., Chen, K., Corrado, G.S., Dean, J.: Distributed representations of words and phrases and their compositionality. In: Advances in Neural Information Processing Systems, pp. 3111–3119 (2013)
21. Hinton, G.E., Srivastava, N., Krizhevsky, A., Sutskever, I., Salakhutdinov, R.R.: Improving neural networks by preventing co-adaptation of feature detectors. arXiv preprint arXiv: 1207.0580 (2012)
22. Kouylekov, M., Magnini, B.: Tree edit distance for recognizing textual entailment: estimating the cost of insertion. In: Proceedings of the Second PASCAL Challenges Workshop on RTE (2006)
23. Kim, M., Goebel, R., Satoh, K.: COLIEE-2015: evaluation of legal question answering. In: Ninth Workshop on Juris-Informatics (JURISIN) (2015)

Lexical-Morphological Modeling
for Legal Text Analysis

Danilo S. Carvalho[1(✉)], Minh-Tien Nguyen[1,2], Chien-Xuan Tran[1],
and Minh-Le Nguyen[1]

[1] School of Information Science, Japan Advanced Institute
of Science and Technology (JAIST), 1-1 Asahidai, Nomi, Ishikawa 923-1292, Japan
{danilo,tiennm,chien-tran,nguyenml}@jaist.ac.jp
[2] Hung Yen University of Education and Technology (UTEHY),
Hung Yen, Vietnam

Abstract. In the context of the Competition on Legal Information
Extraction/Entailment (COLIEE), we propose a method comprising the
necessary steps for finding relevant documents to a legal question and
deciding on textual entailment evidence to provide a correct answer.
The proposed method is based on the combination of several lexical and
morphological characteristics, to build a language model and a set of
features for Machine Learning algorithms. We provide a detailed study
on the proposed method performance and failure cases, indicating that
it is competitive with state-of-the-art approaches on Legal Information
Retrieval and Question Answering, while not needing extensive training data nor depending on expert produced knowledge. The proposed
method achieved significant results in the competition, indicating a substantial level of adequacy for the tasks addressed.

1 Introduction

Answering legal questions has been a long-standing challenge in the Information Systems research landscape. This topic draws progressively more attention, as we experience an explosive growth in legal document availability on the
World Wide Web and specialized systems. This growth is not accompanied by a
matching increase in information analysis capabilities, which points to a severe
under-utilization of available resources and to potential for information quality
issues [1]. As a consequence, increasing pressure has been put into professionals
of law, since having the relevant and correct information is a vital step in legal
case solving and thus is closely tied to the matter of professional ethics and
liability. This problem is often referred as the "information crisis" of law.

The ability to retrieve relevant and correct information given a legal query
has improved over time, with the combination of expert Knowledge Engineering and Natural Language Processing (NLP) methods. However, the ability to
answer questions in the legal domain is of special difficulty, due to the need of

D.S. Carvalho—Supported by CNPq – Brazil scholarship grant.

© Springer International Publishing AG 2017
M. Otake et al. (Eds.): JSAI-isAI 2015 Workshops, LNAI 10091, pp. 295–311, 2017.
DOI: 10.1007/978-3-319-50953-2_21

reasoning over different types of information, such as past decisions, laws and facts. Furthermore, concepts in legal text are often used in a way that differs from common language use, and differences in laws and procedures from each country prevent the creation of comprehensive and coherent international law corpora. Common legal ontologies are among the efforts to facilitate automatic legal reasoning, but have not seen strong development in the past years [2]. In this context, *Textual Entailment Recognition* plays a very important role, as a set of hypothesis presented in a question will certainly have answers in the previously cited types of information (decisions, laws, facts). The Recognition of Textual Entailment (RTE) challenge series[1], although not specific to the legal domain, is a recognized benchmark for methods that can be adapted to legal texts.

To effectively answer legal questions, one fundamental set of information that must be available is the law, presented as the collection of codes, sections, articles and paragraphs that should be unequivocally referenced when a hypothesis is raised as part of a legal inquiry. Therefore, adequate representation of law corpora is the basis of a functional system for legal question answering. The representation problem is often associated with ontologies and other annotated knowledge bases, but these methods are costly and more difficult to automate when compared to fully text-based approaches, such as bag-of-words, n-gram and topic models.

In this work, we propose a fully text-based method for legal text analysis, in the context of the Competition on Legal Information Extraction/Entailment (COLIEE), covering both the tasks of Information Extraction and Question Answering. The goal is the retrieval of relevant law articles to a given yes/no legal question and the use of the retrieved articles to correctly answer the question in a completely automated way. Our contributions in this paper are as follows: (i) a ranking and selection method for legal information retrieval based on a mixed size n-gram model, including an original scoring function for ranking; (ii) an improved adaptation of a Textual Entailment classification method, based on Machine Learning ensembles (Adaboost), including a similarity feature built upon Distributional Semantics (Word2Vec). Lexical and morphological analysis were done on the English translated Japanese Civil Code, comprising tokenization, POS-tagging, lemmatization, word clustering and a set of lexical statistics. A study on success and fail cases is also provided, with common baseline practices and related works used as means of performance comparison. The results of COLIEE are presented as a means of substantiating the experimental evaluation and also discussing the proposed method's perceived shortcomings and improvements.

The remaining of this work is structured as follows: Sect. 2 presents the related works and relevant results; Sect. 3 details the Legal Question Answering problem and the COLIEE competition shared task; Sect. 4 explains our approach to the competition problem; Sect. 5 presents the experimental setting, results and discussion; Finally, Sect. 6 offers some concluding remarks.

---

[1] www.aclweb.org/aclwiki/index.php?title=Recognizing_Textual_Entailment.

2   Related Works

Liu, Chen and Ho [3] presented the three-phase prediction (TPP) method for retrieval of relevant statutes in Taiwan's criminal law, given general language queries. The method was a hierarchical ranking approach to law corpora, featuring a combination of several Information Retrieval techniques, as well as Machine Learning and feature selection ones. Results were evaluated in terms of recall, achieving from 0.52 to 0.91, from the top 3 to 10 retrieved results, respectively.

Inkpen et al. showed one of the first successful models for RTE using SVMs [5]. Later, Castillo proposed a new system for solving RTE using SVMs [6], in which training data includes RTE-3, annotated data set from RTE-4, and the development set of RTE-5. 32 features were used and the training model achieved the best F-measure of 0.69 in two-way and 0.67 in three-way classification task.

Nguyen et al. [7] conducted a study of RTE on a Vietnamese version of RTE-3 [8] translated from Giampiccolo et al. [9]. The author used SVMs trained with 15 features divided in two groups: distance and statistical features, in which the first group captures the distance and the second one represents the word overlapping between two sentences. A voting system combining three classifiers built on three feature groups (distance, statistical, and combined features) was used to judge entailment relation. The method obtained 0.684 of F-measure in two-way task.

In legal text, Tran et al. addressed legal text QA by using inference [10]. The author used requisite-effectuation structures of legal sentences and similarity measures to find out correct answers without training data and achieved 60.8% accuracy on 51 articles on Japanese National Pension Law.

Kim et al. proposed a hybrid method containing simple rules and unsupervised learning using deep linguistic features to address RTE in civil law [11]. The author also constructed a knowledge base for negation and antonym words which would be used for classifying simple questions. To deal with difficult questions, the author used morphological, syntactic and lexical analysis to identify premises and conclusions. The accuracy was 68.36% with easy questions and 60.02 with difficult ones.

This work uses all features in [7], as they apply to the same purpose. Additional features were also included: Word2Vec similarity and term frequency – inverse document frequency (TF-IDF). Our approach differs from [6] in using *Word2Vec* [17] similarity instead of *WordNet*.

3   Legal Question Answering

Legal Question Answering (LQA) consists in finding out and providing *"correct answers"* to a legal question given by users. An overview of LQA is shown in Fig. 1.

LQA can be divided in three tasks: (1) retrieving relevant articles, i.e., the ones containing the answer; (2) finding correct evidence in the relevant articles that allows answering the question; and (3) answering the question. While the first task is a specific case of Information Retrieval (IR), the second can be

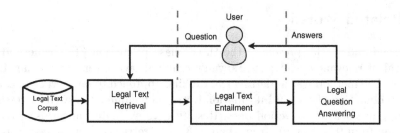

Fig. 1. The model of legal text question answering system

considered as a form of Recognition of Textual Entailment (RTE), in which given a question, the LQA system has to decide whether and how a relevant article can answer the question. The third one is the final result of the two previous tasks, combined with answer formatting.

Legal text is considerably different to other types of text, e.g., news articles, due to their structural and semantic characteristics. Firstly, they have specific logical sentence structures e.g., requisite and effectuation [12]. Secondly, words and writing style are used in a strict form, because law documents require high correctness and should avoid ambiguity. Another aspect is that law documents are written in a high abstraction level [13]; therefore, they often require collection and linking of multiple concept references to enable understanding and answering of a question. The use of concept references leads to a situation in which there are few, or in some cases, even no overlapping words between a law question and its relevant articles.

In this work, LQA tasks are considered into the context of COLIEE, a competition on legal information extraction/entailment which was first held in 2014, in association with Workshop on Juris-informatics (JURISIN). COLIEE 2015[2] is the second competition, consisting of three phases:

- Phase One: retrieving relevant articles from all Japanese Civil Code Articles given a set of YES/NO questions.
- Phase Two: evaluating the entailment relationship between the question and retrieved articles.
- Phase Three: combination of Phase One and Phase Two, the system will retrieve list of relevant articles given a query, and then decide the entailment relationship between retrieved articles and the provided question.

The Japanese Civil Code is composed by a collection of numbered articles, each one containing a set of declarations pertaining to a specific topic of the law, e.g., labor contracts, mortgages.

Information Retrieval Task: Relevance Analysis

The first phase consists on an explicit IR task, for which the goal is to retrieve the relevant articles that can be used to correctly answer a given yes/no question. The challenge in this task is to determine the relative relevance, i.e., Relevance

---

[2] http://webdocs.cs.ualberta.ca/~miyoung2/COLIEE2015/.

Analysis (RA), of an article to the query presented in the question. Different articles dealing with the same topic often have similar wording and it is common for questions not to refer to topic keywords or refer to alternative ones. Furthermore, the restricted size of the Japanese Civil Code means that obtaining reliable linguistic information from articles is difficult and most questions will present new language structures that can range from useful to necessary for answering.

Simple Question Answering Task: Textual Entailment

The goal of Textual Entailment (TE) is to decide whether a legal query/question can be answered by a set of relevant articles retrieved with RA. This task can be accomplished by recognizing textual entailment (RTE), in which the query/question is treated as an hypothesis and relevant articles as evidence. Given a question Q and a set of relevant articles A, ($A = \{a_1, ..., a_n\}$), if Q is answered by a_i ($1 \leqslant i \leqslant n$), then a_i entails Q [9,14]. A pair (Q, a_i) is assigned label YES if a entailment relationship exists, i.e., a_i can answer Q; otherwise, NO.

4   Proposed Approach

In order to be able to perform both Relevance Analysis and Textual Entailment recognition independently in phases one and two, and jointly in phase three, IR and classifier methods were developed separately. First, both the legal corpus and training data are analyzed and combined into representation models. The models are then used to rank articles or classify answers according to the task. The representation model used for Relevance Analysis is a mixed size n-gram collection and the one used for textual entailment are feature vectors for Machine Learning. Figure 2 shows the overall view of the proposed method.

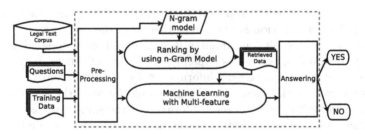

Fig. 2. Model overview

4.1   Relevance Analysis

A detailed analysis of the Civil Code and training data revealed that lexical and syntactic overlapping may vary to a high degree between questions and articles, and also between articles concerning the same topic. However, certain morphological features, such as lemmas, retain a higher level of consistency among

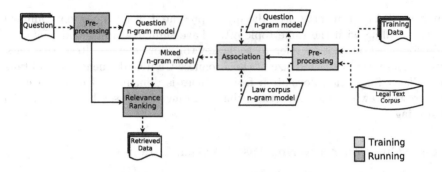

Fig. 3. The process of legal text retrieval

topics. For this reason, the adopted representation model was a mixed size n-gram model, with $n : [1, k]$, i.e., terms made by sequences up to k words, in which the terms are lemmatized. For simplicity, the Relevance Analysis method hereon described was named R_2NC (Ranking Related N-gram Collections). A summarized view of the process is shown on Fig. 3.

The steps to build the model are detailed as follows:

1. Collect the entire content for each article, including section title;
2. Check references between articles and annotate accordingly;
3. Tokenize and POS-tag;
4. Remove stopwords: determiners, conjunctions, prepositions and punctuation;
5. Lemmatize words;
6. Generate n-grams;
7. Expand the n-gram set, by including references n-grams;
8. Associate article number and references;
9. Store the model.

Except for step 4, each step is responsible for adding new information to the model. The information is obtained either from the text, e.g., section title, references, or from morphological analysis, e.g., POS-tags, lemmas. If an article have references, its n-gram set is expanded with the references' n-grams. This is done so that all the necessary information for interpretation of any single article is self-contained. Besides the n-grams, links between the articles are also stored. To include the training data information, the same process is repeated for the questions, and n-gram sets from the trained questions are used to expand the associated articles' n-gram models. Since COLIEE disallowed explicit expert knowledge input, an optional information source was added after the competition, as a way of including expert knowledge in the model when available, and possibly improve system performance. This source consists in a simple term dictionary, where legal terms are associated with other correlated ones. If a given question contains n-grams referred in the dictionary, its n-gram model is expanded with the associated entries. The dictionary was written manually and contains 26 entries that were considered important after analyzing the training data, e.g., "for a third party" → "to others", and extrapolating answers to user defined queries. Tokenization and lemmatization were done using $NLTK^3$ (v. 3.0.2) with the Punkt tokenizer and WordNetLemmatizer modules, respectively. Those modules were used with their unchanged default models and settings, trained with the Punk corpus and WordNet, respectively. POS-tagging was done using *Stanford Tagger*[4] (v. 3.5.2), using the unchanged *english-left3words-distsim* model, which is trained on the part-of-speech tagged WSJ section of the Penn Treebank corpus.

To determine the relative relevance of an article with regard to the content of a question, a ranking approach was adopted. First, the n-gram set of the question is obtained by applying steps 1–6, using the question content instead of article. Then, for each article in the Civil Code, a relevance score is calculated using the following formula:

$$score = \frac{\sum_{\forall t} idf(t)}{I_q \times |q\_ng\_set| + I_{art} \times |art\_ng\_set|}, \quad t \in (q\_ng\_set \cap art\_ng\_set) \quad (1)$$

where *q_ng_set* is the set of n-grams for the question, *art_ng_set* is the set of n-grams for the article in the stored model, I_q is the relative significance of the question n-gram set size and I_{art} is the relative significance of the article n-gram set size. $idf(t)$ is the Inverse Document Frequency for the term t over the articles collection

$$idf(t) = log \frac{N}{df_t} \quad (2)$$

where N is the total number of articles and df_t is the number of articles in which t appears.

The formula (1) is a variation of the traditional *TF-IDF* scoring method, disregarding term frequency and giving different weights for the two types of document being evaluated: articles and questions, according to their size. I_q and I_{art} are parameters to be adjusted according to the corpus characteristics. This formula was developed during the first stages of analysis on the Civil Code corpus, when experiments with a TF-IDF based classifier showed poor results for this task and further observation showed that TF did not contribute for article relevance in many cases. As TF is absent from the formula, document size becomes a more relevant feature and must be considered in the scoring. In the studied corpus, law articles are usually much larger than questions in number of words, hence the different weights to adjust normalization of the score regarding the respective sets.

From this point, the articles are sorted by descending score and the 10 best are selected for filtering. The filtering step consists in fetching the best scoring article and verifying if its score exceeds a parameter threshold *confidence_thresh*. If it does, all the articles in the list that are referred by the first and exceed a parameter threshold *reference_thresh* are also fetched. The fetched articles compose the final list of relevant articles to the input question. Parameter adjustment is described in Sect. 5.

---

[3] www.nltk.org.

[4] nlp.stanford.edu/software/tagger.shtml.

4.2 Textual Entailment

A textual entailment (TE) relation in law domain comprises two levels of information. The first level describes whether or not (YES/NO, respectively) the textual evidence addresses the hypothesis. The second level describes whether the evidence supports or opposes (YES/NO, respectively) the hypothesis. However, due to the time constraint of the competition, only the first level is explored. Therefore, semantic relations such as negation and antonym were not considered in the TE evaluation step.

To detect a TE relation on a pair (Q, a), a similarity-based approach [4] can be used, in which a can answer Q if the similarity is greater than a certain threshold. However, high level inference (see Sect. 3) and the identification of the threshold make these methods more challenging to apply. We, therefore, propose to apply classification for detecting the TE relation with two advantages: (1) use of a rich feature set to represent data characteristics and (2) avoiding to identify the threshold.

This work shares most of the goals presented in Nguyen et al. [7], so all the features in that work were used. However, the corpus size in this case makes it difficult to effectively train Machine Learning algorithms. For this reason, "stronger" features were sought as a way of compensating such problem. An additional Word2Vec feature was added to capture the semantic similarity of a pair (Q, a), as observation of statistical data in Table 1 shows that the lexical overlapping may not be a strong enough feature for the classification on a (Q, a) pair (e.g., cannot capture the similarity of *person* and *manager*). By adding the Word2Vec feature, the model aims to cover the semantic aspect instead of only lexical similarity. Word2Vec was trained by JPN Law corpus: a collection of all Civil law articles of Japan's constitution[5]. It contains 642 cleaned and tokenized articles, with about 13.5 million words in total.

For the classification, the Weka toolset[6] implementation of AdaBoost [18] was used, with *classifier = DecisionStump*.

Table 1. Statistical data observation in phase two

	# pairs	# sentences	# tokens	% uni-gram word overlapping
Training set	267	273	36.562	58.80

The features are shown in Table 2, in which distance features measure distance between a question Q and relevant article a and statistical features capture word overlapping of this pair. After extracting features, a pipeline model was proposed and is shown in Fig. 4.

In Fig. 4, the first step is to preprocess the data from the input files, in which sentences and words are segmented and stopwords[7] are removed. Next, the

---

[5] www.japaneselawtranslation.go.jp.

[6] weka.wikispaces.com.

[7] https://sites.google.com/site/kevinbouge/stopwords-lists.

Table 2. The feature groups; Avg is average; Q is a question, S is a sentence

	Feature	Description
Distance	Manhattan	Manhattan distance from two text fragments
	Euclidean	Euclidean distance from two text fragments
	Cosine similarity	Cosine similarity distance
	Matching coefficient	Matching coefficient of two text fragments
	Dice coefficient	Dice coefficient of two text fragments
	Jaccard	Jaccard distance of two text fragments
	Jaro	Jaro distance of two text fragments
	Damerau-Levenshtein	Damerau Levenshtein distance of two text fragments
	Levenshtein	Levenshtein distance of two text fragments
Statistical	Lcs	The longest common sub string of two text fragments
	Average of TF-IDF	Term frequency-inverse document frequency
	Avg-TF of Q and S	Avg-TF of words in a Q appearing in a S
	Avg-TF of S and Q	Avg-TF of words in a S appearing in a S
	Word overlapping	# word overlapping in a Q appearing in a article
	Average of Word2Vec	Average of word2vec similarity

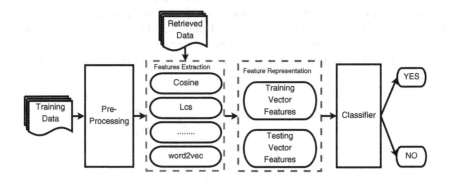

Fig. 4. The process of legal textual entailment recognition

training data is represented in a vector space model by features in Table 2. The retrieved data from relevance analysis is also denoted in the same mechanism. Finally, a classifier was trained on the training data and applied on retrieved data to judge the entailment relation. Note that features in Sect. 4.1 can be also used for this task.

5   Experiments and Results

5.1   Experimental Setup

The dataset was obtained from the published data for the COLIEE shared task (see Footnote 2), consisting in a text file with the Japanese Civil Code and a set of XML files with training and testing data for phases one to three. The training set for the three tasks contains 267 pairs (question, relevant articles). Experiments where divided in phases one and two only, dealing with Information Retrieval and Textual Entailment methods respectively. Each experiment comprised: (i) data analysis, (ii) model and parameter adjustments and (iii) test runs.

5.2   Parameter Adjustment

For R_2NC, parameters I_q, I_{art}, $confidence\_thresh$ (shortened to ct here), $reference\_thresh$ (rt) and also k, the maximum n-gram size, were adjusted empirically on the training data using the following simple procedure:

– Starting with $I_q = 0.8$, $ct = 0.5$, $rt = 0.5$ and $k = 1$,
    1. Increase or decrease a single parameter by step $r = 0.1$ until the F-measure cannot be increased for a leave-one-out test.
    2. Repeat (1) starting from the last obtained value, with $r = 0.01$.
    3. Repeat (1) and (2) for all parameters.

   For k, step was fixed on $r = 1$. I_q and I_{art} respect the constraint $I_q + I_{art} = 1$. The parameters are changed in a specific order: 1. $confidence\_thesh$, 2. k, 3. $reference\_thresh$, 4. I_q. I_q and I_{art} respect the constraint $I_q + I_{art} = 1$. Performance metrics were recorded for the parameter adjustment during the experiments. Figure 5 shows the performance progression on post-competition experiments for the parameters I_q, I_{art}, with the other ones locked into their best respective values. Performance for $k \neq 3$ is negatively affected in both directions $(-, +)$, and no further investigation was conducted for a larger range of values.

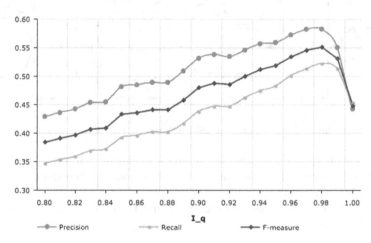

Fig. 5. Performance metrics for phase 1 related to the variation of I_q. $I_{art} = 1 - I_q$.

   Final parameter values used in the competition are $k = 3$, $I_q = 0.965$, $I_{art} = 0.035$, $confidence\_thresh = 0.32$ and $reference\_thresh = 0.2$.
   For the RTE classifier, default parameters from the Weka toolset (see Footnote 6) were used for all the experiments and were not changed. The parameter values are: $iterations = 10$, $seed = 1$, no $re\text{-}sampling$ and $weightthreshold = 100$.

5.3   Baselines

As for the second edition of COLIEE, there is still no definite baseline for the competition dataset. However, common baseline practices and related works could be used for evaluating performance on each task. For phase one, a relationship can be drawn between R_2NC and TPP [3]. For the TE task, the following baselines were used for comparison:

- SVMs: uses Support Vector Machines (SVMs)[8] [19] with Weka. The parameters are $C = 1$, $\gamma = 0$, *kernel Type = radial basis function (RBF)*.
- AdaBoost-SVMs: uses SVMs as weak learners instead of DecisionStump.

5.4   Evaluation Method

Given the limited training data available, leave-one-out validation was used to evaluate the performance of the model in both tasks on the training dataset with three measures: precision (P), recall (R) and F-measure (F) as in Eqs. (3)–(5). In phase two, accuracy (A) measurement is also used as in Eq. (6).

$$P = \frac{Cr}{Rt} \tag{3}$$

$$R = \frac{Cr}{Rl} \tag{4}$$

$$F = \frac{2(P * R)}{P + R} \tag{5}$$

$$A = \frac{Cq}{Q} \tag{6}$$

where Cr counts the correctly retrieved articles for all queries, Rt counts the retrieved articles for all queries, Rl counts the relevant articles for all queries, Cq counts the queries correctly confirmed as true or false and Q counts all the queries.

5.5   Pre-competition Results

Pre-competition experiment results on the shared data are presented in Tables 3 and 4.

The results indicate that R_2NC is expected to be competitive with state-of-the-art approaches to relevance analysis in legal documents, such as *TPP* [3]. However, the proposed method is much simpler when compared to *TPP* and operates with considerably less training data: 266 documents for R_2NC against 1518 documents for *TPP*. R_2NC design also makes it difficult for the model to be overtrained beyond the parameter adjustment, since no training data is counted more than one time and the method is single-shot, as opposed to convergence-based. Experiments were repeated with traditional TF-IDF scoring instead of R_2NC formula, yielding 0.51 F-measure.

---

[8] https://www.csie.ntu.edu.tw/~cjlin/libsvm/.

Table 3. Experiment results for phase one (IR) with R_2NC. In the top 3/10 settings, articles ranked up to 3rd or 10th place are marked as relevant.

	Precision	Recall	F-measure
R2NC	0.568	0.516	**0.54**
R2NC (top 3)	0.27	0.64	0.38
R2NC (top 10)	0.10	0.77	0.17
TPP (top 3)	N/A	0.52	N/A
TPP (top 10)	N/A	0.91	N/A

Table 4. AdaBoost-DecSt (*DecisionStump*) vs. SVMs and AdaBoost-SVMs.

	Precision	Recall	F-measure	Accuracy (%)
AdaBoost-DecSt	0.621	0.614	0.617	**61.42**
SVMs	0.537	0.543	0.539	54.30
AdaBoost-SVMs	0.485	0.491	0.487	49.06

Results of RTE in Table 4 indicate that AdaBoost with a set of appropriate features outperforms the baselines by 7.74% (*SVMs*) and 12.94% (*Adaboost-SVMs*) on F-measure. Moreover, the precision and accuracy of this method also achieve considerable improvements when compared to the baselines. This suggests that the features are expected to be efficient for addressing TE in the legal domain. This conclusion is supported by the accuracy measurements.

Another interesting point is that Word2Vec similarity contributes to improve the performance of RTE. As stated in Sect. 3, legal documents usually require concept linking to understand and answer a question; therefore, semantic similarity from Word2Vec helps to improve the performance. The results also show the efficiency of the lexical features.

The performance of RTE in the law domain, however, is not comparable with the same task in common data i.e., news articles [6,7] due to the characteristics of law dataset, as shown in Sect. 3. The performance was not improved very much even when many features in both phase one and two were combined. This suggests that more sophisticated approaches e.g., semantic inference or semantic rules should be considered in feature construction. Finally, negation and antonym analysis should be considered to improve the quality of the entailment recognition, effectively exploring the second level of entailment information as described in Sect. 4.2.

5.6   Feature Evaluation

Further evaluation of feature impact on TE model was conducted by leave-one-out test. The most effective features are shown in Table 5.

Table 5 shows an indication of contribution from features to the model. Results show that all effective features contribute to the method. Note that both

Table 5. Top 4 influential features, italic is for statistical features. Values are the difference in F-measure between the model with all features and without the single specified feature.

Features	Influential value	Features	Influential value
Euclidean	0.005	*Lcs*	0.0001
Damerau-Levenshtein	0.154	Average of Word2Vec	0.024

Damerau-Levenshtein and *Euclidean* are distance features whereas *the longest common substring* (lcs) is a statistical feature. The results support that in legal texts, there is not much word overlapping between a question and relevant articles. An interesting aspect is that Word2Vec similarity has a big positive impact to the model. This supports the conclusion on similarity stated in Sect. 5.5.

5.7   Competition Results

The method presented in this paper achieved significant results in COLIEE, being ranked 2nd in phase one (IR) and 3rd in phase three (combined IR + TE). It was not well ranked in phase two (TE). The relevant competition results are presented in Table 6 as they were announced in JURISIN 2015.

Table 6. Competition results for phases 1 (IR) and 3 (IR + TE) respectively. First three ranked.

Rank	ID	Prec.	Recall	F-m	Rank	ID	Accuracy.
1	UA1	0.633	0.490	0.552	1	UA1	0.658
2	**JAIST1**	**0.566**	**0.460**	**0.508**	2	Kanolab3	0.620
3	ALV2015	0.342	0.529	0.415	**3**	**JAIST1**	**0.582**

5.8   Post-competition Analysis and Improvements

Post competition analysis pointed us to possible sources of classification problems in phase 2 (TE) and also gave directions of improvement in both tasks.

For R_2NC, the lack of an implicit semantic mapping was an important factor when compared to the top ranked approach. To compensate for that, a term dictionary was included as a new information source for expanding the question n-gram models as described in Sect. 4.1. By using linguistic observations, it was possible to create basic entries in the dictionary (non-expert knowledge), improving phase 1 F-measure on the shared data (Table 3) from 0.54 to 0.55.

In the case of phase 2, over-fitting on training data was deemed the main factor that reduced classification performance. Our system achieved over 61% accuracy (Table 4) when running on the shared data, but only 37.88% reported from the competition results. Phase three results show that accuracy improved when restricting information for the classifier and this is consistent with the over-fitting assumption. Another important point is that a question q and all sentences in an article a were used in building the vector space model. As a result, imbalance of length between the question and the article may have affected feature calculation. This can be addressed by developing a better text segmentation method. Finally, the over-fitting assumption can also be dealt by using other classification approaches e.g., Deep Neural Networks, together with over-fitting avoidance techniques e.g., pruning, dropout.

5.9   Error Analysis and Discussion

An investigation was done on the ranked list obtained with R2NC in phase one (see Sect. 4.1). It revealed that relevant articles ranked 3rd and below had keywords that did not appear in the corresponding question in the corpus. This reinforces the view that the questions are highly directed, albeit in a conceptual level. Relevant articles that ranked lower than 15th (approx. 20%) were found to require a relatively high level of abstraction to obtain an interpretation that could link to the corresponding question. Table 7 shows an example of complex relevance relationship.

Table 8 shows a case in which our system gives correct outputs (ID H18-2-4). In this example, there are several common words from which this approach can correctly judge the TE relation, e.g., *reimbursement.* In addition, several words can be inferred from the questions by using Word2Vec similarity e.g., *person ∼ manager, fees ∼ costs or expenses.* This supports our observation that TE can be addressed by using lexical features and word similarity. For example, in (ID H18-2-4), our system can still predict the TE relation correctly, even with little lexical overlap. This indicates the efficiency of this approach, and especially of the word similarity feature.

Table 7. Example of pair (question, article) with low ranking but high relevance.

ID	Article	Question	Ranked in
H18-2-1	Article 697 (1) A person who commences the management of a business for another person without being obligated to do so (hereinafter in this chapter referred to as "Manager") must manage that business (hereinafter referred to as "Management of Business") in accordance with the nature of the business, using the method that best conforms to the interests of that another person (the principal). (2) The Manager must engage in Management of Business in accordance with the intentions of the principal if the Manager knows, or is able to conjecture that intention	In cases where a person plans to prevent crime in their own house by fixing the fence of a neighboring house, that person is found as having intent towards the other person	424th

Table 8. Examples of entailment judgment; P is predicted and A is annotated

ID	Article	Question	P	A
H18-2-4	(Managers' Claims for Reimbursement of Costs) Article 702 (1) If a Manager has incurred useful expenses for a principal, the Manager may claim reimbursement of those costs from the principal. (2) The provisions of Paragraph 2 of Article 650 shall apply mutatis mutandis to cases where a Manager has incurred useful obligations on behalf of the principal. (3) If a Manager has engaged in the Management of Business against the intention of the principal, the provisions of the preceding two paragraphs shall apply mutatis mutandis, solely to the extent the principal is actually enriched	In cases where a person repairs the fence of a neighboring house after it collapsed due to a typhoon, but the neighbor had intended to replace the fence with a concrete-block wall in the near future, if a separate typhoon causes the repaired sections to collapse the following week, reimbursement of repair fees can no longer be demanded	YES	YES
H18-26-1	(Renunciation of Shares and Death of Co-owners) Article 255 If one of co-owners renounces his/her share or dies without an heir, his/her share shall vest in other co-owners	In cases where person A and person B co-own building X at a ratio of 1:1, if person A dies and had no heirs or persons with special connection, ownership of building X belongs to person B	NO	YES

On the other hand, the pair H18-26-1 exemplifies a case in which the system predicted NO while TE relation was annotated YES even when the question and answer share more common words. This shows the limitation of this feature set in cases where the question and answer are short. In this case, after removing stop words, a few remaining words may not be enough to capture the TE relation. Moreover, the lack of important words e.g., *building*, *connection* or *belong* reveals a big challenge for our system to decide the TE relation. This suggests that a keyword enriching mechanism such as term expansion used in phase one could improve the results.

In order to facilitate the understanding of different error cases and give other people the opportunity to try the system developed for the competition, an online demo system[9] has been made available. In this demo it is possible to input user defined questions or just verify the answers to questions in the COLIEE shared data.

6   Conclusion

This paper explores the challenging issue of building a QA system in the legal domain. We propose a model including three stages: legal information retrieval, legal textual entailment and legal text answering. In the first stage, a mixed size n-gram model built from morphological analysis is used to rank and select relevant articles corresponding to a legal question; next, pairs of questions and retrieved articles are judged by a machine learning algorithm trained on lexical features and Distributional Semantic similarity, to decide whether the questions

---

[9] http://150.65.242.101:3001/.

can be answered positively or negatively by the retrieved articles; and finally, correct answers would be provided for users in the final stage. The contributions of this work in IR and TE task are: (1) a simple, yet effective language model for law corpora coupled with a Relevance Analysis method (R_2NC) capable of exploiting such model; (2) The use of TF-IDF and Word2Vec similarity features for applying Machine Learning algorithms to RTE. With a recall of 0.64 for the top 3 ranked articles, R_2NC appears as competitive when compared to state-of-the-art similar work, in spite of being more simple and applicable with less training data. By combining lexical features and Word2Vec similarity, this approach for LQA also outperformed the baselines by 8.4% (*SVM*) and 11.3% (*Adaboost-SVMs*) on F-measure. Results in the COLIEE competition for the IR task (0.508 F-measure, 2nd place) and the combined IR+TE task (0.582 accuracy, 3rd place) indicate a substantial adequacy to the tasks addressed. The competition also provided important shortcomings of the proposed approach, namely the lack of implicit semantic representation and classifier over-fitting. Those shall be addressed in future work.

Still on future directions, information on a higher abstraction level, e.g., syntactic mappings, could be used to improve the language model for the IR task. In the TE task, since a sentence in a legal article is usually long, a sophisticated method of sentence partition e.g., requisite and effectuation should be considered. In feature extraction, features in IR should be combined with lexical features in TE and investigated to improve the quality of the judgment. Moreover, capturing contradictions in the TE relation by current statistical features is a big challenge. To solve this issue, semantic rules over negation and antonym detection should be defined and incorporated into the feature extraction. Finally, we would like to investigate and apply sentence similarity calculation by Sent2Vec to improve the performance of the TE.

Acknowledgements. This work is supported partly by the grant of NII Research Cooperation and JAIST's Research grant.

References

1. Berring, R.C.: The Heart of Legal Information: The Crumbling Infrastructure of Legal Research. Legal Information and the Development of American Law. Thomson/West, St. Paul (2008)
2. Hoekstra, R., Breuker, J., Di Bello, M., Boer, A.: The LKIF core ontology of basic legal concepts. In: Proceedings of the Workshop on Legal Ontologies and Artificial Intelligence Techniques (LOAIT) (2007)
3. Liu, Y.-H., Chen, Y.-L., Ho, W.-L.: Predicting associated statutes for legal problems. Inf. Process. Manag. **51**(1), 194–211 (2015)
4. Ha, Q.-T., Ha, T.-O., Nguyen, T.-D., Thi, T.-L.N.: Refining the judgment threshold to improve recognizing textual entailment using similarity. In: Nguyen, N.-T., Hoang, K., Jędrzejowicz, P. (eds.) ICCCI 2012. LNCS (LNAI), vol. 7654, pp. 335–344. Springer, Heidelberg (2012). doi:10.1007/978-3-642-34707-8_34

5. Inkpen, D., Kipp, D., Nastase, V.: Machine learning experiments for textual entailment. In: Proceedings of the Second Challenge Workshop Recognising Textual Entailment, pp. 17–20 (2006)

6. Castillo, J.J.: An approach to recognizing textual entailment and TE search task using SVM. Procesamiento del Lenguaje Natural **44**, 139–145 (2010)

7. Nguyen, M.-T., Ha, Q.-T., Nguyen, T.-D., Nguyen, T.-T., Nguyen, L.-M.: Recognizing textual entailment in vietnamese text: an experiment study. In: KSE, pp. 108–113 (2015)

8. Pham, M.Q.N., Nguyen, L.M., Shimazu, A.: Using machine translation for recognizing textual entailment in Vietnamese language. In: RIVF, pp. 1–6 (2012)

9. Giampiccolo, D., Magnini, B., Dagan, I., Dolan, B.: The third PASCAL recognising textual entailment challenge. In: Proceedings of the ACL-PASCAL Workshop on Textual Entailment and Paraphrasing. pp. 1–9. Association for Computational Linguistics (2007)

10. Tran, O.T., Ngo, B.X., Nguyen, M., Shimazu, A.: Answering legal questions by mining reference information. In: Nakano, Y., Satoh, K., Bekki, D. (eds.) JSAI-isAI 2013. LNCS (LNAI), vol. 8417, pp. 214–229. Springer, Cham (2014). doi:10.1007/978-3-319-10061-6_15

11. Kim, M.-Y., Xu, Y., Goebel, R., Satoh, K.: Answering yes/no questions in legal bar exams. In: Nakano, Y., Satoh, K., Bekki, D. (eds.) JSAI-isAI 2013. LNCS (LNAI), vol. 8417, pp. 199–213. Springer, Cham (2014). doi:10.1007/978-3-319-10061-6_14

12. Bach, N.X., Nguyen, L.M., Shimazu, A.: RRE task: the task of recognition of requisite part and effectuation part in law sentences. J. IJCPOL **23**(2), 109–130 (2010)

13. Tran, O.T., Bach, N.X., Nguyen, L.M., Shimazu, A.: Reference resolution in legal texts. In: Proceedings of ICAIL, pp. 101–110 (2013)

14. Dagan, I., Dolan, B., Magnini, B., Roth, D.: Recognizing textual entailment: rational, evaluation and approaches - erratum. Nat. Lang. Eng. **16**(1), 105 (2010)

15. Wang, R.: Intrinsic and extrinsic approaches to recognizing textual entailment, pp. 1-219. Saarland University (2011). ISBN: 978-3-933218-32-2

16. Chelba, C., Mikolov, T., Schuster, M., Ge, Q., Brants, T., Koehn, P., Robinson, T.: One billion word benchmark for measuring progress in statistical language modeling, arXiv preprint arXiv:1312.3005 (2013)

17. Mikolov, T., Sutskever, I., Chen, K., Corrado, G., Dean, J.: Distributed representations of words and phrases and their compositionality. In: Proceedings of NIPS (2013)

18. Freund, Y., Schapire, R.E.: A decision-theoretic generalization of on-line learning and an application to boosting. J. Comput. Syst. Sci. **55**(1), 119–139 (1997)

19. Cortes, C., Vapnik, V.: Support-vector networks. Mach. Learn. **20**(3), 273–297 (1995)

AAA 2015

2nd Workshop on Argument for Agreement and Assurance (AAA 2015)

Kazuko Takahashi[1], Kenji Taguchi[2], Tim Kelly[3], and Hiroyuki Kido[4]

[1] Kwansei Gakuin University, Nishinomiya, Japan
[2] National Institute of Advanced Industrial Science and Technology, Amagasaki, Japan
[3] University of York, York, UK
[4] The University of Tokyo, Tokyo, Japan

In recent decades, argument has been an attractive research topic in artificial intelligence. Research has examined formal models of argumentation defining the semantics of logic programming or nonmonotonic reasoning, the application of persuasion in multi-agent environments, and tools for argumentation analysis or visualization. Using argument is considered an effective approach to resolve inconsistencies and to achieve agreement.

On the other hand, there is also growing interest in assurance cases in safety engineering, where the logical analysis of arguments by Toulmin is much appreciated. Many safety-related standards/guidelines currently mandate the submission of safety cases to certification bodies. Argument is being used in the system development to improve accountability to their customers.

The First Workshop on Argument for Agreement and Assurance was held in 2013, with the goal of deepening a mutual understanding and exploring a new research field involving researchers/practitioners in formal and informal logic, artificial intelligence, and safety engineering working on agreement and assurance through argument. The Second Workshop on Argument for Agreement and Assurance took place at Keio University, Kanagawa, Japan on November 17, 2015. This was held as an international workshop of the seventh JSAI International Symposia on AI (JSAI-isAI 2016), sponsored by The Japan Society for Artificial Intelligence (JSAI).

There were about 20 participants, with eight presentations, one posters-and-demos presentation, three demonstrations and two invited talks.

Ewen Denney gave an invited talk on formalization using argumentation in assurance cases and practical applications based on this. In cooperation with the Ninth International Workshop on Juris-informatics (JURISIN 2015), Phan Minh Dung gave an invited talk on his current research on an argumentation framework based on the strength of inference rules.

The general sessions included a variety of presentations covering material ranging from theoretical work and methodology to demonstrations of practical tools. From these, three were selected as contributions to the post-proceedings. Caminada and Sakama discussed formal criteria for determining whether one agent is more informed than another. Rushby proposed a two-part process using epistemic methods and deductive logic for the interpretation of assurance case arguments. Kido introduced an argumentation framework and its semantics into a decision-tree and presented a method of learning argument acceptability.

We thank all of the reviewers for their valuable comments, all the participants for fruitful discussions, Change Vision Inc. for financial support, and JSAI for giving us the opportunity to hold this international workshop.

Kazuko Takahashi, Kenji Taguchi, Tim Kelly and Hiroyuki Kido
(AAA 2015 organizers)

On the Issue of Argumentation and Informedness

Martin Caminada[1(✉)] and Chiaki Sakama[2]

[1] Cardiff University, Cardiff, UK
CaminadaM@cardiff.ac.uk
[2] Wakayama University, Wakayama, Japan
sakama@sys.wakayama-u.ac.jp

Abstract. In the current paper we examine how to assess knowledge and expertise in an argumentation based setting. In particular, we are looking for a formal criterion to determine whether one agent is more informed than another. Several such criteria are discussed, each of which has its own advantages and disadvantages.

1 Introduction

Formal argumentation theory has been applied to study a whole range of different questions, like "what to accept" [14], "how to come to a joint position" [10] and "how much do different positions differ" [4]. In the current document, we will study a similar question, which can be summarized as "who knows more".

In the context of nonmonotonic reasoning, determining who knows more is far from trivial. Although there are some studies that formulate comparison among nonmonotonic theories ([16,24] for instance), one cannot say that one (nonmonotonic) theory is more informative than another one by simply comparing the sets of entailed conclusions. After all, the fact that the set of entailed conclusions of one agent is bigger than (a superset of) the set of entailed conclusions of the other agent might be due to the fact that the latter agent has information that invalidates one of the inferences made by the first agent. In this case, it would not seem reasonable to claim that the former agent has more knowledge (or is better informed) than the second agent.

Another approach would be to compare not the sets of entailed conclusions, but instead to compare the contents of the knowledge bases. This, however, leads to problems like what to do when two knowledge bases are syntactically different but semantically equivalent. Furthermore, it could very well be that an agent has information that another agent knows to be inapplicable (say, reading a newspaper that another agent knows to be unreliable). Therefore, measuring the raw contents of the knowledge bases is not necessarily appropriate to determine which agent knows more.

Yet, the issue of coming up with a suitable criterion for determining who knows more is an important one. If the aim is for instance to hire an expert or consultant to make a decision, how does one *assure* that the person in question actually possesses expertise? Or, on a more basic level, how even to *define*

© Springer International Publishing AG 2017
M. Otake et al. (Eds.): JSAI-isAI 2015 Workshops, LNAI 10091, pp. 317–330, 2017.
DOI: 10.1007/978-3-319-50953-2_22

expertise or determine how expertise or knowledge of an agent matches up to that of another agent?

In the existing literature on formal logic, issues of reasoning about knowledge (epistemic reasoning) have traditionally been studied using modal logic, in particular by applying modalities that satisfy the KD45 or KT45 axiomatizations. The problem, however, is that where conceptually knowledge stands for "justified true belief", the KD45 and KT45 axiomatization simplify the concept of knowledge to "true belief" since it does not offer the facilities to model the concept of justification. Simplifying the concept of knowledge to "belief that happens to be true" can lead to debatable outcomes, since a lay person who on a particular topic is merely right by accident would be considered to have knowledge, whereas an expert who is wrong due to exceptional and unforeseen circumstances would be considered to have no knowledge. Apart from that, in many cases one cannot simply as an outsider determine whose beliefs are true and whose are not. If two experts disagree about, say, forecasts of climate change, then one cannot simply determine who of them is right (has beliefs that are true) although one might still want to have some idea about who of these experts is better informed. From practical perspective, the concept of "justified belief" is often as more important than that of "true belief".

If one is to take the concept of justified belief seriously, then one is to apply a formal approach that is based not so much on truth (as is the case in modal logic, which essentially models knowledge as true belief) but on justification. The next question then becomes how one could characterize a formal notion of justification, without applying the notion of truth. One possibility would be to interpret justified belief as that which can be defended in rational discussion. In formal argumentation theory, several forms of rational discussion (and the associated argumentation semantics) have been identified. Abstract argumentation theory under grounded semantics, for instance, can be seen as corresponding to persuasion discussion, whereas abstract argumentation theory under (credulous) preferred semantics can be seen as corresponding to Socratic discussion [8].

Furthermore, if one agrees that nonmonotonic aspects are at the heart of many real world reasoning processes, then it makes sense to apply a formalism that is able to deal with nonmonotonicity. Abstract argumentation theory is one of the simplest and most straightforward approaches to nonmonotonic reasoning and has nevertheless been shown to capture a wide range of full-blown nonmonotonic logics [14,18,22,27]. Hence, it makes sense to apply argumentation theory as a starting point for examining the concept of "justified belief", which we will sometimes refer to as "informedness" (to distinguish it from "knowledge", which in the mainstream literature on epistemic logic has come to mean "true belief").[1]

The remaining part of this document is structured as follows. First, in Sect. 2 we briefly summarize some of the key notions in (abstract) argumentation theory. Then, in Sect. 3, we examine several candidate criteria for evaluating relative informedness. We round off in Sect. 4 with a discussion and some issues for further research.

---

[1] Our analysis goes beyond that of for instance [19] in that we aim to consider different grades of justification.

2 Abstract Argumentation Theory

In the current section, we provide a brief overview of some basic notions in abstract argumentation theory. We will focus on the notions that we actually need for examining the concept of argument-based informedness in the next section, and refer to [1] for a more elaborate treatment.

Definition 1. *An* argumentation framework *is a pair* (Ar, att) *in which* Ar *is a (finite) set of arguments, and* $att \subseteq Ar \times Ar$.

Since an argumentation framework is in essence a (finite) graph, we often use a graphical representation of it. Although various approaches (such as [7,14,21,25,27]) have been formulated where arguments do have an internal structure (usually in the form of defeasible derivations), in the current document we will leave the internal structure completely abstract. That is, we will treat argumentation theory at the highest level of abstraction. As for notation, for argumentation frameworks $AF_1 = (Ar_1, att_1)$ and $AF_2 = (Ar_2, att_2)$, we write $AF_1 \sqsubseteq AF_2$ to indicate that $Ar_1 \subseteq Ar_2$ and $att_1 = att_2 \cap (Ar_1 \times Ar_1)$.

The next question becomes what are the reasonable positions one might take based on the conflicting information in the argumentation framework. In the current document, we will apply the approach of *argument labellings* [5,9] for expressing these positions.

Definition 2. *Let* (Ar, att) *be an argumentation framework. An* argument labelling *is a (total) function* $\mathcal{L}ab : Ar \rightarrow \{\mathsf{in}, \mathsf{out}, \mathsf{undec}\}$.

An argument labelling can be seen as a position on which arguments should be accepted (labelled in), which arguments should be rejected (labelled out) and which arguments should be abstained from having an explicit opinion about (labelled undec).

Although argument labellings allow one to express any arbitrary opinion, some opinions can be regarded to be more reasonable than others. When precisely a position can be considered to be reasonable is formally defined by *argumentation semantics*.

Definition 3. *Let* $\mathcal{L}ab$ *be a labelling of argumentation framework* (Ar, att). $\mathcal{L}ab$ *is said to be an* admissible labelling *iff for each argument* $A \in Ar$ *it holds that:*

- *if* $\mathcal{L}ab(A) = \mathsf{in}$ *then for each* $B \in Ar$ *such that* B att A *it holds that* $\mathcal{L}ab(B) = \mathsf{out}$
- *if* $\mathcal{L}ab(A) = \mathsf{out}$ *then there exists a* $B \in Ar$ *such that* B att A *and* $\mathcal{L}ab(B) = \mathsf{in}$

An admissible labelling is called complete *iff it also satisfies:*

- *if* $\mathcal{L}ab(A) = \mathsf{undec}$ *then it is not the case that for each* $B \in Ar$ *such that* B att A *it holds that* $\mathcal{L}ab(B) = \mathsf{out}$, *and it is not the case that there exists a* $B \in Ar$ *such that* B att A *and* $\mathcal{L}ab(B) = \mathsf{in}$

Hence, the idea of an admissible labelling is that one should have sufficient grounds for everything one accepts (because all attackers are rejected) and sufficient grounds for everything one rejects (because it has an attacker that is accepted). A trivial way to satisfy this would be to take an extreme sceptical approach (simply label each argument undec). Therefore, the concept of a complete labelling has the additional requirement on whether one is allowed to abstain from having an explicit opinion on an argument (label it undec). One is only allowed to do so if one has insufficient grounds for accepting it (not all its attackers are rejected) and insufficient grounds for rejecting it (there is no attacker that is accepted).

Based on the concept of a complete labelling, it then becomes possible to define various other argumentation semantics (see [1] for an overview). Furthermore, we have to mention that the labelling approach is equivalent with the traditional extensions approach proposed in [14] (see [5,9] for details).

3   Argument-Based Informedness

We assume the presence of a UAF, a "universal argumentation framework", that serves as the universe of all well-formed arguments. Each individual agent is assumed to have access only to part of the world, therefore the private argumentation framework AF_i of agent i is assumed to be a subgraph of the UAF. Furthermore, if an agent has two arguments at his disposal, then he agrees with the UAF whether one attacks the other. This is in line with the approach of instantiated argumentation [3,7,21,25] where one has access to the internal structure of the arguments and can therefore assess whether they attack each other or not.[2]

Formally, the situation can be described as follows. There exists a $UAF = (Ar_{UAF}, att_{UAF})$, together with n agents ($n \geq 1$), each of which has access to only a subset $Ar_i \subseteq Ar_{UAF}$ of arguments and hence has an argumentation framework $AF_i = (Ar_i, att_{UAF} \cap (Ar_i \times Ar_i))$. That is, $AF_i \sqsubseteq UAF$.

The question we would like to study is "which agent is better informed?" Of course, an easy way to define this would be to use the subgraph relation. That is, agent j is at least as informed as agent i iff $AF_i \sqsubseteq AF_j$ (which in this case simply means that $Ar_i \subseteq Ar_j$). The problem, however, is that for many practical purposes, this characterisation is too strong. If each agent has some private information (like observations that nobody else was able to make) then he will have arguments that are not shared by anybody else. Hence, the resulting partial order will be the empty one, making all agents incomparable.

It does, however, seem reasonable to try to define the "more or equally informed" relation in such a way that an agent with an argumentation framework

---

[2] In terms of instantiated argumentation [3,7,21,25], the UAF consists of the arguments that can be constructed from all the available information in the world (from the "universal knowledge base"). Each agent's private argumentation framework then consists of the arguments that can be constructed from his private knowledge base.

that is a supergraph of that of another agent is automatically more or equally informed. Furthermore, the expression "more or equally informed" seems to suggest at least a partial pre-order. That is, what we are interested in is a relation $\preceq$ that satisfies at least the following properties, for each $i, j, k \in \{1, \ldots, n\}$:

1. if $AF_i \sqsubseteq AF_j$ then $AF_i \preceq AF_j$ (refinement of sub-AF relation)
2. $AF_i \preceq AF_i$ (reflexivity)
3. if $AF_i \preceq AF_j$ and $AF_j \preceq AF_k$ then $AF_i \preceq AF_k$ (transitivity)

Defining a suitable notion of informedness is far from trivial, because one needs to satisfy not only the above stated properties, but also needs to handle a number of examples in a reasonable way that does not deviate too much from what most people's intuitions would be.

To obtain an idea of what the difficulties are, we will now discuss three possible candidates for defining the "more or equally informed" relation. In each of these relations, we focus on a single argument. This is because we believe that different agents can have different competences. One agent may be more informed about, say, climate change and the other more about financial markets. The overall informedness of two experts may be incomparable, but on different topics (or on different particular arguments) it still seems fair to say that one is better informed than the other.

3.1   Informedness Based on Upstream

The first possible criterion is comparing what we call the "upstream" of a particular argument, which consists of all ancestor arguments in a particular argumentation framework. This approach makes sense also because several of the mainstream argumentation semantics, like complete, preferred and grounded, satisfy the principle of *directionality* [2], meaning that for determining the justification status of an argument [15,26] only the upstream is relevant.

Definition 4. *Let $AF = (Ar, att)$ be an argumentation framework and $A \in Ar$. We define $upstream_{AF}(A)$ as the smallest set such that:*

- *$A \in upstream_{AF}(A)$, and*
- *if $X \in upstream_{AF}(A)$ and Y attacks X then $Y \in upstream_{AF}(A)$*

We define a relation $\preceq_{us}^A$ (informedness based on upstream) such that if AF_i and AF_j are subframeworks of the UAF, then $AF_i \preceq_{us}^A AF_j$ iff $upstream_{AF_i}(A) \subseteq upstream_{AF_j}(A)$.

The thus defined notion of informedness, based on upstream, satisfies properties 1, 2 and 3. It satisfies property 1 because a supergraph always has an upstream that is a superset, for any argument. It satisfies properties 2 and 3 because the subset relationship is a partial pre-order.

In spite of these nice formal properties, the upstream-based notion of informedness also has some difficulties, as are for instance illustrated in the situation depicted in Fig. 1.

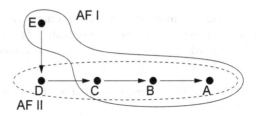

Fig. 1. AF_{II} more informed than AF_I according to the upstream criterion.

In the situation depicted in Fig. 1, we have agent I who has access to arguments A, B, C and E, and agent II who has access to arguments A, B, C and D. We have that $upstream_{AF_I}(A) = \{A, B, C\}$ and $upstream_{AF_{II}}(A) = \{A, B, C, D\}$. Therefore, $AF_I \preceq_{us}^A AF_{II}$. However, one may argue that it is actually agent I who should be more informed, because if one would merge the argumentation frameworks $(AF_I \sqcup AF_{II})$ then agent I would be right about the status of A ("A has to be **in**") whereas the position of agent II ("A has to be **out**") would be wrong.

Furthermore, D could be an argument of which agent I immediately realizes it cannot be warranted. Suppose D is the argument that "Barack Obama was not born in the USA because Bill O'Reilly says so on the Fox News Channel". Agent I, however, understands that Bill O'Reilly is strongly biased and therefore not a reliable source of information, even though he doesn't watch Fox News himself, and hence wasn't even aware of the existence of argument D. However, as soon as agent I learns about the existence of argument D, he is immediately able to construct a counterargument against it (Bill O'Reilly is unreliable, therefore the fact that he says something doesn't necessarily imply that it's actually true). Here, it seems fair to say that it is agent I that is more informed than agent II, which is precisely the opposite as would follow from the upstream criterion.

3.2 Informedness Based on Merging the Argumentation Frameworks

If defining informedness based on upstream is troublesome, then perhaps one should seek for an alternative criterion. The next possibility to be discussed is to define informedness based on whose position would be supported if both agents would share their arguments.

For the definition below, we recall that the *justification status* [26] of an argument consists of the different labels the argument can be assigned to under a particular semantics (for current purposes, we apply complete semantics). For instance, $JS(A) = \{\text{in}, \text{undec}\}$ means there is a complete labelling that labels A **in**, there is a complete labelling that labels A **undec**, but there is no complete labelling that labels A **out**. See [26] for details.

Furthermore, if $AF_i = (Ar_i, att_i)$ and $AF_j = (Ar_j, att_j)$ are the argumentation frameworks of agent i and agent j, respectively, then we write $AF_i \sqcup AF_j$ to denote the argumentation framework $(Ar_i \cup Ar_j, att_{UAF} \cap ((Ar_i \cup Ar_j) \times (Ar_i \cup Ar_j)))$.

Definition 5. *Let* $AF_1, \ldots, AF_n$ *(*$n \geq 1$*) be subargumentation frameworks of the UAF, where for each* AF_i *(*$1 \leq i \leq n$*) it holds that* $AF_i = (Ar_i, att_i)$. *We define a relation* $\preceq^A_{ms}$ *(informedness based on merged status) such that* $AF_i \preceq^A_{ms} AF_j$ *iff* $A \in Ar_i \cap Ar_j$ *and*

- $JS_{AF_i}(A) \neq JS_{AF_j}(A)$ *and* $JS_{AF_i \sqcup AF_j}(A) = JS_{AF_j}(A)$, *or*
- $JS_{AF_i}(A) = JS_{AF_j}(A)$ *and for each* AF_k *(*$1 \leq k \leq n$*) it holds that if* $JS_{AF_i \sqcup AF_k}(A) = JS_{AF_i}(A)$ *then* $JS_{AF_j \sqcup AF_k}(A) = JS_{AF_j}(A)$

The idea of the above definition is as follows. If two agents *disagree* about the justification status of argument A, then one looks at which of their positions would be supported if both agents would share all their information (that is, we look at which position would be supported in $AF_i \sqcup AF_j$). If, however, the two agents *agree* on the justification status of argument A, then one looks at who would be best capable of defending this shared position. For instance, the average newspaper reader may have the same position on whether climate change is going to happen as an expert, but still one would be tempted to say that the expert knows more, because he is better capable of defending his position against criticism. The above definition basically says that if two agents i and j have the same opinion, and for every agent k it holds that if i can convince k then j can also convince k then j is at least as informed as i.

The above definition, although intuitively defensible, does have some undesirable technical properties. Although it satisfies reflexivity, it violates transitivity, as is illustrated in Fig. 2.

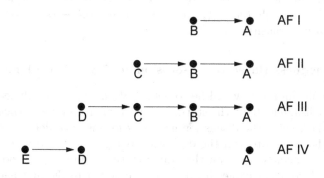

Fig. 2. Informedness based on merged status can violate transitivity

In the example depicted in Fig. 2 we have that AF_I and AF_{II} disagree about the status of A: $JS_{AF_I}(A) = \{\text{out}\}$ and $JS_{AF_{II}}(A) = \{\text{in}\}$. However, if one were to merge AF_I and AF_{II} then the result would be the same as AF_{II}. That is, $JS_{AF_I \sqcup AF_{II}}(A) = \{\text{in}\}$. Hence, $AF_I \preceq^A_{ms} AF_{II}$. For similar reasons, it also holds that $AF_{II} \preceq^A_{ms} AF_{III}$. Transitivity would then require that we also have that $AF_I \preceq^A_{ms} AF_{III}$. However, this is not the case since whereas agent I can maintain his position on A when confronted with agent IV, agent III cannot maintain his position on A when confronted with agent IV. Therefore, $AF_I \npreceq^A_{ms} AF_{III}$, so transitivity does not hold.[3]

It can be observed that property 1 (sub-AF refinement) also does not hold. In the example depicted in Fig. 3. Here, it holds that $AF_I \sqsubseteq AF_{II}$. However, it does *not* hold that $AF_I \preceq^A_{ms} AF_{II}$ because whereas agent I can maintain his position on A when confronted with agent III, agent II cannot.

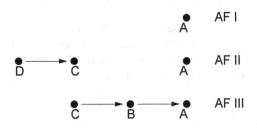

Fig. 3. Informedness based on merged status can violate the sub-AF property

There seems to be no easy way of patching up the definition of $\preceq^A_{ms}$. Omitting the second condition (when $JS_{AF_i}(A) = JS_{AF_j}(A)$) does not help, because it would mean that while $(A) \sqsubseteq (C \to B \to A)$[4] it does not hold that $(A) \preceq^A_{ms} (C \to B \to A)$, hence violating condition 1 (sub-AF refinement). Trivialising the second condition (saying that $AF_i \preceq^A_{ms} AF_j$ and $AF_j \preceq^A_{ms} AF_i$ whenever AF_i and AF_j agree on the status of A) does not work either, since this would imply that $(C \to B \to A) \preceq^A_{ms} (A) \preceq^A_{ms} (B \to A)$ whereas $(C \to B \to A) \npreceq^A_{ms} (B \to A)$, thus violating transitivity.

3.3   Informedness Based on Discussions Using Private Knowledge

A third possible approach would be to define informedness in a dialectical way. If two agents disagree about the status of a particular argument, then let them discuss together, using the discussion games as for instance defined in [6,8,20]. The agent who is able to win the discussion is regarded to be more informed. In case the two agents agree on the status of the argument, we look at what happens if a third agent comes in. If in every case where the first agent is able to convince the third agent, the second agent is also able to convince the third agent, then we say that the second agent is at least as informed as the first agent.

---

[3] The fact that transitivity does not hold can be problematic for applications of the theory. For instance, when the aim is for an agent to select an advisor, he might prefer his advisor to be as informed as possible. Therefore, the agent's preference order would coincide with the informedness order on the possible advisors. As a preference order is usually assumed to be a partial pre-order, one would like the same to hold for the informedness order (which includes satisfying transitivity).

[4] To allow for easy reading, we slightly abuse notation and write for instance $C \to B \to A$ for the argumentation framework $(\{A, B, C\}, \{(C, B), (B, A)\})$.

Definition 6. *Let $AF_1, \ldots, AF_n$ be subargumentation frameworks of the UAF, where for each AF_i ($1 \leq i \leq n$) it holds that $AF_i = (Ar_i, att_i)$. We define a relation $\preceq_{ds}^A$ (informedness based on discussed status) such that $AF_i \preceq_{ds}^A AF_j$ iff $A \in Ar_i \cap Ar_j$ and*

- *$JS_{AF_i}(A) \neq JS_{AF_j}(A)$ and agent j is able to win the relevant discussion game, based on each agent's private argumentation framework, or*
- *$JS_{AF_i}(A) = JS_{AF_j}(A)$ and for each agent k that disagrees with this shared position, it holds that if agent i is able to win the relevant discussion from agent k then agent j is also able to win the relevant discussion from agent k.*

With the "relevant discussion", we mean the discussion in the sense of [20]. With this new definition, we do obtain sub-AF refinement (condition 1). This can be seen as follows. Suppose $AF_I \sqsubseteq AF_{II}$. We distinguish two cases.

- $JS_{AF_I}(A) \neq JS_{AF_{II}}(A)$. Then agent II has a winning strategy for defending his position on A (using the grounded and/or preferred games) which works when applied in AF_{II}. The fact that $AF_I \sqsubseteq AF_{II}$ means that agent I only has access to a subset of countermoves that agent II has already taken into account in his winning strategy. Therefore, the winning strategy of agent II still works when playing against agent I.
- $JS_{AF_I}(A) = JS_{AF_{II}}(A)$. Now, assume the presence of agent III with argumentation framework AF_{III} (which is still a sub-AF of the UAF) who does not agree on the status of A. Then, if agent I has a winning strategy against agent III, then agent II is able to use the same winning strategy, because agent II has access to a superset of possible arguments to move.

Apart from satisfying the sub-AF relation (condition 1), it can easily be verified that $\preceq_{ds}^A$ also satisfies reflexivity (condition 2).

It can be interesting to see how $\preceq_{ds}^A$ deals with the argumentation frameworks of Fig. 1. Here, we have that $JS_{AF_I}(A) = \{\text{in}\}$ and $JS_{AF_{II}}(A) = \{\text{out}\}$. Let us examine what happens if these agents start to discuss (say, using the grounded game as described in [20]).

I: A has to be in

II: but maybe B does not have to be out

I: B has to be out because C has to be in

II: but maybe D does not have to be out (this is where agent I learns about D)

I: D has to be out because E has to be in (after learning about D, agent I realizes that E attacks D, since we assume that whenever an agent is aware of two arguments (like E and D) the agent can also determine whether there exists an attack between them)

So here we see that agent I is able to win the discussion.

The example of Fig. 4, however, is more complicated. Instead of AF_{II} having just one additional argument (as was the case in the example of Fig. 1), it has three additional arguments. For most of the mainstream argumentation semantics, the outcome is not influenced when one substitutes a single argument by a chain of three arguments. The point, however, is that for the criterion of informedness based on discussed status, this substitution does matter. Whereas in the example of Fig. 1 agent I is able to win the discussion, in the example of Fig. 4 it is agent II who is able to win the discussion (this is because when agent II moves argument D, agent I is unable to respond, as he doesn't have access to argument E, and hence loses the discussion).

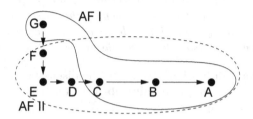

Fig. 4. Informedness based on discussed status can have unexpected results

Another issue that starts to play a role when one takes into account the possibility of dishonesty or strategic behaviour. An example of this is shown in Fig. 5. Here, we have that initially agent I sincerely holds that the status of A has to be out ($JS_{AF_I}(A) = \{\text{out}\}$) whereas agent II sincerely holds that the status of A has to be in ($JS_{AF_{II}}(A) = \{\text{in}\}$). However, during the course of the discussion, agent I learns that his position does not hold, even though he can still successfully defend it. This discussion would look as follows.

II: A has to be in

I: but maybe B does not have to be out

II: B has to be out because C has to be in (this is where agent I learns about argument C)

I: but maybe D does not have to be out (this is where agent II learns about argument D and realizes that with this new information, A is no longer in)

II: D has to be out because E has to be in (agent II realizes that this is not the case, but utters this statement nevertheless, hoping that agent I does not know about argument F)

Since agent I cannot move any more (he doesn't have access to argument F), agent II wins the discussion.

Another problem is that although $\preceq^A_{ds}$ satisfies sub-AF refinement (condition 1) and reflexivity (condition 2), it does not satisfy transitivity (condition 3). A counterexample is provided in Fig. 6.

In the example of Fig. 6, agent I is more informed than agent II on argument A, because they disagree on the status of A and agent I is able to win the discussion. Agent II is more informed than agent III on argument A, because they disagree on the status of A and agent II is able to win the discussion. Transitivity would require that agent I is also more informed than agent III on argument A. However, this is not the case, because agents I and III agree on the status of A, but whereas agent III is able to win the discussion from

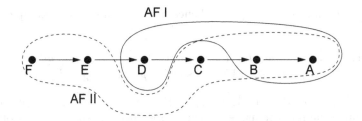

Fig. 5. Informedness based on discussed status and issues of dishonesty

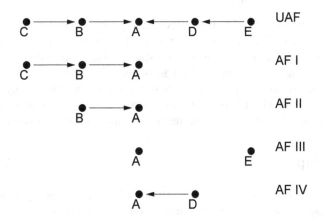

Fig. 6. Informedness based on discussed status can violate transitivity

agent IV, agent I is unable to win the discussion from agent IV. Therefore transitivity does not hold.

3.4   Comparing the Different Types of Informedness

The three different types of informedness, as were introduced above, can be shown to be independent from each other.

Proposition 1. *The informedness relations $\prec_{us}$ and $\prec_{ms}$ are independent from each other. That is, $AF_i \prec_{us}^A AF_j$ does not imply $AF_i \prec_{ms}^A AF_j$ and vice versa.*

Proof. In Fig. 1 it holds that $AF_I \prec_{us}^A AF_{II}$ but $AF_I \not\prec_{ms}^A AF_{II}$, hence providing a counter example against $\prec_{us}$ being subsumed by $\prec_{ms}$. Also, it holds that $AF_{II} \prec_{ms}^A AF_I$ but $AF_{II} \not\prec_{us}^A AF_I$, hence providing a counter example against $\prec_{ms}$ being subsumed by $\prec_{us}$.

Proposition 2. *The informedness relations $\prec_{us}$ and $\prec_{ds}$ are independent from each other. That is, $AF_i \prec_{us}^A AF_j$ does not imply $AF_i \prec_{ds}^A AF_j$ and vice versa.*

Proof. In Fig. 1 it holds that $AF_I \prec_{us}^A AF_{II}$ but $AF_I \not\prec_{ds}^A AF_{II}$, hence providing a counter example against $\prec_{us}$ being subsumed by $\prec_{ds}$. Also, it holds that

$AF_{II} \prec^A_{ds} AF_I$ but $AF_{II} \not\prec^A_{us} AF_I$, hence providing a counter example against $\prec_{ds}$ being subsumed by $\prec_{us}$.

Proposition 3. *The informedness relations $\prec_{ms}$ and $\prec_{ds}$ are independent from each other. That is, $AF_i \prec^A_{ms} AF_j$ does not imply $AF_i \prec^A_{ds} AF_j$ and vice versa.*

Proof. In Fig. 4 it holds that $AF_{II} \prec^A_{ms} AF_I$ but $AF_{II} \not\prec^A_{ds} AF_I$, hence providing a counter example against $\prec_{ms}$ being subsumed by $\prec_{ds}$. Also, it holds that $AF_I \prec^A_{ds} AF_{II}$ but $AF_I \not\prec^A_{ms} AF_{II}$, hence providing a counter example against $\prec_{ds}$ being subsumed by $\prec_{ms}$.

Apart from technical differences between $\prec_{us}$, $\prec_{ms}$ and $\prec_{ds}$, there also exist practical differences. To assess whether agent I is more informed than agent II with respect to upstream ($\prec_{us}$) or merged status ($\prec_{ms}$) one needs to have access to the internal state of the agents. That is, one needs to be able to examine their respective argumentation frameworks. However, to assess whether agent I is more informed than agent II with respect to discussed status ($\prec_{ds}$) no such access is required. Instead, it suffices to examine the discussion between the agents.

4 Roundup

In the current paper, we examined the issue of informedness (justified belief) from the perspective of formal argumentation. We examined different ways of defining informedness, each with its own advantages and disadvantages. Determining, or even defining, which agent is more informed is a non-trivial problem that one could argue is understudied in the literature on formal argumentation and non-monotonic reasoning. Yet, having a proper way to assess agents' informedness can be vital when it comes to selecting the right agents to collaborate with or to source services from.

Some of the proposed criteria for defining informedness make use of dynamic aspects of argumentation, such as discussion or merging argumentation frameworks. It should be mentioned, however, that unlike for instance [11,23] our aim is *not* to study argumentation dynamics by itself, but rather to assess (compare) two static knowledge bases represented as argumentation frameworks. Of course, one way of assessing would be to see what happens when agents with these knowledge bases talk to each other and try to convince each other, but the resulting interactions are purely instrumental to comparing the a-priori (static) knowledge bases, and do not serve as a means for changing them. This makes our work fundamentally different from for instance [11,23].

The current paper should be seen as a first step to opening a research topic. Many research questions are still open, which include the following:

- Is there a reasonable way of defining informedness which satisfies all three conditions, and performs well on the examples stated in the current paper?
- Given a particular way of defining informedness, what are the individual agents' best strategies for determining who is more informed. After all,

the individual agents do not have full access to the UAF, or to the other agent's private AF. It seems most realistic that they try to test the other agent's informedness during some kind of discussion.

– To which extent is the way in which agents try to observe each other's informedness strategy-proof? What are the optimal ways for a particular agent to appear to be more informed than he actually is? To what extent is (undetected) dishonesty possible?[5]

One particular issue is whether one wants to apply the complete justification status [26] for determining the differences of opinion on a particular argument, or whether simpler approaches (such as membership of an admissible set, or membership of the grounded extension) would suffice. Another issue is whether there are more properties (other than condition 1, 2 and 3) that one would like to satisfy, and whether one can formalize the intuitions behind the examples in the current document in the form of postulates, in the same way as is for instance done in [4,7].

References

1. Baroni, P., Caminada, M., Giacomin, M.: An introduction to argumentation semantics. Knowl. Eng. Rev. **26**(4), 365–410 (2011)
2. Baroni, P., Giacomin, M.: On principle-based evaluation of extension-based argumentation semantics. Artif. Intell. **171**(10–15), 675–700 (2007)
3. Besnard, P., Hunter, A.: Constructing argument graphs with deductive arguments: A tutorial. Argument Comput. **5**, 5–30 (2014). Special Issue: Tutorials on Structured Argumentation
4. Booth, R., Caminada, M., Podlaszewski, M., Rahwan, I.: Quantifying disagreement in argument-based reasoning. In: Proceedings of the 11th International Conference on Autonomous Agents and Multiagent Systems (2012)
5. Caminada, M.: On the issue of reinstatement in argumentation. In: Fisher, M., Hoek, W., Konev, B., Lisitsa, A. (eds.) JELIA 2006. LNCS (LNAI), vol. 4160, pp. 111–123. Springer, Heidelberg (2006). doi:10.1007/11853886_11
6. Caminada, M.: A discussion game for grounded semantics. In: Black, E., Modgil, S., Oren, N. (eds.) TAFA 2015. LNCS (LNAI), vol. 9524, pp. 59–73. Springer, Cham (2015). doi:10.1007/978-3-319-28460-6_4
7. Caminada, M., Amgoud, L.: On the evaluation of argumentation formalisms. Artif. Intell. **171**(5–6), 286–310 (2007)
8. Caminada, M., Dvořák, W., Vesic, S.: Preferred semantics as socratic discussion. J. Logic Comput. **26**(4), 1257–1292 (2016). doi:10.1093/logcom/exu005
9. Caminada, M., Gabbay, D.M.: A logical account of formal argumentation. Stud. Logica **93**(2–3), 109–145 (2009). Special issue: new ideas in argumentation theory

---

[5] For instance, Sakama et al. [12,13] provide a formal account of various forms of dishonesty. Their analysis includes Harry Frankfurt's notion of *bullshit* [17], which consists of making claims of which one has no proper knowledge or beliefs. However, their analysis is done in terms of traditional (truth based) multi-modal logic. An open question, therefore, is how to provide an alternative formalisation of this form of dishonesty, one that is not based on truth of belief, but on the notion of argument-based informedness.

10. Caminada, M., Pigozzi, G.: On judgment aggregation in abstract argumentation. J. Auton. Agent. Multi-Agent Syst. **22**, 64–102 (2011). doi:10.1007/s10458-009-9116-7

11. Cayrol, C., Dupin de Saint-Cyr, F., Lagasquie-Schiex, M.: Change in abstract argumentation frameworks: Adding an argument. J. Artif. Intell. Res. **38**, 49–84 (2010)

12. Sakama, C., Caminada, M., Herzig, A.: A logical account of lying. In: Janhunen, T., Niemelä, I. (eds.) JELIA 2010. LNCS (LNAI), vol. 6341, pp. 286–299. Springer, Heidelberg (2010). doi:10.1007/978-3-642-15675-5_25

13. Caminada, M., Sakama, C., Herzig, A.: A formal account of dishonesty. Logic J. IGPL **23**, 259–294 (2015)

14. Dung, P.M.: On the acceptability of arguments and its fundamental role in nonmonotonic reasoning, logic programming and n-person games. Artif. Intell. **77**, 321–357 (1995)

15. Dvořák, W.: On the complexity of computing the justification status of an argument. In: Modgil, S., Oren, N., Toni, F. (eds.) TAFA 2011. LNCS (LNAI), vol. 7132, pp. 32–49. Springer, Heidelberg (2012). doi:10.1007/978-3-642-29184-5_3

16. Fitting, M.: Bilattices and the semantics of logic programming. J. Logic Program. **11**, 91–116 (1991)

17. Frankfurt, H.G.: On Bullshit. Princeton University Press, Princeton (2005)

18. Governatori, G., Maher, M.J., Antoniou, G., Billington, D.: Argumentation semantics for defeasible logic. J. Logic Comput. **14**(5), 675–702 (2004)

19. Grossi, D., van der Hoek, W.: Justified beliefs by justified arguments. In: Proceedings of the Fourteenth International Conference on Principles of Knowledge Representation and Reasoning, pp. 131–140 (2014)

20. Modgil, S., Caminada, M.: Proof theories and algorithms for abstract argumentation frameworks. In: Rahwan, I., Simari, G.R. (eds.) Argumentation in Artificial Intelligence, pp. 105–129. Springer, Heidelberg (2009)

21. Modgil, S., Prakken, H.: The ASPIC+ framework for structured argumentation: A tutorial. Argument Comput. **5**, 31–62 (2014). Special Issue: Tutorials on Structured Argumentation

22. Prakken, H.: An abstract framework for argumentation with structured arguments. Argument Comput. **1**(2), 93–124 (2010)

23. Rienstra, T.: Argumentation in flux: modelling change in the theory of argumentation. Ph.D. thesis, Université du Luxembourg (1999)

24. Sakama, C.: Ordering default theories. Theoret. Comput. Sci. **338**, 127–152 (2005)

25. Toni, F.: A tutorial on assumption-based argumentation. Argument Comput. **5**, 89–117 (2014). Special Issue: Tutorials on Structured Argumentation

26. Wu, Y., Caminada, M.: A labelling-based justification status of arguments. Stud. Logic **3**(4), 12–29 (2010)

27. Wu, Y., Caminada, M., Gabbay, D.M.: Complete extensions in argumentation coincide with 3-valued stable models in logic programming. Stud. Logica **93**(1–2), 383–403 (2009). Special issue: new ideas in argumentation theory

On the Interpretation of Assurance Case Arguments

John Rushby[(✉)]

Computer Science Laboratory, SRI International,
333 Ravenswood Avenue, Menlo Park, CA 94025, USA
Rushby@csl.sri.com

Abstract. An assurance case provides a structured argument to establish a claim for a system based on evidence about the system and its environment. I propose a simple interpretation for the overall argument that uses epistemic methods for its evidential or leaf steps and logic for its reasoning or interior steps: evidential steps that cross some threshold of credibility are accepted as premises in a classical deductive interpretation of the reasoning steps. Thus, all uncertainty is located in the assessment of evidence. I argue for the utility of this interpretation.

1 Introduction

An assurance case provides an argument to justify certain claims about a system, based on evidence concerning both the system and the environment in which it operates. The claims can be about any system property, such as reliability or security, and thereby generalize the previously established notion of a safety case, where the claim is always about safety. Software assurance, in the form this is understood in the DO-178C guidelines for civil aviation [1], provides an important special case: here the top claim is one of correctness with respect to system requirements (safety of those requirements is established separately using guidelines such as ARP-4754A [2] and ARP-4761 [3]).

Assurance cases are standard in many industries (e.g., trains and nuclear power in Europe, and some medical devices in the USA) and are being considered for others, such as civil aircraft, where changes in the operating environment and the pace of that change (e.g., integration of ground and air systems in NextGen, UAVs in civil airspace, and increasingly autonomous flight systems) challenge current methods of assurance. Civil aviation has an exemplary record of safety, so there is interest in achieving greater understanding of both existing and new methods for assurance before making changes. In particular, there is work on reconstructing the argument implicit in DO-178C [4], and exploring whether assurance cases could provide the basis for future evolutions of these and related guidelines [5].

As related in reference [5], modern safety cases developed from methods used in nuclear power, offshore oil, and process industries, where the case was based on a "narrative" about the design and operation of the plant or system and how

M. Otake et al. (Eds.): JSAI-isAI 2015 Workshops, LNAI 10091, pp. 331–347, 2017.
DOI: 10.1007/978-3-319-50953-2_23

its hazards were eliminated or mitigated. Later, as computer control became a larger part of the system, safety cases became more "structured" with an explicit argument organized in a step-by-step manner and often presented in a graphical notation such as CAE (Claims-Argument-Evidence) or GSN (Goal Structuring Notation). This paper is concerned with the *interpretation* of structured assurance case arguments; that is, we ask, what is the meaning of such an assurance case? This question is a necessary precursor to one that will be addressed in a later paper concerning the *evaluation* of assurance cases: that is, how we can tell if an assurance case truly and convincingly justifies its claim.

An assurance case is based on evidence, which is an epistemological concept: that is, it concerns our *knowledge* of the system and its environment. Hence, it seems that the interpretation of a case could or should build, at least in part, on ideas from epistemology, such as those used to formalize scientific theories. On the other hand, an assurance case also employs an *argument*, which is generally viewed as a logical concept. But even within logic, there are different ways of looking at arguments. One perspective focuses on the dialectical, back and forth interpretation of argument; that perspective will be valuable when we turn to the evaluation of assurance cases, but for their basic meaning the classical interpretation of formal logic seems more suitable. However, formal logic deals with deductive validity—that is, truth of the premises must guarantee truth of the conclusion—whereas an assurance case must acknowledge uncertainties in the world and in our knowledge about it, so that truth of the premises may do no more than strongly suggest the conclusion. This is generally referred to as inductive validity (an unfortunate overloading of the term "inductive", which has many other meanings in logic and science) and its interpretation requires a departure from the well-established semantics of classical logic into more contentious areas such as probability logic, fuzzy logic, or evidential reasoning.

Thus, interpretation of assurance cases must reconcile their epistemic and logical aspects, and must acknowledge their inductive character. Furthermore, many industries employ graduated levels of assurance: systems that pose greater risk require greater assurance. If this graduation is framed in terms of assurance cases, then it seems that in addition to the inductive validity (sometimes referred to as the *cogency*) of a case, we must also address the "strength" of that validity. One way in which assurance case arguments may be strengthened is by inclusion of *confidence claims*. These are elements whose falsity would not invalidate the argument but whose truth strengthens it. Clearly, such elements are not part of standard logical interpretations.

Accordingly, some look to rather radical reformulations of the idea of argument, such as Toulmin's treatment [6], or probability logics [7]. In this paper, by contrast, I propose a very simple combination of classical methods and argue for its utility. I present the approach in the following section, provide brief comparison with other methods in Sect. 3, and conclusions in Sect. 4.

2   Structure and Interpretation of Assurance Arguments

As noted in the introduction, an assurance case is composed of three elements: a *claim* that states the property to be assured, *evidence* about the system and its

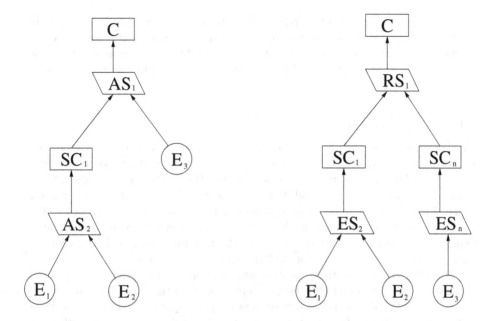

Fig. 1. Converting an argument from free (left) to simple form (right)

environment, and a *structured argument* that the evidence is sufficient to establish the claim. The structured argument is a hierarchical collection of individual argument steps, each of which justifies a local claim on the basis of evidence and/or lower-level subclaims. A trivial example is shown on the left in Fig. 1, where a claim C is justified by an argument step AS_1 on the basis of evidence E_3 and subclaim SC_1, which itself is justified by argument step AS_2 on the basis of evidence E_1 and E_2. The figure is generic and not representative of any specific notation, although its element shapes are from GSN (but the arrows are reversed, as in CAE). Note that a structured argument need not be a tree because subclaims and evidence can support more than one argument step.

Observe that the argument step AS_1 uses both evidence E_3 and a subclaim SC_1. We will see later how to provide an interpretation to such "mixed" argument steps, but it is easier to understand the basic approach in their absence. By introducing additional subclaims where necessary, it is straightforward to convert arguments into *simple form* where each argument step is supported either by subclaims (i.e., a *reasoning step*) or by evidence (i.e., an *evidential step*), but not by a combination of the two; many assurance cases will already have this form—it is natural in GSN, for example. In Fig. 1, the "mixed" or free argument on the left is converted to simple form on the right by introducing a new subclaim SC_n and new evidential argument step ES_n above E_3. Argument steps AS_1 and AS_2 are relabeled as reasoning step RS_1 and evidential step ES_2, respectively.

The key to our approach is that the two kinds of argument step are interpreted differently. Specifically, evidential steps are interpreted epistemically, using ideas grounded in probability, while reasoning steps are interpreted in logic: subclaims supported by evidential steps that cross some threshold of credibility are accepted as premises in a classical deductive interpretation of the reasoning steps. We now consider these two kinds of argument steps in more detail.

2.1   Evidential Steps

Evidential steps are the bridge between our concepts about both the system and its environment, which we express as subclaims, and our observations concerning these, which we document as evidence; in other words, they represent our *knowledge* about the system and its environment. What it means to really know something is the topic of epistemology, a branch of philosophy dating back to the ancient Greeks that provides much insight but no generally accepted treatment. Our focus is a rather more specific than the general theory of knowledge: we want to know what it means for evidence to support a claim.

The intuitive idea is that the evidence in support of a hypothesis or claim should be "weighed" and the claim accepted as a "settled fact" if that weight exceeds some threshold. Modern treatments of this topic derive from the work of I. J. (Jack) Good who framed it in terms of probabilities and Bayesian inference [8]. Good began his work in the codebreaking activity at Bletchley Park during the Second World War and reference [9] recounts some of this history. Recent developments of these ideas are found in Bayesian Epistemology [10] and their application to the theory of science is known as Bayesian Confirmation Theory [11]. Related ideas are developed also in legal theory [12].

When we have evidence E supporting a hypothesis or claim C, it seems plausible that our procedure should be to assess the probability of C given E, $P(C \mid E)$, and to accept C when this probability exceeds some threshold. Unfortunately, assessment of $P(C \mid E)$ poses difficulties. All the quantities under consideration here are subjective probabilities that express human judgement [13] and even experts find it difficult to directly assess a quantity such as $P(C \mid E)$. Furthermore, the significance of $P(C \mid E)$ depends on our prior assessment $P(C)$, which could be one of ignorance (or, in law, prejudice). Rather than attempt directly to assess $P(C \mid E)$, it seems that we should factor the problem into alternative quantities that are easier to assess and of separate significance.

The basic idea of Good and others is that the conditioning should be reversed, so the strength or "weight" of evidence is some function of $P(E \mid C)$. This is related to $P(C \mid E)$ by Bayes' Theorem but seems easier to assess: that is, it seems easier to estimate the likelihood of concrete observations, given a claim about the world, than vice-versa. Furthermore, what we are really interested in is the ability of E to discriminate between C and its negation $\neg C$, so the quantities we should look at are the difference or ratio (or logarithms of these) between $P(E \mid C)$ and $P(E \mid \neg C)$. Such quantities are called *confirmation measures* and may be said to weigh C against $\neg C$ "in the balance" provided by E.

There is no agreement in the literature on the best confirmation measure: Fitelson [14] considers several and makes a strong case for Good's measure

$\log \frac{P(E\,|\,C)}{P(E\,|\,\neg C)}$, Tentori and colleagues [15] perform an empirical comparison and generally approve of Kemeny and Oppenheim's measure $\frac{P(E\,|\,C)-P(E\,|\,\neg C)}{P(E\,|\,C)+P(E\,|\,\neg C)}$, while Joyce [16] argues that different measures serve different purposes.

In criminal law, there is (or was until recently) a reluctance to convict the innocent, even at the price of acquiting some who are guilty, so Gardner-Medwin [17] suggests that appropriate probabilistic criteria for conviction are those that indicate the evidence could very likely have arisen if the defendant is guilty but not if they are innocent—and confirmation measures have this property.

It is a topic for debate whether the criteria for acceptance of evidential steps in assurance cases should use a confirmation measure (so that, as in a criminal trial, we can reduce the chance of accepting a false claim, even at the price of rejecting some good ones) or one that more directly assesses the claim (thereby maximizing utility but possibly accepting some false claims).

My own view is that the final decision is a human judgement that should consider several quantities and measures. It is not necessary to attach numerical estimates to the probabilities nor to actually evaluate the measures, but understanding the basis for their construction can inform our judgement. This judgement is more difficult when several items of evidence are combined to support a subclaim: the items may not be independent so accurate analysis and evaluation requires rather sophisticated probabilistic modeling techniques, such as Bayesian Belief Networks (BBNs) and their tools. Again, informal rather than quantitative modeling and analyses may be used in practice, but it is useful to hone our judgement with numerical experiments that allow sensitivity and "what-if" explorations. An example of such an exploration is presented below.

Bayes' Theorem is the principal tool for analyzing conditional subjective probabilities: it allows a prior assessment of probability to be updated by new evidence to yield a rational posterior probability. It is difficult to calculate over large numbers of complex conditional (i.e., interdependent) probabilities, but usually the dependencies are relatively sparse and can conveniently be represented by a graph (or "net", the term used in BBNs) in which arcs indicate dependencies. An example, taken from [18], is shown above in Fig. 2. This represents a "multi-legged" evidential argument step in which evidence from testing is combined with that from formal verification. The nodes of the graph represent judgments about components of the argument step and the arcs indicate dependencies between these (ignore, for the time being, the arcs associated with A and shown in blue).

The nodes of the graph actually represent random variables but we can most easily understand the construction of the graph by first considering the artifacts from which these are derived. Here, Z is the system specification; from this are derived the actual system S and the test oracle O. Tests T are dependent on both the oracle and the system, while formal verification V depends on the system and its specification. The claim of correctness C is based on both the test results and the formal verification.

For reasons explained later, I think it is best to treat the verification and testing "legs" of the evidence separately, so let us focus on the testing leg alone

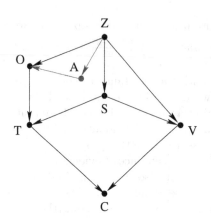

Z: System Specification

A: Assessment of specification quality

O: Test Oracle

S: System's true quality

T: Test results

V: Verification outcome

C: Claim of correctness

Fig. 2. BBN for testing and verification evidence (Color figure online)

(i.e., ignore everything involving V and C); this is shown represented inside the BBN tool *Hugin Expert* [19] in Fig. 3. Here, the interpretation of Z is a random variable representing correctness of the system specification: it has two possible values: correct (i.e., it achieves the requirements established for the system) or incorrect. The assessor must attach some prior probability distribution to these (e.g., 99% confidence it is correct, vs. 1% that it is incorrect).

S is a variable that represents the true (but unknown) quality of the system, stated as a probability of failure on demand (that is, failure wrt. requirements). This probability depends on Z: we might suppose that it is 0.99 if Z is correct, but only 0.5 if it is incorrect.

O is a variable that represents correctness of the test oracle; this is derived in some way from the specification Z and its probability distribution will be some function of the correctness of Z (e.g., if Z is correct, we might suppose it is 95% probable that O is correct, but if Z is incorrect, then it is only 2% probable that O is correct).

T is a Boolean variable that represents the outcome of testing. It depends on both the oracle O and the true quality of the system S. Its probability distribution over these is represented by a joint probability table such as the following, which gives the probabilities that the test outcome is judged successful.

Correct system		Incorrect system	
Correct oracle	Bad oracle	Correct oracle	Bad oracle
100%	50%	5%	30%

In this example, only T is directly observable. Using a BBN tool such as Hugin, it is possible to conduct "what if" exercises on this model to see how prior estimates for the conditional probability distributions of the various nodes are updated by evidence. In particular, Hugin allows the user to manipulate the values of some variables and observe the impact on others. In Fig. 3, we have hypothesized the system is incorrect (indicated by the red bar, and set by double-clicking on the value) and can see that the conditional probability that testing succeeds (i.e., $P(E \mid \neg C)$ for this example) is 33.07%. If the system is assumed correct, the probability that testing succeeds (i.e., $P(E \mid C)$) is 98.53%. Hence, the Kemeny-Oppenheim confirmation measure is 0.49. We can also examine the probability of a correct system, given that testing succeeds (i.e., $P(C \mid E)$), which evaluates to 99.49%, or given that it fails (i.e., $P(C \mid \neg E)$), which is 59.21%.

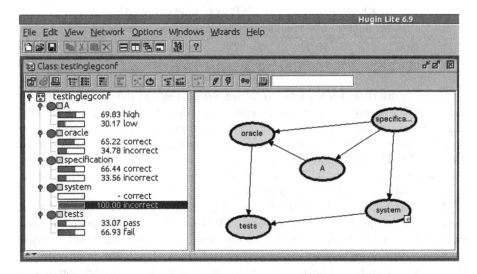

Fig. 3. Hugin analysis of BBN for testing evidence alone (Color figure online)

We see that in this model the assumed prior distributions are such that testing has rather poor evidential weight: it is rather likely that an incorrect system will be accepted or that a rejected system is in fact correct. Further inspection and experimentation will show that part of the explanation is that the modeled test oracle is of low quality. The variable O has strong impact on the test outcome T but is not itself observed or evaluated. We might suppose that reliability of the testing procedure would be improved if we could assess the quality of the test oracle and require this to exceed some threshold. However, it is not easy to see how this artifact can be assessed directly, so an alternative might be to assess the quality of the specification Z, since this has a large impact on the quality of the oracle.

Reasoning similar to this may implicitly underlie some of the DO-178C guide-lines for software assurance in civil aircraft [1]. For the most critical software, DO-178C specifies 71 assurance "objectives" that must be accomplished and sev-eral of these concern the quality of requirements and specifications. For example, its Sect. 6.3.2.d specifies the objective to "ensure that each low-level requirement can be verified". We can introduce this idea into our model as the variable A in Fig. 2 (with dependencies indicated in blue) and similarly in Fig. 3. Here A assesses "confidence" that the specification Z is testable and takes values high and low; we suppose the probability that A is high is 95% when Z is correct and 20% when it is incorrect. There is no arc from A to S because A is not a general evaluation of the specification, just its testability. The probability distribution of O will now depend on both A and Z and we might suppose it takes the following form.

Correct specification		Incorrect specification	
High confidence	Low confidence	High confidence	Low confidence
99%	70%	2%	1%

If we require that A is high before we undertake testing, then we find that the probability of accepting an incorrect system is reduced from 33.07% to 13.33% while the probability of accepting a correct system increases from 98.53% to 99.45%. Hence, the Kemeny-Oppenheim confirmation measure improves from 0.49 to 0.76. The probability the system is correct, given that testing succeeds, improves from 99.49% to 99.85% and, if testing fails, the probability the system is correct reduces from 59.21% to 36.33%.

The probabilities and distributions used in this exercise were "plucked from the air" and cannot be considered realistic. It is possible that experts could provide realistic prior distributions for models such as these, and thereby derive credible posterior estimates. However, I do not think that is the main value in these exercises. Rather, I believe that "what-if" explorations help develop understanding of the relationships among variables and, more particularly, can help guide selection of evidence and supporting confidence claims, and also guide the informal criteria to be used in deciding when the totality of evidential support allows a claim to be regarded as a "settled fact". Thus, although probabilistic modeling provides the underlying semantics and sharpens our understanding, the evidential steps in an assurance case argument may well be comprised of objectives similar (but better justified) to those developed using informal methods in guidelines such as DO-178C.

When all the objectives of an evidential step are satisfied, its subclaim is accepted as a premise in the logical interpretation of the reasoning steps of the argument, as explained in the following section.

2.2   Reasoning Steps

We have seen that in evidential steps, the separate items of evidence are "combined" to justify truth of the claim concerned. This combination may be performed informally or it can use probabilistic modeling with BBNs, as in the example, where we saw testing evidence combined with "confidence" evidence about testability of the specification.

In contrast, I propose that the subclaims appearing in reasoning steps should be *conjoined* to deliver the truth of their parent claim: that is, the claim in a reasoning step is considered true only if all its subclaims are so.[1] This interpretation could be inductive, that is the conjunction of subclaims strongly *suggests* the claim, or it could be deductive, meaning the conjunction *implies* (or entails, or proves) the claim. Let us accept for the time being that logic does provide the appropriate interpretation for reasoning steps and focus on whether this should be deductive or inductive. I maintain it should be deductive and advance two reasons. The first concerns modular reasoning.

Assurance cases are generally very large and cannot truly be comprehended *in toto*: a modular or compositional method is essential. Deductive reasoning steps can be assessed in just such a modular fashion, one step or one claim at a time. First, we check local soundness: that is, for each reasoning step, we must assure ourselves that the conjunction of subclaims truly implies the claim. Second, we must check that claims are interpreted consistently between the steps that establish them and the steps that use them; this, too, is a modular process, performed one claim at a time.

In contrast, the first of these is not modular for inductive steps—for when a step is labeled inductive, we are admitting a "gap" in our reasoning: we must surely believe either that the gap is insignificant, in which case we could have labeled the step deductive, or that it is taken care of elsewhere, in which case the reasoning is not modular.

My second reason for deprecating inductive reasoning steps is that there is no effective way to estimate the size of the gap in our reasoning. We may surely assume that any inductive step is "almost" deductive, or even "as deductive as possible". That is to say, the following generic inductive step

$$p_1 \text{ AND } p_2 \text{ AND } \cdots \text{ AND } p_n \text{ SUGGESTS } c \tag{1}$$

would become deductive if some missing (and presumably unknown) subclaim or assumption a (which, of course, may actually be a conjunction of smaller subclaims) were added, as shown below. (It may be necessary to adjust the existing subclaims p_1 to p_1' and so on if, for example, the originals are inconsistent with a).

$$a \text{ AND } p_1' \text{ AND } p_2' \text{ AND } \cdots \text{ AND } p_n' \text{ IMPLIES } c. \tag{2}$$

If we cannot imagine such a "repair", then surely (1) must be utterly fallacious. It then seems that any estimation of the doubt in an inductive step like (1) must concern the gap represented by a. Now, if we knew anything at all about a it would be irresponsible not to have added it to the argument. But since we did not do so, we must be ignorant of a and it follows that we cannot estimate the doubt in inductive argument steps.

---

[1] Some would allow disjunctions and general logical expressions. My opinion is that these are the hallmarks of evidential—rather than reasoning—steps.

If we cannot estimate the magnitude of our doubt, can we at least reduce it? This seems to be the purpose of "confidence claims", but what exactly is their logical rôle in reasoning steps? One possibility is that confidence claims eliminate some sources of doubt. For example, we may doubt that the subclaims imply the claim *in general*, but the confidence claims restrict the circumstances so that the implication is true *in this case*. But such use of confidence claims amounts to a "repair" in the sense used above: these claims are really assumptions that should be added to the argument as conventional subclaims (see the discussion of assumptions in Sect. 2.4 below), thereby making it deductively sound, or at least less inductive.

The logical rôle of other kinds of confidence claims is less clear; one possibility is that they serve to justify the reasoning involved. Some justification *why* the conjunction of subclaims is believed to suggest (or imply) the claim is, of course required, but I would expect it to take the form of a narrative explanation (as it does in CAE, for example), rather than a claim. On the other hand, Hawkins *et al.* [20] endorse such confidence claims but propose they are removed from the main safety argument and linked to it through assurance claim points (ACPs) at restricted locations to yield "assured safety arguments". However, although this improves the readability of arguments that use confidence claims, Hawkins *et al.* provide no guidance on how to assess their contribution.

Thus, there seems to be no established or proposed method to assess the contribution of confidence claims to inductive reasoning steps. In my opinion, their use opens Pandora's Box, for there is no way to determine that we have "enough" and thus a temptation to employ complex, but still inductive, reasoning steps, buttressed with numerous confidence claims of uncertain interpretation "just in case".

My opinion is that inductive reasoning sets too low a bar and confidence claims do nothing to raise it. Hence, I recommend that reasoning steps should be deductive, for then it is very clear what their evaluation must accomplish: it must review the content and justification of the step and assent (or not) to the proposition that its subclaims truly imply the claim. There is no rôle for confidence claims in deductive reasoning steps and other superfluous subclaims are likely to complicate rather than strengthen the assessment. Hence, the requirement for deductive soundness encourages the formulation of precise subclaims and concise arguments.

An obvious objection to this recommendation is that it may be very difficult to satisfy in practice. I respond to this in the conclusion section, but here I wish to consider a more philosophically motivated objection, which is that science itself does not support deductive theories.

This stems from a controversial topic in the philosophy of science concerning "provisos" (sometime spelled "provisoes") or *ceteris paribus* clauses[2] in statements of scientific laws. For example, we might formulate the law of thermal expansion as follows: "the change in length of a metal bar is directly proportional

---

[2] This Latin phrase is usually translated "other things being equal".

to the change in temperature". But this is true only if the bar is not partially encased in some unyielding material, and only if no one is hammering the bar flat at one end, and .... This list of provisos is indefinite, so the simple statement of the law (or even a statement with some finite set of provisos) can only be inductively true. Hempel [21] asserts there is a real issue here concerning the way we understand scientific theories and, importantly from the assurance perspective, the way we attempt to confirm or refute them. Others disagree: in an otherwise sympathetic account of Hempel's work in this area, his student Suppe describes "where Hempel went wrong" [22, pp. 203, 204], and Earman and colleagues outright reject it [23].

Rendered in terms of assurance cases, the issue is the following. During development of an assurance case argument, we may employ a reasoning step asserting that its claim follows from some conjunction of subclaims. The assertion may not be true in general, so we restrict it with additional subclaims representing necessary assumptions and provisos that are true (as other parts of the argument must show) in the context of this particular system. The "proviso problem" is then: how do we know that we have not overlooked some necessary assumption or proviso? The answer is that the justification for the step should explain this, and we may provide evidential subclaims that cite the methods used (e.g., to show that we have not overlooked a hazard, we will supply evidence about the method of hazard analysis employed) but, lacking omniscience, we cannot be totally certain that some "unknown unknown" does not jeopardize the argument. It can then be argued that the only philosophically sound position is to regard the reasoning step (indeed, all reasoning steps) as inductive.

A counterargument would ask: what does this gain us? If we cannot estimate the size and significance of these "unknown unknowns" (and, indeed, we cannot do this, as explained earlier) then we may gain philosophical purity but no actionable insight, and we lose the benefits of deductive reasoning. Furthermore, there is also a loss in our psychological position: with an inductive reasoning step we are saying "this claim holds under these provisos, but there may be others", whereas for a deductive step we are saying "this claim holds under these provisos, and this is where we make our stand". This alerts our reviewers and raises the stakes on our justification. There may, of course, be unknown provisos, but we are confident that we have identified the only ones that matter. Philosophically, this may be hubris, but for assurance it is the kind of explicit and definitive assertion that encourages effective review.

2.3   Complete Arguments

We have considered evidential steps and reasoning steps separately, now we need to put them together. For arguments that are in simple form, this is easy because they are composed of just those two kinds of steps. The interpretation of a complete argument in simple form is a deductive logical interpretation in which evidentially-supported subclaims are treated as premises that are interpreted epistemically and accepted as true when their weight of evidence is considered to have crossed some threshold (which may assessed either by probabilistic modeling and analysis, or by informal judgment grounded on such modeling). Observe that although the reasoning steps are deductive, the evidential steps admit doubt, and hence the overall argument is inductive. Also note that although confidence claims are not to be used in reasoning steps (because they have no logical interpretation), they may be used in evidential steps and are readily incorporated into their probabilistic interpretation, as illustrated in Sect. 2.1.

We will say that an argument in simple form is *sound* if its reasoning steps are deductively valid and its evidential steps all cross the thresholds established for their claims to be accepted. It seems plausible that the "weight" established for those thresholds could be used to assess the *strength* of a sound argument. We will consider that topic shortly but first turn to arguments that are not in simple form.

2.4   Assumptions, and Arguments Not in Simple Form

I propose two approaches for arguments that are not in simple form. One is to convert them to simple form by the transformation suggested in Fig. 1. The other is to attempt a direct interpretation by treating subclaims appearing in "mixed" argument steps as assumptions. The generic inductive step (1) could be augmented by an assumption a to make it deductive and then be written as

$$p_1 \text{ AND } p_2 \text{ AND } \cdots \text{ AND } p_n \text{ IMPLIES } c, \text{ ASSUMING } a.$$

Assumptions are generally treated as additional premises, so this is interpreted as

$$a \text{ IMPLIES } (p_1 \text{ AND } p_2 \text{ AND } \cdots \text{ AND } p_n \text{ IMPLIES } c),$$

which simplifies under the laws of logic to

$$a \text{ AND } p_1 \text{ AND } p_2 \text{ AND } \cdots \text{ AND } p_n \text{ IMPLIES } c.$$

Thus, we see that any subclaim in a reasoning step can be interpreted as an assumption. This observation might seem trivial, but there are cases where it is useful. In particular, p_1 might only "make sense" if a is true. For example, in some notations a term such as $\frac{y}{x}$ does not "make sense" unless $x \neq 0$, so we should not even inspect the expression corresponding to p_1 until we know the assumption a is true. Since all the subclaims in an assurance case argument must be true if we are to conclude its top claim, we could allow each subclaim to be interpreted under the assumption that all other subclaims are true. However, it can require additional analysis to ensure there is no circularity in this reasoning, so a useful compromise is to impose a left-to-right reading (this strategy is employed effectively in some predicatively-subtyped languages, such as PVS [24]). In concrete terms, this means that each subclaim or item of evidence named in an argument step is evaluated assuming the truth of all subclaims appearing earlier in the argument.

The challenge of "mixed" argument steps such as AS_1 on the left of Fig. 1 is whether to interpret them epistemically, like evidential steps, or logically like reasoning steps. Now that we understand assumptions, my suggestion is that the combination of evidence appearing in the step should be interpreted epistemically, under an assumption comprised of the conjunction of subclaims appearing in the same step. Thus, for example, AS_1 on the left of Fig. 1 would be interpreted as an evidential step in which the evidence E_3 is evaluated under assumptions represented by the subclaim SC_1. This is effectively the same interpretation as for the transformed argument on the right of Fig. 1: there, the evidential step ES_n can use SC_1 as an assumption when interpreting E_3 since it appears earlier in the argument.

2.5 Graduated Assurance

DO-178C recognizes that aircraft software deployed in different functions may pose different levels of risk and it accepts reduced assurance for that which poses less risk. For example, the number of assurance objectives is reduced from 71 for Level A software (that with the potential for a "catastrophic" failure condition) to 69 for Level B, 62 for Level C, and 26 for Level D, and the number of objectives that must be performed "with independence" is likewise reduced from 33 to 21, 8, and 5, respectively. This is an example of *graduated assurance*, and it is found in similar form in many standards and guidelines.

On the one hand, this seems very reasonable, but on the other it poses a serious challenge to the idea that an assurance case argument should be sound. We may suppose that the Level A argument is sound, but how can the lower levels be so when they deliberately remove or weaken some of the supporting evidence and, presumably, the implicit argument associated with them?

There seem to be three ways in which an explicit assurance case argument can be weakened in support of graduated assurance. First, we could simply eliminate certain subclaims or, equivalently, provide no evidence for them. This surely renders the full argument unsound: any deductive reasoning step that employs the eliminated or trivialized subclaim cannot remain deductively sound with one of its premises removed (unless there is redundancy among them, in which case the original argument should be simplified). This approach reduces a deductively sound argument to one that is, at best, inductive, and possibly unsound; consequently, I deprecate this approach.

Second, we could eliminate or weaken some of the evidence supplied in support of selected subclaims. This is equivalent to lowering the thresholds on what constitutes a "settled fact" and does not threaten the soundness of the argument, but does seem to reduce its strength. Intuitively, the *strength* of a sound assurance case argument is a measure of its evidential support—that is the threshold weights that determine settled facts.

It could be argued that there is surely no difference between lowering the threshold for evidential support of a given claim and using that same evidence to provide strong support for a weaker claim, which could then provide only inductive support to those reasoning steps that use it—yet I approve the former and deprecate the latter. My justification is pragmatic: we have a rigorous procedure for evaluating deductive reasoning steps but not inductive ones.

Third, we could completely restructure the argument. Holloway's reconstruction of the argument implicit in DO-178C [4] suggests that this underpins the changes from Level C to Level D of DO-178C. The Low Level Requirements (LLR) and all their attendant objectives are eliminated in going from Level C to Level D; the overall strategy of the argument, based on showing correctness of the executable code with respect to the system requirements, remains the same, but now employs a single step from high level requirements to source code without the LLR to provide an intermediate bridge. This also seems a valid form of weakening. Notice that evidence that is common to Levels C and D could use the same thresholds so the strength of their common parts will be the same; yet it seems clear that Level C is a stronger argument than Level D. Thus, it appears there is more to the strength—or, more accurately, the persuasiveness—of an argument than deductive validity and evidential thresholds. Level C is a bigger argument and has more evidence than Level D, but this is a crude measure; it seems more credible that the persuasive strength of an argument is related to its ability to withstand challenges, which is an idea we will return to briefly in the conclusion.

3   Comparisons with Other Approaches

Other approaches proposed for the interpretation of assurance arguments fall into three classes; I briefly consider each in turn.

The first are probabilistic interpretations: for example, Xeng *et al.* [25], who apply Dempster-Shafer analysis to complete assurance cases, and Denney *et al.* [26], who use BBNs in a similar way. These methods are insensitive to the logical content of reasoning steps so, in effect, they flatten the argument by removing subclaims so that only evidence is left. But this loses the essence of argument-based assurance and takes us back to approaches such as DO-178C, where all we have is a collection of evidence. In contrast, I chose to separate the testing leg of Fig. 3 from the verification leg of the multi-legged case in Fig. 2 because a logical argument for their joint use could consider and mitigate their strengths and weaknesses in a way that a collection of evidence cannot.

A second approach is that of Toulmin [6]. Papers on assurance cases frequently cite Toulmin but do not spell out how his methods should be used. Toulmin's approach is radical and challenges some of the fundamentals of logic: namely, that the validity of our reasoning can be assessed separately from the truth of our premises. This may have some appeal in highly contested areas such as religion or ethics where participants might disagree on basic principles, but seems less appropriate for assurance cases where disagreements concern reasoning, evidence, and the interpretation of these.

A third class of approaches builds on methods from the field of argumentation and agreement [27], including defeasible reasoning and argumentation structures. I am entirely sympathetic to the use of these ideas, in particular the notion of a "defeater", to evaluate the quality or persuasiveness of an assurance case, but I do not think they offer new insight into the basic interpretation of a case.

4   Conclusions

I have proposed a two-part process for interpretation of assurance case arguments: evidential or leaf steps are interpreted epistemically by methods that can be grounded in probability, even if performed informally, while interior or reasoning steps are interpreted in deductive logic. The overall argument is inductive (i.e., admits uncertainty) but all uncertainty is located in assessment of evidence.

A natural objection to this proposal is that it may be very difficult to construct strictly deductive reasoning steps, and still harder to assemble these into a complete argument. One response is that this may accurately reflect the true difficulty of our enterprise—assurance is hard—so that simplifications achieved through inductive reasoning will be illusory. Note that, while the top claim and some of the evidential subclaims may be fixed (by regulation and by available evidence, respectively), we are free to choose the others. Just as formulation of good lemmas can simplify a mathematical proof, so skillful formulation of subclaims may make a deductive assurance argument tractable. I speculate that software assurance cases, where the top claim is correctness, may lend themselves more readily to deductive arguments than other cases, where the top claim is a system property such as safety. Experiments are needed to evaluate these claims.

This approach is simple, even obvious, but I have not seen it explicitly described elsewhere. Haley and colleagues [28] describe a method where reasoning steps are evaluated in formal logic (they call this the *Outer Argument*) while evidential steps are evaluated informally (they call this the *Inner Argument*). They acknowledge that the inner argument concerns "claims about the world" [28, p. 140] but use Toulmin's approach rather than explicitly epistemic methods.

The proposed interpretation provides criteria for assessing the soundness of an assurance case argument and the strength of its evidential support. But it does not provide a means to evaluate whether the total argument is adequately convincing and persuasive. For that, I believe one must assess how well the argument resists challenges; methods from dialectics and defeasible reasoning may be suitable for this purpose, which I plan to address in a subsequent paper.

Acknowledgments. This work was partially funded by NASA under contract NNL13AA00B to The Boeing Company, and by SRI International. I benefited from many suggestions by Michael Holloway, our NASA contract monitor, but the content is solely the responsibility of the author and does not necessarily represent the official views of NASA. Thoughtful comments by the anonymous reviewers improved the presentation of this material.

References

1. RTCA, Washington, DC: DO-178C: Software Considerations in Airborne Systems and Equipment Certification (2011)
2. Society of Automotive Engineers: Aerospace Recommended Practice (ARP) 4754A: Certification Considerations for Highly-Integrated or Complex Aircraft Systems. Also issued as EUROCAE ED-79 (2010)

3. Society of Automotive Engineers: Aerospace Recommended Practice (ARP) 4761: Guidelines and Methods for Conducting the Safety Assessment Process on Civil Airborne Systems and Equipment (1996)

4. Holloway, C.M.: Explicate '78: discovering the implicit assurance case in DO-178C. In: Parsons, M., Anderson, T. (eds.) Engineering Systems for Safety. Proceedings of 23rd Safety-critical Systems Symposium, Bristol, UK, pp. 205–225 (2015)

5. Rushby, J., Xu, X., Rangarajan, M., Weaver, T.L.: Understanding and evaluating assurance cases. NASA Contractor Report NASA/CR-2015-218802, NASA Langley Research Center (2015)

6. Toulmin, S.E.: The Uses of Argument. Cambridge University Press, Cambridge (2003). Updated edition (the original is dated 1958)

7. Adams, E.W.: A Primer of Probability Logic. Center for the Study of Language and Information (CSLI), Stanford University (1998)

8. Good, I.J.: Probability and the Weighing of Evidence. Charles Griffin, London (1950)

9. Good, I.J.: Weight of evidence: a brief survey. In: Bernardo, J., et al. (eds.) Bayesian Statistics 2: Proceedings of the Second Valencia International Meeting, Valencia, Spain, pp. 249–270 (1983)

10. Bovens, L., Hartmann, S.: Bayesian Epistemology. Oxford University Press, Oxford (2003)

11. Earman, J.: Bayes or Bust? A Critical Examination of Bayesian Confirmation Theory. MIT Press, Cambridge (1992)

12. Dawid, A.P.: Bayes's theorem and weighing evidence by juries. In: Swinburne, R. (ed.) Bayes's Theorem. Proceedings of the British Academy, pp. 71–90 (2002)

13. Jeffrey, R.: Subjective Probability: The Real Thing. Cambridge University Press, Cambridge (2004)

14. Fitelson, B.: Studies in Bayesian Confirmation Theory. Ph.D. thesis, Department of Philosophy, University of Wisconsin, Madison (2001)

15. Tentori, K., Crupi, V., Bonini, N., Osherson, D.: Comparison of confirmation measures. Cognition 103, 107–119 (2007)

16. Joyce, J.M.: On the plurality of probabilist measures of evidential relevance. In: Bayesian Epistemology Workshop of the 26th International Wittgenstein Symposium, Kirchberg, Austria (2003)

17. Gardner-Medwin, T.: What probability should a jury address? Significance 2, 9–12 (2005)

18. Littlewood, B., Wright, D.: The use of multi-legged arguments to increase confidence in safety claims for software-based systems: a study based on a BBN analysis of an idealised example. IEEE Trans. Softw. Eng. 33, 347–365 (2007)

19. HUGIN Expert: Hugin home page. http://www.hugin.com/. Accessed 2015

20. Hawkins, R., Kelly, T., Knight, J., Graydon, P.: A new approach to creating clear safety arguments. In: Dale, C., Anderson, T. (eds.) Advances in System Safety: Proceedings of 19th Safety-Critical Systems Symposium, pp. 3–23. Springer, London (2011)

21. Hempel, C.G.: Provisoes: a problem concerning the inferential function of scientific theories. Erkenntnis 28, 147–164 (1988)

22. Suppe, F.: Hempel and the problem of provisos. In: Fetzer, J.H. (ed.) Science, Explanation, and Rationality: Aspects of the Philosophy of Carl G. Hempel, pp. 186–213. Oxford University Press, Oxford (2000)

23. Earman, J., Roberts, J., Smith, S.: *Ceteris paribus* lost. Erkenntnis 57, 281–301 (2002)

24. Rushby, J., Owre, S., Shankar, N.: Subtypes for specifications: predicate subtyping in PVS. IEEE Trans. Softw. Eng. **24**, 709–720 (1998)

25. Zeng, F., Lu, M., Zhong, D.: Using D-S evidence theory to evaluation of confidence in safety case. J. Theor. Appl. Inform. Technol. **47**, 184–189 (2013)

26. Denney, E., Pai, G., Habli, I.: Towards measurement of confidence in safety cases. In: Fifth International Symposium on Empirical Software Engineering and Measurement (ESEM), Banff, Canada, pp. 380–383. IEEE Computer Society (2011)

27. Ossowski, S. (ed.): Agreement Technologies. Law, Governance and Technology Series, vol. 8. Springer, Heidelberg (2013)

28. Haley, C.B., Laney, R., Moffett, J.D., Nuseibeh, B.: Security requirements engineering: a framework for representation and analysis. IEEE Trans. Softw. Eng. **34**, 133–153 (2008)

Learning Argument Acceptability from Abstract Argumentation Frameworks

Hiroyuki Kido[(⊠)]

The University of Tokyo, Hongo 7-3-1, Bunkyo-ku, Tokyo 113-8656, Japan
kido@sys.t.u-tokyo.ac.jp

Abstract. This paper introduces argument-based decision-tree for learning acceptability of arguments. We specifically examine an attack relation existing between arguments, without referring to any contents, either sentences or words, existing in individual arguments. This idea is formalized using decision trees in which their attributes are instantiated by complete, preferred, stable and grounded extensions, respectively, defined by acceptability semantics. This study extracted 38 arguments and 4 utterers from an argument about euthanasia that actually took place on a social media site. Also, 21 training data were collected by asking them to express their attitudes either for or against the individual 38 arguments. By stratifying audiences in accordance with consistency with utterers, leave-two-out cross validation yielded results with a 0.73 AUC value, on average. This fact demonstrates that our argument-based decision-tree learning is expected to be fairly useful for agents who have a definite position on an issue of argument.

Keywords: Acceptability learning · Decision trees · Argumentation

1 Introduction

Many social media sites encourage people to argue about various issues and to explore their valid beliefs. In general, such social media sites have numerous arguments and utterers putting forward their arguments, and audiences reading their arguments. For various reasons such as the scale of issues used for the arguments, time constraints, and lack of expertise, it is difficult for audiences to read every argument and judge their acceptance of opposing arguments whether they agree or disagree with the opinions stated for the argument. Therefore, it is desirable to predict an audience's attitude either for or against every argument based on the information of attitudes related to a few arguments.

Addressing this issue can yield benefits other than those related to social media analysis. A potential application is *argument-based systems assurance.* Growing interest exists in the use of an evidence-based argument that is often called a safety case, assurance case, or a dependability case in safety engineering. Such cases, however, tend to be larger as target systems become increasingly complicated. Our research is expected to help stakeholders to understand cases

© Springer International Publishing AG 2017
M. Otake et al. (Eds.): JSAI-isAI 2015 Workshops, LNAI 10091, pp. 348–362, 2017.
DOI: 10.1007/978-3-319-50953-2_24

and to predict points of conflict and agreements about safety properties of target systems. Another potential application is *argument-based recommendation.* Amazon customer reviews, Facebook comments, and Yahoo comments enable users to post their arguments for or against goods, experiences, and news articles, and also to score the posted arguments in terms of various viewpoints such as approval, fun, and usefulness. If user sentiment related to unseen arguments can be predicted from a certain number of known arguments and their scores, then it is possible to recommend favorable goods, experiences, and news articles from various perspectives.

When regarding argumentation abstractly, we can extract a directed graph in which each node represents an argument and in which each edge represents an attack relation between arguments. Moreover, we can read a certain kind of logic from such a directed graph. For example, it is possible to infer that an audience a agrees with an argument x, i.e., a node in the directed graph, if, although there is an argument y attacking x, y is attacked by another argument z the audience a agrees with. In other words, x is defended by z and a agrees with z.

Based on this observation, we ask *"Is it possible to predict audience's attitude either for or against individual arguments, merely by particularly addressing the attack relation existing between arguments, without referring to any actual contents, i.e., sentences or words existing in arguments?"* We think *yes* and this logical analysis of directed graphs gives a new insight into machine learning. However, little work emphasizes or applies the idea to machine learning, probably because of the distance of logical analysis of argumentation, e.g., the theory of abstract argumentation [1] and machine learning. In fact, sentiment classification techniques [2–7] and feature based opinion mining techniques [8–11] treat information about sentences and words, as existing in each argument rather than information about attack relations existing between arguments.

The key component of our approach is to use decision trees whose attributes are instantiated by complete, preferred, stable, and grounded extensions, defined by the acceptability semantics [1]. Argument data used for our experiment consist of audience expressions either for or against 38 arguments, for all 21 audiences. Our experiment demonstrates that the average area under the ROC curve (AUC) performances on the data in leave-two-out cross validation is measured as 0.61. However, stratification on audiences in accordance with consistency with utterers results in 0.73, on average. These results illustrate that our argument-based decision-tree learning can be expected to be fairly useful for audiences who have a definite position on an issue of the argument.

The contributions of this paper are the following. From machine learning perspectives, this paper presents a proposal of the new idea of decision-tree learning characterized by the theory of abstract argumentation. It characteristically learns from argumentation structure, an attack relation existing between arguments, without referring to sentences or words existing in an argument. From the perspective of the research on argumentation in artificial intelligence, Dung devises the theory of abstract argumentation as a descriptive theory that reformulates consequence notions of nonmonotonic reasoning of various kinds.

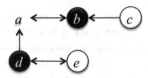

Fig. 1. A vertex labeled directed graph representing an argument graph and audience's attitude either for or against each argument.

This paper contributes to a demonstration that Dung's theory is also useful to interpret agents' actual attitudes related to argument consequences.

2   Motivating Example

This section presents a motivating example of difficulties related to argument-based classification. We consider the following argument on euthanasia (painless death) that actually took place on a social media site: SYNCLON [12].

- *Argument a*: Law should give doctors the right to apply euthanasia because medical treatments should respect a patient's wishes.
- *Argument b*: I disagree with the point that the one who applies euthanasia should be a doctor. Doctors should always consider ways to cure diseases.
- *Argument c*: If a doctor's job is only to consider ways to cure diseases, then doctors can do nothing for patients with untreatable diseases. I think that a medical treatment must face death more. Only doctors can apply euthanasia appropriately because they can assess patients' physical and mental state accurately.
- *Argument d*: The patient's wishes are inadequate to allow doctors to apply euthanasia because no one can have the right to commit murder.
- *Argument e*: We must assign importance to a patient's wishes because costs associated with life extension are severe for the patient and the patient's family.

The directed graph shown in Fig. 1 is an argument graph that structures conflicts and contradictions among the arguments presented above. Each node represents an argument. Each edge from node x to y represents that argument x attacks argument y. A unidirectional attack represents that one argument attacks a premise of the other. A bidirectional attack represents that two arguments attack conclusions or that two arguments are to be regarded as incompatible.

Now, let us assume that an audience agent wants to judge an attitude as either for or against the individual arguments. However, the agent can observe only some arguments for various reasons, e.g., a time constraint or scale of arguments. For example, the agent can observe all arguments except a and then the agent agrees with arguments c and e, but disagrees with remaining arguments b and d. In Fig. 1, the agreed-to arguments are depicted as white circular nodes. The arguments entailing disagreement are depicted as black circular nodes.

Given the vertex labeled directed graph, we ask whether the agent agrees or disagrees with the remaining argument a. Our hypothesis tested in this paper is that it is possible to predict the agent attitude by analyzing an attack relation between arguments, without referring to the actual contents of arguments, e.g., words or inferences. According to our hypothesis, argument a would be classified as an agreed-to argument because it is defended by agreed-to arguments c and e. More precisely, although a is attacked by arguments b and d, there are agreed-to arguments c and e that attack b and d, respectively. As shown in Sect. 4, this idea is formally captured by decision-tree learning characterized by the theory of abstract argumentation [1] rooted in the study of nonmonotonic logic. The theory of abstract argumentation enables us to address a more complicated situation: What happens when the agent agrees with argument b or disagrees with argument c?

3   Preliminaries

Dung's theory of abstract argumentation [1] defines acceptability semantics along with the manner of fixed point semantics. The basic formal notions, with some terminological changes, are defined as follows.

Definition 1 [1]. *An* argumentation framework *is defined as a pair $AF = \langle Args, attacks \rangle$ where Args is a set of arguments, and attacks is a binary relation on Args, i.e. attacks $\subseteq Args \times Args$.*

- $S \subseteq Args$ is conflict-free *if and only if there is no argument $a, b \in S$ such that $(a, b) \in attacks$, i.e., a attacks b.*
- $a \in Args$ is acceptable *with respect to $S \subseteq Args$ if and only if, for all arguments $b \in Args$, if b attacks a then b is attacked by an argument in S.*
- $S \subseteq Args$ is admissible *if and only if S is conflict-free and each argument in S is acceptable with respect to S.*
- $S \subseteq Args$ is a preferred extension *if and only if S is a maximal (with respect to set inclusion) admissible set.*
- $S \subseteq Args$ is a stable extension *if and only if S is conflict-free and every argument not in S is attacked by an argument in S.*
- *A characteristic function $F_{AF} : Pow(Args) \to Pow(Args)$ is defined as follows: $F_{AF}(S) = \{a \in Args \mid a$ is acceptable with respect to $S\}$.*
- $S \subseteq Args$ is the grounded extension *if and only if S is the least fixed point of F_{AF}.*
- $S \subseteq Args$ is a complete extension *if and only if S is admissible and each argument, which is acceptable with respect to S, belongs to S.*

Example 1 [13]. Let $AF = \langle Args, attacks \rangle$ be an argumentation framework where $Args = \{a, b, c, d, e\}$ and $attacks = \{(a, b), (c, b), (c, d), (d, c), (d, e), (e, e)\}$. Figure 2 portrays a directed graph representing AF. AF has the following admissible sets and stable, preferred, complete and grounded extensions.

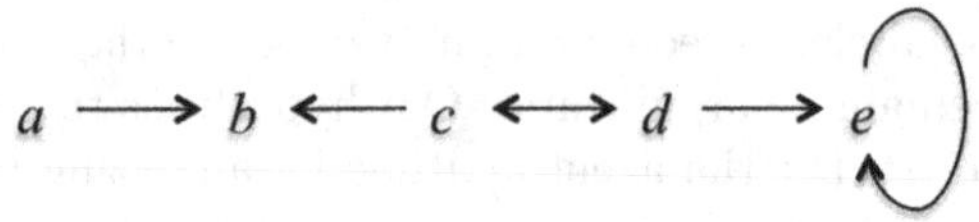

Fig. 2. Directed graph representing argumentation framework $AF = \langle\{a, b, c, d, e\},$ $\{(a, b), (c, b), (c, d), (d, c), (d, e), (e, e)\}\rangle$ [13].

Admissible sets: $\emptyset$, $\{a\}$, $\{c\}$, $\{d\}$, $\{a, c\}$, $\{a, d\}$
Preferred extensions: $\{a, c\}$, $\{a, d\}$
Stable extension: $\{a, d\}$
Grounded extension: $\{a\}$
Complete extensions: $\{a, c\}$, $\{a, d\}$, $\{a\}$

Each extension corresponds to a different attitude for argumentation consequences. For example, rational and skeptical agents would believe arguments in the grounded extension. On the other hand, rational and credulous agents would believe arguments in a preferred extension.

4   Argument-Based Decision-Tree Learning

This section aims to introduce argumentation frameworks and acceptability semantics into machine learning, particularly classification, and define a notion of argument-based learning. Argument-based classification assumes an arbitrary but fixed argumentation framework $AF = \langle Args, attacks \rangle$ where $Args$ is a set of arguments and $attacks$ is an attack relation between arguments in $Args$. Moreover, we assume an audience defined as a partial function $aud : Args \rightarrow \{agree, disagree\}$ that maps an argument to its acceptability status representing the attitude whether the audience agrees with the argument or not: an audience is a vertex label assigning a label to each argument. Given an argumentation framework $\langle Args, attacks \rangle$ and an audience $aud : Args \rightarrow \{agree, disagree\}$, the problem of argument-based classification is to assign a function or rule to classify arguments in $Args$ into an appropriate class in $\{agree, disagree\}$ based on aud.

We are particularly interested in decision-tree learning because it is the most well-used classification technique. It aims at generating a rule corresponding to the propositional logic. Our idea is to use extensions of an argumentation framework as instances of decision tree attributes. The idea derives from the observation that each extension gives a different standpoint on a set of arguments a rational agent believes. We infer that an agent's actual attitude of agreeing or disagreeing with individual arguments can be explained well using a combination of rational beliefs. More precisely, the problem of argument-based decision-tree learning is to find a rule for which conclusion is an acceptability status, i.e., a label, of a target argument. Each premise is a membership relation between the target argument and an extension. Complete, preferred, stable, and

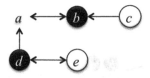

Fig. 3. Visual representation of the argumentation framework and an audience.

Table 1. Examples obtained from Fig. 3.

Example	Attributes		Goal
Argument	PE_1	PE_2	Attitude
b	no	no	disagree
c	yes	yes	disagree
d	no	yes	disagree
e	yes	no	agree

grounded extensions are defined on an argumentation framework independently from a definition of an audience. Therefore, it is possible to imagine at least four argument-based decision trees with attributes instantiated by complete extensions, preferred extensions, stable extensions, and grounded extension of a given argumentation framework.

Example 2 For the argumentation framework and the audience shown in Fig. 3, the argumentation framework is defined as $AF = \langle\{a, b, c, d, e\}, \{(a, b), (b, a), (c, b), (d, a), (d, e), (e, d)\}\rangle$. Here, AF has the following stable extensions, preferred extensions, complete extensions, and grounded extensions.

– Stable extensions: $\{a, c, e\}$, $\{c, d\}$
– Preferred extensions: $\{a, c, e\}$, $\{c, d\}$, denoted respectively by PE_1 and PE_2.
– Complete extensions: $\{c\}$, $\{a, c, e\}$, $\{c, d\}$
– Grounded extension: $\{c\}$

The audience is defined as $aud(e) = agree$ and $aud(b) = aud(c) = aud(d) = disagree$. Table 1 portrays examples of labeled arguments, where the preferred extensions PE_1 and PE_2 are attributes, and the corresponding labels are goals. Each argument has the value *yes* or *no* depending on whether it is a member of each preferred extension, or not.

Figure 4 portrays a decision tree learned from the examples shown in Table 1. The tree shows a rule stating that the audience agrees with an argument if it is a member of PE_1, but is not a member of PE_2. The rule classifies all labeled arguments correctly. Moreover, it classifies the remaining argument a into the class *agree*.

5   Performance Evaluation

5.1   Data Used in Experiment

In this section, we evaluate the learning performance of argument-based decision-tree learning. We extracted 38 arguments related to euthanasia put forward by four utterers. Minor revisions were made to put utterances with the same contents together and get rid of irrelevant utterances. The following are examples of extracted arguments.

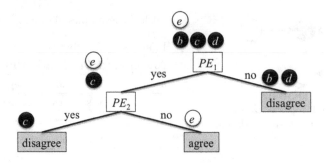

Fig. 4. Decision tree obtained from the examples presented in Table 1.

- Argument a: *Euthanasia should be allowed by law because medical treatments by doctors should respect a patient's wishes.*
- Argument b: *Euthanasia should not be allowed by law. Assume that one who applies euthanasia is a doctor. I doubt a doctor's right to commit a murder.*
- Argument c: *I agree with you that euthanasia should be allowed by law, but disagree with the point that one who applies it is a doctor. Doctors should always consider ways to cure diseases.*
- Argument d: *If a doctor's job is only to consider the way of curing diseases, then doctors can do nothing for patients with untreatable diseases. I think that a medical treatment must face death more. Only doctors can apply euthanasia appropriately because they can assess patients' physical and mental states accurately.*

We defined an attack relation based on the idea that two arguments attack each other if their conclusions mutually conflict or are regarded as incompatible, and one argument (unidirectionally) attacks another if the conclusion of the first argument conflicts with a premise of the second. In the exemplary arguments presented above, we defined arguments a and b as attacking each other, argument c attacks a, and argument d attacks c.

We asked 21 examinees to look at the argumentation framework obtained using the preceding phases and to choose their attitudes either for or against each of 38 arguments. As a result, we got 21 different vertex-labeled directed graphs, i.e., AF and aud_i, for $1 \leq i \leq 21$. Audiences did not know utterers of individual arguments. Figure 5 presents an argumentation framework with some audience.

5.2   ROC Curve Analysis

We used the argumentation system Answer Set Programming Argumentation Reasoning Tool [14] (ASPARTIX) to calculate all complete, preferred, stable, and grounded extensions of abstract argumentation frameworks. It produced 1449 complete extensions, 91 preferred and the same stable extensions, and one grounded extension of AF shown in Fig. 5.

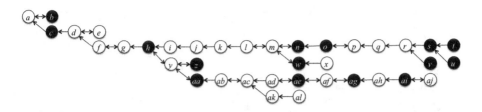

Fig. 5. Visual representation of an argumentation framework with a certain audience.

Decision-tree learning is conducted using machine learning software: Waikato Environment for Knowledge Analysis [15] (WEKA). The classifier we use is J48, generating a pruned or unpruned C4.5 decision tree [16] for which the confidence factor is set to 0.25 and the minimum number of objects is set to 5. The area under the ROC curve, i.e., AUC, obtained by leave-two-out cross validation is used to evaluate learning performance. We follow the common AUC interpretation: excellent (0.9–1.0), good (0.8–0.9), fair (0.7–0.8), poor (0.6–0.7), and failed (0.5–0.6).

Figure 6 presents every examinee's AUC scores with respect to complete, preferred (and stable), and grounded extensions where examinees are arranged in ascending order in terms of AUC scores of the complete extensions. Means of AUC scores of complete, preferred and grounded extensions are 0.61, 0.55, and 0.50, respectively. These results show that decision-tree learning characterized by preferred and grounded extensions is failed and complete extensions is poor, on average. The learning performance of the grounded extension is the same level with random choice because decision trees based on the grounded extension have only one attribute. Therefore, we have no room for improvement for the grounded extension.

However, Fig. 7 presents histograms where the x axis is AUCs based on the complete and preferred extensions, and the y axis is the number of audiences. Bimodality of the histogram implies that stratification of audiences is possible. In fact, eight audiences higher than 0.7 AUC value exist in the case of complete extensions. Six such audiences exist in the case of preferred extensions.

6   Audience Stratification by Utterers

6.1   Stratification Criteria

This section responds to the question of what type of audience is most affected by argument-based decision-tree learning. Our hypothesis is that *audiences with a definite position on a topic of an argument show high performance.* How can we know whether audiences have a definite position or not? We think that it

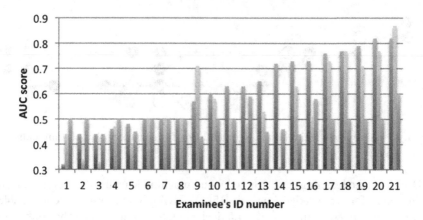

Fig. 6. AUC with respect to extensions of different kinds, for all audiences.

is difficult to measure the related phenomenon directly because of a lack of an objective criterion. Our idea here is that because utterers participating in this argument mutually conflict, it would be safe to infer that if audiences tend to have a consistent position with a particular utterer, then they tend to have a definite position on a topic of an argument. Fortunately, it is possible to assign an objective criterion to ascertain the degree to which utterers have a definite position on utterers. In this subsection, we define accuracy, precision, recall, and the F-measure to evaluate audience consistency with respect to a particular utterer.

We assume an utterer defined as a function $utt : Args \rightarrow \{yes, no\}$ that maps an argument to the fact of whether the utterer put forward the argument or not. The following formulae define true positive, true negative, false positive, and false negative where aud and utt are abbreviated respectively to a and u.

$$tp(a, u) = |\{x \in Args | a(x) = agree, u(x) = yes\}|$$
$$tn(a, u) = |\{x \in Args | a(x) = disagree, u(x) = no\}|$$
$$fp(a, u) = |\{x \in Args | a(x) = agree, u(x) = no\}|$$
$$fn(a, u) = |\{x \in Args | a(x) = disagree, u(x) = yes\}|$$

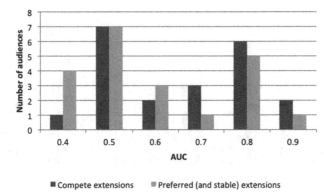

Fig. 7. Bimodal histogram where the x axis shows values of AUC and the y axis shows the number of audiences.

As usual, accuracy, precision, recall, and F-measure are defined as follows.

$$accuracy(a, u) = \frac{tp(a, u) + tn(a, u)}{tp(a, u) + fn(a, u) + fp(a, u) + tn(a, u)}$$

$$precision(a, u) = \frac{tp(a, u)}{tp(a, u) + fp(a, u)}$$

$$recall(a, u) = \frac{tp(a, u)}{tp(a, u) + fn(a, u)}$$

$$F\text{-}measure(a, u) = 2 \cdot \frac{recall(a, u) \cdot precision(a, u)}{precision(a, u) + recall(a, u)}$$

These criteria are used to evaluate how much an audience shares the same view with an utterer. We are interested in an utterer who has the most consistent position with the audience. Therefore, we refer to i'th utterer by u_i and calculate the maximal values from all utterers: $accuracy(a) = \max_{u_i}\{accuracy(a, u_i)\}$, $precision(a) = \max_{u_i}\{precision(a, u_i)\}$, $recall(a) = \max_{u_i}\{recall(a, u_i)\}$, F-$measure(a) = \max_{u_i}\{F\text{-}measure(a, u_i)\}$.

6.2 Audience Stratification

We use the accuracy, precision, recall, and F-measure to sort out audiences who have consistent positions with particular utterers. Figure 8 shows every examinee's accuracy, precision, recall, and F-measure scores where examinee's number is the same as Fig. 6. It is apparent that most recalls are 1.00 because there is an utterer who puts forward only one argument and most audiences agree with it. In fact, four utterers, denoted by utt_1, utt_2, utt_3 and utt_4, were defined as follows in this experiment.

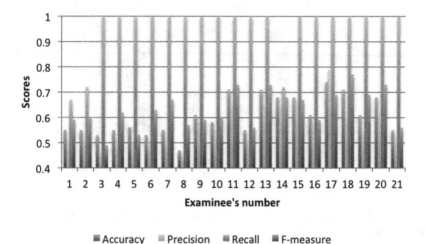

Fig. 8. Audience consistency with a particular utterer.

$$utt_1(x) = \begin{cases} yes & (x = a) \\ no & (otherwise) \end{cases}$$

$$utt_2(x) = \begin{cases} yes & (x = b) \\ no & (otherwise) \end{cases}$$

$$utt_3(x) = \begin{cases} yes & (x = c, f, h, j, l, n, p, r, t, u, w, z, aa, ac, ae, ag, ai, al) \\ no & (otherwise) \end{cases}$$

$$utt_4(x) = \begin{cases} yes & (x = d, e, g, i, k, m, o, q, s, v, x, y, ab, ad, af, ah, aj, ak) \\ no & (otherwise) \end{cases}$$

To use the best criterion for stratifying audiences, we calculate the correlation coefficient to elucidate which criteria have strong association with the performance of argument-based decision-tree learning. Results show that accuracy, precision, recall and F-measure respectively have correlation coefficient values of 0.66, 0.61, 0.37, and 0.55. These facts demonstrate that these criteria and learning performance have more or less mutual correlation.

Figures 9 and 10 show histograms of restricted audiences in terms of accuracy, precision, and F-measure. Recalls are excluded because they show low correlation with the learning performance. These figures show AUC of decision trees whose attributes are the complete and preferred (and stable) extensions. It is apparent that these criteria can reasonably sort out audiences consisting of the right peak in the bimodal histogram shown in Fig. 7. In Fig. 9, the means of AUC restricted to audiences with more than 0.65 accuracy, 0.6 precision, and 0.65 F-measure are 0.73, 0.73, and 0.71, respectively. In Fig. 10, they are all 0.61. These results illustrate that argument-based decision-tree learning with attributes instantiated by complete extensions is fair for the restricted audiences in terms of accuracy,

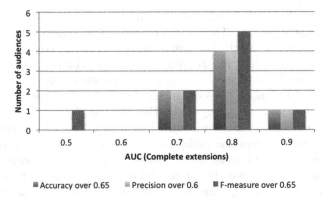

Fig. 9. Stratified histograms on complete extensions where every audience has accuracy greater than 0.65, precision greater than 0.6, and F-measure greater than 0.65.

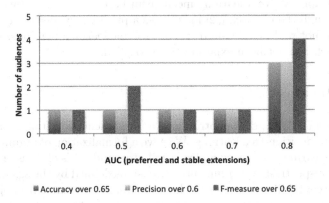

Fig. 10. Stratified histograms on the preferred (and stable) extensions where every audience has accuracy greater than 0.65, precision greater than 0.6, and F-measure greater than 0.65.

precision, and the F-measure. This fact means that argument-based decision-tree learning is expected to be fair for agents who have a definite position on a topic of an argument.

7   Related Work and Discussions

Guilt-by-association techniques [17] such as semi-supervised learning, random walk and belief propagation also infer classes of rest nodes for a given graph, few labeled nodes, and a network effect, i.e., homophily or heterophily. In contrast to our approach, however, such techniques are not aimed at learning a pattern of labeling from a given graph.

Some reports have presented argumentation dealing with machine learning tasks. Amgoud and Serrurier [18] propose an argument-based classification by which classes of examples are argued in an argumentation framework.

Palau and Moens [19] work on argument mining aimed at extracting a conclusion and premises from discourse. They give a classification technique to divide documents into premises or conclusions using training examples stored in the Araucaria corpus [20]. Mozina et al. [21] propose argument-based machine learning (ABML) and describe its benefits compared to classic machine learning techniques. Kido and Ohsawa [22] propose another argument-based classification technique that functions on an argumentation framework.

Their research [18,21] can be regarded as alternative approaches addressing traditional machine-learning tasks. On the other hand, our proposal can be regarded as an instance of decision-tree learning. Moreover, in contrast to their approaches, our argument-based approach is language-independent and domain-independent because our argument-based approach learns from an attack relation of an argumentation framework. The same is true for other research [19] because it applies traditional machine learning techniques to argument mining. They do not aim to give a learning mechanism based on argumentation theory. Finally, although the research [22] introduces a preliminary idea of acceptability learning, it is not well-founded in terms of machine learning methods. Moreover, their approach has not been experimentally verified.

8   Conclusions

As described in this paper, we introduced an argumentation framework and its semantics into decision-tree learning. We gave a formalization of argument-based decision-tree learning where examples, attributes, and goals of decision trees are instantiated respectively by arguments, extensions defined by the semantics, and audience's acceptability attitudes to the individual arguments. We applied it to an actual argument on euthanasia taken from the social media SYNCLON site. We defined accuracy, precision, recall, and the F-measure to evaluate and sort out audiences who have definite positions with respect to particular utterers. We demonstrated that the stratification of audiences based on these criteria increases the learning performance with complete extensions from 0.61 to 0.73 of AUC.

Acknowledgments. This work was supported by JSPS KAKENHI Grant Number 15KT0041. We would like to thank the manager of SYNCLON for the active participation in this work and valuable comments and suggestions.

References

1. Dung, P.M.: On the acceptability of arguments and its funedamental role in non-monotonic reasoning, logic programming, and n-person games. Artif. Intell. **77**, 321–357 (1995)
2. Dave, K., Lawrence, S., Pennock, D.M.: Mining the peanut gallery: opinion extraction and semantic classification of product reviews. In: Proceedings of the 12th International Conference on World Wide Web, pp. 519–528 (2003)

3. Turney, P.D.: Thumbs up or thumbs down?: semantic orientation applied to unsupervised classification of reviews. In: Proceedings of the 40th Annual Meeting on Association for Computational Linguistics, pp. 417–424 (2002)

4. Pang, B., Lee, L., Vaithyanathan, S.: Thumbs up? sentiment classification using machine learning techniques. In: Proceedings of the ACL-02 Conference on Empirical Methods in Natural Language Processing, vol. 10, pp. 79–86 (2002)

5. Kim, S.-M., Hovy, E.: Determining the sentiment of opinions. In: Proceedings of the 20th International Conference on Computational Linguistics (2004)

6. Wiebe, J., Riloff, E.: Creating subjective and objective sentence classifiers from unannotated texts. In: Gelbukh, A. (ed.) CICLing 2005. LNCS, vol. 3406, pp. 486–497. Springer, Heidelberg (2005). doi:10.1007/978-3-540-30586-6_53

7. Wilson, T., Wiebe, J., Hwa, R.: Just how mad are you? finding strong and weak opinion clauses. In: Proceedings of the 19th National Conference on Artifical Intelligence, pp. 761–767 (2004)

8. Liu, B., Hu, M., Cheng, J.: Opinion observer: analyzing and comparing opinions on the web. In: Proceedings of the 14th International Conference on World Wide Web, pp. 342–351 (2005)

9. Hu, M., Liu, B.: Mining and summarizing customer reviews. In: Proceedings of the 10th ACM SIGKDD International Conference on Knowledge Discovery and Data Mining, pp. 168–177 (2004)

10. Popescu, A.-M., Etzioni, O.: Extracting product features and opinions from reviews. In: Proceedings of the Conference on Human Language Technology and Empirical Methods in Natural Language Processing, pp. 339–346 (2005)

11. Ding, X., Liu, B., Yu, P.S.: A holistic lexicon-based approach to opinion mining. In: Proceedings of the International Conference on Web Search and Data Mining, pp. 231–240 (2008)

12. Synclon$_\beta^3$. http://synclon3.com/

13. Besnard, P., Doutre, S.: Checking the acceptability of a set of arguments. In: Proceedings of the 10th International Workshop on Nonmonotonic Reasoning (2004)

14. Egly, U., Gaggl, S.A., Woltran, S.: ASPARTIX: implementing argumentation frameworks using answer-set programming. In: Garcia de la Banda, M., Pontelli, E. (eds.) ICLP 2008. LNCS, vol. 5366, pp. 734–738. Springer, Heidelberg (2008). doi:10.1007/978-3-540-89982-2_67

15. Holmes, G., Donkin, A., Witten, I.H.: Weka: a machine learning workbench. In: Proceedings of the Second Australian and New Zealand Conference on Intelligent Information Systems, pp. 357–361 (1994)

16. Quinlan, J.R.: C4.5: Programs for Machine Learning. Morgan Kaufmann Publishers, San Francisco (1993)

17. Koutra, D., Ke, T.-Y., Kang, U., Chau, D.H.P., Pao, H.-K.K., Faloutsos, C.: Unifying guilt-by-association approaches: theorems and fast algorithms. In: Proceedings of the European Conference on Machine Learning and Knowledge Discovery in Databases, pp. 245–260 (2011)

18. Amgoud, L., Serrurier, M.: Agents that argue and explain classifications. Auton. Agents Multi-Agent Syst. **16**(2), 187–209 (2008)

19. Palau, R.M., Moens, M.-F.: Argumentation mining: the detection, classification and structure of arguments in text. In: Proceedings of the 12th International Conference on Artificial Intelligence and Law, pp. 98–107 (2009)

20. Reed, C., Rowe, G.: Araucaria: software for argument analysis, diagramming and representation. Int. J. Artif. Intell. Tools **13**(4), 961–979 (2004)
21. Moina, M., Abkar, J., Bratko, I.: Argument based machine learning. Artif. Intell. **171**(10–15), 922–937 (2007)
22. Kido, H., Ohsawa, Y.: Defensibility-based classification for argument mining. In: Proceedings of the 4th IEEE International Workshop on Data Mining in Networks, pp. 575–580 (2014)

HAT-MASH 2015

Healthy Aging Tech Mashup Service, Data and People

Ken Fukuda

AI Research Center, National Institute of Advanced Industrial Science
and Technology, Tokyo, Japan
ken.fukuda@aist.go.jp

1 The Workshop

The 1st International Workshop HAT-MASH 2015 (Healthy Aging Tech mashup service, data and people) was successfully held on the 16th and 17th of November, 2015 in Kanagawa, Japan as part of JSAI-isAI 2015, supported by the Society for Serviceology and JST. It was the first international workshop that brings people from healthy aging and elderly care technology, information technology and service engineering all together.

The main objective of this workshop is to provide a forum to discuss important research questions and practical challenges in healthy aging and elderly care support to promote transdisciplinary approaches. The workshop welcomes researchers, academicians as well as industrial professionals of different but relevant fields from all over the world to present their research results and development activities. The workshop will provide opportunities for the participants to exchange new ideas and experiences, to establish research or business networks and to find global partners for future collaboration.

This year, we featured two keynotes and one panel discussion session to help understand the issues and impacts of aging society. We would like to thank Mr. Jukka Lindberg, Client director, Older people, subscriber services, Hämeenlinna City, Finland, and Dr. Masahiro Kanno, CEO, Keiju Healthcare System for their valuable keynote and Ms. Seiko Adachi, Executive Head, Social Welfare Corporation Sinko Fukushikai for the lively discussion as a panelist.

2 Papers

In "Designing Intelligent Sleep Analysis Systems for Auto-mated Contextual Exploration on Personal Sleep-Tracking Data", Zilu Liang, Wanyu Liu, Bernd Ploderer, James Bailey, Lars Kulik, and Yuxuan Li discuss sleep tracking technologies. Although one can find many sleep tracking devices in the consumer market, the problem is they mainly report how well you slept but do not provide actionable information to improve your sleep quality. The paper reports a system called SleepExplorer that visualize not only the quality but also the context of your sleep.

In "Axis Visualizer: Enjoy Core Torsion and Be Healthy for Health Promotion Community Support", Takuichi Nishimura, Zilu Liang, Satoshi Nishimura, Tomoka Nagao, Satoko Okubo, Yasuyuki Yoshida, Kazuya Imaizumi, Hisae Konosu, Hiroyasu Miwa, Kanako Nakajima, Ken Fukuda reports a new trunk torsion model to evaluate the strength of body trunk by smoothness of the muscle movement instead of the muscle mass. The ability to move body trunk muscle smoothly is considered very important for injury prevention, physical strength and beauty.

Acknowledgment

As the organizing committee chair, I would like to thank the committee members, Dr. Taku Nishimura, Dr. Hiroyasu Miwa, Dr. Kentaro Watanabe who were also members of the program committee.

The organizers would like to thank the advisory committee members, Dr. Masaaki Mochimaru, Prof. Takeshi Sunaga, Prof. Hideaki Takeda, the program committee members, Dr. Tom Hope, Dr. Shuichi Ino, Dr. Takahiro KAWAMURA, Dr. Kouji Kozaki, Prof. Noriaki Kuwahara, Prof. Taketoshi Mori, Dr. Yoichi Motomura, Dr. Marketta Niemelä, Dr. Yoshifumi Nishida, Dr. Taku Nishimura, Dr. Yuko Ohno, Dr. Hiroshi Sato, Dr. Atsushi SHINJO, Dr. Marja Toivonen, Dr. Naoshi Uchihira.

The organizers would like to thank Mr. Barada and Ms. Yamada for their support.

The organizers would like to thank again to Mr. Jukka Lindberg, Client director, Older people, subscriber services, Hämeenlinna City, Finland for his keynote lecture and panel discussion and Dr. Masahiro Kanno, CEO, Keiju Healthcare System as keynote speaker, and Ms. Seiko Adachi, Executive Head, Social Welfare Corporation Sinko Fukushikai for participating in the panel discussion as a panelist.

Finally, the organizers would like to thank the Society for Serviceology (http://www.serviceology.org/) and JST (The Japan Science and Technology Agency, http://www.jst.go.jp/EN/index.html) for their support and JSAI for financial support.

Designing Intelligent Sleep Analysis Systems for Automated Contextual Exploration on Personal Sleep-Tracking Data

Zilu Liang[1,3(✉)], Wanyu Liu[1], Bernd Ploderer[1,2], James Bailey[1],
Lars Kulik[1], and Yuxuan Li[1]

[1] Department of Computing and Information Systems, University of Melbourne,
Melbourne, VIC 3010, Australia
[2] Electrical Engineering and Computer Science, Queensland University of Technology,
Brisbane, QLD 4001, Australia
[3] Graduate School of Engineering, The University of Tokyo, Tokyo 113-8654, Japan
z-liang@t-adm.t.u-tokyo.ac.jp

Abstract. There are many sleep tracking technologies in the consumer market nowadays. These technologies offer rich functions ranging from sleep pattern tracking to smart alarm clock. However, previous study indicates that users find these technologies of little use in facilitating sleep quality improvement, as simply making a user aware of how poor his/her sleep is provides no actionable information on how to improve it. Armed with such understanding, we proposed an architecture for designing intelligent sleep analysis systems and developed a prototype called SleepExplorer to help users automatically analyse and visualize the interrelationship of his/her sleep quality and the context (i.e., psychological states, physiological states, lifestyle, and environment). Such contextual information is crucial in helping users understand what the potential reasons for their sleep problems might be. We conducted a 2-week field study with 10 diverse participants, learning that SleepExplorer help users make sense of their sleep-tracking data and reflect on their lifestyle, and that the system has potentially positive impact on sleep behaviour change.

Keywords: Sleep tracking · Self-tracking · Personal informatics · Intelligent systems · Health · Contextual information · Automated data analytics

1 Introduction

Good sleep is essential for personal health, while poor sleep is usually a predictor of other sickness. Existing studies in sleep domain focuses on investigating general relationships between sleep and sleep contextual factors at the population level. However, such relationship may be highly individual, and population-level conclusions may not hold when applying to a specific person. With the prevalence of personal informatics systems [1], movement of Quantified Self (QS), and researches in personal health informatics, sleep technologies such as wearable sleep tracking devices (e.g., Fitbit, Jawbone, etc.) and sleep tracking mobile apps (e.g., SleepAsAndroid, SleepBot, SleepCycle, etc.)

© Springer International Publishing AG 2017
M. Otake et al. (Eds.): JSAI-isAI 2015 Workshops, LNAI 10091, pp. 367–379, 2017.
DOI: 10.1007/978-3-319-50953-2_25

became affordable for everyday use. These technologies has the potential to enable exploration on each individual's sleep affecting factors if combined with system engineering and automated data analytics.

Inspired by the individual-level preventive healthcare framework in [2], we propose an architecture for designing intelligent sleep analysis systems based on sleep-tracking technologies, system engineering and automated analytics. The overarching goal is to guide future researches on the design of intelligent system for preventive and self-managed healthcare. Based on the proposed architecture, we designed and implemented a prototype system called SleepExplorer, which automatically analyzes a user's self-tracking data to identify the factors that are associated to a user's sleep and to visualize such relationships. In order to evaluate the performance of the prototype, we conducted a two-week field study and interviews with 10 users. Our qualitative analysis on the interviews provides rich implications for designing future sleep-tracking and sleep-supporting systems. The contributions of this paper are as follows:

- We proposed a general architecture for designing intelligent sleep analysis systems that empowers individuals to nurture personal data for better sleep and overall health.
- We designed and implemented a prototype system named SleepExplorer to exemplify how the proposed architecture could guide efficient design of intelligent sleep analysis systems in the perspective of system engineering.
- We conducted qualitative study to evaluate SleepExplorer and provided rich implications for designing future sleep-tracking and sleep-supporting systems.

The rest of the paper is organized as follows. In Sect. 2 we summarize related work on sleep research and sleep-tracking technologies. Sections 3 and 4 present the proposed architecture and the implementation of SleepExplorer respectively. In Sect. 5 we describe the design of the field study, the qualitative findings, and the discussions on several issues that are informative for future research. We conclude the paper in Sect. 6.

2   Related Work

2.1   Sleep Health and Sleep Structure

In sleep research, the paradigms for defining sleep health started to shift from those that emphasize the absence of sleep problems to those that focus on how well an individual is doing [3]. Whereas traditional sleep research targets at treatments for disorders and removal of dysfunctions, current research trend focuses on the prevention of sleep problems and the maximization of sleep health.

In practice, sleep health could be measured across multiple levels of analysis such as self-report, behavioral, physiological, and genetic levels of analysis [4]. Within each level of analysis, sleep health can be further characterized along multiple dimensions, such as quantity, continuity, and timing [5]. According to [3], the most important sleep metrics include minutes asleep (MASL), minutes awake (MAWK), number of awakenings (NAWK), minutes to fall asleep (MTFA), and sleep efficiency (SE), subjective sleep quality. It is also emphasized in [3] that sleep health is best understood in the context of individual, social, and environmental demands, as sleep is affected by many contextual factors such as lifestyle, mood, noise.

2.2 Sleep Tracking Technologies

Currently, a large number of commercial sleep technologies are affordable for daily use. The aims of these technologies cover various aspects of sleep: sleep inducing (e.g., White Noise), dream journaling (e.g., Dreamboard), waking (e.g., Smart Alarm Clock), sleep tracking and environment monitoring (e.g., Hello Sense). The platforms of these technologies vary from mobiles (e.g., Sleep Cycle), wearables (e.g., Jawbone) to advanced embedded tracking sensors (e.g., Beddit). These sleep tracking devices and apps provide users with the information about how long they sleep, how well they sleep (sleep quality/ efficiency score), the stages they sleep through, how to fall asleep and wake up with optimized freshness, and how to promote healthy sleep habits through sleep coaching tips.

The mechanism of existing sleep-tracking tools is similar to actigraphy that provides reasonably accurate results for normal, healthy adult populations [6]. However, these tools analyze sleep in isolation from the contextual information, and users of commercial sleep tracking technologies found it difficult to interpret data without being provided with context. Users found no proper tool to integrate data from multiple sources, let alone conducting deeper analysis on the data [7]. In [2], the authors proposed a general framework for individual-level preventive healthcare framework based on self-tracking. Armed with this framework, we propose architecture for designing intelligent sleep analysis systems based on sleep-tracking technologies, system engineering and automatic data analytics. The overarching goal is to inspire the design of future intelligent systems that empower individuals through information technologies to nurture personal data for better sleep. We implement a prototype based on the proposed architecture to illustrate how personal data could generate more values for individuals if they are empowered by proper tools.

3   Proposed Architecture for Intelligent Sleep Analysis Systems

3.1   Overview

We proposed a layered architecture for intelligent sleep analysis systems which is illustrated in Fig. 1. The proposed architecture consists of four complementary layers. Similar functions are grouped into a single layer that provides services to the layer above and receives services from the layer below. From bottom up, the four layers are data collection layer, data integration layer, data analysis layer, and presentation layer. The details of each layer are described in the following sections.

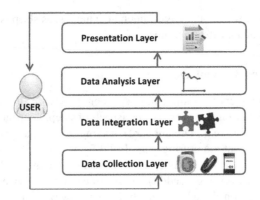

Fig. 1. Proposal of layered structure of intelligent sleep analysis systems.

3.2 Layer 1: Data Collection Layer

This layer defines what data to collect, where to collect the data (i.e., data sources), and how to collect the data (i.e., data collection tools). The data that need to be collected includes targeted sleep metrics and sleep contextual factors. As for sleep metrics, it is recommended to track both objective (e.g., MASL, TTFA) and subjective sleep metrics (e.g., PSQI, Wake-Up Freshness) because the two are only modestly correlated [8, 9] and they assess different aspects of an individual's sleep experience. As for sleep contextual factors, they can be classified into the following four categories:

- *Physiological factors:* blood pressure, body weight, body temperature, etc.
- *Psychological factors:* mood, stress, etc.
- *Life-style factors:* bed time, diet, physical activities, etc.
- *Environmental factors:* temperature, humidity, etc.

Data source refers to the origin of the data. Data source is primary if the data was directly collected by users and is secondary if the data was retrieved from existing database such as electronic health records. Data collection tools include wearable devices, digital medical devices, mobile health apps, spreadsheets, personal journals and so on. A comprehensive summary of the tracking devices can be found in [10] Users utilize these tools to capture data on their sleep and sleep contextual factors such as mood, caffeine consumptions, noise.

3.3 Layer 2: Data Integration Layer

Personal health data collected by using various tracking and monitoring tools are heterogeneous due to the current lack of a common standard for personal health data. Self-tracking data may include numerical data, categorical data, time series data, and unstructured text data. The data integration layer involves unifying data from disparate sources, converting data in various formats into a unified format, and integrating data according to time stamps and sampling rate. This layer functions as a prerequisite for the data analysis layer, as effective data analysis will not be possible until all data are organized under the same format and granularity. This layer retrieves data from tracking/monitoring tools such as Fitbit, personal health data warehouse such as Apple Health, and databases such as electronic health records, and converts unstructured data into structured data sets that will then be analyzed on the upper layer. When the same variables are collected from various sources, data integration layer merges this information together to improve the data quality. This purpose can be achieved by applying e.g. schema mapping [11], record linkage [12] and data fusion [13–15].

3.4   Layer 3: Data Analysis Layer

This Layer is at the heart of an intelligent sleep analysis system. It provides the function and procedural means of extracting meaningful information, i.e. patterns and correlations, from raw data through applying advanced statistical techniques, data mining, and predictive modeling. The operations at this layer create a baseline of the user's sleep pattern, automatically diagnose sleep problems, uncover hidden relations between sleep and contextual factors, analyze how the user deviates from population average, and even predict a user's sleep in the near future. Pre-analysis data processing (e.g. missing data, wrong data) will be conducted at this layer too.

According to [16], there are eight levels of analytical capabilities that any person needs in order to have a panoramic view of their health. In [17], the authors map the eight levels of analytical capabilities to sleep domain and proposed a four-level analytics model for personal sleep informatics. Most existing sleep-tracking tools provide level-1 analysis which answers what is a user's sleep pattern, when did sleep problems happen, and how often did the problems happen. Our implemented SleepExplorer offers level-2 analysis that aims to answers why sleep problems happened, which will be described in detail in Sect. 4. New tools with predictive modelling and optimization functions need to be developed to perform automatic analysis on level-3 and level-4, which answer questions such as "what will happen next?", "how do we do things better?" and "what is the best decision for the problem?"

3.5   Layer 4: Presentation Layer

The Presentation Layer is the closest to the end user. It provides the functions and procedure means of (1) interpreting the output of the data analysis layer, i.e., data analysis results, into plain words that understandable by users; (2) visualizing the analysis results on user interfaces; (3) providing tailored suggestions and recommendations on how to improve sleep quality. This layer plays a critical role in hiding the technical aspects of data processing and analysis from the end users and delivering the insights from data to users on user interfaces.

4 Implementation of SleepExplorer

Based on the proposed architecture in Sect. 3, we implemented a pilot sleep analysis system named SleepExplorer on Microsoft Azure cloud using ASP.NET Model-View-Controller (MVC) framework. The implementation on each layer is described below.

4.1 Layer 1: Data Collection Layer

The sleep metrics collected by SleepExplorer characterize both objective and subjective sleep quality. The objective sleep metrics include minutes asleep (MASL), minutes awake (MAWK), number of awakenings (NAWK), time to fall asleep (TTFA), and sleep efficiency (SE). Fitbit is chosen to track these metrics because it is one of the major vendors in wearable devices market, and it is the most popular wearable device brand in Australia. The subjective sleep metric is wake up freshness and it was tracked using diary every day in the morning. Considering the easiness of tracking the data, SleepExplorer collect the following sleep contextual factors.

- *Physiological factors:* body weight, body temperature, and menstrual cycles.
- *Psychological factors:* mood, stress, tiredness, and dream.
- *Life-style factors:* steps, minutes very active (MVA), minutes fairly active (MFA), minutes lightly active (MLA), calories in, calories out, activity calories, coffee, coffee time, alcohol, electronic devices usage, evening light, nap time, nap duration, social activities, exercise time, and dinner time.
- *Environmental factors:* temperature and humidity.

Among these factors, body weight, steps, MVA, MFA, MLA, calories in/out, activity calories are automatically tracked by Fitbit, the rest are tracked using diary to avoid unnecessary complexity. Users subjectively evaluate each factor against a pre-defined scale. The data source of all sleep metrics and sleep contextual factors are primary, as they are all collected directly by users.

4.2 Layer 2: Data Integration Layer

SleepExplorer collects several types of variables include continuous variables (e.g., body weight, temperature), ordinal variables (e.g., mood, dream), discrete variables (e.g., steps, coffee). Most of the variables are numerical, except time stamps such as exercise time, coffee time and dinner time. In order to study the impact of these time stamps on sleep, these variables are converted to continuous variables. For instance, 18:00 is converted to 1800.

4.3 Layer 3: Data Analysis Layer

On the data analysis layer, SleepExplorer performs the tasks of (1) preprocessing data (e.g. removing missing data, normalizing data, etc.), (2) calculating Spearman correlation coefficient between sleep metrics (e.g., minutes asleep) and contextual factors (e.g., steps, stress, mood, minutes very active, etc.), and (3) calculating significance of the correlation. Spearman correlation is used instead of Pearson correlation coefficient because the former can capture non-linear relationships and can be used on ordinal variables. A C# library named MathNet is used to calculate the correlations. The factors that have moderate and strong Spearman correlation to sleep are passed to the above layer (i.e., presentation layer) for visualization.

4.4   Layer 4: Presentation Layer

The presentation layer, represented by the user interface of SleepExplorer, renders visualization of the data analysis results using a JavaScript library d3.js [18]. A screenshot of the visualization is shown in Fig. 2. Positive and negative related factors were represented by green and red bubbles respectively. For each identified sleep related factor, the strength of correlation were represented by the shades of the corresponding bubbles. Since Ancker and Kaufman's study on health numeracy discovered that many people lack the ability to understand statistical data and have difficulty in interpreting conventional statistical graphs [19], we chose bubbles to visualize the correlations because this presentation is fun and very visually appealing.

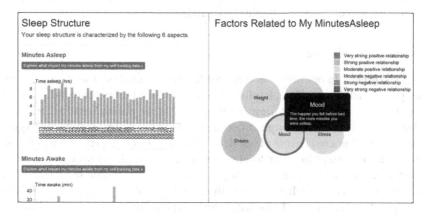

Fig. 2.   A screenshot of SleepExplorer interface.

5   Field Study and Qualitative Findings on SleepExplorer

5.1   Design of Field Study

The aims of the field study were (a) to gain an understanding of current landscape of whether and how users analyze their self-tracking sleep data, (b) to investigate how SleepExplorer could support self-reflection on sleep and other self-tracking data, and (c) to obtain feedback from users for further improvement of the system.

Participants.   We recruited ten diverse participants through university mailing lists. Participants must be at least 18-year-old and must have been using Fitbit to track sleep and at least one contextual factor. Each participant received a $25 book voucher per interview as an appreciation of their commitment to the study.

Data Collection. We collected qualitative data from each participant through two semi-structured interviews conducted at the beginning and at the end of the two-week field study. In the first interview we were interested in their self-tracking experience in general (e.g., tools, motivation, and data management), their sleep quality (quantified using PSQI [20]), and whether/how they analyze their self-tracking data. Users were also asked to use SleepExplorer to analysis the correlation between their sleep data and contextual information that they had tracked using Fitbit. They were encouraged to think out aloud while using the system to give feedback on the usability of SleepExplorer as well as speaking out their interpretation of the visualization on user interface. At the end of the first interview, they selected extra contextual factors from the factor list on SleepExplorer, which would track using the diary function of SleepExplorer in the following two weeks. During the two-week field study, all users received two reminders to fill in their self-tracking diary every day, one in the morning after they got up and one at night before they went to bed. After the two-week field study, we conducted the second interview with participants to discuss their experience and thoughts of the field study. They were also asked to use SleepExplorer to analyze the correlation between sleep data and newly tracked data (including Fitbit data and diary data).

5.2   Qualitative Findings

System Usability. In our first interview, participants pointed out that it was not easy to interpret the bubbles on SleepExplorer interface. Several users mistook the bubbles as an evaluation of their current lifestyle rather than the correlation between the contextual factor and sleep. For example, after analyzing the Fitbit data of P3, the system identified "steps" as a red bubble of her sleep efficiency. Her interpretation was that she should walk more steps in order to eliminate that red bubble. However, the correct interpretation was that her sleep efficiency was worse on days when she walked more steps. This drove us to attach a tooltip with text interpretation to each of the bubbles. Another problem that we found was that sometimes participants were confused whether the bubbles were presented based on general sleep research or based on their own data. In order to emphasize that the visualization was based on each participant's own data and was thus personalized to each participant, we added "my" in the wording on the user interface.

In our second interview, as there was richer contextual information with more factors tracked using the diary function of SleepExplorer, some participants mentioned that there was so much information on the interface that they felt overwhelmed. More information does not necessarily means more value. Therefore, it is imperative to selectively present the most important information rather than rendering all information on the interface, which is one of our future research directions.

Supporting Self-Reflection on Sleep. SleepExplorer helps participants identify the factors that are related to their sleep. Some results validated the subjective perception of participants or brought the interrelationship from back of mind to consciousness, while some were counterintuitive. Both kinds of results support participants' self-reflection on how personal sleep was associated to various aspects of daily routine.

Reflection on Expected Results. Some analysis results confirmed participants' expectation on the relationship between contextual factors and sleep. For example, P2 tended to sleep longer on days when she did more light activities or consumed less alcohol. P3 seemed to have fewer minutes awake during nights when she consumed less coffee in the day or felt more tired before bedtime. Surprisingly, P5 also paid attention to the contextual information that was turned out to be irrelevant to his sleep. He tracked coffee time using the diary function of SleepExplorer but this factor did not appear on the interface. He stated that the result was consistent with his experience, as he had felt no difference whether he drank coffee in the morning or at night.

Reflection on Surprising Results. It seems that participants are curious about why some contextual factors were counterintuitively related to their sleep. When reflect on surprising results, several participants "pealed the onions" to make sense of the results in the context of their lifestyle. When seeing the contextual information that was related to her minutes asleep, P7 provided the following explanation on why exercise was negatively correlated, *"This makes sense. Because when I'll be active, I get up early, and now I'm trying to go to bed earlier as well. But most of the time that I'm using this (Fitbit), I would go to bed at the same time, but then on the days I go to gym in the morning, I had to get up earlier, so I had less sleep."* P10 mentioned that it was possible that the counterintuitive contextual factors were actually related to other factors which eventually led to the seemingly counterintuitive results. Her reflection on why she seemed to sleep better when she had more minutes sedentary was as follows, *"Ok, this could be because there are other factors, for example, I'm sedentary when I work, I'm quite being mentally tired with my work, and I don't drink alcohol at night... So it's not like being still makes me sleep better. It's more like, it happens to be on the days when I'm still, I'm mentally draining, and I don't eat overly at night, and I don't drink alcohol."*

Potential Impact on Behavior Change. Users' mentioned that their intension on behavior change may either due to the self-tracking activity itself or due to the analysis results presented on SleepExplorer user interface.

Behavior Change Due to Self-Tracking. Most participants mentioned that the two-week field study in which they were tracking extra contextual factors increased their awareness of the potential impact of those factors on their sleep. The impact varies. While some participants did not see change in their daily routine, others intentionally adjusted their behaviours either for the purpose of creating variation in data or hoping to see good records. P4 told us that the tracking motivated her to consume less alcohol and to reduce screen time, and she perceived that such change led to her better sleep during the user trial. P5 also reduced screen time before bed, stating that the participation to the user trial is one of the reasons (another reason being the intention of spending more time with his family). However, P5 consumed more alcohol during the user trial just to increase the variation in data.

Behavior Change Due to Analysis Results. Participants' attitude to behavior change varies depending on whether the analysis results validated or contradict their expectations. It seems to be more motivating for participants to adjust their behavior if the analysis results confirmed their previous expectations. When seeing that digital device usage has strongly negative relationship to sleep, which confirmed their subjective perception, P1 and P5 firmed stated that they definitely would adjust their daily schedule to reduce screen time at night. If the analysis result is surprising, however, the decision-making on behavior change becomes more complicated, and people will consider the overall health benefit of such change rather than simply making decision for the sake of sleep. For example, minutes sedentary turned out to have moderately positive relationship to the minutes asleep of P4, which means that she tends to sleep longer if she is more sedentary during the day. P4 said that this is counterintuitive, as her common knowledge told her that it should be exercise rather than sitting that led to better sleep. When asked what action she would take based on this discovery, she told us that she would never try to be more sedentary, because it is not good for cardiovascular system.

Asymmetrical Effects of Positively and Negatively Contextual Factors. Many participants mentioned that they would pay more attention to negatively related contextual factors (red bubbles) than positively related ones (green bubbles). P5 stated that red bubbles offer more actionable information than green ones. This echoes the asymmetrical effects positive and negative events in psychological studies [21], which states that adverse or threatening events trigger strong and rapid physiological, cognitive and emotional responses.

5.3   Discussions

Technical Challenges of Implementing Intelligent Sleep Analysis Systems. Many issues impose great challenges to the design and implementation of reliable automated sleep analysis systems. Three most important issues are data richness and quality, domain knowledge integration, and efficacy of the system.

Data Richness and Quality. Sleep may be affected by a wide range of contextual factors. However, many people do not have the literacy on what contextual factors to track. Additionally, it requires much devotion to track factors such as diet, even though some people are aware of the strong impact of diet on sleep. These reasons may significant reduce the richness of self-tracking data. Data quality is another widely known issue in designing intelligent sleep analysis systems. Missing data, inaccurate data and wrong data all harm the reliability of the final analysis results. Low data quality may either due to the mechanism of tracking tools [22, 23], the misusage of tracking tools [24] or the difficulty in quantifying contextual factors [25]. These problems may be overcome by developing new self-tracking tools or guiding users to use existing devices properly.

Integrating Domain Knowledge into Data Analysis. Different from general data analytics, sleep analysis requires the integration of sleep domain knowledge in order to make sense of the analysis results. It is known that sleep could be affected by a wide variety of factors. However, tracking all potential contextual factors is not feasible in practice, even if it is not impossible. Therefore, it is important to keep the users informed of what the most likely contextual factors are according to sleep research. Moreover, sleep domain knowledge is required in order to interpret the analysis results. For example, what does "you have 12 awakenings last night" mean? Is it normal to fall asleep in 40 min? What does "more minutes very active is associated to shorter time asleep" indicate? It remains a challenge to develop intelligent sleep analysis systems that help users interpret the analysis based on sleep domain knowledge.

Visual Cues for Behavior Change. As mentioned in the previous section, negative contextual factors are stronger visual cues for most participants. However, there are still a few users who are more interested in positive factors, or in stronger factors regardless of whether they are positive or negative. It remains a challenge as to how to adaptively present the most important information to each user based on his/her concern. Also, some users mentioned that too much information on user interface was overwhelming, so it is imperative to strike a balance between informative and concise. We plan to address these issues in our next step.

Social Impact of Intelligent Sleep Analysis Systems. Our qualitative study indicates that users were generally interested in investigating their self-tracking using Slee-pExplorer to gain deeper understanding of their sleep, especially on what factors impact sleep. Tracking and analyzing sleep contextual factors made users more aware of the potential impact of these factors on their sleep quality. It is certain that if empowered by proper tools with automatically data analysis and interpretation functions, users could gain more insights thus higher personal values from self-tracking data.

However, the analysis may not eventually lead to real behavior change. Although many participants expressed their willingness of changing lifestyle based on intuitive analysis results, they also mentioned the barriers for taking such actions in reality. These barriers could be laziness, lacking willpower, tight daily schedule, special events in life (e.g., having a new born baby), bad overall health conditions, and so on. Therefore, further studies are needed to understand the best approach for promoting lifestyle change, and thereby unleash the great potential of intelligent sleep analysis systems in guiding behavior change and in improving sleep quality.

6 Conclusions

In this study, we proposed architecture for designing intelligent sleep analysis systems, and implemented a web application SleepExplorer to exemplify the implementation of the proposed architecture. In order to evaluate SleepExplorer, we conducted a two-week field study with semi-structured interviews before and after. We improved the design of SleepExplorer based on the feedback from participants. The qualitative findings of the field study include: (1) existing sleep tracking technologies rarely integrate contextual information to sleep analysis, but users tend to make connection between sleep and the context in their mind; (2) SleepExplorer helps users make sense of their sleep data through automatic correlation analysis using contextual information and novel visualization for conveying statistical analysis results; (3) SleepExplorer has potentially positive impact in guiding sleep behavior change. We also discussed the challenges and the potential social impact of intelligent sleep analysis systems. In the next step, we intend to address the data quality issue as well as investigate the potential impact of SleepExplorer on users' behavior and sleep quality in long term.

Acknowledgement. This study was supported by Australian Government Endeavour Research Fellowship and Microsoft BizSpark. We would like to thank our study participants for their contribution. We also appreciate feedback and support from Dr. Kathleen Gray, Manal Almalki and our colleagues in Microsoft Center for Social NUI and Natural Language Processing Group, University of Melbourne.

References

1. Li, I., Dey, A., Forlizzi, J.: A stage-based model of personal informatics systems. In: Proceedings of the SIGCHI Conference on Human Factors in Computing Systems, pp. 557–566 (2010)
2. Liang, Z., Chapa-Martell, M.A.: Framing self-quantification for individual-level preventive health care. In: Proceedings of the International Conference on Health Informatics, Lisbon, Portugal, pp. 336–343 (2015)
3. Buysse, D.J.: Sleep health: can we define it? Does it matter? Sleep **37**(1), 9–17 (2014)
4. Carskadon, M.A., Dement, W.C.: Normal human sleep: an overview. In: Kryger, M.H., Roth, T., Dement, W.C., (eds.) Principles and Practice of Sleep Medicine, 4th ed., Elsevier Saunders, Philadelphia, PA, pp. 13–23 (2005)
5. Hall, M.H.: Behavioral medicine and sleep: Concepts, measures and methods. In: Steptoe, A. (ed.) Handbook of Behavioral Medicine: Methods and Applications, pp. 749–765. Springer, New York (2010)
6. Morgenthaler, T., Alessi, C., Friedman, L., et al.: Practice parameters for the use of actigraphy in the assessment of sleep and sleep disorders: An update for 2007. Sleep **30**(4), 519–529 (2007)
7. Liu, W., Ploderer, B., Hoang, T.: In bed with technology: Challenges and opportunities for sleep tracking. In: Proceedings of the Australian Computer-Human Interaction Conference (OzCHI 2015) (2015)
8. Baker, F.C., Maloney, S., Driver, H.S.: A comparison of subjective estimates of sleep with objective polysomnographic data in healthy men and women. J. Psychosom. Res. **47**(4), 335–341 (1999)
9. Coates, T.J., Killen, J.D., George, J., et al.: Estimating sleep parameters: a multitrait-multimethod analysis. J. Consult. Clin. Psychol. **50**(3), 345–352 (1982)
10. Almalki, M., Gray, K., Sanchez, F.M.: The use of self-quantification systems for personal health information: big data management activities and prospects. Health Inf. Sci. Syst. **3**(Suppl 1), S1 (2015)
11. Bellahsene, Z., Bonifati, A., Rahm, E. (eds.): Schema Matching and Mapping. Springer, Heidelberg (2011)
12. Elmagarmid, A.K., Ipeirotis, P.G., Verykios, V.S.: Duplicate record detection: A survey. IEEE Trans. Knowl. Data Eng. **19**(1), 1–16 (2007)
13. Bleiholder, J., Naumann, F.: Data fusion. ACM Comput. Surv. **41**(1), 1–41 (2008)
14. Zhao, B., Rubinstein, B.I.P., Gemmell, J., Han, J.: A Bayesian approach to discovering truth from conflicting sources for data integration. Proc. of the VLDB Endowment **5**(6), 550–561 (2012)

15. Li, Q., Li, Y., Gao, J., et al.: Resolving conflicts in heterogeneous data by truth discovery and source reliability estimation. In: Proceedings of SIGMOD14, Snowbird, UT, USA, pp. 22–27 (2014)

16. Burke, J.: The world of health analytics. In: Kudyba, S.P., (ed.) Healthcare Informatics: Improving Efficiency and Productivity, pp. 161–180. CRC Press (2010). Chapter 8

17. Liang, Z., Ploderer, B., Martell, M.A.C., Nishimura, T.: A cloud-based intelligent computing system for contextual exploration on personal sleep-tracking data using association rule mining. In: Martin-Gonzalez, A., Uc-Cetina, V. (eds.) ISICS 2016. CCIS, vol. 597, pp. 83–96. Springer, Heidelberg (2016). doi:10.1007/978-3-319-30447-2_7

18. D3. http://d3js.org/. Retrieved 31 Dec 2015

19. Ancker, J.S., Kaufman, D.: Rethinking health numeracy: A multidisciplinary literature review. J. Am. Med. Inform. Assoc. 14, 713–721 (2007)

20. Buysse, D.J., Reynolds III, C.F., Monk, T.H., et al.: The Pittsburgh sleep quality index: A new instrument for psychiatric practice and research. Psychiatry Res. 28, 193–213 (1988)

21. Whooley, M., Gray, K., Ploderer, B., Gray, K.: On the integration of self-tracking data amongst quantified self members. In: HCI, pp. 151–160 (2014)

22. Blood, M.L., Sack, R.L., Percy, D.C., Pen, J.C.: A comparison of sleep detection by wrist actigraphy, behavioural response, and polysomnography. Sleep 20(6), 388–395 (1997)

23. Kushida, C.A., Chang, A., Gadkary, C., et al.: Comparison of actigraphic, polysomnograpic, and subjective assessment of sleep parameters in sleep-disordered patients. Sleep Med. 2, 389–396 (2001)

24. Yang, R., Shin, E., Newman, M.W., Ackerman, M.S.: When fitness trackers don't 'fit': end-user difficulties in the assessment of personal tracking device accuracy. In: Proceedings of UniComp 2015, Osaka, Japan, pp. 623–634 (2015)

25. Liang, Z., Liu, W., Ploderer, B., Bailey, J., Kulik, L., Li, Y.: SleepExplorer a visualization tool to make sense of correlations between personal sleep data and contextual factor. Submitted to Pers. Ubiquit. Comput. 20(6), 985–1000 (2016)

Axis Visualizer: Enjoy Core Torsion and Be Healthy for Health Promotion Community Support

Takuichi Nishimura[1(✉)], Zilu Liang[1,6], Satoshi Nishimura[1], Tomoka Nagao[1],
Satoko Okubo[1], Yasuyuki Yoshida[2], Kazuya Imaizumi[3], Hisae Konosu[4],
Hiroyasu Miwa[5], Kanako Nakajima[5], and Ken Fukuda[1]

[1] AI Research Center, National Institute of Advanced Industrial Science
and Technology, Toyota, Japan
takuichi.nishimura@aist.go.jp
[2] Graduate School of Decision Science and Technology, Tokyo Institute of Technology,
Toyota, Japan
[3] Faculty of Healthcare, Tokyo Healthcare University, Toyota, Japan
[4] Japan Dance Sport Federation, Toyota, Japan
[5] Human Information Research Institute, National Institute of Advanced Industrial
Science and Technology, Toyota, Japan
[6] Department of Computing and Information Systems, University of Melbourne,
Melbourne, Australia

Abstract. In Japan, the ratio of people with lifestyle-related diseases has
increased to approximately 30%. Individuals as well as the Nation are getting
more and more health-conscious, and special attention has been made to body
trunk because it is vital for injury prevention, physical strength, and beauty.
Various training methods have been proposed to increase the muscle mass of body
trunk. However, for sports that emphasize somatoform such as dance, the strength
of the trunk is mainly decided by smooth use of the trunk rather than its muscle
mass. In this paper, in order to evaluate the use of the trunk torsion movement,
we proposed a new trunk torsion model for the purpose of evaluating two trunk
torsion standard movements. We also developed a mobile application named
"Axis Visualizer" based on the proposed trunk torsion model analyzing sensor
data in the device. Axis Visualizer generates higher score when a user rotates the
shoulders or hips smoothly with axis fixed and high frequencies. This application
can support trainers and coaches to visualize the use of customers' trunk and to
increase the training effect.

Keywords: Movement modelling · Core strength · Health promotion

1 Introduction

Japan's population is aging rapidly, and tackling this demographic challenge imposes a
heavy burden to the social welfare system. In Japan, the number of old people reached
historical peak 33,000,000 in 2014, which is equivalent to 26.0% of the whole population
[1]. In 2012, the national healthcare cost was more than 39 trillion Japanese Yen, which
increased 626,700,000,000 Japanese Yen compared to that in previous year. In addition,

© Springer International Publishing AG 2017
M. Otake et al. (Eds.): JSAI-isAI 2015 Workshops, LNAI 10091, pp. 380–392, 2017.
DOI: 10.1007/978-3-319-50953-2_26

long-term care insurance benefit sharply increased to 8 trillion Japanese Yen. Moreover, it is expected that the social welfare for nursing and healthcare will roar in 2025 as the baby boom generation ages beyond 75-year-old.

In order to reduce nursing and healthcare cost, it is important to suppress not only public expenditure but also personal medical cost. As nowadays lifestyle-related chronic diseases are more common than diseases caused by pathogenic bacteria, it is imperative to promote healthy lifestyles in order to extend healthy life expectancy. In addition to conventional policy-driven measures, various efforts have been made, including cooperation between health industry (e.g., private sports clubs) and medical institutions or local governments, preventive healthcare measures advocated by health insurance unions, regional comprehensive care system designed for supporting independence of local residents.

Under such background, particular attention has been given to body trunk for the purpose of preventing injuries and increasing physical strength. Various training methods, such as Core Training [2–5], have been introduced to strengthen body trunk. In dance sports, however, it is widely considered that the strength of body trunk is not characterized by the amount of muscle but by the cooperative use of body trunk and limb. Therefore, in order to make it possible for trainers to quantitatively improve customers' trunk strength, we developed a mobile application named Axis Visualizer for evaluating truck torsion movements. A user can get high score if he or she twists shoulders and hips continuously, with axis fixed and high frequency.

The rest of the paper is organized as follows. The strength of trunk is discussed in the next section. In Sect. 3, we present our proposed model of trunk torsion spring model and the evaluation measures of body trunk strength. In Sect. 4, we describe the mobile application Axis Visualizer and an exemplary demo of the evaluation of user's torsion using Axis Visualizer. We then give an overall picture of the health promotion community support that we are working on. The paper is concluded in Sect. 6.

2   What Is "Trunk Strength"?

2.1   General Definition of Trunk Strength

In order to discuss the strength of body trunk, we first describe the definition of posture, movement, motion, action based on kinematics.

- *Posture:* a posture consists of two elements: attitude and position. Attitude represents the relative positional relationship among each part of the body such as head, trunk, limb, and it can be measured through joint angle. On the other hand, position is used to represent the relationship between the body axis and gravity, and it can be indicated by standing, supine (face up), and so on.
- *Movement:* a movement refers to temporal change of posture. In other words, it is described as a change of attitude and position.
- *Motion:* motion is a unit that analyzes the behavior as a task that is specifically carried out by a movement.

- *Action*: action is a unit when taking into consideration the context of the meaning and intention shown by a movement.

Take nodding as an example, the movement is "moving head and neck 10° to the front", the motion is "the flexion of head and neck = nodding", and the action is "agreeing to the other person".

Many services of health promotion and prevention activities put the emphasis on strength training. Body trunk is the torso which has the following three roles: (1) supporting and maintaining postures, (2) the foundation for producing movement, (3) serving as an axis. At present, many people are interested in strengthening body truck, and terms such as core training is becoming very popular. Several methods such as plank have been proposed to increase the mass of surface muscle such as rectus abdominis muscle and deep muscle groups such as transverse abdominal muscle.

It is worth mentioning that the quality of movement is more important than the amount of movement, which is a key point in improving people's healthy life expectancy in the super-aged society in the future. In other words, almost all daily-life movements, such as waking-up, standing, walking, are antigravity movements. Depending on what posture a movement is performed in, not only the major muscle used but also muscle contraction patterns are changed. It is imperative to focus on the point of change. Bad motion patterns caused by bad postures may lead to musculoskeletal pain syndrome. Based on this rationality, the MSI (Movement System Impairment) approach was proposed, which reduces mechanical stress by correcting the movements and motions and thereby enables the prevention and treatment of pain [9]. The prevention and treatment of pain is directly related to the activity in everyday life, therefore it is a very effective and efficient means for health promotion and preventive care. Moreover, system theory can also be applied to this field. When trying to achieve a certain action, human body has a functional mechanism that reduces the freedom of movement through the cooperation of various parts of the body. This mechanism was called synergy by Bernstein. With respect to the interaction of external force such as gravity and the various initial conditions of the body, it is likely that motion control may be distributed among mutually coordinated systems (for example, musculoskeletal system, nervous system, etc.). Incorporating various other motion control theories, Shumway-Cook and Woollacott proposed a system theory which advocates that movement is not simply the result of muscle-specific exercise program or uniform reflection, but rather the result of the dynamic interaction among perception system, cognition system, and musculoskeletal system [10]. System theory has been applied in rehabilitation and has been proved effective in practice. For instance, instead of applying approaches that target at the body dysfunction (e.g., muscle weakness) itself, practicing motion control through repeated movement such as walking was applied. The result demonstrated that even though muscle strength was not really improved, the capability of walking independently was increased.

Therefore, it is considered that system theory as well as MSI approach should be applied to health promotion and preventive medicine. In other words, rather than evaluating single function, such as muscle strength and range of motion, it is important establish an easy and proper approach to evaluate and visualize the quality of movement pattern in a comprehensive manner by considering nerve-muscle coordination, joint sense, and so on. In this way, it will become possible to learn and acquire the optimal movement pattern without injury or secondary dysfunction.

2.2 "Trunk Strength" in Dance Sport

In physically expressive sports such as dance sports, it is widely known that smooth usage from deep muscles to the surface muscles is more important than muscle mass. As is described in technique books of World DanceSport Federation [11], the core stability is the result of motion control and the muscle function of the Lumbar spine, pelvis, and hip joint complex. If the trunk is twisted while it is stretched by deep muscle group, it will return to the original position like a spring. The torque that pulls the trunk back to the original position will be generated naturally, which makes the dynamic torsion movement more efficient. By consciously linking toe to chest center, it is possible to make continuous and smooth movements at minimum muscle strength. Depending on the intensity of the movement, surface muscle group may be used to reinforce the movement in addition to deep muscle group.

2.3 Trunk Movement in Dance Sports

According to the WDSF technique book "Rumba" [11], there are four types of trunk movement.

(1) *Left and right horizontal movement of pelvis*. While keeping the pelvis and shoulder line horizontal, move pelvis left and right horizontally.
(2) *Left and right tilt of pelvis* (Fig. 1 left). The left side of the body is compressed vertically while the right side of the body is stretched. Keep the left shoulder close to the left hip and the right shoulder away from the right hip. The right side is the same.

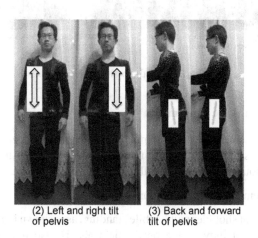

(2) Left and right tilt   (3) Back and forward
of pelvis   tilt of pelvis

Fig. 1. Examples of trunk movements in dance sports.

(3) *Back and forward tilt of pelvis* (Fig. 1 right). For forward tilt, tilt the pelvis forward by moving the upper part of the pelvis forward and the bottom part backward, and vice versa for back tilt.

(4) *Trunk torsion.* While keeping the pelvis and shoulder line horizontal, rotate around the vertical axis of the trunk.

The trunk can also make various movements by opening and closing the chest. In the case of standard dance, many movements involve the conjunction move of limbs and head. In the case of Latin dance, many movements only move the lower part of the body smoothly while keeping the upper part stable.

In this paper, we focus on trunk torsion that may frequently occur in walking and running motion, which in characterized by the angle change in the vertical axis of the shoulder line and the hip line.

3   Proposal of Trunk Torsion Spring Model and Evaluation Method

In this section we present a method for evaluating the movement of the trunk. First, we propose the spring model for trunk movement. We selected the standard movements of trunk torsion as shown in Fig. 2 for easy evaluation on trunk movements. Second, we propose a method to use accelerometer on mobile phone for evaluating trunk movement. This method could be used more widely in comparison with motion capture systems or floor reaction force systems that are only available in laboratories.

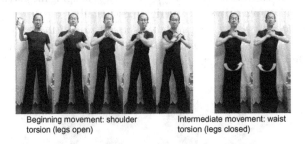

Beginning movement: shoulder   Intermediate movement: waist
torsion (legs open)   torsion (legs closed)

Fig. 2. Basic body torsion movements.

3.1   Trunk Torsion Basic Movements

We selected the following three types of movements for easy evaluation of trunk torsion. All these movements require stretching the body vertically by using abdominal deep muscle group and rotating trunk around the vertical axis naturally.

- Beginning movement: shoulder torsion (legs open)

Separate the legs at shoulder width while standing, as is shown in Fig. 2 (left). Rotate the shoulder while not moving the head (the waist is also rotated naturally). The shoulders and waist are almost in the same phase when moving slowly but are almost in the opposite phase when moving rapidly.

- Intermediate movement: waist torsion (legs closed)

Close the legs while standing, as is shown in Fig. 2 (right). Rotate the waist so as not to move the upper chest. In practice the chest is also rotating slightly and is almost in the opposite phase as the waist.

- Advanced movement: Kuka Racha action

Separate the legs at shoulder width while standing, and shift the center of gravity left and right. In order to move the part from chest to toe constantly and as if they were connected, move the pelvis in the shape of character 8. Since this motion can only be performed smoothly by advanced dancers, we will not cover it in this paper.

3.2 Proposed Trunk Torsion Spring Model

Our proposed trunk torsion spring model is described by Eq. (1).

$$T = -k\theta \tag{1}$$

Where,

T: the torque around the vertical axis.

θ: torsion angel (rotating angle of the shoulder line to the pelvis line)

k: spring constant ($k_p + k_a$).

k_p: the constant that characterizes the passive return of the trunk torsion to its original posture.

k_a: the constant that characterizes the active force around the trunk axis generated by deep muscles, which is in the reverse direction of the rotation force.

The rationality of this model is that smooth body torsion movement is generated by the rotational force of deep muscle group along the body axis rather than the linear force of surface muscle group. The movement is supported by the passive restoring force caused by the torso, which is stretched vertically by deep muscle group. It is considered that k_p increases with the activities of deep muscle group, and we plan to measure it in the future. We will modify the model if the relationship was non-linear. The k_a increases with torsion force that accelerates the rotation of the trunk. We plan to improve the accuracy of the model based on measurement using motion capture systems and floor reaction force meter.

The trunk is usually twisted repeatedly less than 90° as shown in Fig. 3. In this model, oscillation frequency can be calculated using Eq. (2).

$$f = 1/2\pi \sqrt{(k/M)} \tag{2}$$

where M is the rotation moment. The Axis Index in the proposed model is defined as follows:

$$Axis\ Index: f(Naturalness,\ Elasticity,\ Position) \tag{3}$$

- *Naturalness:* if only deep muscle group was used and surface muscle group was not used, the movement satisfies Eq. (1) and is close to sine wave. It becomes coordinated motion and the body can move in harmony from chest to foot. In the implemented mobile app Axis Visualizer which will be described in the next section, peak ratio (the power of peak divided by the total power) is used to approximate *Naturalness*.
- *Elasticity:* refers to the oscillation frequency, or the square root of the restoring force (spring constant). The unit is [Hz] or [s−1]. High frequency of torsion movement indicates stronger force to restore. Since it is in proportion to the square root of the value obtained by dividing the spring constant against the rotational moment of the trunk, stronger force is required if the body is large.
- *Position:* depending on the posture, the torsion center could shift forward or backward. This can be understood by plotting the trajectory of the sensor.

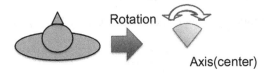

Fig. 3. The sinusoidal motion is around the axis, center of rotation. The frequency of front-back movement is twice compared to that of right-left motion.

3.3 Measurement Method

The body motion during trunk torsion can be measured by motion capture systems and floor reaction force systems. Based on the measurement, we can calculate the parameters mentioned in the previous section and therefore further measure the shake of the body axis or the connection of the trunk and the limbs. However, these systems are usually not available in daily training situation. As an alternative, we take advantage of the imbedded sensors in mobile terminals. In the initial version of the app, the mobile terminals were worn on the sacrum using low back pain belt in order to see the movement of trunk. As we further developed the app, it becomes possible for users to keep the mobile terminals against chest or shoulders by hand.

4   Implementation of Axis Visualizer

4.1 Overview

Axis Visualizer is a measurement application for trunk strength based on the proposed trunk torsion spring model described in Sect. 3. This app works on iOS and uses the imbedded accelerometer of mobile terminals to measure trunk strength. This app has the following three main characteristics.

- It adopts the measurement method based on trunk torsion spring model.
- It realizes easy measurement without special devices.
- It provides straightforward visualization of the measurement results.

The procedure of measurement is as follows: (1) physical condition check, (2) practice of movement, (3) practice of measuring method, (4) measurement (12 s), (5) visualization of results, and (6) data export. In particular, the measurement method in step 4 can use two of the methods described in Sect. 3.1. If one becomes familiar with those methods, step 2 and 3 can be omitted. Similarly, step 6 is optional. In the following subsections, we will describe in detail the design of application in step 4 and the data visualization in step 5.

4.2 Measurement with Sound Feedback

We implemented a function of enhancing natural movements by implementing sound feedback during the measurement. When the acceleration of the movement is greater than a fixed threshold value, the sound of a bell will be played. We used the following equations to determine the upper and lower bound of the threshold value.

$$\text{Upper bound } \theta_{max} = min[0.5, a_{max} - \left(a_{max} - a_{min}\right) \cdot \theta_r] \tag{4}$$

$$\text{Lower bound } \theta_{min} = max[0, a_{min} + \left(a_{max} - a_{min}\right) \cdot \theta_r] \tag{5}$$

where a_{max} and a_{min} are the maximum and minimum acceleration during the past N seconds ($N = 2$ in the implementation). The sound will is played when the acceleration is higher than θ_{max} or lower than θ_{min}. We also validated that $\theta_r = 0.2$ for this implementation. Figure 4 presents a screenshot of the application. Users simply need to input nickname. The measurement starts when users tap the start button and stops in 12 s.

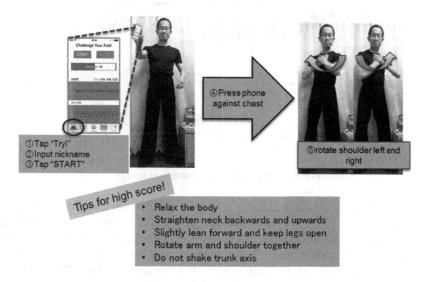

Fig. 4. The use of axis visualizer.

4.3   Result: Peak Ratio * Peak Frequency [Hz]

The application presents the value obtained by multiplying the peak frequency to the peak ratio of the movements as evaluation results. During measurement, the sampling rate, FFT and sampling time are set to 50 Hz, 512 taps, and 10.24 s respectively. Figure 5 shows screenshots of measurement results. The percentage value shown on the orange bar is a value obtained by multiplying the peak frequency to the peak ratio, which makes it easy to understand the measurement results. The graph in the center in Fig. 5 shows the acceleration. The red line indicates the acceleration in the left-right direction (X-axis), and the blue line indicates the acceleration in the back-forth direction (Z-axis). The value of the peak ratio and the peak frequency is presented below it. The graph at the bottom shows the frequency analysis results, illustrating which frequency components are often obtained during measurement. The fewer the peaks are, the more stable the torsion movement is. Peak ratios are obtained by dividing the peak power over full power.

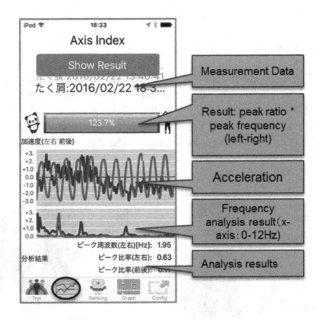

Fig. 5. Screenshots of measurement results.

As shown in Fig. 3, the evaluation is only in terms of whether the movement is sinusoidal based on the peak ratio of left-right movement. In future, we further utilize the peak ratio of back-forth movement for more accurate evaluation based on the arcuate movement about the axis of the trunk. Since the period of back-forth movement is twice of that of left-right movement, the frequency is twice the peak ratio (Fig. 5 below). Therefore based on the frequency analysis of left-right acceleration, it is possible to calculate the peak ratio of back-forth movement.

4.4 Ranking Function

As is shown in Fig. 6, we also implemented a ranking function for users to compare results with others. A ranking is generated based on the descending order of the scores stored in local memory. This function can help users (1) track the change in their quality of movement, (2) understand the difference of between their movement and others', and (3) be motivated for further improvement in core stability.

Fig. 6. Image of ranking results.

4.5 Visualization of Analysis

The simple visualization function allows users to select two axis out of the X-axis, Y-axis, and Z-axis, and plot a graph of either acceleration or gyros on a two dimensional plane. The visualization helps users understand their movements straightforwardly. Figure 7 illustrates a plot of acceleration on a two-dimensional plan with X-axis and Z-axis. This function not only helps users understand their own movements but also makes it possible to compare to previous measurement or the measurement results of others.

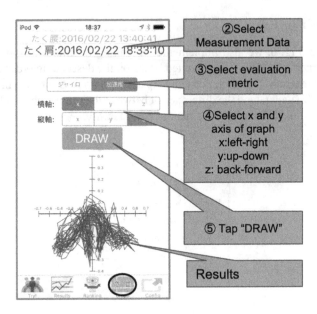

Fig. 7. Two dimensional trajectory of accelerometers.

5   Health Promotion Community Support

In this section, we describe our on-going health promotion community support and shows the positioning of the implemented Axis Visualizer. As is shown in Fig. 8, health promotion community support refers to the repetitive cycle of performing physical activities within the organization, measuring, analyzing, visualizing, and re-designing (making strategic decision on tailored behavior change for better activity and better health state) the activity of community members. Health promotion communities can stay active by repeating this cycle. Furthermore, the measurement data and the insight of redesign can be aggregated in a database. After necessary processing such as anonymization, it is possible to share health community information/knowledge with other organizations. As a result, the useful information and knowledge obtained in one of the community can be utilized in other communities, and therefore the overall quality of health promotion community can be improved nationwide.

Within the framework of such health promotion community support, Axis Visualizer is positioned as a tool for measuring the intensity of the trunk of the participants. Trunk strength is one of the indicators to measure whether the activities were carried out without injury and whether the activities were carried out effectively in a wide variety of physical activities such as dance sports. Axis Visualizer makes it possible to autonomously measure trunk strength in each community and thereby contributes to the prosperity of the entire health promotion community.

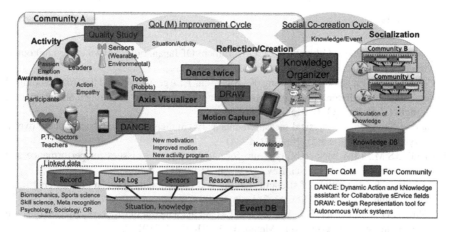

Fig. 8. Health promotion community support

6   Conclusions

In this paper, we proposed standard trunk torsion movements and a trunk torsion model. Based on the proposed model, we implemented and preliminarily evaluated a trunk strength evaluation application called Axis Visualizer and we conducted a preliminary evaluation of the application.

In the next step, we will refine the trunk torsion model based on measurement using motion capture systems or floor force plate systems, as well as assessing the model against users' subjective evaluation. We also plan to improve the real-time biofeedback in Axis Visualizer and to measure the actual spring constant k_p and k_a, so that the system can be used by trainers. We will continue improve the trunk torsion model, the visualization of the analysis results, and the way to save and share the results after applying our system in practical use. In addition, we will build the modeling and evaluation techniques for other trunk movements and by doing so we will eventually promote the health promotion community support

Acknowledgment.   This study was partly supported by Japanese METI's "Robotic Care Equipment Development and Introduction Project", "Future AI and Robot Technology Research and Development Project" commissioned by the New Energy and Industrial Technology Development Organization (NEDO) and JSPS KAKENHI Grant Numbers 24500676 and 25730190. We would also like to thank the member of the health promotion project in Odaiba and Tsukuba for their kind support.

References

1. White Paper in 2015, Cabinet Office, Government Of Japan. http://www8.cao.go.jp/kourei/whitepaper/w-2015/html/zenbun/index.html [in Japanese]
2. Cowley, P.M., Fitzgerald, S., Sottung, K., Swensen, T.: Age, weight, and the front abdominal power test as predictors of isokinetic trunk strength and work in young men and women. J. Strength Cond. Res. **23**(3), 915–925 (2009)
3. Biering-Sørensen, F.: Physical measurements as risk indicators for low-back trouble over a one-year period. Spine **9**(2), 106–119 (1984)
4. McGill, S.M.: Low back stability: from formal description to issues for performance and rehabilitation. Exerc. Sport Sci. Rev. **29**(1), 26–31 (2001)
5. Baechle, T.R.: Essentials of strength training and conditioning. In: National Strength and Conditioning Association (NSCA), 3rd edn. (2008)
6. Behm, D.G., Drinkwater, E.J., Willardson, J.M., et al.: The use of instability to train the core musculature. Appl. Physiol. Nutr. Metab. **35**(1), 91–108 (2010)
7. Panjabi, M.M.: The stabilizing system of the spine. J. Spinal Disorder **5**(4), 390–397
8. Majewski-Schrage, T., Evans, T.A., Ragan, B.: Development of a core-stability model: a delphi approach. J. Sport Rehabil. **23**(2), 95–106 (2013)
9. Sahrmann, S.A.: Diagnosis and Treatment of Movement Impairment Syndromes. Mosby, St. Louis (2002)
10. Shumway-Cook A, Marjorie Woollacott, Motor Control: Translating Research into Clinical Practice, 2nd edn (1999)
11. WDSF Latin Technique Books and DVD, The World DanceSport Federation. https://www.worlddancesport.org/WDSF/Academy/Technique_Books
12. Nakamura, Y., Yamane, K., Suzuki, I., Fujita, Y.: Somatosensory computation for man-machine interface from motion capture data and musculoskeletal human model. IEEE Trans. Robot. (2004)

TSDAA 2015

Workshop on Time Series Data Analysis and its Applications (TSDAA 2015)

Basabi Chakraborty[1], Goutam Chakraborty[1], Masafumi Matsuhara[1], and Tetsuji Kuboyama[2]

[1] Graduate School of Software and Information Science, Iwate Prefectural University, 152-52 Aza Sugo, Takizawamura, Iwate 020-0693, Japan
{basabi, goutam,masafumi}@iwate-pu.ac.jp
[2] Computer Center, Gakushuin University, 1-5-1 Mejiro, Toshima ku, Tokyo 171-8588, Japan
ori-isAI2015@tk.cc.gakushuin.ac.jp

1 The Workshop

The Workshop on Time series Data analysis and its Applications (TSDAA 2015) was held on November 17, 2015 in Sosokan Building, Yagami Campus of Keio University, Yokohama, Kanagawa, Japan as a part of the JSAI International Symposia on AI 2015 (JSAI-isAI 2015), sponsored by The Japanese Society of Artificial Intelligence (JSAI). Around 20–25 persons from eight different countries attended this one day workshop.

The main objective of this workshop was to provide an interdisciplinary forum for discussion of different approaches and techniques of time series data analysis and their implementation in various real life applications. As time series data is abundant in nature, an unifying approach is needed to bridge the gap between traditional multivariate time series data analysis with state-of-the-art methodologies of data mining from real life time series data (numerical and text) in various applications ranging from medical and health related, biometrics or process industry to finance or economic data analysis or weather prediction.

The workshop comprised of three sessions with 13 presentations including 3 invited lectures. The first session was on Confidence in Time-Series Prediction with 3 presentations and an invited lecture by Prof. Maciej Huk from Department of Computer Science and Management, Wroclaw University of Technology, Wroclaw in Poland. Prof Huk talked about the problem of prediction disbelief of time series applications and proposed methods of avoiding it. The second session was on Applications of Time series Analysis in Medical Data with 3 presentations and an invited lecture by Prof. Keun Ho Ryu from College of Electrical and Computer Engineering, Chungbuk National University, Cheongju, South Korea. Prof. Ryu delivered lecture on Geosemantics knowledge mining model framework in social network media. In the third and final session on Uncertainties in Time series data acquisition, Prof. Cedric Bornand from University of Applied Sciences, Western Switzerland HES-SO, delivered invited lecture on careful simultaneous acquisition of time series data from heterogeneous information sources. In addition, there were 4 presentations in the final session. All the presentations were published by JSAI in The Proceedings of International Workshop on Time Series Data Analysis and its Applications (TSDAA 2015) ISBN : 978-4-915905-71-1 C3004(JSAI).

2   The Post-Workshop Proceedings

Four papers out of 10 contributory papers presented in the workshop has been selected for publication in this post-workshop proceedings after careful review by 3 PC members. One of the four papers deal with similarity measure for time series classification or clustering. Two papers presented applications of time series analysis in medical domain while the last one represented application of text time series processing in social media.

Yoshida et al. presented a comprehensive comparison of popular similarity measures for time series classification or clustering regarding computation time and classification accuracy by simulation experiment on 43 bench mark data sets. They then proposed a technique for improving the computation time of presently available popular algorithm DTW. Finally a new measure has been proposed for improvement of classification accuracy and its efficiency has been judged by simulation experiment. Nagayama et al. proposed an algorithm for extraction of infection propagation pattern from bacterial culture data in medical facility by using exhaustive search. Kamiyama et al. proposed an efficient algorithm for anomaly detection of periodic bio signals like ECG which can work well in real-time on computationally weak platforms like smart phones. The simulation experiments with MIT-BIH data set proves the efficiency of the new concept introduced in this paper. Ramamonjisoa proposed an algorithm for analyzing and summarizing comments regarding some events, disasters, political turmoil etc. over social media for a certain period of time. The proposed technique of analysis was found to be able to reveal some interesting hidden fact behind the event.

Acknowledgement

Let us take this opportunity to convey our sincere acknowledgement to all the members of organizing committee of JSAI-isAI 2015 for their sponsorship and various supports in organizing the workshop. We would also like to acknowledge the great help of all the PC members of our workshop in reviewing the papers. We would like to convey our heartiest thanks to the invited speakers for coming all the way to Japan and delivering excellent talks and finally thanks to all the speakers for whom the workshop could achieve such a success.

A Comparative Study of Similarity Measures for Time Series Classification

Sho Yoshida and Basabi Chakraborty[(✉)]

Graduate School of Software and Information Science, Iwate Prefectural University,
152-52 Aza Sugo, Takizawamura, Iwate 020-0693, Japan
g231n033@s.iwate-pu.ac.jp, basabi@iwate-pu.ac.jp

Abstract. Time series data are found everywhere in the real world and their analysis is needed in many practical situations. Multivariate time series data poses problem for analysis due to its dynamic nature and traditional machine learning algorithms for static data become unsuitable for direct application. A measure to assess the similarity of two time series is essential in time series processing and a lot of measures have been developed. In this work a comparative study of some of the most popular similarity measures has been done with 43 benchmark data set from UCR time series repository. It has been found that, on the average over the different data sets, DTW performs better in terms of classification accuracy but it has high computational cost. A simple processing technique for reducing the data set to lower computational cost without much degradation in classification accuracy is proposed and studied. A new similarity measure is also proposed and its efficiency is examined compared to other measures.

1 Introduction

Now-a-days univariate or multivariate time series (MTS) data is abundant in nature and in real life events. The example includes on line handwritten signature data or human gait data in the area of biometric authentication, stock market or exchange rate fluctuation in the area of financial analysis, EEG or ECG data in medical domain or temperature, humidity time series in weather pattern recognition. Observations of several parameters or features over a period of time constitute multivariate time series. The analysis of multi variate time series is essential for mining, prediction, classification or clustering of data in variety of domains and there are lot of techniques available for time series analysis [1]. Due to high volume and dynamical nature of MTS data, their analysis is a challenging task.

Due to importance for classification of time series data, various approaches have been developed ranging from Neural and Bayesian networks to Genetic Algorithms, Support Vector Machines and Characteristic Pattern Extraction [2]. Traditional classification techniques like Bayesian classifier or decision tree are modified for MTS data and temporal naive Bayesian model (T-NB) and temporal decision tree (T-DT) are developed [3]. In [4] MTS data is transformed to a

© Springer International Publishing AG 2017
M. Otake et al. (Eds.): JSAI-isAI 2015 Workshops, LNAI 10091, pp. 397–408, 2017.
DOI: 10.1007/978-3-319-50953-2_27

lower dimensional compact representation by extracting characteristic features to facilitate the use of classical machine learning algorithms for classification. Now for any classification task, pair-wise similarity measure for grouping the time series is the most important. Euclidean distance is widely used as the simple similarity measure. Dynamic time Warping (DTW) and its various variants are considered to be the most successful similarity measure. Among other measures, Longest Common Subsequence (LCSS) and edit distance are quite popular [5]. An algorithm for classification of MTS data is proposed by author in [6] with an earlier proposed similarity measure based on time delay embedding in [7].

In this work, a comparative study of popular similarity measures (including the proposed measures) for time series classification has been done with 43 bench mark data set from UCR repository [8]. DTW is found to be the most effective one regarding classification accuracy but it has high computational cost. The result also is consistent with other similar studies [9,10]. A simple preprocesing technique has been proposed to reduce the computational cost with minimum degradation in classification accuracy. A new similarity measure also has been proposed and its efficiency has been examined. In the next two sections, existing approaches for time series classification and popular similarity measures are presented in brief.

2   Approaches for Time Series Classification

Existing approaches for time series classification can be broadly classified into 3 categories [11]

1. *Feature based classification* in which a multidimensional time series is transformed into a feature vector and then classified by traditional classification algorithms. The choice of appropriate features plays an important role in this approach. A number of techniques has been proposed for feature subset selection by using compact representation of high dimensional MTS into one row to facilitate application of traditional feature selection algorithms like recursive feature elimination (RFE), zero norm optimization etc. [3,12]. Time series shapelets, characteristic subsequences of the original series, are recently proposed as the features for time series classification [13]. Another group of techniques extract features from the original time series by using various transformation techniques like Fourier, Wavelet etc. In [14], a family of techniques have been introduced to perform unsupervised feature selection on MTS data based on common principal component analysis (CPCA), a generalization of PCA for multivariate data items where all the data items have the same number of dimensions. Any distance metric is used for classification of the feature based representation of the time series data.

2. *Model based classification* in which a model is constructed from the data and the new data is classified according to the model that best fits it. Models used in time series classification are mainly statistical, such as Gaussian, Poisson, Markov and Hidden Markov Model (HMM) or based on neural networks.

Naive Bayes is the simplest model and is used in text classification [15]. Hidden Markov models (HMM) are successfully used for biological sequence classifications. Some neural network models such as recurrent neural network (RNN) are suitable for temporal data classification. Probabilistic distance measures are generally suitable for model based classification of time series.

3. *Distance based classification* in which a distance function which measures the similarity between two time series is used for classification. Similarity or dissimilarity measures are the most important component of this approach. Euclidean distance is the most widely used measure with 1NN classifier for time series classification. Though computationally simple, it requires two series to be of equal length and is sensitive to time distortion. Elastic similarity measures such as Dynamic Time Warping (DTW) [16] and its variants overcome the above problems and seem to be the most successful similarity measure for time series classification in spite of high computational cost.

3   Existing Similarity Measures

Some similarity measures popularly used for multivariate time series analysis for classification/clustering, prediction or mining problems are listed below.

3.1   Euclidean Distance

Euclidean measure is the simplest and the most popular dissimilarity measure.

The dissimilarity $D(x, y)$ between two time series x and y using any L_n norm is defined as

$$D_{ec}(x, y) = \left(\Sigma_{i=1}^{M}(x_i - y_i)^n \right)^{\frac{1}{n}} \tag{1}$$

where n is a positive integer, M is the length of the time series, x_i and y_i are the i-th elements of x and y time series respectively. For $n = 2$, we obtain Euclidean distance. This measure is difficult to use for time series of different lengths and having a time lag.

3.2   Fourier Coefficient Measure

Instead of comparing the raw time series, comparison can be done between the ith Fourier coefficients of the time series pair after the Discrete Fourier Transform. This measure falls under the category of feature based classification. The equation is given as

$$D_{fc}(x, y) = \left(\Sigma_{i=1}^{\theta}(\hat{x}_i - \hat{y}_i)^2 \right)^{\frac{1}{2}} \tag{2}$$

where $\hat{x}_i$ and $\hat{y}_i$ represent i-th Fourier coefficients of x and y time series and $\theta = \frac{M}{2}$, M is the length of the time series.

3.3 Auto Regression Coefficient Measure

This distance measure falls under the category of model based classification and uses the model parameters for calculating similarity values. Auto regression coefficients of two time series are calculated beforehand from AR(Auto Regressive) models and the distance between corresponding coefficients is taken as the dissimilarity measure. The number of AR coefficients is controlled by a parameter in this model and directly affects the speed of the similarity calculation.

3.4 Dynamic Time Warping (DTW) Distance Measure

Dynamic Time Warping (DTW) is a classic approach for computing dissimilarity between two time series. DTW belongs to the group of elastic measures and works by optimally aligning the time series in temporal domain so that the accumulated cost of the alignment is minimal. The accumulated cost can be calculated by dynamic programming, recursively applying

$$D_{i,j} = f(x_i, y_j) + \min\left(D_{i,j-1}, D_{i-1,j}, D_{i-1,j-1}\right) \tag{3}$$

for $i = 1 \ldots M$ and $j = 1 \ldots N$ where M and N are the length of the time series x and y respectively and $f(x_i, y_j) = \sqrt{(x_i - y_j)^2}$.

Currently DTW is the main benchmark against any promising new similarity measure, though its computational cost is quite high.

3.5 Edit Distance on Real Sequences

Edit distance on real sequences or EDR is an extension of original edit or Levensthein [17] distance to real valued time series. Computation of EDR formalized by dynamic programming is similar to DTW but $f(x_i, y_j)$ is different as follows:

$$m(x_i, y_j) = \Theta(\epsilon - f(x_i, y_j)) \tag{4}$$

where Θ is the Heaviside step function, such that $\Theta(z) = 1$ if $z \geq 0$ and 0 otherwise.

3.6 Time-Warped Edit Distance

Time-warped edit distance or TWED is an extension and combination of DTW and EDR [18]. TWED uses a penalty parameter λ and a stiffness parameter ν. For uniformly sampled time series, the formulation of TWED is as follows:

$$D_{i,j} = \min\left(D_{i,j} + \Gamma_{x,y}, D_{i-1,j} + \Gamma_x, D_{i,j-1} + \Gamma_y\right) \tag{5}$$

for $i = 1 \ldots M$ and $j = 1 \ldots N$ where
$\Gamma_{x,y} = f(x_i, y_j) + f(x_{i-1}, y_{j-1}) + 2\nu|i - j|$
$\Gamma_x = f(x_i, x_{i-1}) + \nu + \lambda$
$\Gamma_y = f(y_j, y_{j-1}) + \nu + \lambda$

3.7 Earlier Proposed Similarity Measure: CTE

A new approach for time series classification has been proposed by author in [6]. In the proposed approach, time series is represented by a multidimensional delay vector (MDV) by delay coordinate embedding which is a standard approach for analysis and modeling of nonlinear time series [19]. The similarity between two time series is measured by the proposed similarity measure *Cross Translation Error*(CTE) [20] based on MDV representation of time series.

A deterministic time series signal $\{s_n(t)\}_{t=1}^{T_n}(n = 1, 2, \ldots, N)$ can be embedded as a sequence of time delay co-ordinate vector $v_{s_n}(t)$ known as experimental attractor, with an appropriate choice of embedding dimension m and delay time τ for reconstruction of the original dynamical system as follows:

$$v_{s_n}(t) \equiv \{s_n(t), s_n(t+\tau), \ldots, s_n(t+(m-1)\tau)\}, \tag{6}$$

Now for correct reconstruction of the attractor, a fine estimation of embedding parameters (m and τ) is needed. There are variety of heuristic techniques for estimating those parameters [21].

Cross Translation Error (CTE) has been proposed in [20] for calculating similarity between two time series. The algorithm is described below, the details can be found in [20].

1. Multi-dimensional delay vector $v_{s_n}(t)$ can be generated from time series $\{s_n(t)\}_{t=1}^{T_n}(n = 1, 2, \ldots, N)$ based on Eq. (6). $(m + 1)$ dimensional vector $v_{s_n}(t)'$ including the normalized time index t/T_n is defined as follows;

$$v_{s_n}(t)' \equiv \{s_n(t), s_n(t+\tau), \ldots, s_n(t+(m-1)\tau), t/T_n\}. \tag{7}$$

2. Let $v_{s_i}(t)$ and $v_{s_e}(t)$ denote m-dimensional delay vectors generated from time series $s_i(t)$ and time series $s_e(t)$ respectively. $v_{s_i}(t)'$ and $v_{s_e}(t)'$ denote the corresponding $(m+1)$-dimensional vector including the normalized time index t/T_n.

3. A random vector $v_{s_i}(k)$ is picked up from $v_{s_i}(t)$. Let the nearest vector of $v_{s_i}(k)$ from $v_{s_e}(t)$ be $v_{s_e}(k')$. The index k' for the nearest vector is defined as follows;

$$k' \equiv \arg \min_t \|v_{s_i}(k)' - v_{s_e}(t)'\| \tag{8}$$

4. For the vectors $v_{s_i}(k)$ and $v_{s_e}(k')$, the transition in the each orbit after one step are calculated as follows;

$$V_{s_i}(k) = v_{s_i}(k+1) - v_{s_i}(k), \tag{9}$$
$$V_{s_e}(k') = v_{s_e}(k'+1) - v_{s_e}(k'). \tag{10}$$

5. Cross Translation Error (CTE) e_{cte} is calculated from $V_{s_i}(k)$ and $V_{s_e}(k')$ as

$$e_{cte} = \frac{1}{2}\left(\frac{|V_{s_i}(k) - \bar{V}|}{|\bar{V}|} + \frac{|V_{s_e}(k') - \bar{V}|}{|\bar{V}|}\right), \tag{11}$$

where $\bar{V}$ denotes average vector between $V_{s_i}(k)$ and $V_{s_e}(k')$.

6. e_{cte} is calculated for L times for different selection of random vector $v_{s_i}(k)$ and the median of e_{cte}^i $(i = 1, 2, \ldots, L)$ is calculated as

$$M(e_{cte}) = Median(e_{cte}^1, \ldots, e_{cte}^L). \tag{12}$$

The final cross translation error E_{cte} is calculated by taking the average, repeating the procedure Q times to suppress the statistical error generated by random sampling in the step (3).

$$E_{cte} = \frac{1}{Q} \sum_{i=1}^{Q} M_i(e_{cte}). \tag{13}$$

Cross translation error is a distance metric, so lower value of E_{cte} represents higher similarity. For multivariable time series, each dimension is considered separately as a single time series and represented by a multidimensional vector.

4   Comparative Study

The efficiency of a time series similarity measure is commonly evaluated by the classification accuracy it achieves with a distance based classifier. The most commonly used, effective and simple classifier is 1NN (nearest neighbour) classifier. Some works on time series classifier suggest that the best results of time series classification are achieved by nearest neighbor classifier [9].

4.1   Data Set Used

The benchmark data sets consisting of 43 different time series data from University of California, Riverside (UCR) time series repository [8] are used for the simulation experiments. Table 1 shows the data set description in brief. The training data is used as labeled data for classifier and classification accuracy is calculated on the test set. The average classification accuracy for twenty trials are noted for all the data sets.

4.2   Simulation Results

The simulation results of classification accuracies with different similarity measures are shown in Table 2 respectively. CTE(1) and CTE(2) represent two implementations of the measure. In CTE(1), CTE is calculated with fixed m and τ where m is taken as 3 and $\tau = 2$. In CTE(2), CTE is calculated with different m and τ for different time series. m and τ for each time series is calculated using popular method of mutual information [21]. Table 3 represents the average classification accuracy and computational cost over 43 data sets. From Table 2 it is found that no similarity measure exists which performs the best for all the data sets. From Table 3 it seems that on the average TWED and DTW are the best two measures though their computational cost is much higher than Euclid

Table 1. Data set used

Serial No.	Data set name	No. of class	Size of train set	Size of test set	Data length
1	50words	50	450	455	270
2	Adiac	37	390	391	176
3	Beef	5	30	30	470
4	CBF	3	30	900	128
5	ChlorineConcentration	3	467	3840	166
6	CinC_ECG_torso	4	40	1380	1639
7	Coffee	2	28	28	286
8	Cricket_X	12	390	390	300
9	Cricket_Y	12	390	390	300
10	Cricket_Z	12	390	390	300
11	DaitomSizeReduction	4	16	306	345
12	ECG200	2	100	100	96
13	ECGFivedays	2	23	861	136
14	FaceAll	14	560	1690	131
15	FaceFour	4	24	88	350
16	FacesUCR	14	200	2050	131
17	Fish	7	175	175	463
18	Gun_Point	2	50	150	150
19	Haptics	5	155	308	1092
20	InlineSkate	7	100	550	1882
21	ItalyPowerDemand	2	67	1029	24
22	Lighting2	2	60	61	637
23	Lighting7	7	70	73	319
24	MALLAT	8	55	2345	1024
25	MedicalImages	10	381	760	99
26	MoteStrain	2	20	1252	84
27	Oliveoil	4	30	30	570
28	OSULeaf	6	200	242	427
29	SonyAiboRS	2	2	601	70
30	SonyAiboRS2	2	27	953	65
31	StarLightCurves	3	1000	8236	1024
32	SwedishLeaf	15	500	625	128
33	Symbols	6	25	995	398
34	Synthetic_control	6	300	300	60
35	Trace	4	100	100	275
36	Two_Patterns	4	1000	4000	128
37	TwoLeadECG	2	23	1139	82
38	uWGL_X	8	896	3582	315
39	uWGL_Y	8	896	3582	315
40	uWGL_Z	8	896	3582	315
41	wafer	2	1000	6164	152
42	WordsSynonyms	25	267	638	270
43	yoga	2	300	3000	426

Table 2. Classification accuracy with different similarity measures

Data set Name	No. of class	Classification accuracy with								
		Euclid	Fourier	AR	DTW	EDR	CTE(1)	DTE	CTE(2)	TWED
50 words	50	0.67	0.63	0.21	0.71	0.72	0.48	**0.78**	0.64	0.76
Adiac	37	0.6	**0.65**	0.29	0.59	0.54	0.51	**0.65**	0.60	0.60
Beef	5	0.5	0.6	0.53	0.5	0.57	0.7	**0.87**	0.60	0.50
CBF	3	0.89	0.75	0.53	**0.99**	0.98	0.32	0.47	0.72	**0.99**
Cl. conc	3	0.62	0.67	0.68	0.62	0.66	0.64	**0.72**	0.64	0.63
CinCECGtorso	4	**0.94**	0.88	0.54	0.89	**0.94**	0.70	0.73	0.79	0.76
Coffee	2	0.75	**0.93**	0.86	0.82	0.5	0.61	0.90	0.79	0.75
CricketX	12	0.59	0.51	0.30	**0.78**	0.62	0.2	0.74	0.41	0.69
CricketY	12	0.67	0.49	0.22	0.76	0.64	0.18	0.72	0.39	**0.77**
CricketZ	12	0.64	0.53	0.28	**0.78**	0.53	0.18	0.71	0.41	0.75
DaiSizeRed	4	0.93	0.93	0.77	**0.96**	0.94	0.81	0.94	0.92	0.95
ECG200	2	0.89	**0.9**	0.79	0.79	0.89	0.85	0.86	0.78	0.85
ECG5days	2	0.78	0.74	0.74	0.79	**0.94**	0.78	0.86	0.92	0.80
FaceAll	14	0.72	**0.80**	0.36	0.77	**0.80**	0.57	0.74	0.58	0.78
Face4	4	0.84	0.73	0.42	0.84	**0.90**	0.5	0.16	0.49	0.88
FacesUCR	14	0.80	0.72	0.34	0.94	0.93	0.58	0.88	0.84	**0.95**
Fish	7	0.79	0.79	0.35	0.86	0.91	0.44	0.91	0.85	**0.94**
GunPt	2	0.95	0.92	0.79	0.88	0.97	0.86	**1.00**	0.97	0.98
Haptics	5	0.36	0.38	0.30	0.36	0.36	0.33	0.37	**0.40**	**0.40**
InlineSkate	7	0.35	0.30	0.35	0.37	0.3	0.23	**0.49**	0.28	0.43
ItalyPD	2	**0.96**	0.95	0.61	0.95	0.94	0.73	0.92	0.85	0.95
Lighting2	2	0.82	0.66	0.67	0.80	0.80	0.64	0.46	0.52	**0.84**
Lighting7	7	0.71	0.53	0.37	**0.77**	0.66	0.33	0.14	0.42	**0.77**
MALLAT	8	**0.92**	0.90	0.49	0.91	0.87	0.60	0.80	0.86	0.90
MedImage	10	0.70	0.69	0.5	**0.76**	0.66	0.53	0.73	0.61	0.74
MoteStrain	2	0.86	0.86	0.57	0.89	0.87	0.81	**0.93**	0.85	0.88
Oliveoil	4	0.83	0.83	0.50	**0.87**	**0.87**	0.6	0.8	0.83	0.83
OSULeaf	6	0.55	0.53	0.44	0.64	0.61	0.42	**0.88**	0.60	0.81
SonyAiboRS	2	0.69	0.69	**0.93**	0.73	0.71	0.64	0.43	0.82	0.68
SonyAiboRS2	2	**0.87**	0.84	0.86	0.84	0.85	0.80	0.63	0.84	**0.87**
StarLightC	3	0.85	0.82	0.83	0.89	0.87	0.78	**0.92**	0.87	0.88
SwedishLeaf	15	0.79	0.75	0.60	0.80	0.87	0.55	**0.88**	0.78	0.87
Symbols	6	0.90	0.87	0.74	0.95	0.95	0.82	0.24	0.92	**0.97**
Syncontl	6	0.88	0.79	0.51	**0.99**	0.92	0.31	0.48	0.66	0.96
Trace	4	0.76	0.80	0.84	**0.99**	0.68	0.46	0.97	0.87	0.97
TwoPattern	4	0.96	0.78	0.23	**1.00**	0.98	0.29	0.25	0.64	0.99
TwoLeadECG	2	0.74	0.78	0.67	0.95	0.79	0.81	0.84	**0.98**	0.97
uWGLX	8	0.74	0.73	0.26	0.73	0.64	0.41	0.12	0.55	**0.76**
uWGLY	8	**0.67**	0.63	0.29	0.64	0.48	0.34	0.12	0.41	0.66
uWGLZ	8	0.65	0.64	0.27	0.66	0.45	0.39	0.12	0.51	**0.67**
wafer	2	0.99	0.99	0.99	0.98	0.99	0.92	**1.00**	0.99	0.99
WordsSyn	25	0.63	0.59	0.21	0.67	0.54	0.46	**0.75**	0.58	0.70
yoga	2	0.83	0.83	0.61	0.84	0.51	0.71	**0.88**	0.79	0.86

Table 3. Average classification accuracy and computational cost

	Average values with								
	Euclid	Fourier	AR	DTW	EDR	CTE(1)	DTE	CTE(2)	TWED
Classification accuracy	0.76	0.73	0.59	0.79	0.75	0.61	0.67	0.69	0.81
Computation cost	$O(M)$	$O(M^2)$	$O(M^2)$	$O(M^2)$	$O(M^2)$	$O(M)$	$O(M^2)$	$O(M)$	$O(M^2)$

distance or CTE(1)/CTE(2). CTE(2) has improved efficiency over CTE(1). The bold figures in Table 2 represents the best value for each data set.

In closer examination of the data set characteristics, it is found that Auto regression measure works quite well for data sets with less number of classes as ECG200, ECGFivedays or Gunpoint with number of classes 2, but performance rapidly decreases for data sets with high number of classes like Words Synonyms, Adiac or 50Words. It has also been found that data sets with less time lag between the two samples of time series as in CinCECGtorso, Euclid distance measure produces highest classification accuracy while the data sets with more time lags between samples as in cricket_X, Cricket_Y or Cricket_Z, DTW produces the highest classification accuracy.

5   Proposals for Performance Improvement

On the average, for most of the data sets, DTW performs quite satisfactorily, but its computational cost is high. For the improvement of the performance of the similarity measure, to achieve high classification accuracy with low computational cost, a simple preprocessing technique is proposed here.

5.1   A Simple Preprocessing Method

The objective of this proposal is to reduce computational cost with minimum degradation of classification accuracy. So to lower the computational cost, the length of the time series for a particular data set is shortened by deleting some points.

The consecutive increasing stretch of values or consecutive decreasing stretch of values in the time series are examined and intermediate points in the increasing or decreasing stretches are removed as shown in Fig. 1. It is done by calculating the gradients of the consecutive points. With this processing, the number of points in the time series data set samples can become unequal in length for different samples and Euclid measure can not be applied. But by reducing with this pre processing method, the data points in time series samples can be reduced a lot and computational cost can be effectively lowered. The average classification accuracy over 43 data sets using DTW is obtained as 0.76 (compared to 0.79 for original data sets) while the computational cost can be reduced by 85% as is shown in Table 4. The classification accuracy obtained is same as Euclid measure of original time series with much lesser computational cost.

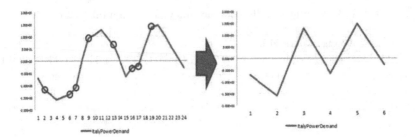

Fig. 1. Example of time series point reduction

Table 4. Performance with reduced data set

	Normal time series	Reduced time series
Average classification accuracy with DTW	0.79	0.76
Average computation time in sec	294707	43497

5.2   A New Similarity Measure

It is found that CTE is computationally light and can be used for comparing time series of unequal lengths unlike Euclid distance but average classification accuracy is poor compared to DTW. A new measure combining DTW and CTE is proposed here as DTE (Dynamic Translational Error) in which the calculation of distance in DTW is done by CTE in order to improve classification accuracy. The measure DTE (with m = 3 and $\tau = 2$) is used for classification of 43 time series and the results are shown in Tables 2 and 3. It seems that DTE outperforms DTW in many data sets and comes out to be the best measure in 13 data sets while DTW come out to be the best measure in 10 data sets, though on the average, DTW outperforms DTE. But the computational cost of DTE is high as it basically uses DTW algorithm.

6   Conclusion

Time series classification techniques are very important in processing practical data in various fields. For grouping or classifying samples of time series, it is necessary to define a good similarity or distance measure for calculation of similarity between two time series. As there are a lot of similarity measures available in literature, the performance of the measures need to be known before applying them to a particular application.

It is found that no similarity measure achieves consistently the best performance for all sorts of data set. The Euclid distance is the most simple and computational cost is the minimum (of the order of $O(m)$). So in many cases Euclid measure is taken as the baseline measure [10] and new measures are compared against Euclid measure in terms of computational cost. Dynamic time warping based measure achieves the best performance on the average over all the data sets but the computational cost is high.

A new similarity measure (DTE) combining DTW with earlier developed Cross translational error (CTE) has been proposed and its efficiency has been studied. Here we considered the parameters of DTE as we have done for CTE(1) i,e fixed m and τ. There is a scope for further investigation regarding the optimization of the parameters. Also a simple modification for achieving comparable results with DTW to the baseline Euclid measure has been proposed by reducing the data points in the time series judiciously. We would also like to use the data reduction method in conjunction with DTE in our future study. The results has been found to be promising. The study on the proposal of DTE presented here is elementary and currently an extensive evaluation of the performance of DTE and measure of reducing its computational cost is undertaken.

References

1. Kantz, H., Schreiber, T.: Nonlinear Time Series Analysis. Cambridge University Press, Cambridge (2004)
2. Buza, K., Nanopoulos, A., Schmidt-Thieme, L.: Time series classification based on individual error prediction. In: IEEE International Conference on Computational Science and Engineering (2010)
3. Chakraborty, B.: Feature selection and classification techniques for multivariate time series. In: Proceedings of ICICIC 2007, September 2007
4. Zhang, X., Wu, J., et al.: A novel pattern extraction method for time series classification. Optimiz. Engin. **10**(2), 253–271 (2009)
5. Buza, K., Nanopoulos, A., Schmidt-Thieme, L.: Fusion of similarity measures for time series classification. In: Corchado, E., Kurzyński, M., Woźniak, M. (eds.) HAIS 2011. LNCS (LNAI), vol. 6679, pp. 253–261. Springer, Heidelberg (2011). doi:10.1007/978-3-642-21222-2_31
6. Chakraborty, B.: A proposal for classification of multi sensor time series database on time delay embedding. In: Proceedings of 8th International Conference on Sensing Technology ICST, Liverpool, England, pp. 31–35, September 2014
7. Chakraborty, B., Manabe, Y.: An efficient approach for person authentication using online signature verification. In: Proceedings of SKIMA 2008, pp. 12–17 (2008)
8. Keogh, E., Zhu, Q., Hu, B., Hao, Y., Xi, X., Wei, L., Ratanamahatana, C.A.: The UCR Time Series Classification/Clustering (2011). www.cs.ucr.edu/~eamonn/time_series_data/
9. Serra, J., Arcios, J.: An empirical evaluation of similarity measures for time series classification. Knowl. Based Syst. **67**, 305–314 (2014)
10. Giusti, R., Batista, G.: An empirical comparison of dissimilarity measures for time series classification. In: Brazilian Conference on Intelligent Systems (BRACIS 2013) (2013)
11. Xing, Z., Pei, J., Keogh, E.: A brief survey on sequence classification. ACM SIGKDD Explor. Newslett. **12**(1), 40–48 (2010)
12. Lal, T.N., et al.: Support vector channel selection in BCI. IEEE Trans. Biomed. Eng. **51**(6), 1003–1010 (2004)

13. Ye, L., Keogh, E.: Time series shapelets: a new primitive for data mining. In: Proceedings of ACM SIGKDD International Conference of Knowledge Discovery and Data Mining (2009)

14. Yoon, H., Yang, K., Sahabi, C.: Feature subset selection and feature ranking for multivariate time series. IEEE Trans. Knowl. Data Eng. **17**(9), 1186–1198 (2005)

15. Kim, S.B., Han, K.S., et al.: Some effective techniques for naive bayes text classification. IEEE Trans. Knowl. Data Eng. **18**(11), 1457–1466 (2006)

16. Wang, X., Mueen, A., Ding, H., Trajcevski, G., Scheuermann, P., Keogh, E.: Experimental comparison of representation methods and distance measures for time series data. Data Min. Knowl. Discov. **26**, 275–309 (2013)

17. Levenshtein, I.: Binary codes capable of correcting deletions, inssertions and reversals. Sov. Phys. Dokl. **10**, 707–710 (1966)

18. Marteau, F.: Time warp edit distance with stiffness adjustment for time series matching. IEEE Trans. PAMI **31**, 306–318 (2009)

19. Alligood, K., Sauer, T., Yorke, J.A.: Chaos: An Introduction to Dynamical Systems. Springer, New York (1997)

20. Manabe, Y., Chakraborty, B.: Identity detection from online handwriting time series. In: Proceedings of SMCia08, pp. 365–370 (2007)

21. Aberbanel, H.D.I.: Analysis of Observed Chaotic Data'. Springer, New York (1996)

Extracting Propagation Patterns from Bacterial Culture Data in Medical Facility

Kazuki Nagayama[1], Kouichi Hirata[4(✉)], Shigeki Yokoyama[2], and Kimiko Matsuoka[3]

[1] Department of Artificial Intelligence,
Graduate School of Computer Science and Systems Engineering,
Kyushu Institute of Technology, Kawazu 680-4, Iizuka 820-8502, Japan
nagayama@dumbo.ai.kyutech.ac.jp

[2] KD-ICONS Co. Ltd., Ohmori-minami 4-6-15-303, Ohta, Tokyo 143-0013, Japan
yokoyama@kd-icons.co.jp

[3] Osaka Prefectural General Medical Center,
Bandai-higashi 3-1-56, Sumiyoshi, Osaka 558-8558, Japan

[4] Department of Artificial Intelligence, Kyushu Institute of Technology,
Kawazu 680-4, Iizuka 820-8502, Japan
hirata@ai.kyutech.ac.jp

Abstract. In this paper, we formulate *propagation patterns* as the pairs of records in the same bacterial culture occurring within a fixed span in bacterial culture data. Then, we design the exhaustive search algorithm to extract all of the propagation patterns from bacterial culture data based on the *extended principle* of the *2-dimensional career map* to determine whether two records in bacterial culture data belong to the same bacterial culture or the different ones. In particular, we focus on *infectious propagation patterns*, in which two patients are not identical, and they are in the same room and/or treated by the same physician. Finally, we give the experimental results to extract all of the propagation patterns and analyze them.

1 Introduction

In recent years, since hospital-acquired infection becomes a big social problem, it is important to predict and prevent hospital-acquired infection in medical facility. In our previous works [3–5], we extract the time-related rules such as episodes [6] or temporal patterns [1,8] representing *replacements of bacteria* and *changes for drug resistant*, which are regarded as the factors of hospital-acquired infection, from *bacterial culture data*. Here, the bacterial culture data consists of ID, patient ID, date, sample, detected bacterium and the antibiograms for more than 100 antibiotics. The values of antibiograms are one of empty (ε), susceptible (S), intermediate (I) and resistant (R).

This work is partially supported by Grand-in-Aid for Scientific Research 24300060, 24240021, 26280085 and 15K12102 from the Ministry of Education, Culture, Sports, Science and Technology, Japan.

ⓒ Springer International Publishing AG 2017
M. Otake et al. (Eds.): JSAI-isAI 2015 Workshops, LNAI 10091, pp. 409–417, 2017.
DOI: 10.1007/978-3-319-50953-2_28

Whereas our previous works have succeeded to extract such time-related rules on some level, it remains a future work to extract them by adding *place data* consisting of ID, room number, bed number and physician number. First, in this paper, we incorporate the bacterial culture data with the place data.

In our previous works [3–5], we assume that all of the records with the same detected bacterium in bacterial culture data belong to the same bacterial culture. Then, in order to extract more accurate and reliable rules, it is necessary to distinguish records belonging to the different bacterial cultures.

However, it is too expensive in general to determine whether two records belong to the same bacterial culture or the different ones exactly, for example, a pulsed-field gel electrophoresis (PFGE), in molecular epidemiology. Furthermore, when dealing with passed bacterial culture data, we cannot apply such a method to them in order to identify bacterial cultures.

The 2-*dimensional career map* (*2DCM*) [2] is an alert system for hospital-acquired infection with observing antibiograms in bacterial culture data. The principle of the 2DCM is to determine that two records belong to the different bacterial cultures if there exists an antibiotic for which value of the antibiograms in a record is S and that in another record is R.

Since we can apply the principle of the 2DCM to passed bacterial culture data, we can regard it as an alternative and approximate method to identify bacterial cultures from antibiograms. On the other hand, when applying the principle of the 2DCM to our bacterial culture data directly, it implies complex identification of bacterial cultures [7]. The reason is that our bacterial culture data is too sparse to apply just the principle of the 2DCM.

Hence, in this paper, we extend the principle of the 2DCM applicable to our bacterial culture data by adding two rules, which we call the *extended principle* of the 2DCM. The first rule is to exclude the record such that every value in the antibiograms is either ε or S from bacterial culture data. The second rule is to regard that two records sharing k or more antibiotics for which value of the antibiograms in a record is ε and that in another record is one of S, I or R belong to the different bacterial cultures.

In this paper, we formulate *propagation patterns* as the pairs of records in the same bacterial culture occurring within a fixed span in bacterial culture data. Then, based on the extended principle of the 2DCM, we design the exhaustive search algorithm to extract all of the propagation patterns within a fixed span from bacterial culture data. In particular, we focus on *infectious propagation patterns*, in which two patients are not identical, and they are in the same room and/or treated by the same physician, which implies that they are possible to cause the hospital infection. Finally, we give the experimental results to extract all of the propagation patterns and analyze them.

2   Extended Principle of 2-Dimensional Career Map

In this paper, we deal with data provided from Osaka Prefectural General Medical Center from 1999 to 2007. In order to extract the propagation patterns of the same bacterial culture, first we incorporate the *bacterial culture data* consisting of 56,478 records with the *place data* consisting of 78,853 records.

The bacterial culture data consists of the following attributes, where the value of antibiograms is one of empty (ε), susceptible (S), intermediate (I) and resistant (R):

ID, patient ID, date, sample, detected bacterium and antibiograms for 108 antibiotics.

On the other hand, the place data consists of the following attributes:

ID, room No., bed No. and physician ID.

By incorporating the bacterial culture data with the place data through ID, we obtain the data consisting of 31,300 records in the following forms.

ID, patient ID, date, room No., bed No., physician ID, sample, detected bacterium and antibiograms for 108 antibiotics.

In this paper, we call the above data *bacterial culture data*.

The *2-dimensional career map* (*2DCM*) [2] is an alert system for hospital-acquired infection with observing antibiograms in bacterial culture data. In this paper, we adopt the principle of the 2DCM as an alternative and approximate method to determine whether two records with the same detected bacterium in bacterial culture data belong to the same bacterial culture or the different ones from antibiograms. The principle of the 2DCM is to determine that:

Two records belong to the different bacterial cultures if there exists an antibiotic for which value of the antibiograms in a record is S and that in another record is R.

For example, consider the bacterial culture data consisting of 8 records illustrated in Table 1. Here, "bac" denotes the detected bacterium and a_i denotes an antibiogram ($1 \leq i \leq 10$).

Table 1. An example of bacterial culture data.

ID	$\cdots$	bac	a_1	a_2	a_3	a_4	a_5	a_6	a_7	a_8	a_9	a_{10}
1	$\cdots$	SA	ε	ε	S	S	R	ε	I	R	I	S
2	$\cdots$	SA	ε	ε	S	S	S	ε	S	S	S	S
3	$\cdots$	SA	ε	ε	S	S	I	ε	S	S	S	S
4	$\cdots$	SA	ε	ε	S	S	R	ε	S	R	S	I
5	$\cdots$	PA	S	R	ε	S	ε	S	I	R	I	S
6	$\cdots$	PA	S	I	ε	S	ε	S	R	R	I	S
7	$\cdots$	PA	S	S	ε	S	ε	S	I	R	S	S
8	$\cdots$	PA	ε	S	ε	I	ε	ε	ε	ε	ε	ε

Then, by using the principle of the 2DCM, for the bacterium SA, the group of the records 1 and 4 and the group of the records 2 and 3 belong to the different bacterial cultures, because of the value of a_8. For the bacterium PA, the records 5 and the group of the records 7 and 8 belong to the different bacterial cultures, because of the value of a_2.

However, it is known that the principle of the 2DCM implies the complex identification of bacterial culture [7]. The reason of the complex identification is that our bacterial culture data is too sparse to apply just the principle of the 2DCM. Hence, in this paper, we extend the principle of 2DCM as follows.

1. We exclude the record such that every value in the antibiograms is either ε or S from bacterial culture data.
   (In Table 1, we exclude the record 2 for the bacterium SA.)
2. We regard that two records sharing k (called a *sharing threshold*) or more antibiotics for which value of the antibiograms in a record is ε and that in another record is one of S, I or R belong to the different bacterial cultures.
   (In Table 1, for $k = 6$, we regard that the bacterium PA in the record 8 and that in the records 5, 6 and 7 belong to the different bacterial cultures.)

We call the principle of the 2DCM with the above two rules an *extended principle* of the 2DCM. We regard two records not belonging to the different bacterial cultures as those belonging to *the same bacterial culture*. In Table 1, for the bacterium SA, we divide 3 records into two bacterial cultures of $\{1, 4\}$ and $\{3\}$. For the bacterium PA, we divide 4 records into three bacterial cultures of $\{5\}$, $\{6, 7\}$ and $\{8\}$.

3   Extracting Propagation Patterns

In this paper, we call the pairs of records in the same bacterial culture occurring within a fixed span in bacterial culture data *propagation patterns*. Then, we design the algorithm to extract all of the propagation patterns from bacterial culture data.

Let D be the set of all the records in bacterial culture data and B the set of all the bacteria. For a bacterium $b \in B$, let $D_b \subseteq D$ be the set of records of which bacterium is b. We denote the number of antibiotics in antibiograms by n (in this paper, $n = 108$). For an antibiotic a_i ($1 \leq i \leq n$) and a record $r \in D$, we denote the value of antibiograms for a_i in r by $a_i(r)$, that is, $a_i(r) \in \{\varepsilon, S, I, R\}$.

For $r, r' \in D$, we define the following functions δ, σ and γ. Here, $|S|$ denotes the cardinality of a set S.

1. $\delta(r, r') = 1$ if there exists an antibiotic a_i ($1 \leq i \leq n$) such that either (1) $a_i(r) = S$ and $a_i(r') = R$ or (2) $a_i(r) = R$ and $a_i(r') = S$; 0 otherwise.
2. $\sigma(r) = |\{a_i \mid a_i(r) \in \{\varepsilon, S\}\}|$.
3. $\gamma(r, r') = \left| \left\{ a_i \left| \begin{array}{l} (1) \ a_i(r) = \varepsilon \text{ and } a_i(r') \in \{S, I, R\} \text{ or} \\ (2) \ a_i(r) \in \{S, I, R\} \text{ and } a_i(r') = \varepsilon \end{array} \right\} \right|\right.$.

```
    procedure PROPPAT(D)
        /* D: the set of all the records in bacterial culture data */
        /* B: the set of all the bacteria in bacterial culture data */
        /* k: a sharing threshold (k = 6), s: a fixed span */
1       PP ← ∅;
2       foreach b ∈ B do
3           foreach r ∈ D_b s.t. σ(r) < n do
4               PP(r) ← ∅;
5               foreach r' ∈ D_b − {r} s.t. σ(r') < n and |d(r) − d(r')| ≤ s do
6                   if δ(r, r') = 0 and γ(r, r') < k then
7                       PP(r) ← PP(r) ∪ {(r, r')};

8               PP ← PP ∪ PP(r);

9       output PP;
```

Algorithm 1. PROPPAT.

By using the above functions, we design the exhaustive search algorithm PROPPAT in Algorithm 1 to extract all of the propagation patterns within a fixed span s. Here, we denote the date of r by $d(r)$ and set a sharing threshold k to 6.

In the algorithm PROPPAT in Algorithm 1, the condition that $\delta(r, r') = 0$ in line 6 represents the principle of the 2DCM, the conditions that $\sigma(r) < n$ in line 3 and $\sigma(r') < n$ in line 5 represent the first rule in the extended principle of the 2DCM and the condition that $\gamma(r, r') < k$ in line 6 represents the second rule in the extended principle of the 2DCM. Also, the condition that $|d(r) − d(r')| \leq s$ in line 5 guarantees that both r and r' occur within a fixed span s.

We divide into 8 kinds of propagation patterns whether a patient is same or different, a room is same or different and a physician is same or different, respectively. We refer the 8 propagation patterns to sss, ssd, sds, sdd, dss, dsd, dds and ddd for a patient, a room and a physician, where s and d mean "same" and "different", respectively. In particular, we focus on the following patterns:

1. dss: the different patient, the same room and the same physician;
2. dsd: the different patient, the same room and the different physician;
3. dds: the different patient, the different room and the same physician.

The patterns dss, dsd and dds claim that either a physician or a room, a room and a physician are possible to cause the hospital infection, respectively. We call them *infectious propagation patterns*.

4   Experimental Results

In this section, we give the experimental results to extract all of the propagation patterns of the same bacterial culture and analyze them.

Table 2 illustrates the computation time to extract all of the propagation patterns by using the algorithm PROPPAT in Sect. 2. In this paper, we set a fixed span s to 5 days, 10 days, 15 days, 20 days, 25 days and 30 days, because no propagation of bacteria occur over 30 days. The computer environment is that

Table 2. The computation time to extract all of the propagation patterns.

Span (days)	5	10	15	20	25	30
Time (ms)	7017	8082	8624	9262	10287	11337

CPU is Intel(R) Core(TM) i7-3770 3.40 GHz, RAM is 8 GB, OS is Windows 7 Home Premium (64bit) and programming language is Borland C++ 5.5.1.

Table 2 shows that the computation time increases almost linearly (but not exponentially) when the span increases 5 days.

In the remainder of this section, we focus on the following 12 bacteria, that are top 12 bacteria whose number of extracted propagation patterns is large. We refer them to Bx ($1 \leq x \leq 12$).

B1: Staphylococcus aureus,
B2: Pseudomonas aeruginosa,
B3: Klebsiella pneumoniae,
B4: Escherichia coli,
B5: Enterococcus faecalis,
B6: Enterobacter cloacae,

B7: Candida albicans,
B8: Candida glabrata,
B9: Serratia marcescens,
B10: Staphylococcus epidermidis,
B11: Streptococcus pneumoniae,
B12: Enterococcus faecium.

Tables 3 and 4 illustrate the number of all of the propagation patterns (denoted by #), the ratio (%) of propagation patterns and the ratio (%) of infectious propagation patterns of dss, dsd and dds (denoted by †) within 5 days, 10 days and 15 days (Table 3) and 20 days, 25 days and 30 days (Table 4), respectively.

Tables 3 and 4 claim the following statements.

1. For all of the 12 bacteria, the number of propagation patterns and the ratio of the pattern ddd tends to increase when the span s increases. Also the pattern ddd is the most frequent propagation pattern.
2. For the 11 bacteria except B11, the pattern sss is the second most frequent propagation pattern and the ratio of the infectious propagation patterns is less than 6% for every span.
3. For Streptococcus pneumoniae (B11), the infectious propagation pattern dds is the second most frequent propagation pattern. Also the infectious propagation pattern dsd is the fourth most frequent propagation pattern within 5 days and the third most frequent propagation pattern within other spans.
4. The ratio of the pattern sss for the bacteria B5, B6, B9, B10 and B12 is more than 5% for every span. In particular, Enterococcus faecium (B12) have the largest ratio of not only the pattern sss (more than 12%) but also the patterns ssd and sds (more than 7%) and the smallest ratio of the pattern ddd (less than 66%) for every span.
5. Streptococcus pneumoniae (B11) have the largest ratio of infectious propagation patterns which is about 15%. Staphylococcus epidermidis (B10) have the second largest ratio of infectious propagation patterns which is about 4%. On the other hand, Candida albicans (B7) within 5 days and 15 days, Candida glabrata (B8) within 10 days and 20 days and Enterococcus faecalis (B5) within 25 days and 30 days have the smallest ratio of infectious propagation patterns.

Table 3. The number of all of the propagation patterns (denoted by #), the ratio (%) of propagation patterns and the ratio (%) of infectious propagation patterns (denoted by †) for the 12 bacteria within 5 days (upper), 10 days (center) and 15 days (lower).

5	B1	B2	B3	B4	B5	B6	B7	B8	B9	B10	B11	B12
#	29507	6315	2654	2451	1944	1708	1428	615	567	513	367	370
sss	5.67	8.23	5.35	6.32	14.35	14.40	6.02	4.88	17.11	17.93	6.81	26.22
ssd	2.46	4.61	3.01	3.96	7.46	9.25	2.87	3.90	9.35	7.02	1.63	11.08
sds	0.95	0.79	0.94	1.10	1.75	3.16	1.54	1.63	3.17	3.12	0.54	7.03
sdd	1.01	1.05	1.24	1.39	2.78	2.93	0.98	0.98	1.23	3.12	0.54	3.78
dss	0.08	0.03	0.11	0.08	0.05	0.00	0.00	0.00	0.00	0.00	1.63	0.00
dsd	0.32	0.38	0.41	0.61	0.36	0.12	0.21	0.33	0.18	0.00	5.18	0.00
dds	2.52	2.50	2.45	1.63	1.80	2.63	1.47	1.63	1.76	3.31	7.90	2.43
ddd	86.98	82.41	86.47	84.90	71.45	67.51	86.90	86.67	67.20	65.50	75.75	49.46
†	2.92	2.91	2.98	2.33	2.21	2.75	1.68	1.95	1.94	3.31	14.71	2.43
10	B1	B2	B3	B4	B5	B6	B7	B8	B9	B10	B11	B12
#	58147	12342	5189	4833	3630	3183	2771	1144	1002	884	651	602
sss	4.34	6.64	4.28	4.28	11.07	10.78	4.47	4.28	13.87	13.12	4.76	21.10
ssd	2.12	4.01	2.27	3.50	7.19	7.67	2.49	3.41	8.08	5.32	1.08	9.80
sds	1.15	1.14	1.19	1.49	2.20	3.46	1.62	1.66	3.69	3.73	0.31	8.14
sdd	1.18	1.56	2.00	1.63	3.25	3.83	1.23	2.36	2.50	3.85	0.61	3.82
dss	0.10	0.07	0.10	0.08	0.06	0.00	0.04	0.00	0.20	0.23	1.69	0.17
dsd	0.43	0.55	0.46	0.48	0.39	0.41	0.25	0.26	0.20	0.34	4.92	0.17
dds	2.38	2.54	2.37	1.78	1.87	2.48	1.91	1.75	2.40	3.28	8.60	3.16
ddd	88.30	83.49	87.32	86.76	73.97	71.38	87.98	86.28	69.06	70.14	78.03	53.65
†	2.91	3.16	2.93	2.34	2.31	2.89	2.20	2.01	2.79	3.85	15.21	3.49
15	B1	B2	B3	B4	B5	B6	B7	B8	B9	B10	B11	B12
#	85000	18203	7616	7028	5127	4527	3921	1615	1413	1231	947	785
sss	3.55	5.47	3.37	3.60	8.84	8.77	3.52	3.53	11.89	9.83	3.59	18.47
ssd	1.79	3.33	1.93	3.07	6.16	5.90	2.01	2.66	6.65	4.22	0.84	8.79
sds	1.17	1.19	1.06	1.39	2.34	3.56	1.43	1.92	4.10	3.41	0.21	8.92
sdd	1.21	1.49	1.67	1.91	3.02	3.80	1.10	2.11	2.62	3.82	0.42	3.57
dss	0.11	0.10	0.11	0.07	0.12	0.11	0.08	0.00	0.28	0.16	1.58	0.38
dsd	0.52	0.73	0.50	0.57	0.51	0.66	0.41	0.62	0.35	1.06	4.75	0.25
dds	2.27	2.53	2.23	1.72	1.74	2.56	1.81	1.86	2.69	3.49	9.08	2.68
ddd	89.39	85.17	89.13	87.66	77.28	74.64	89.65	87.31	71.41	74.00	79.51	56.94
†	2.90	3.36	2.84	2.36	2.36	3.34	2.30	2.48	3.33	4.71	15.42	3.31

Table 4. The number of all of the propagation patterns (denoted by #), the ratio (%) of propagation patterns and the ratio (%) of infectious propagation patterns (denoted by †) for the 12 bacteria within 20 days (upper), 25 days (center) and 30 days (lower) respectively.

20	B1	B2	B3	B4	B5	B6	B7	B8	B9	B10	B11	B12
#	107153	22849	9554	8936	6360	5575	4893	2025	1717	1517	1177	914
sss	3.00	4.71	2.91	3.00	7.36	7.46	2.96	3.06	10.37	8.17	2.97	16.41
ssd	1.57	2.89	1.67	2.70	5.31	4.88	1.70	2.17	6.17	3.56	0.68	7.99
sds	1.13	1.12	0.93	1.26	2.25	3.44	1.19	1.73	3.79	3.03	0.17	7.77
sdd	1.17	1.36	1.48	1.76	2.72	3.41	0.94	1.83	2.74	3.63	0.42	3.28
dss	0.11	0.13	0.10	0.08	0.16	0.13	0.08	0.00	0.23	0.13	1.61	0.33
dsd	0.56	0.84	0.61	0.67	0.58	0.72	0.47	0.69	0.58	1.25	4.42	0.77
dds	2.22	2.49	2.20	1.84	1.86	2.55	1.90	1.73	3.03	3.63	9.09	2.74
ddd	90.24	86.46	90.10	88.70	79.76	77.42	90.76	88.79	73.09	76.60	80.63	60.72
†	2.89	3.45	2.91	2.59	2.59	3.39	2.45	2.42	3.84	5.01	15.12	3.83
25	B1	B2	B3	B4	B5	B6	B7	B8	B9	B10	B11	B12
#	133012	28232	11821	11013	7770	6673	6115	2512	2112	1846	1457	1056
sss	2.56	4.08	2.49	2.57	6.24	6.56	2.40	2.47	9.09	6.72	2.40	14.49
ssd	1.38	2.49	1.36	2.41	4.49	4.12	1.37	1.79	5.40	2.98	0.55	7.67
sds	1.06	1.09	0.83	1.22	2.07	3.21	1.03	1.43	4.17	2.65	0.14	7.58
sdd	1.10	1.27	1.31	1.79	2.64	3.22	0.78	1.47	2.60	3.03	0.41	3.31
dss	0.11	0.12	0.14	0.08	0.14	0.12	0.11	0.00	0.19	0.16	1.65	0.28
dsd	0.62	0.93	0.74	0.62	0.60	0.87	0.70	0.80	0.57	1.19	4.67	0.66
dds	2.18	2.50	2.27	1.85	1.79	2.64	1.91	1.87	3.27	3.47	8.51	2.65
ddd	90.98	87.52	90.87	89.47	82.02	79.26	91.68	90.17	74.72	79.79	81.67	63.35
†	2.91	3.56	3.14	2.55	2.54	3.63	2.73	2.67	4.02	4.82	14.82	3.60
30	B1	B2	B3	B4	B5	B6	B7	B8	B9	B10	B11	B12
#	157930	33432	14092	13183	9089	7826	7105	2933	2431	2139	1738	1199
sss	2.24	3.61	2.19	2.26	5.40	5.78	2.14	2.15	8.02	5.89	2.01	12.84
ssd	1.23	2.20	1.16	2.19	3.89	3.53	1.20	1.57	4.81	2.57	0.52	7.01
sds	1.01	1.06	0.77	1.12	2.05	2.84	0.93	1.30	3.99	2.34	0.12	7.09
sdd	1.05	1.14	1.18	1.73	2.40	3.03	0.70	1.30	2.43	2.85	0.40	3.17
dss	0.11	0.11	0.12	0.10	0.13	0.12	0.10	0.00	0.16	0.28	1.55	0.58
dsd	0.66	0.97	0.78	0.71	0.61	0.88	0.77	0.85	0.62	1.17	4.78	0.75
dds	2.16	2.54	2.38	1.84	1.77	2.75	1.93	1.84	3.29	3.23	8.52	3.25
ddd	91.54	88.37	91.42	90.05	83.75	81.09	92.23	91.00	76.68	81.67	82.11	65.30
†	2.93	3.61	3.28	2.65	2.51	3.74	2.80	2.69	4.07	4.68	14.84	4.59

5   Conclusion

In this paper, we have designed the exhaustive search algorithm PROPPAT to extract all of the propagation patterns from bacterial culture data based on the extended principle of the 2DCM. Then, we have analyzed the propagation patterns and, in particular, the infectious propagation patterns such that a patient is different and either a room or a physician or both is same.

It is a future work to design a more efficient algorithm than the exhaustive search in this paper. It is also a future work to introduce the (infectious) propagation patterns with more useful forms than pairs of records and to design the algorithm to extract them. Furthermore, it is a future work to analyze the infectious propagation patterns from the viewpoint of not only the frequency but also the attributes.

Concerned with the statement 4 in the last of Sect. 4, the distribution of the ratios in Tables 3 and 4 for Enterococcus faecium is different from other bacteria. Then, it is a future work to investigate the medical reason.

Concerned with the statements 3 and 5 in the last of Sect. 4, for Streptococcus pneumoniae, the cavity of nose is the sample increasing the ratio of infectious propagation patterns. On the other hand, Streptococcus pneumoniae are known not to infect by airborne infection. Hence, it is a future work to investigate the medical reason why the ratio is much larger than others.

Acknowledgment. The authors would like to thank anonymous referees of TSDAA 2015 for valuable comments to revise the previous version.

References

1. Allen, J.: Maintaining knowledge about temporal intervals. Commun. ACM **26**, 832–843 (1983)
2. Fujimoto, S.: Control for hospital acquiared infection by automatic classification of antibiogram and 2 dimensional career map. IASR **32**, 9–10 (2011). (in Japanese)
3. Hirata, K., Motoyama, K., Yokoyama, S., Matsuoka, K.: Temporal interrelations of bacteria based on the occurrence of time. In: Proceedings of CME 2010, pp. 188–192 (2010)
4. Katoh, T., Hirata, K., Arimura, H., Yokoyama, S., Matsuoka, K.: Extracting sequential episodes representing replacements of bacteria from bacterial culture data. In: Proceedings of CME 2009, 4 p. (2009)
5. Katoh, T., Hirata, K., Arimura, H., Yokoyama, S., Matsuoka, K.: Aligned bipartite episodes between the genera of bacteria. In: Proceedings of CME 2010, pp. 193–197 (2010)
6. Mannila, H., Toivonen, H., Verkamo, A.I.: Discovery of frequent episodes in event sequences. Data Min. Knowl. Discov. **1**, 259–289 (1997)
7. Nagayama, K., Eto, K., Haraguchi, N., Hirata, K.: Inferring bacterial culture from identification graph of bacteria for antibiogram. In: Proceedings of Hinokuni Symposium, IPSJ Kyushu (2014). (in Japanese)
8. Papapetrou, P., Kollios, G., Sclaroff, S., Gunopulos, D.: Mining frequent arrangements of temporal intervals. Knowl. Inf. Syst. **21**, 133–171 (2009)

Real-Time Anomaly Detection of Continuously Monitored Periodic Bio-Signals Like ECG

Takuya Kamiyama[1](✉) and Goutam Chakraborty[2]

[1] Graduate School of Software and Information Science,
Iwate Prefectural University, Takizawa, Iwate, Japan
g231m006@s.iwate-pu.ac.jp
[2] Department of Software and Information Science,
Iwate Prefectural University, Takizawa, Iwate, Japan
goutam@iwate-pu.ac.jp

Abstract. In this paper, we proposed an efficient heuristic algorithm for real-time anomaly detection of periodic bio-signals. We introduced a new concept, "mother signal" which is the average of normal subsequences of one period length. Their number is overwhelmingly large compared to anomalies. From the time series, first we find the fundamental time period, assuming the period to be stable over the whole time. Next, we find the normal subsequence of length equal to time-period and call it the "mother signal". When the distance of a subsequence of same length is large from the mother signal, we identify it as anomaly. While calculating the distance, we ensure that it is not large due to time shift. To ensure that, we shift-and-rotate the subsequence in step of one slot at a time and find the minimum distance of all such comparisons. The proposed heuristic algorithm using mother signal is efficient. Results are compared and found to be similar to that obtained using brute force comparisons of all possible pairs. Computational costs are compared to show that the proposed method is more efficient compared to existing works.

Keywords: Periodic time series · Anomaly detection · Fundamental period · Clustering

1 Introduction

Recently, systematic collection, storing and analysis of medical data is getting adapted everywhere and being used for personalized health-care, medical informatics, drug testing and a plethora of applications. There exist healthcare systems for mobile devices which is used on a daily basis.

In this work, we analyze bio-signals for healthcare applications. The main motivation of this work is to detect anomaly in real-time, on computationally weak platforms like smart phones. Anomalies in bio-signal can detect/predict heart disease, pulse failure or other kinds of life-threatening situations.

M. Otake et al. (Eds.): JSAI-isAI 2015 Workshops, LNAI 10091, pp. 418–427, 2017.
DOI: 10.1007/978-3-319-50953-2_29

2   Related Works

There are many mobile applications to record bio-signals. The recorded data can be uploaded on a PC for off-line analysis. There are studies on anomaly detection, both on periodic and non-periodic signals. Various techniques are used, such as HMM [1] based, prediction based [2], similarity based [3], window based [4] and segmentation based [5]. Depending on the algorithm used, anomaly location and the length of anomaly would vary. Unfortunately, the ground truth could be understood and identified only by the domain expert, in the case of bio-signals by a health professional. Anomaly detected on the basis of algorithmic analysis may not tell the ground truth. In fact, some signal anomaly could be of no health concern.

Previous related works have some pitfalls. They are computationally heavy, and are not able to identify all anomalies present, if there are more than one. As we assume weak computational platform, the proposed algorithm has to be light in computation and memory requirement. In addition, we need to be able to locate multiple anomalies, if they are present.

3   Definition of Time Series Discords

The anomaly subsequences in the bio-signal are called discords. Time series discords are subsequences of a longer time series that are maximally different to all the rest of the subsequences of the whole sequence. Discords could be detected by comparing every pair of sub-sequences (also called windows) and identifying the one with largest distances from their nearest (least distant) neighbors. We can find such a discord using brute force method which is computationally heavy with time complexity of $O(n^2)$, where n is the total number of subsequences possible out of the whole time series. Brute force method can list such subsequences in order of distances, and thus is able to detect all discords.

Let us consider a discrete time series consisting of T time-slots. Let us also consider subsequences of length m time-slots where $m \ll T$. Thus, the original signal consists of $n = (T - m + 1)$ such subsequences. i^{th} subsequence starts at i^{th} slot, where $0 \leq i \leq (T - m)$. In previous works, the length of the subsequence was user defined [4]. In our work, we set the length m equal the fundamental period of time series [6]. As $m \ll T$, $m \ll n$.

We propose a new concept we named "mother signal". We consider only periodic signals and m is set equal to the fundamental period. Physically, mother signal is the average of subsequences of length m, which are normal (not discords) and therefore their number is overwhelmingly large compared to discords. Once mother signal is known discords are detected more efficiently. Even if we use exhaustive comparison with mother signal, the complexity is $O(mn)$. As $n \gg m$, $O(n^2) \gg O(mn)$. In other words, comparisons with mother signal will be much more efficient compared to brute force comparison of every possible pairs. The largest discord is the one whose distance is highest from the mother signal. Multiple discords are defined as subsequences whose distances with mother signal exceed a pre-defined threshold. Otherwise, we can identify and list the discords in order of their distances from the mother signal.

3.1 Mother Signal

From a given time series, first its fundamental period is determined. That will be the value of m. In a periodic time series, every normal subsequence of length equal to time period would be similar, if the subsequence is shift-and-rotated by proper length for maximum match. Only in case of discords the distance from mother signal would be large even when all possible shifts and rotations are tried.

At the outset, our assumption about the mother signal is verified. From a given time-series we randomly pick up 30 subsequences which are not discords. Each subsequence is normalized to mean 0 and standard deviation 1. Following normalization, we shift-and-rotate all subsequences to maximally match with the first subsequence. For computational efficiency, we segment a subsequence (length m) into a few divisions, say M, where m is divisible by M. Over each segment, m/M slots of data are averaged. Thus, a subsequence is reduced from m dimensional vector to M dimensional vector.

Our algorithm is based on the assumption that normal subsequences form a compact cluster, with near normal distribution and small variance. We discarded the anomaly region from a known signal and selected 30 samples from random locations. We tested the distribution of these 30 subsequences for normality. The results is shown in Fig. 1, confirming our assumption that they follow Gaussian distribution [7].

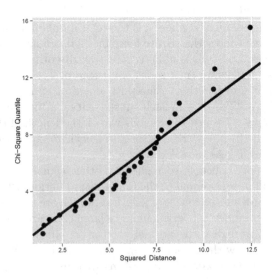

Fig. 1. Testing the Gaussian distribution of normal subsequences.

4   Creation of the Mother Signal

By comparing with the mother signal, to find the discord, we could reduce the complexity from $O(n^2)$ to $O(mn)$. In Sect. 3, we described the physical meaning of the mother signal. Here, we describe how the mother signal is created. The whole procedure is explained in Fig. 2.

In the first step, we extract subsequences from the whole time series shown in Fig. 2(a), at $m/2$ intervals. A few samples are shown, in Fig. 2(b). Thus, instead of $(n - m + 1)$ subsequences, we get much less, $(2n/m - 1)$ subsequences. The reduction in number of   subsequences is by a factor of $(m/2)$. We will show

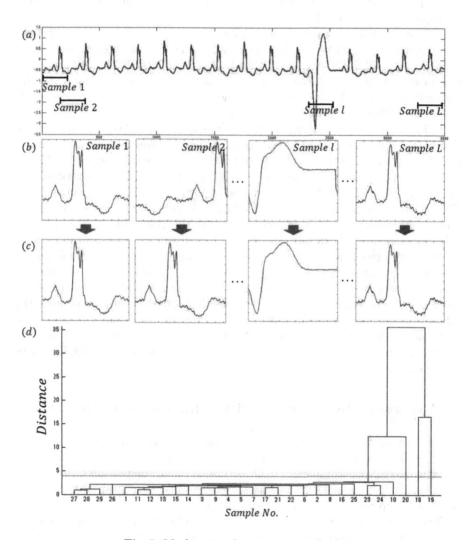

Fig. 2. Mother signal generation method.

that though this improves efficiency, the obtained mother signal is almost the same. Next, we shift-and-rotate each subsequences in step of one slot, to a point where the matching with leading subsequence is maximized. Thus subsequences in Fig. 2(b) are changed to Fig. 2(c). Finally, mutual distances of all subsequences were calculated and a dendrogram is drawn for clustering.

The original signal is shown in Fig. 2(a). Its time period of m-slots is determined. We take subsequences of window size equal to m in Fig. 2(b). The window is shifted by $m/2$ slots at a time and different subsequence samples are drawn. In Fig. 2(c), we show how all the subsequences are shifted and rotated (shifted on the right and that portion is carried back to the beginning of the subsequence) to an appropriate number of slots, such that the matching with the first subsequence is maximized.

All the normal subsequences look alike. For the discord it is very different. If we cluster them, where the distance metric is the Euclidean distance, normal subsequences will fall in one group with largest membership. Discord sequences will have large distances from the normal ones. This is the basic idea of forming mother signal, which will represent the normal shape. Subsequences which deviate from the normal one are discords.

In Fig. 2(d), the dendrogram using $(2\,n/m-1)$ sample subsequences is shown. The dotted line determines the minimum distance between clusters. As we can see, the subsequences are divided into 4 clusters. The mother signal is defined as the mean of members of the cluster with largest cardinality. We verified that mother signal created by this method, and by using all subsequences are similar. This was true for all the data we tried. In Fig. 3, the four clusters with member subsequences drawn overlapped, are shown. We see that cluster 2, 3 and 4 have only one member each. In fact, the sets with cardinality 1 are abnormalities.

We investigated how the mother signal changes, when we used all normal subsequences (shifting subsequence by 1-slot) and just $(2n/m-1)$ subsequences shifting in steps of $m/2$ for efficiency. The bottom line of Fig. 3 depicts the distances of cluster mean from the true mother signal formed by shifting the subsequence one-slot at a time. We see that even after reducing the number of subsequences by a large factor, we still obtain very similar results. The distance of this efficiently generated mother signal and the true one is only 2.59. Distances of different discords are much larger.

5   Proposed Heuristic Algorithm for Anomaly Detection

In Sect. 3, we discussed that the computational complexity for anomaly detection using brute force method is $O(n^2)$ and all possible comparisons with mother signal will reduce the complexity to $O(mn)$. As $n \gg m$, comparison using mother signal is much more efficient. In this section, we propose a heuristic algorithm which will improve the efficiency even further.

The basic idea of proposed heuristic for reducing computation is as follows. Suppose, for i^{th} subsequence, to find the minimum distance (i.e., maximum match) with mother signal, we need to shift-and-rotate the ith signal by p slots.

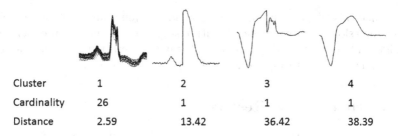

Cluster	1	2	3	4
Cardinality	26	1	1	1
Distance	2.59	13.42	36.42	38.39

Fig. 3. Four clusters of subsequences, their cardinality and distances between cluster average and mother signal.

In that case, for comparing $(i+1)^{th}$ subsequence with the mother signal, we need to shift-and-rotate $(i+1)^{th}$ subsequence by $(p+1)$ slots, because $(i+1)^{th}$ subsequence is only 1-slot shifted from the ith subsequence. In that way, the number of shift-rotate-compare will be reduced from m to 1. We can reduce the computation for distance comparison by m times. Through experiment, we found that 97% of times, this 1-slot shift gave best match. But, it does not always happen. First, we shift $(i+1)^{th}$ subsequence by p-slots, $(p+1)$-slots and $(p+2)$-slots, and calculate the distance with mother signal. Ideally, $(p+1)$-slots shift will give the best match. If that does not happen, we need more elaborate searching for proper shift that would achieve maximum match with the mother signal. An efficient heuristic algorithm for this more elaborate searching is proposed and explained below.

In Fig. 4, we show the distance when a normal signal is compared with the mother signal. The x-axis represents the shift and the y-axis is the distance. The change of distance with shift is smooth, with a prominent minimum at shift around 115 slots. This shape of this minimum is a symmetric wide valley, not a deep narrow gorge. Because of this shape, it is easy to find the minimum without comparing distances for all possible m shifts. We calculate distances for a few equally spaced shifts, and can easily converge to the minimum point (maximum match) in a few trials.

The heuristic search algorithm to find minimum distance is explained in Fig. 5. First, we find distances for α number of equidistant shifts each of $(\lfloor m/\alpha \rfloor)$ in the first iteration. We may not find the minimum in the first iteration, which may lie between two shift points, as shown in Fig. 5(a).

The minimum here lies between 9th and 10th of the m/α shift points where the distances are calculated. We sort the distances corresponding to α points. With a reasonably high value of α (here we use $\alpha = 20$), and due to the symmetric shape of distance around the minimum, we can assure that the minimum and the second minimum points are on two sides of the actual minimum, one on the positive slope side and the other on the negative slope side. We divide the space between first and second minimum, and first and third minimum equally, and calculate the distances at those middle points, as shown in Fig. 5(b). This is the first iteration. We again take the minimum 3 points out of 5, and repeat.

We continue this till the minimum and second minimum are consecutive points, i.e., one time slot apart, when the algorithm stops. In fact, in all our experiments with eight different sets of data, we needed only two iterations. The computation is thus reduced from m to $\alpha + \epsilon$ where ϵ is just 4 or 6, and $\alpha \ll m$.

6   Experiments and Results

We used data from MIT-BIH Database [8], the detail of which is shown in Table 1. The algorithm complexity to find anomaly is basically the number of times we need to calculate distance between two subsequences. Distance function calculates Euclidean distance between two subsequences. Computation time for different algorithms, are shown in Fig. 6. For the proposed algorithm, time to generate mother signal is included.

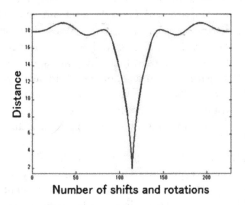

Fig. 4. A subsequence similar to mother signal is shifted one-slot at a time and distance with the mother signal is plotted.

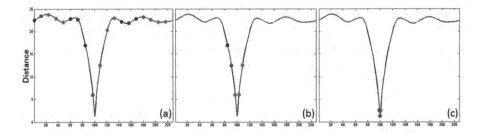

Fig. 5. The heuristic search algorithm to find the minimum.

Table 1. Data used for experiments

Serial No.	T	m	n
1	3751	228	3524
2	3750	163	3588
3	3750	251	3500
4	3750	143	3608
5	5400	351	5050
6	5401	292	5110
7	5400	399	5002
8	16500	370	16131

In Fig. 7, we show the average of maximum F-measure for all 8 data sets using our proposed algorithm and related work reported in [4].

Figure 8 is presented to visually evaluate the quality of result. The original data is shown in Fig. 8(a). Distances of subsequences calculated using brute force method is shown in Fig. 8(b), and that using the proposed mother signal with heuristic is shown in Fig. 8(c). The strong resemblance between Fig. 8(b) and (c) confirms the correctness of the proposed algorithm.

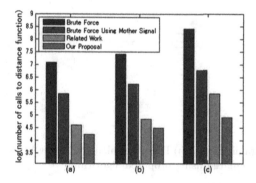

Fig. 6. The number of calls to dist function. (a) Average of results while data 1 to data 4 of Table 1 are used. (b) Average results of data 5 to data 7. (c) Result of data 8.

It is usual to evaluate any pattern recognition algorithm by precision and recall. In our case, the ground truth is not known. The decision of anomaly is solely based on the distance from the normal trend - both for the proposed method as well as brute force method. We, therefore, subjectively compared the results in Fig. 8, as we can not judge what really the discords are, what discords are biologically significant.

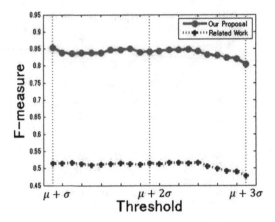

Fig. 7. The average of maximum F-measure for all data using our proposed algorithm and work reported in [4].

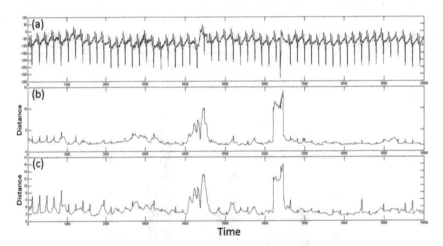

Fig. 8. (a) The Original Signal. (b) Distance calculated using brute force method, and (c) is the distance calculated using proposed algorithm.

7   Conclusion

We could improve the efficiency of discord detection in periodic signals. We need to compare our algorithm to other kind of algorithms [9] proposed recently. The memory requirement is not analyzed. The parameters used in our algorithm, like M, α etc., determine both the efficiency as well as the quality of the result. To find their optimum values for maximum efficiency without sacrificing quality is a challenge. We will investigate how the result changes with parameter values, for certain type of bio-signals. Finally, the most important aspect is to compare the results with ground truth. We hope to obtain supervised data with discords, identified by health experts. Only then we can conclude how useful the result is or modify the algorithm to detect discords for the supervised data. Our algorithm works for periodic signals. It is true that the period changes over time, like pulse rate varies with the level of exhaustion or emotional state. We need to modify the algorithm so that it could adapt such changes. Moreover, discords with different waveforms have different meanings. For practical application we need to analyze the discords to provide the user with their level of emergency and suggestions for action to be taken.

References

1. Sha, W., Zhu, Y., Huang, T., Qiu, M., Ming, Z., Zhu, Y., Zhang, Q.: A Multi-order markov chain based scheme for anomaly detection. In: Computer Software and Applications Conference Workshops (COMPSACW), pp. 83–88 (2013)
2. Ma, J., Perkins, S.: Online novelty detection on temporal sequences. In: KDD 2003, Proceedings of the Ninth ACM SIGKDD International Conference on Knowledge Discovery and Data Mining, pp. 613–618 (2003)
3. Rebbapragada, U., Protopapas, P., Brodley, C.E., Alcock, C.: Finding anomalous periodic time series. Mach. Learn. J. **74**, 281–313 (2008)
4. Keogh, E., Lin, J., Ada Waichee, F., Van Herle, H.: Finding the unusual medical time series, algorithms and applications. IEEE Trans. Inform. Technol. Biomed. **10**(3), 429–439 (2006)
5. Salvador, S., Chan, P.: Learning states and rules for detecting anomalies in time series. Appl. Intell. **23**(3), 241–255 (2005)
6. Basha, R., Ameen, J.R.M.: Unusual sub-sequence identifications in time series with periodicity. Int. J. Innov. Comput. Inform. Control (IJICIC) **3**, 471–480 (2007)
7. Korkmaz, S., Goksuluk, D., Zararsiz, G.: MVN: an R package for assessing multi-variate normality. R J. **6**(2), 151–162 (2014)
8. Goldberger, A.L., Amaral, L.A.N., Glass, L., Hausdorff, J.M., Ivanov, P.C., Mark, R.G., Mietus, J.E., Moody, G.B., Peng, C.-K., Stanley, H.E.: PhysioBank, PhysioToolkit, and PhysioNet: components of a new research resource for complex physiologic signals. Circulation **101**(23), e215–e220 (2000)
9. Sugiyama, M., Borgwardt, K.M.: Rapid distance-based outlier detection via sampling. In: NIPS 2013

Aggregating and Analyzing Articles and Comments on a News Website

David Ramamonjisoa[✉]

Faculty of Software and Information Science, Iwate Prefectural University,
IPU, Takizawa, Sugo 152-52, Japan
david@iwate-pu.ac.jp

Abstract. In the top news stories, the commenting activity is rising and falling until it stops. In some ongoing news stories such as disasters like the disappearance of flight MH370, global warming or climate change, political turmoil or economic crisis, this commenting activity cycle can repeat and last many years. To our knowledge, a study and analysis of those data does not exist up to now. There is a need to separate facts, opinions and junk within those comments data. In this paper, we present our framework for supporting readers in analyzing and visualizing facts, opinions and topics in the comments and its extension with comments aggregation and summarization for comments within several news articles for the same event. We added a time-series analysis and comments features such as surprising comments and a preferential threads attachment model.

Keywords: Articles and comments analysis · Comments visualization · Network analysis · Text mining

1 Introduction

User comments are a kind of user-generated content. Their purpose is to collect user feedback, but they have also been used to form a community discussing about any piece of information on the internet (news article, video, live talk show, music, picture, and so on). The commenting tool becomes social gathering software where commenters share their opinions, criticism, or extraneous information. Recent research consists of assisting the end-user to reading comments by providing succinct summary or useful comments according to some algorithms [1–3].

In the top news stories, the articles and commenting activity are rising and falling until they stop. In some unresolved events in the news such as disasters like the disappearance of flight MH370, global warming or climate change, political turmoil or economic crisis, this commenting activity cycle can repeat and last many years. To our knowledge, a study and analysis of those data has not been done. There is a need for an overview of the articles on the events and to separate facts, opinions and junk in the comments data. However, some studies have investigated the challenges in summarizing jointly the articles and comments covering the event, and generating socially-informed timelines [4]. The timelines are built by maximizing topic cohesion between the article and the comment summaries. The maximization uses an optimization algorithm that

© Springer International Publishing AG 2017
M. Otake et al. (Eds.): JSAI-isAI 2015 Workshops, LNAI 10091, pp. 428–441, 2017.
DOI: 10.1007/978-3-319-50953-2_30

allows the generation of a high quality article summary and comment summary via mutual reinforcement. A similar study on this method is described in the dual wing factor graph model between Web documents and their associated social contexts to generate summaries [5].

In this paper, we present our framework to support readers in analyzing and visualizing facts, topics, hidden structure in the comments and its extension with comment aggregation and summarization of comments within several news articles on the same event. A time-series analysis and comments specific features such as surprising comments and a preferential threads attachment model are also introduced. The results obtained from the experiments showed that the lift measure used for the surprising event feature combined with clustering and interesting features visualization can summarize and synthetize the event.

Firstly, we describe the comments modeling and system framework. Then, we present the features used for analysis such as clustering, classifying, summarizing and finding surprising comments. Thirdly, we detail the analysis models. Finally, we demonstrate the use of the models with a test data and discuss the results.

2 Aggregating News Articles and Comments

An event in the news (top story) can generate several articles and on the order of tens of thousands of comments on a single news website. The goal of this process is to gather all articles and comments from the event published in a single news website.

Let E_k be some event, A_k the associated articles and T_k the topic category of the event. Then we have $E_k = \{T_k, A_k, C_k\}$.

$$T_k = \text{TopStories|World|Local|Entertainment|SciTech}$$
$$\text{|Business|Politics|Sports|Health|Products}$$

$A_k = \{a_{1k}, a_{2k}, \dots, a_{mk}\}$ where a_{ik} is any article concerning the event

All comments are noted C_k (as the comment corpus for the event). The model of the comment collection is described below:

$$C_k = \{c_i\}$$

where

$$c_i = (commentID_i, time_i, author_i, title_i, content_i, rating_i) \tag{1}$$

$$content_i = \begin{cases} \{s_{1i}, \dots, s_{ji}, \dots, s_{li}\} | \\ < reply - to > \\ < quotation > \end{cases} j \in [1..l] \tag{2}$$

s_{ji} is the j-th sentence of the comment c_i as a sequence of words $(w_1, .., w_k)$.

From the Eqs. 1 and 2, we can build a data matrix, a network graph or time series data from the comment corpus. We can also extend the equations to model comment properties such as the lift measure or topic model as described in this paper.

The aggregating module is part of the comments analysis framework depicted in Fig. 1. Comments are pre-processed by selecting and extracting relevant features. They can be summarized, compared or statistically examined. Pre-processed data are then used to discover some interesting patterns or regularities and then to visualize those findings for the end user.

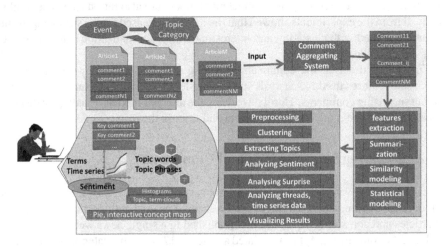

Fig. 1. Framework for the articles and comments aggregation and analysis.

3   Features on Articles and Comments

The feature selection and extraction module uses patterns of statistic relevance, sentence writing style relevance, natural language processing (NLP) techniques relevance and topic/user/sentiment modeling relevance. For example, in the work on social context summarization described in [4, 5] or on comments information retrieval in [2, 6], more than 10 features have been selected and applied. In general, those features are used to build a model for comments classification or clustering, then to summarize or visualize the performance of the model. Particularly, a system can use a filter to visualize the comments data by selecting the features in interest [7]. Those features in text data can be mixed with time series data to build a specific analyzer for financial data and news together [8].

In our framework, we are using the common features (statistics, NLP) and comments specific features. The common features are for example the number of words in each article or comment, the number of sentences, the readability, the similarity with articles, and the divergence with comment or article. In this paper, we focus on features that are used for the specific modeling described in Sect. 4 and other important ones for the framework such as preferences on specific threads and lift measure.

3.1   Vector Space Models

Each comment is transformed into a set of n-gram terms. The process requires the natural language processing on the original comment text. For each sentence in the comment, we tokenize it and remove unnecessary punctuation marks connected with the word. A heuristic process is used to remove redundant character to restore the original word. All upper-case letters are changed to lower-case letters. N-gram terms are extracted and selected to represent the comment. They are called the Bag of Words (BOW). Any 1-g, 2-g, or 3-g terms composed entirely with stopwords are deleted.

3.2   Lift to Measure Surprise Comments

The vector space model cannot provide information about the association between words. The Lift model is used on pairs of words and to compare the likelihood of the co-occurrence of words happening by chance to that observed in the actual corpus.

The best estimate of the probability of a word occurring in a corpus is given by the observed number of comments that contain that word divided by the total number of comments

$$P(w_i) = \frac{NumContaining(C, w_i)}{\#AllComments} \tag{3}$$

Lift can be generalized to all words $\{w_1, .., w_n\}$ in the comment.

$$Lift_1(\{w_1, \ldots, w_n\}) = \frac{P(w_1 \cap w_2 \ldots w_i \ldots \cap w_n)}{\prod_{i=1}^{n} P(w_i)} \tag{4}$$

$Lift_2$ assumes that the co-occurrence of pairs of words is independent.

$$Lift_2(\{w_1, \ldots, w_n\}) = \frac{\prod_{\forall i \neq j} P(w_i \cap w_j)}{\prod_{i=1}^{n} P(w_i)^{n-1}} \tag{5}$$

$Lift_3$ is the logarithm of the $Lift_2$ and useful when the number of comments is large.

$$Lift_3(\{w_1, .., w_n\}) = \left| \sum_{\forall i \neq j} log(P(w_i \cap w_j)) \right| - (n-1) \left| \sum_{\forall i} log(P(w_i)) \right| \tag{6}$$

$Lift_3$ calculates the surprise of a comment as the combination of the mutual information value of every pair of words in that comment using the corpus as a whole to calculate the relevant probabilities. This model is introduced and used for the twitter analysis in [9].

4 Analysis Models

The analysis of the event from articles and comment data is concerned with the detection of the burst period, clustering and synthetizing the topics in each burst to create a visual representation of the summarized data. In the following, we describe techniques used to realize those analyses.

4.1 Clustering Comments and Multi-dimensional Reduction Scaling

We used K-means clustering algorithm and the multi-dimensional reduction scaling (MDS) to implement this module. The similarity measure is the cosine similarity of the tf-idf vectors. The representation of those clusters in two-dimensional space is performed with the MDS.

4.2 Topic Model

Topics are the thematic summary of the comment collection. In other words, it answers the question what themes are those comments discussing.

The topic modeling is used to extract T topics out of the comments collection. That is, we have a set of comments $C_k = \{c_i\}i\varepsilon[1 \dots N]$ and a number of topics $T_k = \{t_i\}i\varepsilon[1 \dots m]$. A comment c_i can be viewed by its topic distribution. For example, $\Pr(c_1 \in t_1) = 0.50$ and $\Pr(c_1 \in t_2) = 0.20$ and so on. The default topic modeling based on LDA is a soft clustering. It can be modified into hard clustering by considering each comment as belonging to a single topic (cluster) t_r,

$$r = argmax_r P(t_r|c) = argmax_r P(c|t_r)P(t_r) \tag{7}$$

where r is the number of topics that has the maximum likelihood for each comment. Hence, the output of the LDA based topic clustering approach is an assignment from each comment to a cluster.

Key comments selection within clusters of comments is important for summarizing the contents.

For each topic obtained by the topic modeling, a set of comments are associated. We define the key comment as the top of the comments by ranking them within their clusters. The ranking method is realized by comparing each comment vector (a bag of words) to the list of words which form the topic vector. We use cosine distance for the comparison. The most similar to the topic is the key comment. The experiment on this analysis is first described in [10].

4.3 Preferential Attachment (PA) Model for Comments and Commenting an Article

A statistical analysis and comparison of the structure and evolution of the different discussion threads associated to the news websites and Wikipedia comments have shown that the threads popularity network is following the preferential attachment model or the Yule process [11]. That is, a new comment is added to one of the existing comments with a probability proportional to their existing in-degree. The initial comment in the thread is most like or dislike one by the readers in the sense that the reaction and activity is high.

4.4 Visualization Techniques

We want to represent only the most important information in the results to the end-user. A quick overview for understanding the event timeline and story is necessary. Article titles contain the most concise information on the event, so we gather them and process the terms using cloud conversion. The structure of the conversion can also be visualized at glance with a specific network graph database tool.

For the topic model, we use the topic explorer tool [12].

5   Test Data and Experiments

5.1 Text Data Content

In this paper, we present an analysis of articles and comments on the unexplained disappearance of Malaysia Airlines Flight MH370, that occurred last March 2014, on the Guardian news website. This event was also studied in the work of Wang Lu et al. on timeline summary generation from different news websites such as CNN, NYT (New York Times) and BBC [4]. Their dataset included articles and comments published during the period from March 2014 until June 2014. Our analysis of the event started on March 2014 and ended at the end of September 2015. During that time, there are 368 articles, videos and audio clips published on the Guardian news website[1]. The total number of comments during the experiment period is 77900 and there are 5780 unique comments. The number of commenters is 1780. In Fig. 2, we can observe the comment ratings and user participation activity during the period. We only plotted unique comments by removing redundant comments cross-posted in several articles. Unique comments are selected on their first appearance chronologically in the comment corpus. Therefore, the interval between two time stamps in the horizontal axis, representing the comment time, is not linear. The time stamps in the horizontal axis indicate the vertical dotted line in the grid as the first comment is posted on '2014-3-10 18:41:00'. The peaks are during March and June 2014 and then decreasing in intensity by July 2015.

We can observe in Fig. 3 the repeating bursting and decaying of the article data during this event. The word cloud set to 100 terms on the top left in Fig. 3 represents the initial month of the event. There were 185 articles during the first month on the news website. Following the second Malaysian Airline Flight MH17 explosion, there were 6000 jobs cut at the company during August 2014 represented by the word cloud on the bottom left. The finding of the flaperon debris is shown by the word clouds on the right.

---

[1] http://www.theguardian.com/world/malaysia-airlines-flight-mh370.

Some experts on aviation, specifically the Boeing 777 s, commented in the early stages of the finding (July 2015) that there was no doubt that the debris was from the missing airplane. It took two months for the investigators to give the same conclusion.

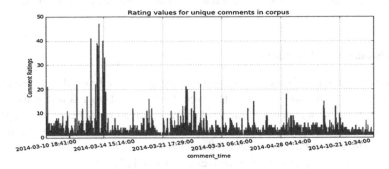

Fig. 2. Comment Ratings by users during the one and half year period

Figure 3 shows the number of articles each month and the generated keywords in important periods.

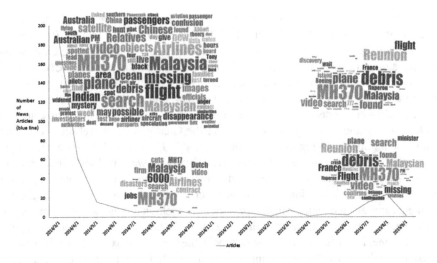

Fig. 3. Time series data representing the frequency of articles on the disappearing Malaysian Flight MH370 from March 2014 to September 2015, on the Guardian news website and the word clouds (top left: March 2014, bottom left: August 2014, top right: July 2014, bottom right: August 2014) are depicted.

Many unrelated comments were posted although the Guardian news website is moderated. The discussions on the comment board are analyzed with our tool from our previous research [10].

5.2   Clustering and Multidimensional Scaling (MDS) Results

In Figs. 4 and 5, the 368 articles in 10 clusters (represented by 10 colors) are plotted and respectively in five clusters during the 18 month period (18 points). The legend in each plot represents the main keywords as the top terms per cluster. The MDS results show the distance between clusters in a two dimensional space and clearly separate them. Clusters with the term 'Malaysia' are close to each other in the center and clusters with the terms 'search, plane, missing' are at the periphery in the 10 clusters plot. In the 5 clusters plot, the cluster with the term 'SES' (State Emergency Service) is singled out on the right, the cluster with the terms 'debris, Reunion' is at the bottom center and the cluster with terms 'MH17, jobs' is at the top center.

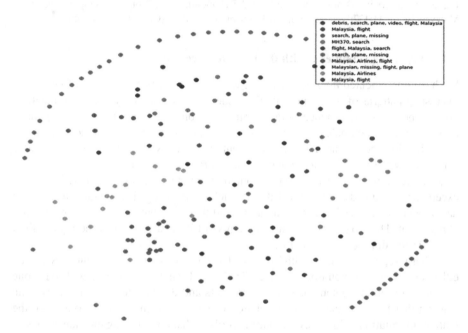

Fig. 4.   10 clusters in 2D plots obtained from the 368 articles on the disappearance of Air Malaysia Flight MH370 from March 2014 to September 2015. (Color figure online)

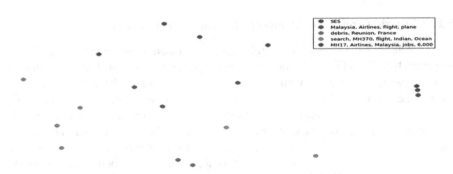

Fig. 5. 5 clusters in 2D plots obtained from the 18 months of data on the disappearance of Air Malaysia Flight MH370 from March 2014 to September 2015. (Color figure online)

5.3   Surprising Comments with the Lift Measure

We have implemented the surprise model with our own specifications and found some interesting patterns of comments with the model. For the simplicity of the computation, we assume that each sentence in the comment is composed only by the noun phrases (nps). Hence, we compute the occurrence of each pair of these to obtain the lift measure for each unique comment in the text comment data. The extraction of the noun-phrases is realized with the parsing and noun phrases extractor in NLTK [13].

Comments with great Lift3 value contain many sentences (num_sent) in general except the top ranked comment in Table 1 which is surprising. The top ranked comment has only six sentences but upon reading it, find that this comment has the most noun-phrases in it. The top ranked comment in terms of Lift measure is the itinerary of Flight MH370 according to Inmarsat [14].

Otherwise, the comments with $Lift_3$ equal to zero show one or two sentences and are either empty or one noun-phrases (see Table 2). Usually, comments with only one sentence are replies to some previous comments and their contents are brief. We can confirm that $Lift_3$ calculates the correlation of terms in comments by comparing the mutual information value of every pair of words in that comment against the corpus as a whole to calculate the relevant probabilities. The mutual information value of two random variables X and Y is zero if X is independent of Y. Figure 6 shows the values of the Lift3 measure during the period of our experiment.

Table 1. Top 5 of Lift3 computation results on the comments from "Malaysian Airline flight MH370" event and their related comment records

	comment	lift3	num_sent	nps
4447	\nTimeline of events, by the hour:\n16:30 UTC ...	17614.684420	6	[hour interval, acars, penang, strait, radar l...
3895	\nLets Get Real Here People... If They Are "Na...	16821.676339	21	[looks, ...where, global scale your capabili...
4699	\nElectrical problems developed first from 17:...	15018.411232	32	[uniquely, fire self, radio compass, acars, vo...
4488	\nThe Cambodian Connection (cont. - Part 2)\nM...	11902.477066	36	[complete contradiction, aircraft 's whereabou...
3272	\nI don't believe it "diverted around Indonesi...	11888.181240	33	[2:03am local, low altitude, radio compass, us...

comment_time	comment_type	comment_id	comment_recommend	comment_content	comment_author	commented_article	lift3
2014-04-11 15:52:00	{}	comment-34245297	2	\nLets Get Real Here People... If They Are "Na...	hogan412	3z9mt	16821.676339
2014-05-03 20:02:00	{}	comment-35166215	7	\nThe Cambodian Connection (cont. - Part 2)\nM...	CliveG7	3zpnt	11902.477066
2014-05-04 13:10:00	{}	comment-35191013	6	\nTimeline of events, by the hour:\n16:30 UTC ...	EnoughIsEnoughUS	3zpnt	17614.684420
2014-06-28 18:05:00	{u'reply_ref': u'#comment-37531499', u'reply_t...	comment-37540502	4	\nI don't believe it "diverted around Indonesi...	Sy Gunson	3qejk	11888.181240
2014-10-05 08:58:00	{u'reply_ref': u'#comment-41801533', u'reply_t...	comment-41804323	1	\nElectrical problems developed first from 17:...	Sy Gunson	426bm	15018.411232

Table 2. Last 15 of Lift3 computation results on the comments

	comment	lift3	num_sent	nps
0	\nThey did not have, nor do they need visas. T...	4.401002	3	[beijing, kind, chinese government]
2	\nTerrorists (and spies) use stolen passports....	7.875785	6	[live, fact, terrorists]
3	\nYou can assume whatever you like - after all...	0.000000	1	[]
4	\nIt's a roaming thing, you call out from your...	1.520780	2	[network tries, ca n't]
5	\nBecause by understanding what has happened t...	2.365878	3	[aviation industry, good safety record]
9	\nThat is an excellent question.. There is no ...	9.329782	2	[excellent question.., 's database, interpol]
11	\nAnd refused to allow outside assistance unti...	2.515640	2	[google, expendable means]
13	\nI was surprised to hear that an aircraft can...	2.842999	4	[mid air, dangerous parts]
14	\nmust have been a group who went shopping and...	0.000000	1	[vip]
15	\nLooks like you might be right. So where does...	0.000000	2	[looks]

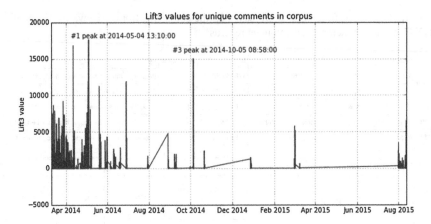

Fig. 6. The variation of the Lift measure during the experiment period and the peak during the Inmarsat data publication on May 4th 2014.

5.4   Visualization of the Comments Network Graph

Figure 7 depicts the structure of conversations and the preferential attachments to some users' or comment posts. They are realized with Neo4j[2]. Neo4j is a tool which we can use to build a property graph from a database. The property graph contains connected entities (the *nodes*) which can hold any number of attributes (key-value pairs) and *relationships* which give meaning to the connection between two-node entities. A relationship always has a direction, a type, a start node and an end node. It can also have properties. Neo4j implements the property graph and optimizes the graph database to output query results faster independent of the size of the database. The comment corpus is transformed into a comment database where each comment is a record containing the properties described in Eq. (1). The modeling of this database in Neo4j consists of defining the nodes and the relationships. In our comments visualization, we want to observe users, comments and their relationships. Therefore, we define users as entities such as people with names and identifications. The Guardian news commenting facility does not allow anonymous people to post comments. Hence, each user must have a single user name and user identification. Comments are also entities in the graph with all properties they have in the record. We define the relationships that can be modeled as the interactions between users and comment entities. After the data modeling, we put the comment database in the Neo4j server and build the graph network describing the interactions of those entities.

---

[2] http://neo4j.com.

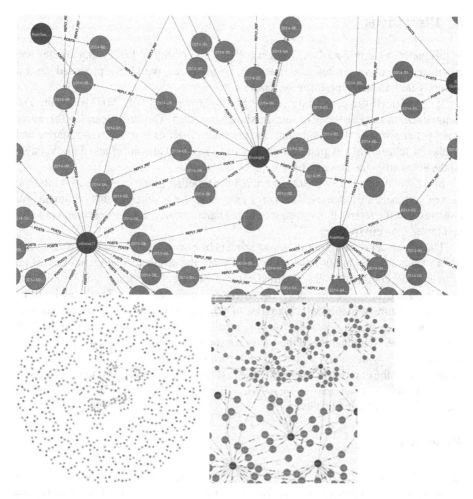

Fig. 7. The comment graph network on the comments data with minimum rating equal to 3 (Color figure online)

In Fig. 7, users (red nodes) are labeled with their usernames. They can post several comments. The most active users are centered on the graph with the highest number of comments and replies. Comments are labeled with the time stamp (blue nodes). The three types of relationship are users posting comments, comments that are replies to other comments and comments quoted other users. By clicking a comment node, we can read its properties such as subject, content, and rating score. The graph can be zoomed in or out and queried in SQL-like format.

6   Discussions

In this paper, we introduced our framework for aggregating and analyzing articles and comments on news websites after events occur in the real world. We proposed several methods and models to perform the analysis.

We presented the case study on the Malaysia Airline Flight MH370 event. The summarization and clustering results are shown in Sect. 5. The truth regarding this event is not yet known. We did text mining and comment analysis in order to summarize and synthetize (clustering and graph network analysis) the vast amount of text data available on the event from the news website.

Surprisingly the Lift3 measure extracted the timeline of the event based on the top ranked comment in this experiment. By analyzing the features in this comment, we observed that it contains the highest number of name entities, a lower sentiment polarity and lower subjectivity tone.

The PA model is valid for the commenting behavior on articles discussing the event. Most of the articles do not have comments. Only the articles already commented on present active comments unless the moderator closed the commenting tool. Usually, the most popular articles are the ones most shared and commented on. This model will be used in the future to predict which articles to monitor using our comment analysis tool.

In the future, we will consider the optimization of the program to compute the $Lift_3$ values because computations of the joint probabilities of two terms are very expensive for comments with many sentences and corpuses with more than ten thousands comments for the duration of the event. The real-time summarization as in the IncreSTS [1] should also be taken into account. We will apply our framework to other domains such as finance data and text events.

References

1. Liu, C.Y., Tseng, C.Y., Chen, M.S.: IncreSTS: towards real-time incremental short text summarization on comment streams from social network services. IEEE Trans. Knowl. Data Eng. **27**(11), 2986–3000 (2015). http://doi.ieee.org/10.1109/TKDE.2015.2405553. IEEE Press, New York
2. Momeni, E., Sageder, G.: An empirical analysis of characteristics of useful comments in social media. In: Proceedings of the Conference on Web Sciences, WebSci 2013, Paris, France, 02–04 May 2013, pp. 258–261. ACM, New York (2013). http://doi.acm.org/10.1145/2464464.2464490
3. Schinas, M., Papadopoulos, S., Kompatsiaris, Y., Mitkas, P.A.: Visual event summarization on social media using topic modelling and graph-based ranking algorithms. In: Proceedings of the 5th ACM International Conference on Multimedia Retrieval, ICMR 2015, Shanghai, China, 23–26 June 2015, pp. 203–210. ACM, New York (2015). http://doi.acm.org/10.1145/2671188.2749407
4. Wang, L., Cardie, C., Marchetti, G.: Socially-informed timeline generation for complex events. In: The Proceedings of the 2015 Annual Conference of the North American Chapter of the Association for Computational Linguistics: Human Language Technologies, Denver, Colorado, USA, 31 May–5 June 2015, pp. 1055–1065. ACL Anthology publisher (2015). http://www.aclweb.org/anthology/N15-1112

5. Yang, Z., Cai, K., Tang, J., Zhang, L., Su, Z., Li, J.: Social context summarization. In: The Proceedings of the 34th International ACM SIGIR Conference on Research and Development in Information Retrieval, SIGIR 2011, Beijing, China, 24–28 July 2011, pp. 255–264. ACM, New York (2011)

6. Potthast, M., Stein, B., Loose, F., Becker, S.: Information retrieval in the commentsphere. ACM Trans. Intell. Syst. Technol. (TIST) 3(4), 68 (2012). ACM, New York

7. Wanner, F., Ramm, T., Keim, D.A.: ForAVis – explorative user forum analysis. In: Proceedings of the International Conference on Web Intelligence, Mining and Semantics, WIMS 2011, Sogndal, Norway, 25–27 May 2011, pp. 1–10. ACM, New York (2011)

8. Wanner, F., Schreck, T., Jentner, W., Sharalieva, L., Keim, D.A.: Relating interesting quantitative time series patterns with text events and text features. In: Proceedings of the IS&T/SPIE Volume 9017, Electronic Imaging, International Society for Optics and Photonics, Visualization and Data Analysis 2014, San Francisco, USA, 2 February 2014, pp. 1–15. SPIE, USA (2014). doi:10.1117/12.2039639

9. Rooney, S.: Using lift as a practical measure of surprise in a document stream. In: The EBW 2015, Proceedings of the 3rd International Conference on E-Technologies and Business on the Web, Paris, France, 26–28 March 2015, pp. 7–12. SDIWC, Hong Kong (2015)

10. Ramamonjisoa, D., Murakami, R., Chakraborty, B.: A framework for comments analysis and visualization. Int. J. Digit. Inf. Wirel. Commun. (IJDIWC) 5(3), 179–187 (2015). SDIWC, Hong Kong

11. Gomez, V., Kappen, H.J., Litvak, N., Kaltenbrunner, A.: A likelihood-based framework for the analysis of discussion threads. World Wide Web 16((5-6)), 645–675 (2013). Social Networks and Social Web Mining

12. Murdock, J., Allen, C.: Visualization techniques for topic model checking. In: Proceedings of the 29th Conference on Artificial Intelligence, AAAI 2015, Austin Texas, USA, 25–30 January 2015, pp. 4284–4285. AAAI Press, USA (2015

13. Bird, S., Klein, E., Loper, E.: Natural Language Processing with Python, 1st edn. O'Reilly, Sebastopol (2009)

14. Ashton, C., Bruce, A.S., Colledge, G., Dickinson, M.: The search for MH370. J. Navig. 68(1), 1–22 (2015). doi:10.1017/S037346331400068X. The Royal Institute of Navigation 2014

ASD-HR 2015

Intervention of Children with Autism Spectrum Disorders Using a Humanoid Robot (ASD-HR 2015)

Yuichiro Yoshikawa[1], Yoshio Matsumoto[2], Hirokazu Kumazaki[3,4],
and Yukie Nagai[5]

[1] Graduate School of Engineering Science, Osaka University/JST ERATO
Ishiguro Symbiotic Human-Robot Interaction Project, 1-3 Machikane-yama,
Toyonaka, Osaka 560-8531, Japan
yoshikawa@irl.sys.es.osaka-u.ac.jp
[2] AIST, 1-1-1 Umezono, Tsukuba, Ibaraki 305-8560, Japan
yoshio.matsumoto@aist.go.jp
[3] Kanazawa University, 13-1 Takaramachi, Kanazawa, Ishikawa 920-8640 Japan
kumazaki@tiara.ocn.ne.jp
[4] Department of Neuropsychiatry, Keio University School of Medicine,
35 Shinanomachi, Shinjuku-ku, Tokyo 160-8582 Japan
kumazaki@tiara.ocn.ne.jp
[5] Graduate School of Engineering, Osaka University, 2-1 Yamadaoka, Suita,
Osaka 565-0871, Japan
yukie@ams.eng.osaka-u.ac.jp

The First International Workshop on Intervention of Children with Autism Spectrum Disorders using a Humanoid Robot (ASD-HR 2015) associated with JSAI International Symposia on AI 2015 (IsAI-2015) was held in Yagami/Hiyoshi Campus, Keio University, Kanagawa, Japan on 17th and 18th November, 2015.

As appeared in the statistics in the recent reports, the necessity of treatment and education for children and adolescents with autism spectrum disorders (ASD) has been widely recognized. Researchers have recently considered using humanoid robots to treat ASD-associated deficits in social communication. In the workshop, presentation of the studies in the interdisciplinary field of research on this topic including both engineering and medical sides, namely, 16 oral presentations including five invited talk were given. To facilitate communication between researchers in engineering and psychological/medical fields, we asked five pioneering and/or representative researchers in many fields to give us talks on their activities; Dr. Hideki Kojima who is the pioneering researcher on robotic therapy of children with ASD, Dr. Nilanjan Sarkar and Dr. Zackary Warren who leads the group of Vanderbilt University in the engineering and psychological sides, respectively, which is the one of the world-leading group on the field of research of children with ASD and humanoid robot, Dr. Taro Muramatsu from the department of psychiatry in University of Keio as well as Dr. Masutomo Miyao from Donguri clinic for children with developmental disorder. On the same purpose, we also planned the session of "robot demonstration" where participants brought their robots that are used for the studies on this filed. We had five exhibitions of robots; Keepon developed by Dr. Kozima from Miyagi University, an android robot Actroid-F from Advanced Institute of Science and Technology (AIST), table-top humanoid robot CommU from Osaka University, animal type robot Paro from Tokushima University and AIST, and huggable communication media Hugvie from ATR.

Three of five accepted papers in this workshop were given by researchers who have been mainly working in engineering field while the rest two were given by medical doctors in child psychiatry. Engineering papers describes how their robots were constructed and/or evaluated by individuals with ASD. Yoshikawa et al. reported how adolescents with ASD responded to an android robot, namely Actroid, by using their gaze. Jimenez et al. concerns the feasibility of collaborative learning between children with ASD and an educationally supportive robot, Ifbot. Sumioka et al. described their investigation of needs and interests for robots in a special school by demonstrating three different communicational media to the teachers: a telecommunication robot with minimal design of humanity, i.e., Telenoid, a huggable telecommunication device, i.e., Hugvie, and small mechanically-looking humanoid robots, i.e., M3-Synchy.

The rest two papers were given by medical doctors who were the experts of developmental disorders, which were also targeted by the engineering researchers who contributed to the workshop. Kumazaki et al. gave a case report of intervening to a child where the child had chances to communicate with others though an android robot by letting it speak as the child typed. Nakadoi reported cases of introducing an animal typed communication robot, Paro, into his psychiatric ward.

As discussed in the workshop, through the above interdisciplinary five papers, we could discuss not only about whether such robots give positive effect for children with ASD but also about the analysis on the necessary features of robots or tendency or types of children with ASD for such possible positive effects. Such a discussion with data including ones obtained in the real fields is expected to contribute on the future progress in this field.

Positive Bias of Gaze-Following to Android Robot in Adolescents with Autism Spectrum Disorders

Yuichiro Yoshikawa[1(✉)], Yoshio Matsumoto[2], Hirokazu Kumazaki[3,4], Yujin Wakita[2], Sakiko Nemoto[5], Hiroshi Ishiguro[1], Masaru Mimura[4], and Masutomo Miyao[5]

[1] Graduate School of Engineering Science, Osaka University/JST ERATO Ishiguro Symbiotic Human-Robot Interaction Project, 1-3 Machikane-Yama, Toyonaka, Osaka 560-8531, Japan
{yoshikawa,ishiguro}@irl.sys.es.osaka-u.ac.jp
[2] The National Institute of Advanced Industrial Science and Technology (AIST), 1-1-1 Umezono, Tsukuba, Ibaraki 305-8560, Japan
{yoshio.matsumoto,wakita}@aist.go.jp
[3] Fukui University, 23-3 Matsuoka-Shimoaizuki, Eiheiji-cho, Yoshida-gun, Fukui 910-1193, Japan
kumazaki@tiara.ocn.ne.jp
[4] Department of Neuropsychiatry, Keio University School of Medicine, 35 Shinanomachi, Shinjuku-ku, Tokyo 160-8582, Japan
mimura@a7.keio.jp
[5] Donguri Psycho Developmental Clinic for Developmental Disorder, 14-5, 4 Minami-Karasuyama, Setagaya, Tokyo 157-0062, Japan
miyaomncchd@gmail.com

Abstract. We investigated whether it is easier for adolescents with ASD to establish communication using eye-gaze with an android than a human. Two-days-experiment was conducted to measure eye-gaze patterns of subjects during conversation with two types of interlocutor, a female type android robot and a human female, where the interlocutors used their gaze to identify what they were mentioning to. Fixation bias on the target object that the interlocutor was referring to by using her eye gaze showed that the adolescents with ASD followed the gaze of android more than human's although the sample size was still small.

1 Introduction

As appeared in the statistics in the recent reports (ex. [1]), the necessity of treatment and education for children and adolescents with autism spectrum disorders (ASD) has been widely recognized. It has been reported that persons with ASD less attend to the area of eyes in the pictures of a human face than persons with typically development [2]. Accordingly, they often fail to establish eye contact as mentioned in Diagnostic and Statistical Manual of Mental Disorder, Fifth Edition (DSM-5) [3], which is one of the most important social communication cues. Researchers have recently considered using humanoid robots to treat ASD-associated deficits in social communication [4–6]. Although small humanoid robots have been programmed to teach social

© Springer International Publishing AG 2017
M. Otake et al. (Eds.): JSAI-isAI 2015 Workshops, LNAI 10091, pp. 447–453, 2017.
DOI: 10.1007/978-3-319-50953-2_31

cues such as head-gaze and hand-pointing [5], they have not been generalized for interactions with humans.

One possible reason for it might be stemmed from insufficiency of human likeness of the robot used as the interaction partner for children with ASD. Another possible reason might be insufficiency of repetition of the interaction in daily conversation. It is necessary to find a sufficiently acceptable and influential design of robot as well as program of treatment and education for children and adolescents who might need supports of such a robot. In this study, therefore, we have started basic investigations of how adolescents with ASD respond to a special type of robots called android with much humanity. An "android" robot is another type of humanoid robot that has the appearance of a real person and have recently focused on as an influential information media for humans [7]. Due to their similar appearance to humans, it is expected that they could be useful partners for adolescents with ASD to learn social interaction with humans.

It was reported that the positive bias of eye-contact, namely the tendency of looking at the upper region of the face including eyes, appeared in adolescents with ASD when they face with an android interlocutor, which did not appear with a human interlocutor [8]. However, the potential of the android for generalizability has not been confirmed. One possible reason might come from the artificial setting of the interaction where participants did not have to attend to eyes of the interlocutor to follow the context in the conversation. In other words, even though the opportunity to attend to eye-region was increased in the setting of the previous study, it did not sufficiently contribute on letting children recognize any meaning or usefulness of attending to eyes. Therefore, it seems a feasible approach to modify the contents of interaction to involve necessities to attend to eyes and examine the effect of the android again. As the first step for this approach, it should be examined whether it is easier for adolescents with ASD to establish communication using eye-gaze, namely gaze-following, with an android than a human. In this paper, we report the result of a preliminary experiment to analyzed eye-gaze patterns of subjects during conversation with an android and a human where the interlocutor used their eyes to identify what they were mentioning to (see Fig. 1 for the experimental setup).

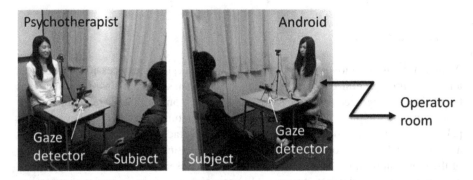

Fig. 1. Schematic representation of human and android rooms for the conversational experiment

2 Method

2.1 Subject

Four participants were recruited from a child mental clinic well known in Japan for specializing in developmental disorders and related conditions. Inclusion criteria were being aged 15 to 18 years and having a previous diagnosis of ASD. Informed consent was obtained from all individual participants and their guardians included in the study. All participants had received a clinical diagnosis of ASD based on the DSM-5 (APA 2013) and were further diagnosed through the consensus of a clinical team comprised of experienced professionals (child and adolescent psychiatrists, clinical psychologists, and pediatric neurologists). The team assessments were made following a detailed clinical examination on the first visit, follow-up observations, and through an evaluation of the answers provided in response to a questionnaire related to the development and symptoms of participants as completed by guardians. Clinical psychologists collected information from guardians concerning developmental milestones (including joint attention, social interaction, pretend play, and repetitive behaviors, with onset prior to 3 years of age) and episodes (e.g., how the individual with ASD behaved at kindergarten and school). Additional professionals, such as teachers, provided further background based on their detailed observations of interactions with people (particularly non-family members), repetitive behaviors, obsessive/compulsive traits, and stereo-typed behaviors. The third and eighth authors confirmed existing diagnoses by using both diagnostic instruments and screening questionnaires, including the Pervasive Developmental Disorder–Autism Society Japan Rating Scale (PARS), a diagnostic interview scale for ASD developed in Japan (PARS Committee 2008) [9]. Sub and total scores of this scale correlate with the domain and total scores of the Autism Diagnostic Interview-Revised (ADI-R) [10, 11].

2.2 Apparatus

Two experimental booths were adjacently settled: the human room for communication with a female person and the android room for one with a female type android robot named Actroid-F (Kokoro Co. Ltd.). The human and the android talked to participants by following interaction scripts prepared in advance. In each script, they asked questions to the participants, waited for a while, and then gave comments on the participants' answers. The questions and comments in the scripts were so carefully chosen that they can maintain consistency of conversation after receiving varieties of possible adolescents' answers. In other words, participants' experiences were designed to be interactive as well as equivalent among participants. The system to control the android and one to support the human interlocutor to produce behavior as scheduled in the interaction script were installed in the space beside these rooms and concealed from participants by using wall partitions.

Android is a kind of humanoid robots which has completely the same appearance of a real individual by making a plaster cast of this person and behaving like humans by using pneumatic actuators to silently and rapidly move its skin. It has eleven degrees of freedom: neck (3), eyeballs (2), eyelids (1), cheek (1), lip (1), eye-blow (2), bow (1). It performs communication with the participant by uttering sentences written in the interaction script when a tele-operator judged the timing for the next utterance. The utterances were produced by playing voice sound from a speaker located close to it. Note that the voice was pre-recorded from a person who also took the role of interlocutor in the human room. Meanwhile, it produced spontaneous eye-blinking behavior and mouth open-close movement synchronized with its utterance. Furthermore, it produced gaze and facial expressions on the scheduled timing in the interaction script. The gaze expression implemented based on the head movement to keeping facing with the participants or looking at predefined locations in the room (see Fig. 2).

Fig. 2. The android that is looking at an object in front of her

The human interlocutor wore an earphone and concealed it within her long black hair. The operator outside the human room played voice sound via the earphone to let her know what she should utter and where she should look at.

In each booth, an eye-tracker device (Tobii, X2-60) was settled to detect fixation points of the participants during the conversations. Before starting the trials, each of them was calibrated so as to output the participant's fixation points on the virtual screen located on the position of the human's or the android's face, which corresponded to the image plane captured by a video camera behind the participant.

2.3 Procedure

Participants attended to two-days-experiment in each of which days they alternately talked to a female person and a female-type android six times totally. They were instructed that they would alternately and repeatedly communicate with a female person and a female type android robot and saw their appearances for habituation. Before starting the series of sessions in each day, the gaze detection devices in both rooms were calibrated for each participant by an experimenter.

The participant first attended to the warm up session in the human room for one or two minute. After the warm up session, he moved to the android room and attended to what we call a gaze session with the android where the android used its gaze to identify what it was mentioning to. He then came back to the human room and attended to the similar gaze session with the person. After that, he attended to further gaze sessions in the android and human rooms where the interlocutors exhibited some gaze expressions and mentioned to the expressions. Finally, the participants attended to the farewell session in the android room where the android told the finish of the sessions and asked the impression of the participant. Therefore, each participants totally attended to six sessions in one day: the warm-up session with the human, the gaze sessions with the android and the human, the further gaze sessions with them, the farewell session with the android. The four gaze sessions took about one or two minute. The same process after calibration was repeated in the second day.

Note that the structures of the interaction contents in both types of gaze sessions were prepared so that they matched between sessions in the human and android rooms. Especially, the times of opportunities for the participant to answer and the times of gaze expressions of the interlocutors were matched between them.

3 Result

The first gaze sessions with the android and the human were analyzed. In the first half period of these sessions, the interlocutors talked about either object placed on the table between them and the participant while looked at it three times. To check whether the participant looked at the same object with the interlocutor, in other words followed the gaze of the interlocutor, the degree of matched looking D_m was calculated by the following equation:

$$D_m = \frac{T_c}{T_c + T_o} \tag{1}$$

where T_c is time spent for looking at the correct side while T_o is time spent for looking at the opposite side. T_c and T_o was calculated by integrating time when the fixation points stayed on the predefined regions on the object on the correct and opposite side, respectively.

The average D_m in the session with the android was 0.67 and 0.81 in the first and second day, respectively while one with the human was 0.38 and 0.52. Although the there was no significant difference between the android and the human in the first day (t(3) = 2.15, *n.s*), one with the android was significantly higher than one with human in the second day (t(3) = 3.89, p < .05). There was no significant differences in the average D_m between the human sessions of the first and the second day (t(3) = 0.62, *n.s*).

4   Discussion and Conclusion

Since the degree of matched looking in the session with the android was higher than one with the human in the second day, it is considered that the android robot could promote the tendency of gaze following of adolescents with ASD than human interlocutor. Supposing the positive bias of the android for eye-contact reported in the previous work [8], the finding in the current paper encourages the future application of androids for improving social skills related to gaze signals in conversation. However, since the current result is still limited because the number of subjects is small, the experiment should be reexamined with larger samples in the future.

Another concern on the current result is about generalizability. Although we found that the participants could have experiences of joint attention, i.e., namely looking at the same side of object that the interlocutor was mentioning to, with the android more with the human, the significant difference appeared only in the second day. Therefore, it is difficult to say that the android in the first day could provide participants with sufficiently promoted experiences of joint attention. Rather, adolescent with ASD might need some training sessions to be promoted for joint attention even with the android. Accordingly, there was no significant difference in the tendency of gaze following in the sessions with human between the first and second days, which was considered to be the measure of generalization. Therefore, it is worth examining the effect in the experiment with more repetitions for so longer days that participants can have chances to communicate with humans after experiences of gaze following are promoted with the android.

Meanwhile, it is important to identify relationships between effective aspects of robot appearance and behavior and subjects' symptoms that underlie these tendencies. Through understanding such relationships, the design of the robot and program for treatment and education for adolescents with ASD should be developed and tested in the real fields. To consider such a treatment and education, it is necessary to carefully consider who is willing to participate and can successfully receive positive effect of it without any stress. Therefore, it is worth evaluating the stress from this treatment and education by using not only subjective evaluation but also objective one like bio-marker like stress hormone.

Acknowledgement. This work was supported by JSPS KAKENHI Grant Numbers 24680022, 15K12117, 25220004.

References

1. Baio, J., National Center on Birth Defects and Developmental Disabilities, CDC; Prevalence of Autism Spectrum Disorder Among Children Aged 8 Years —— Autism and Developmental Disabilities Monitoring Network, 11 Sites, United States (2010)

2. American Psychiatric Association; Diagnostic and Statistical Manual of Mental Disorders Fifth Edition (DSM-V)
3. Pelphrey, K.A., et al.: Visual scanning of faces in autism. J. Autism Dev. Disord. **32**(4), 249–261 (2002)
4. Kozima, H., Michalowski, M.P., Nakagawa, C.: Keepon - a playful robot for research, therapy, and entertainment. Int. J. Soc. Robot. **1**, 3–18 (2009)
5. Warren, Z.E., Zheng, Z., Swanson, A.R., Bekele, E., Zhang, L., Crittendon, J.A., Weitlauf, A.F., Sarkar, N.: Can robotic interaction improve joint attention skills? J. Autism Dev. Disord. **45**, 3726–3734 (2013). 1–9
6. Wainer, J., Dautenhahn, K., Robins, B., Amirabdollahian, F.: Collaborating with Kaspar: using an autonomous humanoid robot to foster cooperative dyadic play among children with autism. In: 2010 IEEE-RAS International Conference on Humanoid Robots Nashville, TN, USA, December 6–8 (2010)
7. Ishiguro, H.: Scientific issues concerning androids. Int. J. Robot. Res. **26**(1), 105–117 (2007)
8. Yoshikawa, Y., et al.: Positive Bias for Eye Contact in Adolescents with Autism Spectrum Disorders During Conversation with an Android Robot, Annual Meeting of INSAR (2015)
9. PARS Committee. Pervasive Developmental Disorders Autism Society Japan Rating Scale. Spectrum Publishing Company, Tokyo (2008)
10. Ito, H., Tani, I., Yukihiro, R., Adachi, J., Hara, K., Ogasawara, M., et al.: Validation of an interview-based rating scale developed in Japan for pervasive developmental disorders. Res. Autism Spectr. Disord. **6**, 1265–1272 (2012)
11. Lord, C., Rutter, M., Couteur, A.L.: Autism diagnostic interview-revised: a revised version of a diagnostic interview for caregivers of individuals with possible pervasive developmental disorders. J. Autism Dev. Disord. **24**, 659–685 (1994)

Feasibility of Collaborative Learning and Work Between Robots and Children with Autism Spectrum Disorders

Felix Jimenez[1]([✉]), Tomohiro Yoshikawa[1], Takeshi Furuhashi[1],
Masayoshi Kanoh[2], and Tsuyoshi Nakamura[3]

[1] Graduate School of Engineering, Nagoya University, Nagoya, Aichi, Japan
felix@cmplx.cse.nagoya-u.ac.jp, {yoshikawa,furuhashi}@cse.nagoya-u.ac.jp
[2] School of Engineering, Chukyo University, Nagoya, Aichi, Japan
mkanoh@sist.chukyo-u.ac.jp
[3] Graduate School of Engineering, Nagoya Institute of Technology,
Nagoya, Aichi, Japan
tnaka@nitech.ac.jp

Abstract. With the growth of robot technology, educational support robots have received increasing attention. Most of the previous studies have focused on collaborative learning effects between educational support robots and healthy children. However, few studies have addressed the use of educational-support robots for children with autism spectrum disorders. Because the number of students in primary school with autism spectrum disorders has increased, the need for educational support robots developed to assist autistic children becomes more imperative. Therefore, this study investigates the practicality of collaborative learning and work with educational support robots and reports the feasibility of collaborative learning between educational-support robots and children with autism spectrum disorders.

Keywords: Collaborative learning · Collaborative work · Educational-support robots · Children with autism spectrum disorders · Learning by teaching

1 Introduction

The growth of robot technology has prompted an increasing interest in robots that assist learning. For example, an educational support robot can support students throughout their school life [1] or help them learn English [2]. Koizumi [3] developed a robot-directed series of Lego-block building classes intended to promote spontaneous collaboration among children. The robots facilitated collaborative learning among children and established positive social relationships with these children by praising their efforts. These experimental results suggest that, in addition to stimulating spontaneous collaboration, robots can enhance children's enthusiasm for learning.

© Springer International Publishing AG 2017
M. Otake et al. (Eds.): JSAI-isAI 2015 Workshops, LNAI 10091, pp. 454–461, 2017.
DOI: 10.1007/978-3-319-50953-2_32

Most existing studies report on the learning effects of collaborative learning between educational support robots and healthy children. However, the number of children in primary schools who have been diagnosed with autism spectrum disorders has increased [4]. These children have difficulty in communicating with other children. Therefore, the development of educational support robots for children with autism spectrum disorders is an important undertaking. Among the studies focusing on children with autism, one proposed a support methodology that was based on medical knowledge about autism spectrum disorders [5]. Another study developed a learning support system for autistic children [6]. A recent study developed a robot that teaches children with autism spectrum disorders and reported that the robot prompted effective learning [7].

We researched the use of a robot to support learning as a collaborative partner. However, few reports have addressed collaborative learning between educational support robots and children with autism spectrum disorders. Therefore, we do not know how the behavior of a robot affects collaborative learning with these children. Children with autism tend to receive peer teaching from healthy children in their school. In addition, autistic children might have low self-esteem and might suffer from depression [8].

To increase the self-esteem of autistic children, we think that learning by teaching, i.e., learning by teaching another learner, is a better method. This learning method promotes self-esteem and improves learning motivation. Tanaka's study [9] suggested that a robot that answers a question incorrectly and says things such as "Please teach me", prompts learners to learn by teaching in a collaborative learning environment. We think that a robot could help children with autism spectrum disorders to learn by teaching and to participate in collaborative learning with the robot.

Therefore, this study investigated the feasibility of collaborative learning between children with autism spectrum disorders and a robot that encourages them to learn by teaching. To encourage collaborative learning, the robot is designed to be unable to answer questions correctly and say things such as "Please teach me". To foster collaborative work, the robot is designed to make statements such as "Please explain the piece of work that you built". We believe that the robot can help children with autism spectrum disorders to learn by teaching collaboratively. We conducted two experiments. In one experiment, we conducted a collaborative learning session between a robot and one child with an autism spectrum disorder. In another experiment, we conducted a collaborative work session in which three children with an autism spectrum disorder built a Lego-bloc structure with the help of a robot.

2   Overview of the Robot

The robot used in this study was an Ifbot (Fig. 1). Ifbot is a conversational robot that can be used to support learning English conversation skills and promote effective learning [10]. This robot has a limited number of expressions and does not move its arm and body. The experiment reported in this study was conducted

Fig. 1. Ifbot

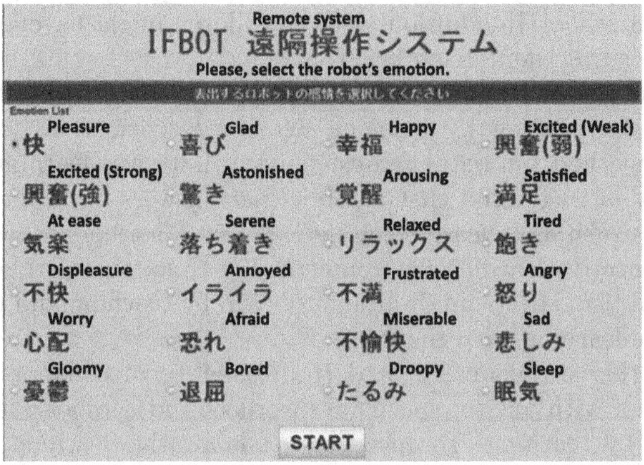

Fig. 2. Remote system to control the robot's expression

using the Wizard of Oz method; that is, the experimenter's voice is projected as the robot's voice through a personal computer. We used a remote system (Fig. 2) to control the robot's expressions, which were limited to happy and unhappy expressions. Each expression has 12 patterns.

A previous study reported that a robot can prompt a learner to learn by teaching when it answers a question incorrectly [9]. For example, the robot responds to questions by saying, "This question is difficult for me. The answer is XXXX". or "Please teach me the right answer". Moreover, when learners answer a question correctly, the robot praises the learner and says thing such as "It's amazing". or "Yes, the answer was right. Very good".

Therefore, the robot was designed to answer the question incorrectly or ask questions, such as "What is it?" Moreover, the robot says "Please teach me".

In this study, the robot does not have the ability to learn. Therefore, it always answers a question incorrectly.

3 Experiment of Collaborative Learning

3.1 Method

We conducted the experiment at Hikari Kids that was established by Gifu Emergent Research, a general incorporated association. One child with an autism spectrum disorder (boy, age:11) and one teacher from Hikari Kids participated in this experiment of collaborative learning. Two researchers operated the robot from a separate room using a remote system. This experiment was conducted for 50 min, and the experimental procedure was as follows:

(1) Introductions: 10 min
   The robot and the child took turns to introduce themselves.
(2) Collaborative learning: 40 min
   The autistic child and the robot engaged in collaborative learning while the teacher observed. The child and robot learned the history of Japan from a comic book, "Doraemon no syakaika omoshiro kouryaku nihon no rekishi 2" (in Japanese) [11]. The teacher asked the child and the robot alternately to read one page of the comic book aloud. The robot was designed to misread a sentence. For example, if the comic includes the phrase "Oda Nobunaga", the robot misreads it as "Oda Shintyou". After the robot misreads the sentence, it said statements such as "This sentence is difficult for me. Please teach me", which prompted the child to "teach" the robot. We consider that this response of the robot can prompt a learner to learn by teaching in a collaborative learning.

3.2 Results

The experimental conditions are shown in Fig. 3. During the introduction, the child played a Japanese word game with the robot and the teacher. The child named the robot "XXX kun" and communicated with it as it would with a human friend. During the collaborative learning time, we found that the child and the robot alternately read one page of the comic book aloud. Moreover, when the robot misread the sentence, e.g., saying "Oda Shintyou", the child said statements such as "That is not correct. It should be Oda Nobunaga". The child voluntarily explained who "Oda Nobunaga" was. This indicates that the robot, which misread the sentence, can help a child with autism spectrum disorder to learn through collaborative learning by prompting the child to teach. Therefore, it is possible that collaborative learning can occur between a robot and children with autism spectrum disorders.

However, when the robot repeatedly misread the sentence after being corrected by the child, the child was reluctant to correct the robot again.

(a) They play Japanese word-chain game.   (b) The child teaches the robot.

(c) The child ignores the robot.

Fig. 3. Experimental conditions 1

After the experiment, the child said statements such as "The robot could not remember what was being taught by me. Thus, teaching the robot was bothersome". This indicates that the child did not teach the robot again when it repeatedly misread the sentence after being corrected. A previous study reported that the same phenomenon occurred in collaborative learning between a robot and healthy children [12].

4   Experiment of Collaborative Work

4.1   Method

We conducted another experiment at Hikari Kids. Three children (boys) with an autism spectrum disorder and one teacher from Hikari Kids participated in this experiment of collaborative work. Two children was eight years old. One children was 11 years old. Two researchers operated the robot from a separate room using a remote system. This experiment was conducted for 50 min, and the experimental procedure was as follows:

(1) Introductions: 10 min
   The robot and the children took turns to introduce themselves.

(2) Collaborative work: 40 min

The autistic children and the robot engaged in collaborative work while the teacher observed. The children and robot built a Lego-block structure known as "Lego education story starter" [13]. The teacher gave a Lego-block task to each child. The task were "Home", "Garden", and "Christmas Day". The robot was designed to say statements such as "What is it?" or "Please explain that". If the child explained the Lego-block to the robot, the robot responded with "It is very good". We consider that the robot's response can help a student to learn in a collaborative work environment by prompting the child to reach.

4.2   Result

The experimental conditions are shown in Fig. 4. The Lego-block structures that were built by the children are shown in Fig. 5. During the introduction time, the children played a Japanese word game with the robot and the teacher. During the collaborative work time, the children said "I'm building a home for the robot" and "I'm building the chute for the robot" because the robot does not move its

d) The children and the robot play a Japanese word-chain game. (e) The children build the Lego block.

(f) The children explain the Lego block structure to the robot.

Fig. 4. Experimental conditions 2

Fig. 5. Lego-block structure built by the children

arm and body. Moreover, when the robot says "What is it?" or "Please explain the Lego-block structure that you built", the children explained to the robot how they built the structure. It is possible that collaborative work can occur between a robot and three children with autism spectrum disorders.

5   Conclusion

This study investigated the feasibility of collaborative learning and work between children with autism spectrum disorders and a robot that prompts them to learn by teaching. For collaborative learning, the robot was designed to answer questions incorrectly and make statements such as "Please teach me", to collaborative learners. In collaborative work, the robot was designed to say "Please explain a piece of work that you build" and similar statements.

From the collaborative learning experiment, we found that a child with autism spectrum disorder communicated with the robot as though it were a human friend and taught the robot the answer when it made a mistake. Moreover, the child voluntarily explained the meaning of the correct answer. This suggests that a robot that answers questions incorrectly can help an autistic child through collaborative learning by prompting a child to learn by teaching. Therefore, collaborative learning between a robot and children with autism spectrum disorders is possible. However, when the robot repeatedly misread a sentence taught by the child, the child did not teach it the correct sentence again.

Moreover, form the collaborative work experiment, we found that the children built the Lego-block structure for the robot. Moreover, when the robot said "What is it?" or "Please explain the Lego-block structure that you built", the children explained to the robot how they built the structure. Therefore, collaborative work between a robot and children with autism spectrum disorders is possible.

In future works, a longer-term experiment will be conducted to investigate the psychological and learning effects of collaborative learning between a robot and children with autism spectrum disorders.

References

1. Kanda, T., Hirano, T., Eaton, D., Ishiguro, H.: Interactive Robots as Social Partners and Peer Tutors for Children: A field trial. Hum-Comout. Interact. **10**(1), 61–84 (2004)
2. Kwon, O.H., Koo, S.Y., Kim, Y.G., Kwon, D.S.: Telepresence robot system for English tutoring. In: IEEE Workshop on Advanced Robotics and its Social Impacts, pp. 152–155. IEEE Press (2010)
3. Koizumi, S., Kanda, T., Miyashita, T.: Collaborative learning experiment with social robot. J. Robot. Soc. Jpn. **29**(10), 902–906 (2011). (in Japanese)
4. Ministry of Education, culture, sports, science and technology - Japan: Tsuuzyou no gakkyou ni zaiseki suru hattatsu syougai no kanousei noaru tokubetsu na kyouikuteki sien wo hitsuyou tosuru zidou seito ni kansuru tyousa kekka nituite (in Japanese). http://www.mext.go.jp/a_menu/shotou/tokubetu/material/1328729.htm
5. Handa, K.: Short-term effects of social skills training including self-monitoring for children with developmental disorders. Behav. Res. Ther. **40**(3), 177–187 (2014). (in Japanese)
6. Miura, T.: Hattatsusyougaizi no Soukiryouikusisutemu niokeru Hoken to Sosyaruwa-ku: Syokusyumonrenkei to Hattatsusien tositeno Ketugou. Jpn. Soc. Hum. Sci. Health-Soc. Serv. **19**(2), 55–57 (2013). (in Japanese)
7. Standen, P.J., Brown, D.J., Hedgecock, J., Roscoe, J., Galves Trigo, M.J., Elgajiji, E.: Adapting a humanoid robot for use with children with profound and multiple disabilities. In: 10th International Conference on Disability, Virtual Reality and Associated Technologies, pp. 205–211 (2014)
8. Hiruta, G.: Cognitive rehabilitation in children with ADHD. Cogn. Rehabil. **16**(1), 8–14 (2011)
9. Tanaka, H.: Social robotics research and its application at early childhood education. J. Robot. Soc. Jpn. **29**(1), 19–22 (2011). (in Japanese)
10. Ichiryu, A., Shizuka, H.: Kansei communication robot with English conversation. Jpn. Soc. Mech. Eng. **1P1-F08**(4) (2009)
11. Fujiko, F.F., Hamagakuen: Doraemon no Syakaika Omoshiro Kouryaku Nihon no rekishi 2 Kamakurazidai - Edozidai Zenhan (Doraemon no Gakusyu Siri-zu). Syougakukan (2013)
12. Matsuzoe, S., Tanaka, F.: The difference of excellence in educational-support robots affects children's learning english vocabularies. J. Jpn. Soc. Artif. Intell. **28**(2), 170–178 (2013)
13. Lego Education Story Starter. https://education.lego.com/ja-jp/preschool-and-school/upper-primary/7plus-storystarter

Teachers' Impressions on Robots for Therapeutic Applications

Hidenobu Sumioka[1,3]([✉]), Yuichiro Yoshikawa[2,3], Yasuo Wada[4],
and Hiroshi Ishiguro[1,2,3]

[1] Hiroshi Ishiguro Laboratories, Advanced Telecommunication Research
Institute International, Seika, Japan
sumioka@atr.jp
[2] Graduated School of Engineering Science, Osaka University, Suita, Japan
yoshikawa@irl.sys.es.osaka-u.ac.jp
[3] ISHIGURO Symbiotic Human-Robot Interaction Project,
ERATO, JST, Bunkyō, Japan
[4] Kyoto Prefectural Minamiyamashiro Support School, Sōraku, Japan
ishiguro@sys.es.osaka-u.ac.jp

Abstract. We examine whether teachers in special school accept communication robots as new educational tools and what they expect by introduction of them into class activities. We asked the teachers to experience demonstrations of three different types of robots, Telenoid, Hugvie, and M3-Synchy, which are available to assist teachers in educational situations and to fill in questionnaire about their impression on the robots and possible applications. The results of the questionnaire showed that more than half of teachers recognized that the robots show positive effect on their students and are useful tools not only in group activities but also in independent activities. Surprisingly, almost all teachers considered Hugvie, which is a cushion-like communication medium, have educational effects on their students, comparing it with the other robots. We also discuss what kind of robots are required by teachers.

Keywords: Human-robot interaction · Mediated interaction · Therapeutic robots · Education

1 Introduction

Recently, robotic technologies have been utilized to support children who have special educational needs due to severe learning difficulties, physical disabilities or behavioral problems. For example, research has shown the potential of treatment with robots for individuals with Autism Spectrum Disorder (ASD). Many studies report that children with ASD display the social behaviors, such as joint attention, eye contact, and imitation, to robots that are not observed while interacting with their peers, caregivers, and therapists [4,13]. This implies that the interaction with robots helps children with special needs learn new skills and

© Springer International Publishing AG 2017
M. Otake et al. (Eds.): JSAI-isAI 2015 Workshops, LNAI 10091, pp. 462–469, 2017.
DOI: 10.1007/978-3-319-50953-2_33

knowledge from other people since social behaviors are important to facilitate communication with others.

Some social behaviors can be fostered through repetition of appropriate training as shown by studies with applied behavior analysis [9,10]. Kasari *et al.* showed that 8-week interventions by caregivers improved responsiveness of toddlers with ASD to joint attention and this effect was maintained one year later [7]. Therefore, it is expected that long intervention with robots facilitates improvement of social behaviors and learning.

When we consider increasing opportunities of interaction between children with special needs and robots, we cannot ignore school life because children spend as much time in school as they spend at home. If teachers in special schools introduce robots into their daily class to support their students with special needs, the improvement of social behaviors and learning would be accelerated. However, to the best of our knowledge, there is no research that investigates whether they expect positive effect of such robots on students and in which situation they assume to utilize the robots. There, in this paper, we examine the potential of communication robots for application in special school by demonstrating different robotic systems to teachers in special school.

2   Experiment

2.1   Robots

We used three different types of robots that have different levels of human-likeness: Telenoid, Hugvie, and M3-Synchy (hereafter Synchy) because we aim to contribute to a student's improvement of social behaviors and learning and robot's human-likeness is supposed to influences interaction between a student and a robot as described in [13]. Telenoid is designed as a human-like robotic medium with minimal human appearance characteristics [12] (Fig. 1). Thanks to its teleoperation system, its movement and voice are synchronized with teleoperator's ones. Some studies show Telenoid has positive effect on communication with elderly people and group activities with children [15]. Hugvie is a simpler version of Telenoid and designed as a cushion-like communication device that focuses on the hugging experience. By putting a hands-free mobile phone inside a pocket of its "head", people can call and talk while hugging the Hugvie. Studies with Hugvie show its stress reduction [14] and the enhancement of children listening to others [11].

While these two robots were developed as a proxy of a remote person to facilitate telecommunication between people, Synchy was designed as an independent and autonomous agent like other humanoids such as Robovie [6], Nao (ALDEBARAN), and Pepper (Softbank) although it was often controlled by Wizard-of-Oz who remotely operates it in another room because of insufficient functions. Synchy was developed to study social interaction among multiple people and robots [5]. It has seventeen DoFs to express gaze and gestural behaviors and fifteen red LEDs mainly on its head and cheek for more impressive expression of its internal state such as emotion. Synchy has been used for studies about social interaction [1] and treatment of children with autism [8].

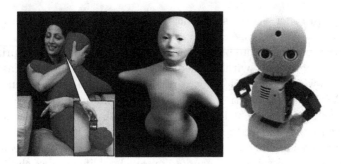

Fig. 1. Robots used in the experiment: Hugvie (left), Telenoid (center), and M3-Synchy (right)

2.2 Procedure

The experiment was conducted as an event where we introduce the robots to teachers at Kyoto prefectural Minamiyamashiro Support School. About sixty teachers (about forty female teachers and twenty male ones) for elementary school section, junior high school one, or senior high school one participated in the event in total though some of them partially joined it. Before the event starts, all participants were informed that they have to fill in a questionnaire about their impression of the robots after the event. This study was approved by the ethics committee of the Advanced Telecommunications Research Institute International (Kyoto, Japan).

Whole procedure is shown in Fig. 2. At the beginning of the event, the experimenters who are familiar with both the robots themselves and studies with them explained abilities of each robot and results confirmed in the previous studies to the teachers about for forty minutes. After that, the teachers were divided into two groups: Group A and Group B. The person who has the expert knowledge of Telenoid and Hugvie gave demonstrations of these robots to Group A at one room. Group B was given Synchy's demonstrations by experts of the robot at another room. After 25-min demonstration, each group moved to the other demonstration.

In the first half of the demonstrations of Telenoid and Hugvie, the experimenter explained teleoperation system of Telenoid and operated it. Then, he asked some participants to operate Telenoid or to have conversation with the operator through it. The others could observe the conversations. In the second half, all participants experienced Hugvie. They listened to the explanation about how Hugvie works from the experimenter with the storytelling system proposed in [11], holding Hugvie. They also listened to a recorded story through it. After all explanations by the experimenter, the participants could freely use those robots and asked questions to the experimenter.

In Synchy demonstrations, three Synchys were presented on a desk in front of participants. The participants observed or experienced four different types of social interactions among robots and participants who were interested.

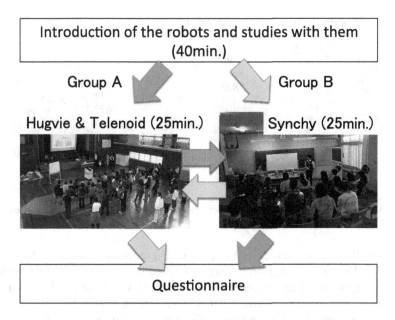

Fig. 2. The experimental procedure

First, they observed social gaze behaviors among the robots, such as making eye contact and averting one's gaze from another robot without verbal interaction. Next, one volunteer among the participants joined a conversation with the robots that actually talked according to a script without speech recognition. After that, they listened to the scripted conversation among the robots which teaches students to say thank you when they got help from other people. Finally, some participants had conversation with a Synchy that spoke according to text input from an experimenter. After all demonstrations, the participants could ask questions to the experimenter.

After the event finished, we asked the participants who experienced all demonstrations to complete questionnaires that consist of three questions: (Q1) Do you think each robot has educational effects on students? (Q2) In which situation do you think each robot is useful? (Q3) Do you want to use the robots in class activities? The participants selected one from three answers ("yes", "no", and "don't know") for Q1 and Q3 and some from three activities mainly performed in the classroom (morning meeting, independent activities, coursework) and "don't know" for Q2. They also wrote down other potential activities as "other suggestions" if they have them.

3 Result

We collected answers from twenty-seven participants and found that 60.5% of participants expected positive effect of the robots on students in total while

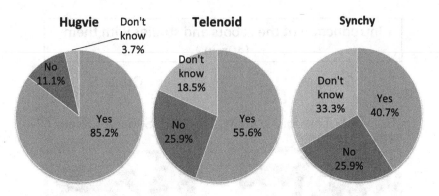

Fig. 3. Answers to Q1 (Do you think each robot has educational effects on students?)

there were 21.0% ones who do not expect it. Figure 3 shows results of teachers'
impressions for each robot. Surprisingly, 85.2% of teachers expected educational
effect of Hugvie on their students while about only half of teachers did that
for Telenoid and Synchy. Telenoid had higher rate of "yes" than Synchy. This
difference was caused by higher rate of those who had no idea about how to use
Synchy because the rates of "no" between Telenoid and Synchy were the same.

Table 1 displays the results of Q2. The teachers showed that they mainly
expect to use each robot in independent activities. Interestingly, they expected
to use Hugvie in morning meeting and coursework while the other robots were
not expected. Telenoid was also attractive for coursework. As is the case in Q1,
many teachers showed no idea about the introduction of Synchy into their class.
As other suggestions, teachers suggested playtime for all robots, sex education
for Telenoid, comprehensive study for Synchy. Hugvie was suggested to be used
in leisure activity, storytelling, and class activity.

According to the result of Q3 (Fig. 4), 59.3% of teachers answered that they
want to use these robots in their class while 11.1% did not want to use them.
There were 29.6% who have no idea.

Table 1. Answers to Q2 (In which situation do you think each robot is useful?)

	Hugvie	Telenoid	Synchy
Morning meeting	10	2	1
Independent activities	13	10	7
Coursework	8	7	1
Don't know	3	8	10

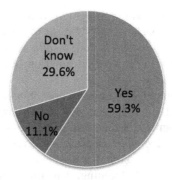

Fig. 4. Answers to Q3 (Do you want to use the robots in class activities?)

4   Discussion and Conclusion

The results showed that about 60% of teachers in special school expected educational benefits of the robots. Especially, 85.2% of them expected that positive effect of Hugvie on their students. A possible reason is that they could easily imagine the introduction of Hugvie into their class because Hugvie can be used for simultaneous support to many students as well as individual support. This is also supported by the fact that many teachers recognized usefulness of Hugvie in morning meeting and coursework as well as in independent activities as shown in Table 1. This result implies that robots to assist students with special needs should be available for both a group and an individual.

Another possibility is that the teachers confirmed Hugvie's effect of the enhancement of listening to other people talk through the demonstration and expected it on their students. Obviously, listening is a fundamental skill for students to learn new skill and knowledge in a classroom since it was reported that they spend at least 45% of the school day to be engaged in listening activities [3]. In fact, a study showed a possibility that speech perception of students with Down syndrome improves when sound condition improves in classroom [2]. Therefore, many teachers would have students who have severe problem on listening and consider that Hugvie is effective to support them.

Telenoid and Synchy seem to be expected to support an individual since many teachers answered independent activities as their application. These robots are also helpful in a lesson that human teachers have difficulty in giving such as sex education. Although several applications were presented in Synchy's demonstration, it seems difficult for the teachers to imagine how they use Synchy or an autonomous robot in their classroom according to the results of Q1 and Q2. This might be because teachers could not estimate the difficulty in producing educational materials with using Synchy. Therefore, when we introduce autonomous robots into a classroom, we should provide its application in practical situation for teachers.

In results of Q2, the rate of possible application of Telenoid was similar to Hugvie except for its application to morning meeting. The low rate of morning meeting would be because a morning meeting provides teachers for health checks of their students. In fact, a teacher said, *"I want to see and talk with all students at least in morning meeting to check their health conditions"*. Teachers would use a robot in morning meeting if it has a function of health checks or it serves as an assistant of the teachers.

In summary, we examined the impressions of teachers on communication robots as educational tools in a special school. The results showed that more than half of teachers recognized educational benefits of the robots. Especially, the teachers expected positive effect of Hugvie on their students. Since numerous studies suggest that it is possible for people with special needs to partially acquire some social skills by daily training, it is important to develop the system which is available at special school as well as home and a clinic. We have started to introduce Hugvie, which was the most acceptable for the teachers, into class activities in a special school, hypothesizing that Hugvie facilitates students' concentrating on listening their teacher and reduces their problem behaviors in their classroom. We hope that this introduction shed light on the effect of Hugvie on students with special needs and some problems of its introduction on school education.

Acknowledgments. This research was supported by JST, ERATO, ISHIGURO symbiotic Human-Robot Interaction Project. We thank to teachers in Kyoto prefectural Minamiyamashiro Support School for their participation.

References

1. Arimoto, T., Yoshikawa, Y., Ishiguro, H.: Nodding responses by collective proxy robots for enhancing social telepresence. In: Proceedings of the Second International Conference on Human-Agent Interaction, pp. 97–102. ACM (2014)
2. Bennetts, L., Flynn, M.: Improving the classroom listening skills of children with down syndrome by using sound-field amplification. Down Syndr. Res. Pract. **8**(1), 19–24 (2002)
3. Berg, F.S.: Facilitating Classroom Listening: A Handbook for Teachers of Normal and Hard of Hearing Students. College Hill Pr, Boston (1987)
4. Diehl, J.J., Schmitt, L.M., Villano, M., Crowell, C.R.: The clinical use of robots for individuals with autism spectrum disorders: a critical review. Res. Autism Spectr. Disord. **6**(1), 249–262 (2012)
5. Ishiguro, H., Minato, T., Yoshikawa, Y., Asada, M.: Humanoid platforms for cognitive developmental robotics. Int. J. Humanoid Rob. **8**(3), 391–418 (2011)
6. Kanda, T., Ishiguro, H., Imai, M., Ono, T.: Development and evaluation of interactive humanoid robots. Proc. IEEE **92**(11), 1839–1850 (2004)
7. Kasari, C., Gulsrud, A.C., Wong, C., Kwon, S., Locke, J.: Randomized controlled caregiver mediated joint engagement intervention for toddlers with autism. J. Autism Dev. Disord. **40**(9), 1045–1056 (2010)
8. Kumazaki, H., Yoshikawa, Y.: Preliminary study for the effect of intervention to joint attention with autism spectrum disorder using a robot. Hum. Dev. Res. **29**, 169–174 (2015). In Japanese
9. Lovaas, O.I.: Behavioral treatment and normal educational and intellectual functioning in young autistic children. J. Consult. Clin. Psychol. **55**(1), 3 (1987)

10. McEachin, J.J., Smith, T., Lovaas, O.I.: Long-term outcome for children with autism who received early intensive behavioral treatment. Am. J. Mental Retard. **97**, 359–359 (1993)
11. Nakanishi, J., Sumioka, H., Sakai, K., Nakamichi, D., Shiomi, M., Ishiguro, H.: Huggable communication medium encourages listening to others. In: Proceedings of the 2nd International Conference on Human-Agent Interraction, pp. 249–252 (2014)
12. Ogawa, K., Nishio, S., Koda, K., Balistreri, G., Watanabe, T., Ishiguro, H.: Exploring the natural reaction of young and aged person with telenoid in a real world. J. Adv. Comput. Intell. Intell. Inform. **15**(5), 592–597 (2011)
13. Ricks, D.J., Colton, M.B.: Trends and considerations in robot-assisted autism therapy. In: IEEE International Conference on Robotics and Automation, pp. 4354–4359. IEEE (2010)
14. Sumioka, H., Nakae, A., Kanai, R., Ishiguro, H.: Huggable communication medium decreases cortisol levels. Sci. Rep. 3, No. 3034 (2013)
15. Sumioka, H., Nishio, S., Minato, T., Yamazaki, R., Ishiguro, H.: Minimal human design approach for sonzai-kan media: investigation of a feeling of human presence. Cogn. Comput. **6**(4), 760–774 (2014)

An Intervention for Children with Social Anxiety and Autism Spectrum Disorders Using an Android Robot

Hirokazu Kumazaki[1,2(✉)], Yuichiro Yoshikawa[3], Yoshio Matsumoto[4],
Masutomo Miyao[5], Hiroshi Ishiguro[3], Taro Muramatsu[2], and Masaru Mimura[2]

[1] Kanazawa University, Kanazawa, Japan
kumazaki@tiara.ocn.ne.jp
[2] Department of Neuropsychiatry, Keio University School of Medicine,
Shinjuku-ku, Tokyo, Japan
muramatsu@a8.keio.jp, mimura@a7.keio.jp
[3] Graduate School of Engineering Science, Osaka University, Suita, Japan
yoshikawa@irl.sys.es.osaka-u.ac.jp
[4] Robot Innovation Research Center, National Institute of Advanced Industrial Science
and Technology (AIST), Tsukuba, Japan
yoshio.matsumoto@aist.go.jp
[5] Donguri Psycho Developmental Clinic, Setagaya-ku, Tokyo, Japan
miyaomncchd@gmail.com

Abstract. Some children with social anxiety and autism spectrum disorders (ASD) are characterized by deficits in social communication consisting of a lack of speech in specific social situations in which there is an expectation of speaking, although these children do speak in other situations. Comorbid social anxiety and ASD is considered hard to treat. A variety of therapeutic and educational approaches have been developed, which have had educational benefits for some children with social anxiety and ASD. Thus, there is an urgent need for the development and application of novel and more efficacious treatment strategies. We introduced a "taking note system," in which a subject operates an android robot and communicates with others through it. Post intervention, the participant showed a significant decrease in social anxiety. Based on our case report, an intervention using a "taking note system" is effective for decreasing stress. This finding suggests that this approach is promising and warrants further study.

Keywords: Animal-assisted therapy · Robot therapy · PARO · Autism spectrum disorder

1 Introduction

1.1 Social Anxiety

Autism spectrum disorder (ASD) is a developmental disorder that can cause lifelong challenges [1]. As adolescents with ASD grow more conscious of their own social difficulties, many of them develop symptoms of social anxiety [2–4]. Some children with social anxiety and autism spectrum disorders (ASD) are characterized by deficits

M. Otake et al. (Eds.): JSAI-isAI 2015 Workshops, LNAI 10091, pp. 470–477, 2017.
DOI: 10.1007/978-3-319-50953-2_34

in social communication involving a lack of speech in specific social situations in which there is an expectation of speaking, although these children do speak in other situations. Social anxiety is an emotion characterized by discomfort or fear when in a social interaction that involves a concern of being judged or evaluated by others. It is typically characterized by an intense fear of what others are thinking about one, which results feelings of insecurity and being not good enough for other people, and/or the assumption of automatic rejection by peers. Developmental social anxiety occurs early in childhood as a normal part of the development of social functioning and is a stage that most children outgrow; however, it may persist or resurface and grow into chronic social anxiety during the teenage years or even adulthood. The causes of social anxiety in specific children are not always clear; however, social anxiety is known to be associated with an autistic profile. Many factors are related to comorbid social anxiety and autism spectrum disorders, such as psychological factors, neurobiological factors, and environmental factors.

1.2   Treatment for Social Anxiety

Treatment for social anxiety consists of two primary domains: psychotherapeutic-and medication-based interventions. Within the non-medication-based or psychotherapeutic approaches, psychodynamic therapy, behavioral therapy, and family therapy are among the most common. A medication-based approach may also play a role in treating social anxiety, with SSRIs being most commonly recommended.

Some forms of social anxiety are considered hard to treat; as a result, a variety of therapeutic and educational approaches have been developed. Educational benefits may be seen in some but not all children with social anxiety and ASD. Considering the present situation, there is an urgent need for the development and application of novel and more efficacious treatment strategies [5]. Given recent rapid technological advances, it has been argued that technology could be effectively harnessed to provide innovative clinical treatments for individuals with social anxiety.

2   Outline of the Study

There are few reports about therapy using a robot for social anxiety comorbid with ASD. Thus, we proposed an intervention using robotics. Researchers investigating robots as communication therapy tools often report increased engagement, levels of attention, and social communication, along with amelioration of social anxiety [5–9]. Much consideration has been given to the type and form of robots used in interventions as a variety exist [10]. A humanoid robot, particularly an android robot, can be used as a tele-communication medium for distant inter-human communication. Android robots look like humans. Thus, they may have the greatest potential for generalization. Individuals with ASD often find it difficult to process abstract information (Baron-Cohen 1991; Ropar and Mitchell 2002; Tager-Flusberg 1985; Weeks and Hobson 1987) and prefer concrete features (Ropar and Peebles 2006) such as color and shape, which is indeed the android model with strong visual human-likeness. Given this factor, it is possible that individuals with ASD prefer android robot.

2.1 ACTROID-F

The android robot used in this study was ACTROID-F (Kokoro Co. Ltd.), a female humanoid robot with an appearance similar to that of a real person (see Fig. 1). The artificial body of the ACTROID-F has the same proportions, facial features, hair color, and hairstyle that a human might have. At first sight and from a distance, it is difficult to distinguish this android robot from a real person. There are advantages that interlocutors can be absorbed in conversation and operators can experience real conversation. Our preliminary experiment suggested that ACTROID-F might help improve nonverbal communication and decrease social anxiety of children.

Fig. 1. Geminoid F (android robot)

2.2 Introduction for "Taking Note System"

We introduced "taking note system," in which the subject operates the android robot and can communicate with others through the android robot. For the "taking note system," the operator input words into the computer, and ACTROID-F read aloud. Moreover, ACTROID-F can have a conversation naturally by transferring the motions of the operator measured through a motion capturing system. The operator can watch the look of ACTROID-F through camera image. Our preliminary experiment suggested that appearance of Actroid-F made operators as if they spoke themselves. There are many differences between speech and taking notes (Fig. 2). In terms of vector, speech is direct, while taking notes is indirect. In terms of durability, speech does not remain, while taking notes remains. In terms of revisions, speech cannot be revised, while taking notes can be revised. In terms of time to respond, speech is difficult to adjust, while taking notes is easy to adjust. Considering these differences, these elements of "taking note system" makes communication easier for children with social anxiety comorbid with ASD whose working memory and processing speed have been reported to be impaired.

Differences between speech and taking notes

	Speech	Taking notes
Vector	direct	indirect
Durability	does not remain	remains
Revision	can't revise	can revise
Time to respond	difficult to adjust	easy to adjust

Fig. 2. Differences between speech and taking notes

Thus, we proposed the development and initial application of a robotic intervention system capable of targeted interaction for children with social anxiety and ASD, to improve their communication skills.

3 Case Report

The developmental milestones of the 18-year-old female (Subject A) were normal until about two and a half years of age, when she was noticed to have diminution in the volume of her speech. She refused to talk in class, which contributed to her poor performance at school. Her diagnosis was ASD. The author confirmed existing diagnoses using both diagnostic instruments and screening questionnaires, including the Pervasive Developmental Disorder–Autism Society Japan Rating Scale (PARS), a diagnostic interview scale for ASD developed in Japan [11]. Sub and total scores of this scale correlate with the domain and total scores of the Autism Diagnostic Interview-Revised (ADI-R) [12, 13]. There was no history of seizure disorder or any suggestion of neurological disorder.

The experiment was conducted in a familiar room often used with children for various activities (see Fig. 3). As the subject typed on the keyboard, ACTROID-F spoke.

ACTROID-F was positioned at one end of the room within plain view of the participants. In addition, a camera image was placed in front of the subject for her to watch.

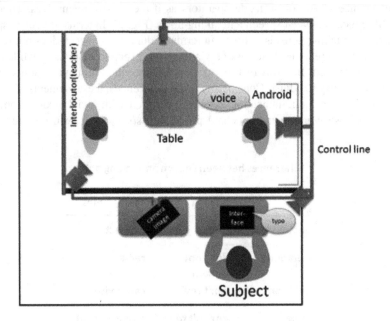

Fig. 3. The experimental room

The procedure consisted of three sessions. In the first session, communication of ACTROID-F was controlled by Subject A and a teacher in a separate room. In the second session, ACTROID-F was controlled by Subject A and two teachers in a separate room. In the third session, ACTROID-F was controlled by Subject A and a teacher in the same room.

Before and after the intervention, the Stress Response Scale (SRS-18) [14] was completed by Subject A. All statements were rated on a 4-point scale ranging from 1 (not at all) to 4 (very much). The SRS-18 consists of 18 items ("become irritable," "in a sad frame of mind," "worried about something," "feel angry," "feel like crying," "can't control of the passion," "have a frustrating thought," "feel bad," "sink deeply," "irritated," "don't have confidence in lots of things" "mind everything" "think negatively," "ideas and behaviors haven't taken shape," "want to be cheered up," "feel like being by myself," "cannot concentrate on something"). The total SRS-18 score is obtained by summing each item score and, thus, ranges from 18 to 72.

In each session, Subject A could respond and type in any situation. Although the experimental settings seems to become difficult as session processes, the time after the question until typing began decreased, while utterance action, type action, and open question letter increased. The results of the intervention are presented in Table 1. For the SRS-18, the items "become irritable," "mind everything," "sink deeply," and "cannot concentrate on something" improved after the intervention. The total score on the SRS-18 decreased from 40 to 32. In an analysis of the videotape, when she typed on the keyboard her expression became soft. Before she pressed "enter," she was stiff, but after she pressed "enter," she became relieved. She also seemed to be deep in the conversation because she concentrated on utterance from a teacher and devoted herself completely to type. For some questions, she typed and then pressed "enter." For other questions, she typed but could not press "enter" (see Fig. 4). It seemed to be difficult to press "enter" for those questions. She showed a significant decrease in social anxiety when she pressed "enter" and ACTROID-F spoke. Therefore, it is possible that an intervention using a "taking note system" is effective for decreasing stress.

Table 1. Transition of the response of subject for intervention.

	Time after question untill beginning to type (S)
One to one (separate)	5.30
One to one (separate)	1.18
One to one (same room)	1.05

One to one (separate): Communication between ACTROID–F controlled by A and a teacher in a separate room.
One to one (separate): ACTROID–F controlled by A and a teacher in a separate room.
One to one (same room): ACTROID–F controlled by A and a teacher in a same room.

Questions that she could answer

Do you want to go with your classmate?

Please tell me who is the most interesting fellow in this class.

Do you have a colleague who is an understanding person?

Please tell me which teacher you seem to be talking with.

Questions that she could not answer

What kinds of things about jet coasters do you find fascinating?

Aren't you embarrassed about talking with the teachers?

Fig. 4. Examples of questions

4   Conclusion

Due to the situational nature of social anxiety comorbid with ASD, we hypothesized that children would show increased speaking behavior primarily in private situations. The goal of investigating robots for children with social anxiety comorbid with ASD is to develop systems that elicit these positive and productive interactions. Despite limited effectiveness of the intervention, it is suggested that an intervention using ACTROID-F could be used as a tool for the development of communication. Generalization of the skills learned in these sessions has not been observed outside of the experimental room. This initial finding suggests that this approach is promising and warrants further study. Future research needs to examine this intervention with a larger, more diverse sample, and using multiple levels of analysis of behavioral change.

Ethical Approval. All procedures performed involving human participants were in accordance with the ethical standards of the institutional and/or national research committee and with the 1964 Helsinki declaration and its later amendments or comparable ethical standards. We have no competing interests.

Acknowledgement. This work was supported in part by Grants-in-Aid for Scientific Research from the Japan Society for the Promotion of Science (15H01577). We sincerely thank the participants and all the families who participated in this study.

References

1. Kaboski, J.R., Diehl, J.J., Beriont, J., Crowell, C.R., Villano, M., Wier, K., Tang, K.: Brief report: a pilot summer robotics camp to reduce social anxiety and improve social/vocational skills in adolescents with ASD. J. Autism Dev. Disord. **45**, 3862–3869 (2014)
2. Gillott, A., Furniss, F., Walter, A.: Anxiety in high-functioning children with autism. Autism **5**, 277–286 (2001)
3. Sterling, L., Dawson, G., Estes, A., Greenson, J.: Characteristics associated with presence of depressive symptoms in adults with autism spectrum disorder. J. Autism Dev. Disord. **38**, 1011–1018 (2008)
4. Shtayermman, O.: Peer victimization in adolescents and young adults diagnosed with asperger's syndrome: a link to depressive symptomatology, anxiety symptomatology and suicidal ideation. Issues Compr. Pediatr. Nurs. **30**, 87–107 (2007)
5. Warren, Z.E., Zheng, Z., Swanson, A.R., Bekele, E., Zhang, L., Crittendon, J.A., Weitlauf, A.F., Sarkar, N.: Can robotic interaction improve joint attention skills? J. Autism Dev. Disord. **45**, 3726–3734 (2013)
6. Bekele, E., Crittendon, J.A., Swanson, A., Sarkar, N., Warren, Z.E.: Pilot clinical application of an adaptive robotic system for young children with autism. Autism **18**, 598–608 (2013)
7. Huskens, B., Palmen, A., Van der Werff, M., Lourens, T., Barakova, E.: Improving collaborative play between children with autism spectrum disorders and their siblings: the effectiveness of a robot-mediated intervention based on lego® therapy. J. Autism Dev. Disord. **45**, 3746–3755 (2014)
8. Goodwin, M.S.: Enhancing and accelerating the pace of autism research and treatment: the promise of developing innovative technology. Focus Autism Other Dev. Disabil. **23**, 125–128 (2008)
9. Dautenhahn, K., Werry, I.: Towards interactive robots in autism therapy: background motivation and challenges. Pragmatics Cogn. **12**, 1–35 (2004)
10. Ricks, D.J., Colton, M.B.: Trends and considerations in robot-assisted autism therapy. In: 2010 IEEE International Conference on Robotics and Automation (ICRA) (2010)
11. PARS Committee. (1997). Pervasive Developmental Disorders Autism Society Japan Rating Scale. Spectrum Publishing Company, Tokyo (2008). (in Japanese)
12. Ito, H., Tani, I., Yukihiro, R., Adachi, J., Hara, K., Ogasawara, M., Inoue, M., Kamio, Y., Nakamura, K., Uchiyama, T., Ichikawa, H., Sugiyama, T., Hagiwara, T., Tsujii, M.: Validation of an interview-based rating scale developed in Japan for pervasive developmental disorders. Res. Autism Spectr. Disord. **6**, 1265–1272 (2012)

13. Lord, C., Rutter, M., Le Couteur, A.: Autism diagnostic interview-revised: a revised version of a diagnostic interview for caregivers of individuals with possible pervasive developmental disorders. J. Autism Dev. Disord. **24**, 659–685 (1994)
14. Suzuki, S., Shimada, H., Miura, M., Katayanagi, K., Umano, R., Sakano, Y.: Development of a new psychological stress response scale (SRS-18) and investigation of the reliability and the validity. Japan. J. Behav. Med. **4**, 21–29 (1997)

Usefulness of Animal Type Robot Assisted Therapy for Autism Spectrum Disorder in the Child and Adolescent Psychiatric Ward

Yoshihiro Nakadoi[✉]

Department of Child Psychiatry, Shikoku Medical Center for Children and Adults,
Senyuu-Cyo 2-1-1, Zentsuji City, Kagawa Pref 765-8507, Japan
y-nakadoi@umin.ac.jp

Abstract. Objectives: The purpose of this study is to assess the effectiveness of PARO assisted therapy for Autism Spectrum Disorder (ASD) in the child and adolescent psychiatric ward. Methods: In the Child and Adolescent Psychiatry Ward of Shikoku Medical Center for Children and Adults, we put PARO near the door of the nurse station and told the inpatients to play with it freely in the hall of the ward after getting permission from the staff. The interaction between patients and PARO was observed. Results: It was thought to be useful for same patients with ASD in developing good communication or in reduction of impulsive behaviors or anxiety. However, others hated it for some features such as its big eyes or a slight drive noise. Conclusions: Before animal type robot-assisted therapy is introduced as a tool for the patients with ASD, the kind of patients who are benefitted by this approach and how the approach would work in the treatment must be clarified.

Keywords: Animal-Assisted Therapy · Robot therapy · PARO · Autism Spectrum Disorder

1   Introduction

Animal-Assisted Therapy (AAT) is a type of psychotherapy that works with animals, such as dogs and horses. It has been suggested as a treatment practice for Autism Spectrum Disorder (ASD) in child and adolescent area [1–4]. It is expected to have three effects; psychological effect (e.g., relaxation, motivation), physiological effect (e.g., improved vital signs, stress reduction), and social effect (e.g., stimulation of communication among in-patients and caregivers).

PARO is a baby harp seal type therapeutic robot that was developed by Japan' National Institute of Advanced Industrial Science and Technology [5]. PARO has five sensors: tactile, light, audition, temperature, and posture sensors. With the light sensor, PARO can recognize light and dark. With tactile sensor, PARO feels being stroked and beaten. By posture sensor, PARO feels being held, and can also recognize the direction of sound with its audio sensor. PARO assisted therapy is expected to be able to provide therapy similar to AAT for patients in a variety of clinical settings. Unlike AAT, it does

© Springer International Publishing AG 2017
M. Otake et al. (Eds.): JSAI-isAI 2015 Workshops, LNAI 10091, pp. 478–482, 2017.
DOI: 10.1007/978-3-319-50953-2_35

not cause any allergies, infections, bites and scratches and can be introduced at lower cost. To date, there are some studies of its effectiveness for elderly people or dementia patients [6–10]. However, the effectiveness of it in the treatment of ASD remains unclear.

The purpose of this study was to assess the effectiveness of the use of animal type robot PARO for ASD in the child and adolescent psychiatric ward.

2 Methods

Nine patients (four boys and five girls, aged 8–19 years) with ASD were enrolled. They were all admitted to the Child and Adolescent Psychiatry Ward of Shikoku Medical Center for Children and Adults and they had any contacted with PARO. This study was conducted in the Ward from November in 2014 to January in 2015. We put PARO near the door of the nurse station and told the inpatients to play with it freely in the hall of the ward after getting permission from the staff. The interaction between patients and PARO was observed. The procedure was explained and informed consent was obtained from the patients and their parents. This research was approved by the Ethical Committee of Shikoku Medical Center for Children and Adults (Fig. 1).

Fig. 1. PARO was put near the door of the nurse station and told the inpatients to play with it freely in the hall of the ward after getting permission from the staff.

3   Results

Some patients with ASD treated PARO like a real animal. Especially, it was thought to be effective for two patients.

First, it was thought to be useful for 10-year-old girl with ASD to develop good communication. She was identified as having delayed in motor and language development. However, she was sensitive to small change around her. She got often excited because of her difficulties dealing with emotional and psychological transitions. She liked to play by herself and often escaped and ran into the street without awareness of the traffic. At the age of six, she was diagnosed with ASD and Attention Deficit Hyper-activity Disorder (ADHD), and prescribed risperidone and valproic acid for her irrita-bility and abnormal EEG. At the age of ten, she refused to go to an elementary school and suffered from auditory hallucinations, such as music or nasty things being said about her and from delusion of persecution. After that, she was referred to our hospital. She was diagnosed with ASD, ADHD and acute and transient psychotic disorders. She was admitted to our hospital. After admission, psychiatric symptoms (auditory hallucinations and delusions of persecution) has gone, but she had trouble associating with others in the ward. She often talked excessively and suddenly became excited. One day, she became interested in PARO. She liked to hold PARO and talked about PARO with other patients or staffs in the day room of the ward almost every day. She often said "PARO is so cute". She seemed to treat PARO like a living animal. We saw a pleased expression on her face when some patients got interested in PARO and gathered around her and PARO. Two weeks later, though her autistic trait remained completely unchanged, she gradually started to talk in a relaxed way.

Second, it was also considered to be useful for 16-year-old boy with ASD and moderate mental retardation in reduction of impulsive behaviors or anxiety. He was identified as having difficulties in speech and communication. At the age of four he was diagnosed with autism and intellectual disability. At elementary school, he attended a special needs education class, and he went to a special needs education junior high school and high school. At the age of fifteen, he refused to go to high school and suffered from visual hallucinations, a fear of animals, and delusions of persecution. He was referred to with our hospital because he got physically uncontrollable at home. He was diagnosed with ASD, moderate intellectual disability and acute and transient psychotic disorders. He was admitted to our hospital. After admission, the visual hallucinations and delusions of persecutions have gone, but he got physically uncontrollable and repeatedly asked the same question, "When can I go back home?" We explained to him that he could go back home when he no longer not easily excited. We also tried to establish a token economy, but he did not understand it. We assumed that this was due to his intellectual disability and autistic persistence. One day, we gave him PARO instead of an antipsy-chotic drug when he had the impulse to go back home. He liked PARO, and held PARO with pleasure in the day room of the ward almost every day though we could not find deep interaction between him and other patients. He often said hello to PARO. He seemed to treat PARO like a living animal. He sometimes went to sleep with PARO, which was an exceptional allowance for him. Two weeks later, we saw a calm expression on his face.

On the other hand, 12-year-old boy with high functioning ASD did not like PARO because of some features like big eyes or a slight drive noise. He said to us, "PARO is noisy. It is a machine!". Moreover, 8-year-girl with high functioning ASD and ADHD who had to do with lack of normal attachment often attacked PARO.

4   Discussion

Patients with ASD in the Child and Adolescent Psychiatry Wards have various comorbid disorders and diverse causes such as other developmental disorders (e.g. ADHD, learning disorder), mood disorder, psychosis, panic disorder, anxiety disorder, dissociative disorder, obsessive compulsive disorder, eating disorder, posttraumatic stress disorder, substance dependence and maltreatment. So we should treat the patients by using combination methods, including PARO assisted intervention. PARO is effective for some autistic children who have impaired communication abilities. PARO helps to promote their communication abilities or feel calm and relaxed with its simple emotional expressions; for example, PARO barks happily in response to being petted. However, others hated PARO for some features such as its big eyes or a slight drive noise, because they have a strong tendency to put undue focus on details. Before animal type robot-assisted therapy is introduced as a tool for the patients with ASD, the kind of patients who are benefitted by this approach and how the approach would work in the treatment must be clarified.

Moreover, some hospital staffs said that they felt relaxed with PARO, though we did not use the objective index. PARO might be useful for good communication between staffs and patients.

This study had some important limitations. First, the positive changes in the successful cases might be just from the temporal changes in motivation or emotional state. We confirmed the lasting positive change by gross observation, though we did not use the objective index. Second, the positive changes might be caused by the other hospital treatment. At least, we did not change in medication during observation period.

Further interventional studies using the objective index are needed for children with ASD in order to show the effectiveness of PARO assisted therapy.

References

1. Martin, F., Farnum, J.: Animal-assisted therapy for children with pervasive developmental disorders. West. J. Nurs. Res. **24**, 657–670 (2002)
2. Bass, M.M., Duchowny, C.A., Llabre, M.M.: The effect of therapeutic horseback riding on social functioning in children with autism. J. Autism Dev. Disord. **39**, 1261–1267 (2009)
3. Viau, R., Arsenault-Lapierre, G., Fecteau, S., Champagne, N., Walker, C.D., Lupien, S.: Effect of service dogs on salivary cortisol secretion in autistic children. Psychoneuroendocrinology. **35**, 1187–1193 (2010)
4. Gabriels, R.L., Pan, Z., Dechant, B., Agnew, J.A., Brim, N., Mesibov, G.: Randomized controlled trial of therapeutic horseback riding in children and adolescents with autism spectrum disorder. J. Am. Acad. Child Adolesc. Psychiatry **54**, 541–549 (2015)

5. Shibata, T., Tanie, K.: Influence of a priori knowledge in subjective interpretation and evaluation by short-term interaction with mental commit robot. In: IEEE/RSJ International Conference on Intelligent Robots and Systems, 31 October–5 November, Takamatsu, Japan (2000)

6. Takayanagi, K., Kirita, T., Shibata, T.: Comparison of Verbal and Emotional Responses of Elderly People with Mild/Moderate Dementia and Those with Severe Dementia in Responses to Seal Robot, PARO. Front. Aging Neurosci. 6(257) (2014)

7. Sung, H.C., Chang, S.M., Chin, M.Y., Lee, W.L.: Robot-assisted therapy for improving social interactions and activity participation among institutionalized older adults: a pilot study. Asia Pac. Psychiatry 7, 1–6 (2015)

8. Robinson, H., MacDonald, B., Broadbent, E.: Physiological effects of a companion robot on blood pressure of older people in residential care facility: a pilot study. Australas. J. Ageing 34, 27–32 (2015)

9. Robinson, H., MacDonald, B., Kerse, N., Broadbent, E.: The psychosocial effects of a companion robot: a randomized controlled trial. J. Am. Med. Dir. Assoc. 14, 661–667 (2013)

10. Robinson, H., MacDonald, B., Kerse, N., Broadbent, E.: Suitability of healthcare robots for a dementia unit and suggested improvements. J. Am. Med. Dir. Assoc. 14, 34–40 (2013)

SKL 2015

International Workshop on Skill Science

Tsutomu Fujinami

Japan Advanced Institute of Science and Technology, 1-1 Asahidai, Nomi, Ishikawa 923-1292, Japan

1 Aims and Scope

Human skills involve well-attuned perception and fine motor control, often accompanied by thoughtful planning. The involvement of body, environment, and tools mediating them makes the study of skills unique among researches of human intelligence. The symposium invited researchers who investigate human skill. The study of skills requires various disciplines to collaborate with each other because the value of skills is not determined solely in terms of efficiency, but calls for consideration of quality. Quality resides in person and often needs to be transferred through apprentice systems. The procedure of validation is strict, but more complex than scientific activities, where everything needs to be described by referring to data. We are keen to discussing the theoretical foundations of skill science as well as practical and engineering issues in the study.

2 Topics

We invited wide ranges of investigation into human skills, from science and engineering to sports, art, music, craftsmanship, and whatever concerns cultivating human possibilities. Thirteen pieces of work were presented at the workshop, including one invited lecture. Four selected pieces of work are included in the issue from our workshop.

The article titled "The Trend in the Frontal Area Activity Shift with Embodied Knowledge Acquisition during Imitation Learning of Assembly Work", written by Asaka and co-authors, reports how they identified motor learning process in the brain using near-infrared spectroscopy. Another article titled "Identifying Context-dependent Modes of Reading", written by Fuyama and Hidaka, proposes a method to detect processes running in subject's mind while reading books. The third article titled "The Cognitive Role of Analogical Abduction in Skill Acquisition", written by Furukawa and his colleagues, employs a logic approach to studying cellist's skill to propose a cognitive model of skill acquisition. The other article titled "Whole-body coordination skill for dynamic balancing on a slackline", written by Kodama and his colleagues, reports how they traced the coordination developed in the whole body while subjects learnt to balance themselves on a slackline.

The workshop organizer is honored to present these reports, which deal with varieties of issues from theoretical to practical problems. He hopes that the reader will also find them interesting and will be interested to study human skills.

The Trend in the Frontal Area Activity Shift with Embodied Knowledge Acquisition During Imitation Learning of Assembly Work

Yusuke Asaka[1(✉)], Keiichi Watanuki[1,2,3], and Lei Hou[3]

[1] Graduate School of Science and Engineering, Saitama University,
255 Shimo-Okubo, Sakura-Ku, Saitama-Shi, Saitama 338-8570, Japan
s15dr001@mail.saitama-u.ac.jp
[2] Brain Science Institute, Saitama University, Saitama, Japan
[3] Institute of Ambient Mobility Interfaces, Saitama University, Saitama, Japan
{watanuki,hou}@mech.saitama-u.ac.jp

Abstract. This paper discusses the relationship between brain activity and improvement of skills during the process of embodied knowledge acquisition by imitation. Study subjects watched a video clip of a working procedure and then executed the same series of actions. Each experiment was conducted twice. After the first experiment, we set up three practice trials. Using near-infrared spectroscopy, we found that the trend in oxy-hemoglobin levels during the observation task shifted toward a low-level increase in the dorsolateral prefrontal area and a low-level decrease in the frontal lobe with improvement in performing the skill. In the execution task, the trend in oxy-hemoglobin shifted toward an increase in the dorsolateral prefrontal area and toward a decrease in the frontal pole with improvement in skill performance. These results suggest that activity in the frontal area changes during the process of embodied knowledge acquisition.

Keywords: Body intelligence · Brain activity · Near-Infrared spectroscopy · Knowledge acquisition

1 Introduction

Explicit knowledge can be expressed in a variety of ways, such as text, figures, and tables. On the other hand, knowledge that is not easily expressed is called tacit knowledge [1, 2]. In this study, we define tacit knowledge to include embodied knowledge, which is a set of skills based on experiences and intuitive sense, as seen in performing an art, sport, craft, or other skilled task. Embodied knowledge cannot be easily communicated and shared because of the difficulty in evincible expression. For this reason, the efforts of both the learner and the instructor are important for acquisition of embodied knowledge.

In core manufacturing industries, embodied knowledge includes skills pertinent to making products, and effective sharing of knowledge is an important issue in developing human resources. Instructors with a lot of embodied knowledge can provide advice to learners regarding the quality of their work and their products. However, from a practical

© Springer International Publishing AG 2017
M. Otake et al. (Eds.): JSAI-isAI 2015 Workshops, LNAI 10091, pp. 485–498, 2017.
DOI: 10.1007/978-3-319-50953-2_36

standpoint, ensuring that learners have the time to practice with instructors is difficult [3]. In addition, because of the problem with evincible expression, the learner's level of embodied knowledge is difficult to evaluate. An example of a method to evaluate the level is a practical exam taken by the learner. Exam monitors need to be experts in the skill, and they evaluate the learner's level based on their own experience. Due to the nature of embodied knowledge and the realities of the practical worksite, learners need to objectively evaluate their level of embodied knowledge in order to acquire it on their own. This involves a high level of information manipulation, such as internalization of tacit knowledge. Measuring brain activity is an appropriate method for objective evaluation of the level of internalization.

In our study, we used near-infrared spectroscopy (NIRS) to investigate the relationship between brain activity and embodied knowledge during the process of embodied knowledge acquisition (EKA). Evaluation of the level of knowledge acquisition by monitoring brain activity can be an objective indicator of the learner's degree of skill progression and can enable prediction of improvement with modeling and options for more effective methods of learning. Therefore, our ultimate goal was to construct a new learning model with the use of NIRS to improve learning efficiency.

In this study, we targeted procedural memory and imitation learning. In core manufacturing industries, the process of EKA plays an important role in remembering the operative procedures of the process machinery. To construct the learning model, we conducted the same experiment twice. After the first experiment, we set up three practice times and evaluated the influence of the improvement of the task skill on brain activity.

2   Measurement of Frontal Area Activity by NIRS

2.1   Optical Brain Imaging System

When neural activity occurs in the brain, blood flow increases in the tissue near the active neurons, and the rate of oxygenated and deoxygenated hemoglobin (oxyHb, doxyHb) levels in the blood changes. Near-infrared light (700–900 nm) is harmlessly transmitted through the human body, and hemoglobin characteristically changes following near-infrared absorbance, depending on the oxygen level. These properties enable non-invasive measurement of brain activity. Another advantage of NIRS is that it allows subjects to move, unlike other brain function measurement techniques. NIRS has relatively high temporal resolution, and the device is small and portable. Thus, in this study of the process of learning embodied knowledge, NIRS is a valid measurement technique.

2.2   Measurement of the Frontal Area and Removal of Artifact Due to Biofunction

The dorsolateral prefrontal area of the brain is closely related to working memory, since it establishes long-term memory [4]. The ability to remember words later is predicted by the amplitude of activation in the left prefrontal and temporal cortices during word encoding [5]. In a previously conducted experiment, in which the subjects remembered a set of simple body actions by imitation learning and then performed the task, the oxyHb levels increased in the dorsolateral prefrontal area and decreased in the frontal pole [6].

In the current study, we measured activity in the frontal lobe and analyzed the same areas, including the right prefrontal area (Channel 20), the frontal pole (Channel 23), and the left prefrontal area (Channel 26) (Fig. 1).

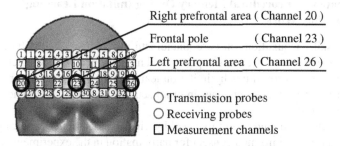

Right prefrontal area (Channel 20)

Frontal pole   (Channel 23)

Left prefrontal area   (Channel 26)

○ Transmission probes
○ Receiving probes
□ Measurement channels

Fig. 1. Regions of NIRS measurement

With the use of NIRS, optical fibers were placed on the subject's scalp, based on the international 10–20 system. Because of the contact of the fibers with the scalp, the measurements were affected by the subject's body motion, metabolism, and breathing. With the assumption that these artifacts are similar in all brain regions, we applied global average references [7]. In their study, they verified their effectiveness for static tasks such as listening to music, reading text, solving puzzles, or other tasks.

This hypothesis can be applied to such experiments involving body motion. For each trial, the results were standardized to the measurement result at rest before the task (Pre-Rest) with Eq. (1):

$$\Delta oxy(t)_{Z-SCORE} := \frac{\Delta oxy(t)_{raw} - \mu_{pre-rest}^{\Delta oxy}}{\sigma_{pre-rest}^{\Delta oxy}} \tag{1}$$

where $\Delta oxy(t)$ denotes the measured value of oxyHb at time t on each channel; $\mu_{pre-rest}^{\Delta oxy}$ is the average oxyHb change in the Pre-Rest time; and $\sigma_{pre-rest}^{\Delta oxy}$ is the standard deviation for the Pre-Rest time. Then, the standardized measurement result was averaged for the times of each of the 32 channels. Finally, this result was subtracted from the standardized measurement result at each point with Eq. (2):

$$\Delta oxy(t)_{GR} := \Delta oxy(t)_{Z-SCORE} - \frac{\sum_{l=1}^{n} \Delta oxy_l(t)_{Z-SCORE}}{n} \tag{2}$$

where n denotes the total number of channels. In this experiment, n is defined as 32.

3   Measurement of Brain Activation During the Learning of Work Procedures

3.1   Formation of Procedural Memory During Imitation Learning of Assembly Work

Simulating skill acquisition in core manufacturing industries involves acquiring embodied knowledge, which leads to procedural memory of a skill. Subjects remember and execute a set of procedures by imitation learning. The purpose of our experiment was to measure the trends of changes in brain activity with the development of a skill by imitation learning.

The subjects in our study were five healthy, right-handed males (age range 21–24), all of whom provided informed consent for participation in the experiment. The experiment ran for two days. One experiment was composed of three trials, and one trial was composed of three periods. The portion of the trial during which the subject carried out the experimental task lasted 180 s (the task time). The rest time before and after the task time lasted 30 s (the rest time). In the rest time, the subjects rested with their eyes open. This trial was executed three times by each subject.

The subjects were seated at a desk so that they could see a display in front of them (Fig. 2). Before the experiment, the experimental procedure and instructions to remember the procedures, as distinct from the finished product, were explained. To carefully handle the assembly parts, subjects assumed a bent-forward posture, which

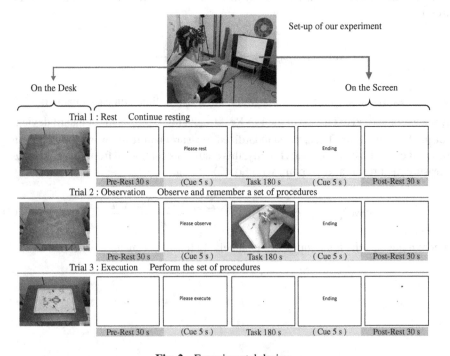

Fig. 2. Experimental design

they were instructed to maintain. Instructions for the task were shown on the display for 5 s after the task started, and the instructions ended 5 s before the end of the task. During the rest time, the subjects were told to rest without thinking. During the first trial, the subjects rested with their eyes open (the rest trial). In the second trial, the subjects observed and remembered a set of procedures shown on the display, which was a video clip of assembly work (the observation trial). The assembly work lasted 157 s, and then the clip of the finished product lasted 13 s. The parts for assembly were placed on the desk after the trial. In the third trial, the subject executed the procedures that they remembered in the observation trial (the execution trial). The end of the task was indicated on the display and signaled with a buzzer only in the execution trial.

After the first trial, the subjects practiced the tasks to improve their skill, without undergoing NIRS (Table 1). The subjects practiced the observation task and the execution task once after measurement by NIRS on the first day and then performed these tasks twice before measurement by NIRS on the second day.

Table 1. Experimental schedule.

	First day		Second day	
	Measurement by NIRS	Practice after measurement by NIRS	Practice before measurement by NIRS	Measurement by NIRS
Task 1 Rest	1[a]	–	–	1
Task 2 Observation	1	1	2	1
Task 3 Execution	1	1	2	1

[a] The numerals indicate the subjects' performance counts.

3.2 Accuracy of the Remembered Procedures

We evaluated the procedures that each subject executed in the trial for accuracy, which indicated the degree of task improvement. In general, an expert in the skill determines the skill level. However, in this experiment, we determined the skill level by limiting the assembly work procédures, using the following evaluative standard. The procedures were broken down into 12 numerically ordered sections. First, if the parts which use in the section had assembled correctly, the section would be considered to have been completed. Second, if the number of the completed sections was higher than that of the previous section, it would score 1 point. This method allowed a score of up to 11 points. The assembly parts were distinguished with asymmetrical shapes and colors.

Evaluation of the executed procedures is shown in Table 2. On the first day that the subjects were instructed to perform the assembly work, their scores were low (~ 0–2 points). On the second day, they all scored 10 points. This result suggests that all subjects improved their skills during the trials.

Table 2. Evaluation results of the executed procedures.

Subjects		Completed sections												Scores
A	First day	1ª	4	2	–	–	–	–	–	–	–	–	–	1
	Second day	1	2	3	4	5	6	7	9	10	11	12	–	10
B	First day	2	1	–	–	–	–	–	–	–	–	–	–	0
	Second day	1	2	3	4	5	6	7	9	10	11	12	–	10
C	First day	1	4	2	–	–	–	–	–	–	–	–	–	1
	Second day	1	2	3	4	6	7	8	9	10	11	5	12	10
D	First day	1	2	4	–	–	–	–	–	–	–	–	–	2
	Second day	1	2	3	4	5	6	7	8	9	10	11	–	10
E	First day	2	–	–	–	–	–	–	–	–	–	–	–	0
	Second day	1	2	3	4	5	6	7	8	10	9	11	12	10

ª The numerals indicate the number of sections.

3.3 The Relationship Between Brain Activity and Accuracy of Actions During the Process of Embodied Knowledge Acquisition by Imitation

The results of the oxyHb levels in the rest trial were compared with those in the observation trial (Fig. 3). In the rest trial, the oxyHb levels stabilized at low levels on both days, whereas in the observation trial they increased in Channels 20 and 26 and decreased in Channel 23. These tendencies subsequently stabilized at a low level, and the confidence interval of the oxyHb levels narrowed on the second day. These results indicate that the frontal area activity during formation of procedural memory by imitation learning tended to stabilize at a low level. One explanation for this may be that practice of the tasks resulted in a reduced amount of information to remember from the video clip, and that frontal area activity shifted to a similar tendency as in the rest trial. Previous research has reported that it is possible that these tendency shifts are involved in the inverse models which are neural representations of how, enable us to recognize a familiar object, comprehend what we hear quickly, and carry out complex movements so easily and accurately in the cerebellum [8, 9]. In this experiment, NIRS measurement showed that the inverse model had received enough information to execute the task on the second day, which is why brain activity stabilized at a low level in the observation task.

The results of the oxyHb levels in the rest trial were compared with those in the execution trial (Fig. 4). The oxyHb levels in the rest trial were the same as those shown in Fig. 3. The results on the first day showed a similar tendency as those in the rest trial. The oxyHb in the execution trial increased in Channels 20 and 26 and decreased in Channel 23. The differences in Channels 23 and 26 on the second day were significant. This indicates that the trend in frontal area activity was amplified during the execution of remembered procedures, due to the effect of practice. The subjects had more information about the assembly work on the second day, and this increase was related to the functions of retrieving the memory, executing the remembered procedures, or both.

Comparing the results of the first day with those of the second day, the accuracy rate was improved by imitation learning, which may have increased the skill levels. Limited acquisition of embodied knowledge due to procedural memory produced an increase in the oxyHb levels in Channels 20 and 26 and a decrease in Channel 23 in the observation trial during the early stage. These tendencies were mitigated due to a reduction in the amount of information to remember. In the execution trial, the oxyHb levels did not show a trend that depended on the task during the early stage. The increase in oxyHb in Channels 20 and 26 and the decrease in Channel 23 were due to an increased amount of information to execute.

We compared the total amounts of the adjusted oxyHb levels during the task time for each subject on the first day with those on the second day to focus on the change in the trend (Fig. 5). The total amounts in the rest trial were low on both days. In the observation trial, the total amounts of oxyHb levels showed positive high levels in Channels 20 and 26 and a high negative level in Channel 23, and these subsequently shifted to a lower level. In the execution trial, the total amounts were low in all three channels and shifted to positive high levels in Channels 20 and 26 and a high negative level in Channel 23.

These results suggest that frontal area activity changed during EKA, which improved the accuracy of the assembly work. Assuming these activity shifts are applicable to more high-level skills, we can predict a degree of improvement based on brain activity.

With NIRS, oxyHb levels or total Hb levels (which are sums of oxyHb and deoyHb), can be indices of brain activity, and deoxyHb levels behavior is complex [10]. Assuming that artifacts are similar in all brain regions, we employed global average references. Thus, we considered the deoxyHb levels for confirmation of the effectiveness of the global average references. If the results of the deoxyHb levels were similar in trend to the oxyHb levels, the global average references were deemed to be not applicable, and the previous trends were thought to be caused by artifacts, such as body motions.

The results of the deoxyHb levels in the rest trial were compared with those in the observation trial (Fig. 6). The deoxyHb levels in the rest and observation trials were stable at low levels on both days. The deoxyHb results in the rest trial were also compared with those in the execution trial (Fig. 7). On the first day, the deoxyHb in the execution trial increased in Channels 20, 23, and 26. On the second day, the deoxyHb decreased in Channels 20 and 26. To focus on the change in the trend, we compared the sum total amounts of the adjusted deoxyHb levels during the task time for each subject on the first day with those on the second day (Fig. 8). The total amounts in the rest and observation trials were low on both days. The amounts in the execution trials were at high levels and showed different trends for each subject. There was no trend shift with EKA in the deoxyHb. These results suggest that the previous trends were not caused by artifacts.

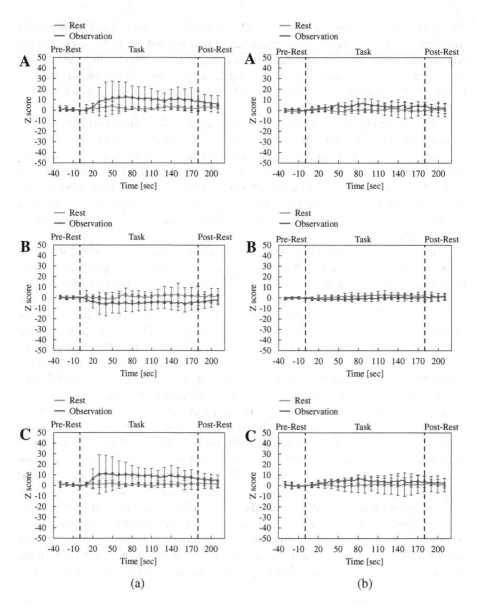

Fig. 3. Comparison of the mean oxyHb variation and the 95% confidence interval (n = 5) in the observation trial with those in the rest trial. (a) First day. (b) Second day. (A) The right dorsolateral prefrontal area (Channel 20, in the near F8 of the international 10–20 system). (B) The frontal pole (Channel 23, in the near Fpz of the international 10–20 system). (C) The left dorsolateral prefrontal area (Channel 26, in the near F7 of the international 10–20 system).

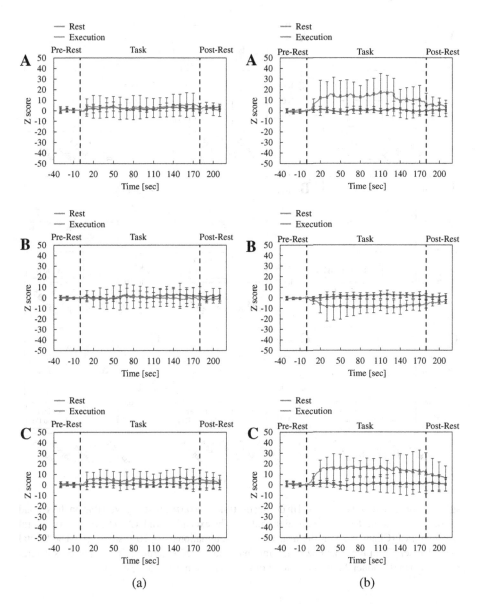

Fig. 4. Comparison of the mean oxyHb variation and the 95% confidence interval (n = 5) in the execution trial with those in the rest trial. (a) First day. (b) Second day. (A) The right dorsolateral prefrontal area (Channel 20, in the near F8 of the international 10–20 system). (B) The frontal pole (Channel 23, in the near Fpz of the international 10–20 system). (C) The left dorsolateral prefrontal area (Channel 26, in the near F7 of the international 10–20 system).

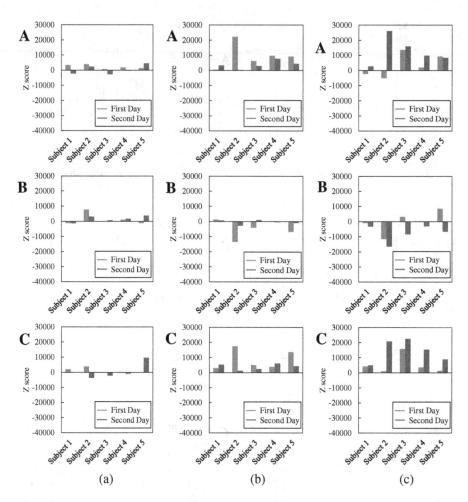

Fig. 5. Comparison of the total oxyHb during the task time on the first day with that on the second day. (a) The rest trial. (b) The observation trial. (c) The execution trial. (A) The right dorsolateral prefrontal area (Channel 20, in the near F8 of the international 10–20 system). (B) The frontal pole (Channel 23, in the near Fpz of the international 10–20 system). (C) The left dorsolateral prefrontal area (Channel 26, in the near F7 of the international 10–20 system).

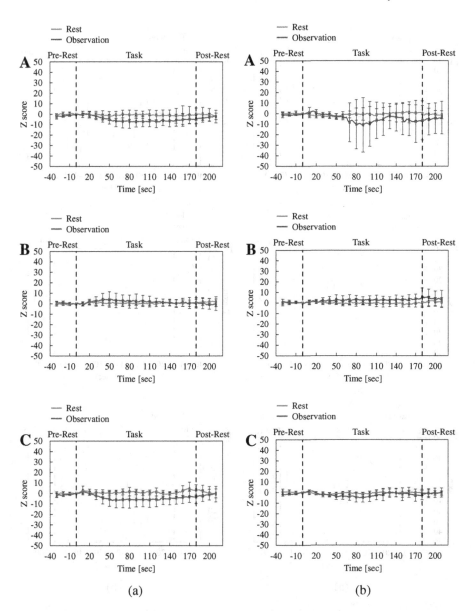

(a)   (b)

Fig. 6. Comparison of the mean deoxyHb variation and the 95% confidence interval (n = 5) in the observation trial with that in the rest trial. (a) First day. (b) Second day. (A) The right dorsolateral prefrontal area (Channel 20, in the near F8 of the international 10–20 system). (B) The frontal pole (Channel 23, in the near Fpz of the international 10–20 system). (C) The left dorsolateral prefrontal area (Channel 26, in the near F7 of the international 10–20 system).

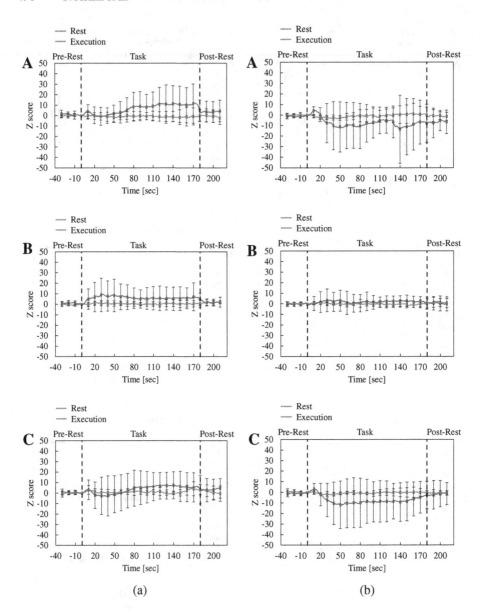

Fig. 7. Comparison of the mean deoxyHb variation and the 95% confidence interval (n = 5) in the execution trial with those in the rest trial. (a) First day. (b) Second day. (A) The right dorsolateral prefrontal area (Channel 20, in the near F8 of the international 10–20 system). (B) The frontal pole (Channel 23, in the near Fpz of the international 10–20 system). (C) The left dorsolateral prefrontal area (Channel 26, in the near F7 of the international 10–20 system).

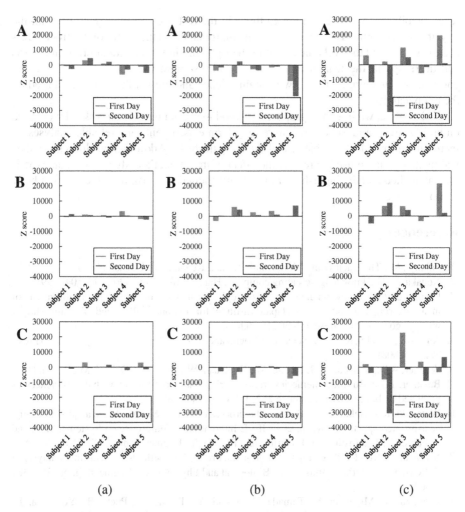

Fig. 8. Comparison of the total deoxyHb during the task time on the first day with that on the second day. (a) The rest trial. (b) The observation trial. (c) The execution trial. (A) The right dorsolateral prefrontal area (Channel 20, in the near F8 of the international 10–20 system). (B) The frontal pole (Channel 23, in the near Fpz of the international 10–20 system). (C) The left dorsolateral prefrontal area (Channel 26, in the near F7 of the international 10–20 system).

4   Conclusion

In this paper, we analyzed the relationship between brain activity and accuracy of actions during the process of EKA by limiting the acquisition of procedural memory through imitation learning.

The oxyHb levels increased in the right and left dorsolateral prefrontal areas and decreased in the frontal pole during the process of observing and remembering the procedures. With improvement in the scores of the procedure's accuracy, the oxyHb levels stabilized to a level similar to those in the rest trial. In the execution trial, the oxyHb levels stabilized at a low level. With improvement in the scores of the procedure's accuracy, the oxyHb levels increased in the right and left dorsolateral prefrontal areas and decreased in the frontal pole. These results suggest the possibility that we can construct a new learning model with brain science and predict the skill level based on the brain activity.

In the future, we intend to propose a method to predict the learner's skill level based on brain activity and to perform more long-term experiments and reliable analyses by increasing the number of subjects and changing the tasks. Additionally, although analytical methods can be used to reduce the effect of the subject's body motions, an experimental method that distinguishes brain activity elicited by a task from the effect of body motions is needed.

References

1. Polanyi, M.: The Tacit Dimension. Routledge and Kegan Paul, London (1966)
2. Nonaka, I.: The knowledge-creating company. Harvard Bus. Rev. **69**(6), 96–104 (1991)
3. Results of the survey on new business deployment and human resource cultivation of manufacturing enterprise, The Japan Institute for Labour Pollcy and Training, http://www.jil.go.jp/institute/research/2014/126.html
4. Amari, S., Tanaka, K.: Brain Science of Cognizance and Behavior. University of Tokyo Press, Tokyo (2008)
5. Wagner, A.D., Schacter, D.L., Rotte, M., Koutstaal, W., Maril, A., Dale, A.M., Rosen, D.B., Buckner, R.L.: Building memories: remembering and forgetting of verbal experiences as predicted by brain activity. Science **282**, 1188–1191 (1998)
6. Watanuki, K., Asaka, Y.: Analysis of the process of embodied knowledge acquisition using near-infrared spectroscopy. In: 2012 IEEE International Conference on Systems, Man, and Cybernetics, The Institute of Electrical and Electronics Engineers, pp. 2693–2699 (2012)
7. Nozawa, T., Kondo, T.: Comparison of artifact reduction methods for real-time analysis of fNIRS data. In: 24th Symposium on Biological and Physiological Engineering, pp. 381–384 (2009)
8. Imamizu, H., Miyauchi, S., Tamada, T., Sasaki, Y., Takino, R., Puetz, B., Yoshioka, T., Kawato, M.: Human cerebellar activity reflecting an acquired internal model of a new tool. Nature **403**, 192–195 (2000)
9. Ito, M.: Internal model visualized. Nature **403**, 153–154 (2000)
10. Fukuda, M.: Near-infrared spectroscopy in psychiatry. Equilib. Res. **69**(1), 1–15 (2010)

The Cognitive Role of Analogical Abduction in Skill Acquisition

Koichi Furukawa[1]([✉]), Keita Kinjo[2], Tomonobu Ozaki[3], and Makotoc Haraguchi[4]

[1] Keio University, Minato, Japan
kfurukawa@kaetsu.ac.jp
[2] College of Economics and Environmental Policy, Okinawa International University,
Ginowan, Japan
keita.kinjo@okiu.ac.jp
[3] College of Humanities and Sciences, Nihon University, Tokyo, Japan
tozaki@chs.nihon-u.ac.jp
[4] Graduate School of Information Science and Technology, Hokkaido University, Sapporo, Japan
makoto@kb.ist.hokudai.ac.jp

Abstract. In this paper, we discuss the cognitive role of analogical abduction in skill acquisition. Abductive inference makes it possible to find missing links that explain a given knack in achieving a skillful task. We introduced meta level abduction to realize rule abduction which is mandatory in finding intermediate missing links to be added in knack explanation. Analogical abduction can be achieved by adding analogical inference rules to causality rules within meta level abduction. We have applied our analogical abduction method to the problem of explaining the difficult cello playing techniques of spiccato and rapid cross strings of the bow movement. Our method has constructed persuasive analogical explanations about how to play them. We have used a model of forced vibration mechanics as the analogy base world for spiccato, and the specification of the skeletal structure of the hand as the basis for the cross string bowing technique. We also have applied analogical abduction to show the effectiveness of a metaphorical expression of "eating pancake on the sly" to achieve forte-piano dynamics, and successfully identified an analogical explanation of how it works. Through these examples, we show the effectiveness of analogical abduction in skill acquisition. Furthermore we discuss the importance of meta level representation as a basis for providing rich human cognitive paradigm such as causality, analogy and metaphor. Finally we propose a cognitive architecture which gives a possible structure for realizing accommodation on our analogical abduction schema.

Keywords: Rule abduction · Analogical abduction · Cello playing · Accommodation · Cognitive architecture

1 Introduction

In acquiring skills in such activities as sports, playing instruments, drawing picture and so on, it is essential to get some sort of "knack" to perform those activities. The notion

© Springer International Publishing AG 2017
M. Otake et al. (Eds.): JSAI-isAI 2015 Workshops, LNAI 10091, pp. 499–513, 2017.
DOI: 10.1007/978-3-319-50953-2_37

of achieving a knack refers to some kind of unexplained but necessary skill component, without which performance is lacking. In acquiring professional skill, it is said that we need continuous daily training or practice something like 10,000 h. However it is very hard to spend such long period of time for ordinary people, like amateur athletes or musicians. For those people, the key strategy is to acquire some critical knack for achieving those skills. There are many possibilities in acquiring a knack, e.g., observing professionals' performance, being taught by trainers, conducting trial and error by themselves and so on. Such training methods have two important features in their processes; encountering a knack and assimilating and/or accommodating the knack. Trainers' suggestions are quite useful to encounter key points which play essential roles in understanding the knack. Observation of professionals' performance sometimes makes it possible to acquire an ideal form of performance which may give a solution for achieving the given task. Trial and error is useful to finding key points to realize the task and to get the knack by themselves. It is always important for the players to consider how the performance task is related to possible activities that can achieve the goal. For such mental activities, abduction and analogy play central roles in deepening the thinking that relates the problem with various activities which may not always be directly related to the problem domain. In Particular, analogical reasoning is quite useful to expose relationships which may not be directly related in the performance domain in question.

Knacks play crucial roles in acquiring artistic or sports skills. Generally knacks themselves are hard to understand. This is the reason why we call the secret as knack to perform difficult tasks. Abduction is a kind of synthetic reasoning used to construct explanatory hypotheses about knacks i.e. surprising observations. In this paper, we show how we have succeeded in applying abductive inference to provide explanation structure about how to perform difficult cello playing techniques, by exposing "hidden secrets" behind a given "knack" for achieving a difficult task.

Furthermore, we try to give proper explanation of a knack by employing analogical abduction. The role of abduction is to find explanation structure i.e. missing links in the explanation, whereas that of analogy gives understandable explanation to either the knack itself or the introduced links. In realizing the analogical abduction engine, we integrate abduction and analogy on the basis of meta level expression of causality and analogy.

In Sect. 2, we discuss aspects of skill discovery in skill acquisition, focusing on two approaches; "meta cognitive verbalization" and "analogical abductive reasoning". In Sect. 3, we give formulation of skill acquisition by abduction. In Sect. 4, we augment abduction by analogy. In Sect. 5, we discuss other possibilities for explaining knacks. In Sect. 6, we propose a cognitive architecture based on analogical abduction. Lastly, in Sect. 7, we conclude our paper by giving discussion and future work.

2   Aspects of Skill Discovery in Skill Acquisition

In acquiring any kinds of skill, an essential point is the mental activity of trying to discover a knack to perform a given difficult task. Knack discovery is essential in skill acquisition. Previously, we found the importance of closing one's right arm to increase sound volume in playing the cello as a case study [1, 2]. This is an example of a knack. Later, we discovered another knack to increase the sound by tilting the bow to touch to the string by the edge of the bow hair, which we call the "edge bowing method". These knacks provided significant improvement in achieving the given task.

The verbalization of a knack helps one to be more confident about acquired skills, both to deliver them to other people and to make them more objective. Among several approaches of skill verbalization, we especially notice two methods; "meta cognitive verbalization" [3, 4] and "analogical abductive reasoning" [5]. Meta cognitive verbalization tries to memorize one's physical status during performance in terms of notions which appeared in one's mind by self-reflection. By accumulating those memos for a long period of time, one can discover important facts within the change of vocabulary usage patterns, which reflects skill development.

On the other hand, analogical abductive reasoning tries to extract possible explanations how to perform given hard tasks by selecting adequate combinations of candidate hypotheses in a repertoire of body movement actions. For example, in our experimental study, we tried to find methods to perform "traverse between two strings repeatedly with bow direction change" and discovered a hypothesis "activate right forearm muscles strongly".

Skill acquisition has many issues to be addressed. Some are listed as follows;

  i. finding a knack for skillful performance,
 ii. finding missing links (secrets behind a knack) in skill explanation,
iii. identifying the role of a surprising fact (a knack) in skill discovery, and
 iv. accommodating the new skill.

Interestingly, most of the issues listed above can be properly treated in the analogical abductive reasoning framework. In this paper, we focus on the skill of playing the cello. A player often exercises some basic methods at the first step of training. In some later steps, however, the player may face a passage which s/he cannot play by using only acquired methods. In such case, none of the acquired methods can be applied to the passage, so new methods are required. Typically, the passage in question contains compound tasks to be achieved simultaneously. In that case, simple adoption of component basic skills does not work properly; we need to invent a new skill to avoid potential inconsistency amongst the compound tasks: we call this process *skill abduction*. The new skill is called an abduced skill. The solution may be unexpected and hard to achieve. Our goal is to aid the player and/or the trainer to find a solution by analyzing the goal task, basic skills and relevant physical constraints.

3   Formulating Skill Acquisition by Abduction

3.1   Generating Hypotheses by Abductive Reasoning

Although abductive reasoning does not necessarily derive the right answer, it produces plausible hypotheses to explain observation and therefore useful in hypotheses generation. The philosopher Pierce first introduced the notion of abduction. In Pierce [6] he identified three forms of reasoning.

Deduction.   An analytic process based on the application of general rules to particular cases, with the inference of a result.

Induction.   Synthetic reasoning which infers the rule from the case and the result.

Abduction.   Another form of synthetic inference, but of the case from a rule and a result.
   Peirce further characterized abduction as the "probational adoption of a hypothesis" as explanation for observations (results), according to known laws. "It is however a weak kind of inference, because we cannot say that we believe in the truth of the explanation, but only that it may be true" [6]. We omit formal definition of abductive inference to avoid complexity [7]. The essence of abductive inference is to augment missing facts or rules to derive the given surprising observation (the knack). Therefore an abductive inference engine is synonymous with a theorem prover augmented by a mechanism of finding missing links in deriving the given problem (a knack).

3.2   Logical Explanation of a Knack by Abduction

Knacks are target-dependent and are expressed by such phrases as "if you want to achieve a target exercise A, you should do an action B." But it is typically difficult to explain why the action B works for achieving the exercise A because of either the existence of "hidden secrets" behind the knack or the lack of proper knowledge to understand the given knack. In this section we solve the former problem by applying abduction. The latter problem is solved in Sect. 4.
   A knack is usually a surprising observation and therefore hypotheses generation by abduction can help in finding candidates for the "secret" prerequisite for achieving the given exercise. To elaborate, we try to abduce missing hypotheses to achieve (explain) the goal (exercise) A under the assertion of the fact (action) B. Since B appears at the leaf of the proof tree, the abduction procedure has to find hypotheses in between the goal A and the leaf B, identified as a (set of) rule(s). We refer to this abductive procedure as rule abduction. Note that rule abduction itself is realized in the framework of ATMS (Assumption based Truth Maintenance System) [8]. In this paper, we select logic programming approach because it is simpler and more expressive than ATMS. However, rule abduction cannot be achieved by standard Abductive Logic Programming (ALP) [7], because "abducibles" (predicates which are allowed to appear in the hypotheses to be generated) are limited only to "facts" in ALP. It means that generated hypotheses are simple (unknown) facts. A simple example of fact abduction is to explain the lack of a person's alibi by hypothesizing that he is a criminal. This limitation is due to the difficulty

of handling rule abduction. To resolve this difficulty, we developed a rule abduction method using meta level abduction [2] where causality relations between predicates are expressed by a meta predicate "caused(X, Y)" which represents that the goal X is caused by an action Y. Note that we restrict the logical implication to causality. The detail of the meta level representation is described in the next section.

There may be a situation where a (set of) intermediate proposition(s) is necessary to fill a gap between the premise B of the knack and its goal A, in which case we need to invent a new node (predicate) between them. This ability is called as "predicate invention" in Inductive Logic Programming (ILP) community, which has been claimed to be very hard to realize. We found that SOLAR was equipped with this function naturally by virtue of the ability to produce hypotheses having variables with existing quantifier [2]. An example having his feature is shown later in Subsect. 4.3.2.

3.3 Meta Level Representation for Rule Abduction

A weakness of available abductive inference engines such as PrologICA [9] is that we can only abduce facts but not rules. As explained in the last section, we need rule abduction to explain why knacks work. Our approach to overcome this problem is to introduce meta level representation to express rules as atoms by introducing causality relations between predicates such as *caused(spiccato, bow_support_with_ringfinger)*, which states that spiccato is caused by supporting the bow with the ring finger. This representation allows us to state a rule "spiccato is caused by supporting the bow with the ring finger" in terms of a meta level atom *caused(spiccato, bow_support_with_ringfinger)*. Since we can abduce meta level atoms with a predicate *connected* (which represents a direct causality relation) by applying conventional abductive engines, we succeed in obtaining a rule "*spiccato ← bow_support_with_ringfinger*." Formally, the predicate *caused* is defined recursively as follows:

$$caused(X, Y) \leftarrow connected(X, Y) \tag{1}$$

$$caused(X, Y) \leftarrow connected(X, Z), \ caused(Z, Y) \tag{2}$$

Here, the predicates *connected* and *caused* are both meta-predicates for object-level propositions X, Y and Z. From now on, we refer to this representation of causality relations as Meta Level (ML) representation of causality.

4 Augmenting Abduction by Analogy

4.1 Why Analogical Abduction?

Our rule abduction alone is insufficient to obtain meaningful missing prerequisites in the real application domain of skill acquisition. For example, consider this example of a knack: "you should bend the thumb joint to realize crossing strings quickly." In this example, a possible missing rule is the knack itself; that is, "to achieve crossing strings quickly, bend the thumb joint" is a rule to be hypothesized by rule abduction. But it is easy to see that this rule is essentially useless because it does not explain why it works effectively. Here we introduce an analogical abduction system which makes it possible to give a suitable explanation to the proposed knack. To show the effectiveness of the knack, we need to identify a hidden reason. The hidden reason is typically provided by analogical reasoning which gives a possible explanation of the knack by means of an argument in an underlying analogical domain associated with the original vocabulary of the abducible rules.

Abductive reasoning generates possible hypotheses to prove a given knack to achieve a given difficult task. However it proposes only a possible proof (explanation) structure, i.e., the identification of missing links in the proof tree. It remains the user's task to give an appropriate meaning to generated hypotheses. Analogical reasoning is a possible way to automatically identify potential meanings of generated hypotheses. For example, to give an explanation of "spiccato is directly caused by *bow_support_with_ringfinger*," we use an analogy to the dynamics of forced vibration which is known to be analogous to spiccato, a fast jumping staccato. Furthermore we know that the forced vibration is directly caused by both supplying energy to the system with appropriate timing (just after the point of maximum amplitude) and absorbing shock at the point of energy supply. It is quite persuasive if we find a correspondence of *bow_support_with_ring-finger* to shock absorbing in forced vibration. We try to extract this correspondence automatically by incorporating analogical reasoning into an abduction engine SOLAR [10, 11].

4.2   Incorporating Analogical Reasoning to Abduction

In this section, we incorporate analogical reasoning into our ML framework. We refer to the world under consideration as the target world and the corresponding analogical world as the base world. Analogical reasoning is achieved by introducing a base world similar to the target world, where we conduct inference [12]. Analogical reasoning can be formulated as logical inference with equality hypotheses [13]. We achieve analogical abduction by extending our ML based rule abduction framework.

We modify the causality relationship Formulas (1) and (2) to deal with causalities in the different worlds separately as follows:

$$
\begin{aligned}
&t\_caused(X, Y) \leftarrow t\_connected(X, Y) \\
&t\_caused(X, Y) \leftarrow t\_connected(X, Z), \ t\_caused(Z, Y) \\
&b\_caused(X, Y) \leftarrow b\_connected(X, Y) \\
&b\_caused(X, Y) \leftarrow b\_connected(X, Z), \ b\_caused(Z, Y)
\end{aligned}
\tag{3}
$$

where the prefix "t" represents a predicate in the target world and "b" in the base world. Although the predicate "b_caused" does not appear in following examples, we define it because of the symmetry with "t_caused" for possible future use. We also introduce a predicate "$similar(X, Y)$" to represent similarity relations between an atom X in the target world and a corresponding atom Y in the base world.

Now we have to define the predicate "$t\_connected$," for which we have to consider three cases to show the connectedness in the target world as follows:

$$t\_connected(X, Y) \leftarrow connected\_originally(X, Y) \tag{4}$$

$$t\_connected(X, Y) \leftarrow connected\_by\_abduction(X, Y) \tag{5}$$

$$t\_connected(X, Y) \leftarrow connected\_by\_analogy(X, Y) \land$$
$$print\_connected\_by\_analogy(X, Y) \tag{6}$$

The first case is that the connectedness holds from the beginning, (4); the second case is that it holds by abduction as a solution of abductive inference, (5); and the third case is that it is derived by analogy, (6). Definition (6) contains an auxiliary predicate *"print connected by analogy(X, Y)"* which indicates that it is to be *"printed"* as a part of an abduced hypothesis to provide evidence that the analogical connection is actually used to show the *"t_connected"*ness. Since analogical reasoning can be achieved without any defects in the inference path, we need to prepare an artificial defect atom *"print_connected_by_analogy(X,Y)"* on the reasoning path. This printing in turn is defined by specifying the predicate *"print_connected_by_analogy"* as an abducible predicate.

We have to further define three predicates; *"connected_originally"*, *"connected_by_abduction"* and *"connected_by_analogy"*. The predicate *"connected_originally"* is used in the assertion of facts representing the original connection; *"connected_by_abduction"* is introduced as an abducible predicate. Finally, the definition of *"connected_by_analogy"* is given by the following analogy axiom which plays a central role in analogical abduction.

Analogy Axiom

$$connected\_by\_analogy(X, Y) \leftarrow b\_connected(XX, YY),$$
$$similar(X, XX), similar(Y, YY) \tag{7}$$

This axiom states that the nodes X and Y in the target world can be linked by the predicate *"connected_by_analogy(X, Y)"* because of the base relationship *"b_connected(XX, YY)"* between XX and YY which are similar to X and Y, respectively, as shown in Fig. 1. Note that there may be more than one similarity candidates. In this

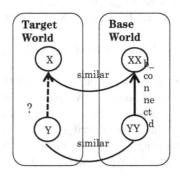

Fig. 1. A schema representing the analogy axiom.

paper, we assume that the user provides some of the initial similarities, and that the abductive inference engine will compute any remaining possible similarity hypotheses to explain an observation.

4.3   Giving Analogical Explanation to Generated Hypotheses

4.3.1   Interpreting a Causal Link by Analogy

We first start with an example of a simple analogical abduction. The problem is how to explain the effectiveness of the ring finger.

% Observation (G):

t_caused(spiccato, support_bow_with_ringfinger).

% Abducible predicates (Γ):

abducibles([connected_by_abduction/2, similar/2,   (8)

print_connected_by_analogy/2]).

% Background Knowledge(B):

%%% Base world:

b_connected(forced_vibration, shock_absorber).   (9)

%%% Target world:

:- connected_by_abduction(spiccato, support_bow_with_ringfinger).   (10)

% Similarity:

similar(spiccato, forced_vibration).   (11)

%Axioms:
b_caused(X; Y):-b_connected(X, Y).
b_caused(X, Y):-b_connected(X, Z), b_caused(Z, Y).
t_caused(X, Y):-t_connected(X, Y).
t_caused(X, Y):-t_connected(X, Z), t_caused(Z, Y).
t_connected(X, Y):-originally_connected(X, Y).
t_connected(X, Y):-connected_by_abduction(X, Y).
t_connected(X, Y):-connected_by_analogy(X, Y), print_ connected_by_analogy
(X, Y).
connected_by_analogy(X, Y):-b_ connected(XX, YY), similar(X, XX), similar
(Y, YY).

In this program, the goal (observation) to be satisfied is "t_caused(spiccato, support_bow_with_ringfinger)" (Clause 8). We provide the following two facts: (1) "shock_absorber" is one of the possible causes to achieve the forced_vibration (Clause 9), and (2) spiccato is analogous to the forced_vibration (Clause 11). In addition, we provide a negative clause asserting that the direct connection from "support_bow_with ringfinger" to "spiccato" cannot be hypothesized (Clause 10). In one of our SOLAR experiments, the number of obtained hypotheses is 7 when the maximum search depth is set to 10 and the maximum length of produced clauses is 4. One plausible hypothesis is:

print_connected_by_analogy (spiccato, support_bow_with_ringfinger)∧

similar(support_bow_with_ringfinger, shock_absorber)

which indicates that the support of the bow with the ring finger in spiccato is analogous to the shock absorber in the forced vibration as shown in Fig. 2.

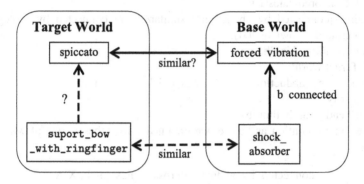

Fig. 2. Analogical abduction for achieving spiccato playing.

The dotted lines are to be computed as a hypothesis.

4.3.2 Interpreting a Newly Invented Predicate by Analogy
In this subsection, we consider the problem of showing the effectiveness of bending the thumb to achieve the quick crossing of strings (cross strings quick). We use the skeletal structural linkage of the knuckle (of the first four fingers) and the thumb (b_ connected(knuckle, thumb)) as a counterpart of a functional linkage of bending the knuckle and bending the thumb (t connected(knuckle bend, thumb bend)) in the analogy

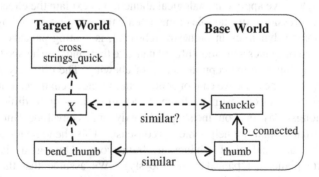

Fig. 3. Analogical abduction with predicate invention. A predicate X is introduced by abduction in Target World. An analogical reasoning is conducted to give an interpretation of X as similar to "knuckle" in the base world.

setting. Note that we define the similarity only between "bending thumb" and "thumb" without providing the predicate "bend knuckle", which is to be invented by abductive reasoning. In this example, we discover missing similarities and invent a predicate at the same time. The problem structure is shown in Fig. 3.

The abduction program for this problem is shown as follows (axiom clauses are omitted here):

% Observation(G):
t_caused(cross_strings_quick, bend_thumb).
% Abducible predicates(Γ):
abducibles([connected_by_abduction/2, similar/2,print_connected_by_analogy/2]).
% Background Knowledge(B):
b_connected(knuckle, thumb),
%%% Target world:
:-connected_by_abduction(cross_strings_quick, bend thumb).
% Similarity:
similar(bend_thumb, thumb).

Under the same condition as before, we obtained 7 hypotheses, one of which is the following:

$$connected\_by\_abduction(cross\ strings\ quick, X) \wedge$$
$$similar(X, knuckle) \wedge$$
$$print\_connected\_by\_analogy(X, bend\ thumb)$$

This hypothesis accurately represents the structure shown in Fig. 3. We further conducted our experimental study by deleting the similarity relation "similar(bend_thumb, thumb)" from the above program and then succeeded in recovering this link as well.

4.4   Explaining the Effectiveness of Metaphorical Expression

To show the applicability of our approach to different kinds of problems other than mechanical models, we apply our analogical abduction to explain the effectiveness of a metaphorical expression. An example of metaphorical expression, issued by a trainer to achieve forte-piano dynamics in orchestra rehearsal, is "eating pancake on the sly," which means that one takes a big mouthful of pancake first, and then he/she tries to make it secret by a motion of imperceptible action of chewing. The difficulty of achieving such dynamics arises because we cannot control our muscle strength accurately because of an inability to precisely estimate force. In addition, it is quite difficult to attain consensus amongst players about the shape of the dynamics envelope. But a metaphorical expression can sometimes help achieve a consensus. This phenomenon is formalized in terms of our analogical abduction framework. Our goal is to prove "t_caused(forte_piano, eat_pancake_on _the_sly)". We assume that the expression "eating pancake on the sly" induces a sequence of motor control commands indicating a big action followed by an imperceptible action in the brain, which arises within the metaphorical base world (see Fig. 4). The analogical abductive reasoning is shown as follows:

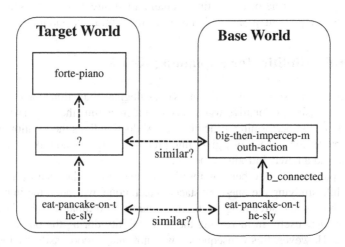

Fig. 4. Metaphorical expression of "eating pancake on the sly" to achieve forte-piano.

% Observation(G):
t_caused(forte piano; eat pancake on the sly).
% Abducible predicates (Γ):
abducibles ([connected by abduction/2, similar/2, print_connected_by_analogy/2]).
% Background Knowledge (B):
%%% Base world:
b_connected(big_then_impercep_action, eat pancake on the sly).
%%% Target world:
:-connected_by_abduction(forte_piano, eat_pancake_on_the_sly).

Under the same condition as before, we obtained 6 hypotheses, one of which is the following:

connected_by_abduction(forte_piano, X) ∧
similar(X, big_then_impercep_action) ∧
similar(eat_pancake_on_the_sly, eat_pancake_on_the_sly) ∧
print connected by analogy(X, eat_pancake_on_the_sly)

Note that the third atom has the form "similar(X,X)". Namely we regard the same thing as similar.

The entire problem structure of this analogical abduction is almost the same as our previous predicate invention example shown in Fig. 3 except for the treatment of the similarity relation at the bottom; it is abduced in the metaphorical analogy case whereas it is given from the beginning in Fig. 3. The key characteristic of the metaphorical analogy is that the same analogical expression appears in both the base and the target worlds. Since a metaphorical expression directly induces an emotional feeling to encourage the production of adequate motor control commands for achieving the given goal, it should be included in the target world. Alternatively, the same metaphorical expression triggers a similar motion in the eating action which means that it should be in the base world. Another remark is that the metaphorical expression of "eating pancake on the sly" plays the role of converting a quantitative direction of the sound volume adjustment into a qualitative one, which is much more intuitive and understandable.

5   Other Possibilities for Explaining Knack

In the last section, we showed the usefulness of analogical abduction as a promising way to produce persuasive explanative arguments for understanding the reason why the given knacks work well in performing difficult tasks. Abduction finds the location of missing links in the proof tree and analogy gives interpretation of the found links including both a causal link and a newly introduced predicate.

This chapter discusses other possibilities for explaining the idea of a knack. For example, while studying the one-bow staccato technique, we found the importance of holding the bow while stretching the thumb contrary to ordinal bow-holding. In fact, this knack is very useful in increasing the bow stability during the one-bow staccato performance. However, this consequence was not understood easily by the learners before observing a performance video showing virtuoso technique of the one-bow staccato. By looking the video, most of the learners suddenly understood the role of the new bow-holding way which can be expressed as pinching by the thumb and other fingers. This experience supports the usefulness of observing skillful videos to understand the key points of the knack.

Another experience supports the importance of metaphorical expression for delivering a sense of musicality in ensemble performance. We introduced the example metaphorical expression "eating pancake on the sly," where we claimed that such an expression sometimes helps achieve a consensus among players. Precisely speaking, this situation is not a knack explaining problem. However it provides all the players a common musical feature how to play the given note having the "forte-piano" sign. Therefore it is a musicality explaining problem which is closely related to knack explaining. Furthermore, we succeeded in formulating this "forte-piano" expression problem in terms of our analogical abduction framework.

6   Proposal of Cognitive Architecture

We have discussed the feasibility of our analogical abduction in skill acquisition. In acquiring skills, we need to understand adequate knacks to achieve given difficult performance tasks like spiccato or rapid cross strings of bow movement in cello playing. There are two kinds of activities required to obtain knacks: to encounter knacks and to assimilate and/or accommodate them to their own knowledge. The problem of encountering knacks is achieved in various ways: being taught by teachers, by watching good performance, by trial and error by themselves and so on. A possible scientific support for this encountering is physical meta cognition [3, 4]. In this paper, we focused on the accommodation aspect in knack acquisition. We discussed the importance of knack explanation to achieve the accommodation problem. Analogical abduction plays an essential role in this mental processes, since we need a precise explanation why a given knack is useful in achieving the given performance task. Analogical abduction gives an explanatory argument to achieve a task by showing the validity of the knack as a proof in causality links and analogical arguments.

Let us show how the accommodation process proceeds by our spiccato example. The essence of this example is to find the reason why supporting the bow with the ring finger makes it possible to achieve the spiccato. Since spiccato is modelled by the forced vibration, we understand that we need to absorb the shock possibly brought by adding outer force to keep the continuous vibration. However, in case of the spiccato, we apt to achieve the shock absorption by adjusting the force of the index finger which is used to give force to the bow and therefore it is far from our default thinking to use the ring finger for this purpose. On the other hand, "supporting the bow by the ring finger" is suggested by a professional cellist, one of the co-authors of this paper. Furthermore, the analogical abduction process to which the background knowledge of the forced vibration and the suggestion of the professional cellist were given, proposed the similarity between the ring finger usage and the shock absorption as one of the solutions. Given this proposal, we human cellists began to think the validity of using the ring finger to realize the shock absorption and found the proper explanation of the ring finger's shock absorption as a result of lifting the bow by the ring finger, but not pushing it by the index finger. Then, we tried to practice spiccato with conscious use of the ring finger as shock absorption and succeeded in achieving the skill. This entire process lead our deep understanding of the reason why supporting the bow by the ring finger achieved the shock absorption.

This example shows the role of the analogical abduction in the accommodation process. The whole process achieves the accommodation of the ring finger usage in adjusting the bow force, which had been believed to be achieved only by the index finger. This is the moment when the accommodation occurred. In this process, the analogical abduction triggered the change of the belief on the bow force adjustment. The valid explanation of the ring finger's role was given by us. The knack given by the professional cellist was "to use the ring finger in holding the bow to achieve the spiccato" and it was provided as an input to the abductive analogical engine to obtain a hypothesis relating it to shock absorption. Therefore the knack is a kind of surprising observation to be explained and our abductive analogical engine succeeded in producing an appropriate hypothesis. Furthermore we can identify the proposed hypothesis as a new theorem and its deep explanation by human as the proof. That is to say, the analogical abduction proposes a new theorem and human gives its proof.

Furthermore, we notice the importance of ML representation of causality and analogical reasoning. At first, we introduced the ML representation to realize rule abduction. Later we succeeded in realizing analogical reasoning by adding an analogy axiom with the predicate "$connected\_by\_analogy(X, Y)$" as well as the similarity predicate "$similar(X, Y)$". Note that both predicates are meta predicates both of whose arguments are propositions. In a sense, the ML representation made it possible to concisely augment the functionality of analogical reasoning to our rule abduction system. It is interesting to note that rule abduction and analogical reasoning are important aspects of human cognitive functions. This leads an important suggestion that ML representation may work as a key role in human thinking. The handling of metaphor is another evidence of this conjecture.

These considerations suggest a promising cognitive architecture which realizes accommodation, based on our analogical rule abduction as shown in Fig. 5.

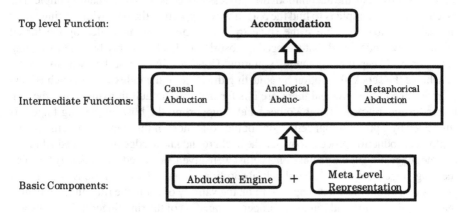

Fig. 5. A proposed cognitive architecture for achieving "accommodation" on an analogical abduction engine.

7   Discussion and Future Work

In order to achieve more realistic analogical abduction, there is a fundamental issue to be addressed. In this paper, we explicitly provide a base world analogous to the target world. In real problems for discovering or explaining skills, we may need to find an appropriate base world itself, before being able to conduct analogical reasoning, or to find and extract similar sub-worlds adequate for analogical abduction from the given target and base worlds. To deal with these problems, we have to provide detailed attributes to the components of each world and compute the degree of similarity for each pair of subset to find analogous pairs [14].

In our approach, we put an abduction engine in the center and tried to add analogical reasoning on top. However, there are other possibilities to generalize our approach further to find better integration of abduction and analogy, including metaphor. One viewpoint is to make analogical reasoning propose adequate abducibles for abduction. This should be realized by strengthening abductive reasoning engine by adding the feature of automatic preparation of abducibles supported by analogical reasoning. Another viewpoint is to use abduction to propose appropriate similarity relations to establish analogical reasoning, which has been reported here. In other words, abduction and analogy are supporting each other. An ideal implementation of a complementary abduction-analogy system is future research work.

Our final remark is on physical meta cognition, in other words, meta-level verbalization. In Sect. 6, we identified its role as a possible scientific support in knack encountering to achieve a difficult performance. However the meta-level verbalization covers the process of accommodation discussed in this paper. We leave the problem of further consideration on this issue for future work.

Acknowledgement. We express our special thanks to Professor Randy Goebel from Alberta University for his suggestions and fruitful discussions on how to incorporate analogy into abduction.

References

1. Furukawa, K., Masuda, T., Kobayashi, I.: Abductive reasoning as an integrating framework in skill acquisition. J. Adv. Comput. Intell. Intell. Inf. **15**(8), 954–961 (2011)
2. Inoue, K., Furukawa, K., Kobayashi, I., Nabeshima, H.: Discovering rules by meta-level abduction. In: Raedt, L. (ed.) ILP 2009. LNCS (LNAI), vol. 5989, pp. 49–64. Springer, Heidelberg (2010). doi:10.1007/978-3-642-13840-9_6
3. Suwa, M.: Metacognitive verbalization as a tool for acquiring embodied expertise (in Japanese). J. Jpn. Soc. Artif. Intell. **20**(5), 525–532 (2005)
4. Suwa, M.: A cognitive model of acquiring embodied expertise through meta-cognitive verbalization. Trans. Jpn. Soc. Artif. Intell. **23**(3), 141–150 (2008)
5. Furukawa, K., Kinjo, K., Ozaki, T., Haraguchi, M.: On skill acquisition support by analogical rule abduction. In: Kawtrakul, A., Laurent, D., Spyratos, N., Tanaka, Y. (eds.) Information Search, Integration, and Personalization, vol. 421, pp. 71–83. Springer International Publishing, Cham (2014). doi:10.1007/978-3-642-13840-9_6
6. Peirce, C.S.: Collected papers of charles sanders peirce, 2. In: Hartshorn, C., et al. (eds.) Harvard University Press, pp. 1931–1958
7. Kakas, A.C., Kowalski, R.A., Toni, F.: The role of abduction in logic programming. In: Handbook of Logic in Artificial Intelligence and Logic Programming, vol. 5, pp. 235–324. Oxford University Press (1998)
8. Reiter, R., de Kleer, J.: Foundation of assumption-based truth maintenance systems: preliminary report. In: Proceedings of AAAI87, pp. 183–188 (1987)
9. Ray, O. and Kakas, A.C.: ProLogICA: a practical system for abductive logic programming. In: Proceedings of the 11th Non Monotonic Reasoning Workshop, pp. 304–314 (2006)
10. Inoue, K.: Linear resolution for consequence finding, Artif. Intell. **56**(2/3), 301–353 (1992). Elsevier
11. Nabeshima, H., Iwanuma, K., Inoue, K.: SOLAR: a consequence finding system for advanced reasoning. In: Cialdea Mayer, M., Pirri, F. (eds.) TABLEAUX 2003. LNCS (LNAI), vol. 2796, pp. 257–263. Springer, Heidelberg (2003). doi:10.1007/978-3-540-45206-5_22
12. Haraguchi, M., Arikawa, S.: A formulation of analogical reasoning and its realization. J. Jpn. Soc. Artif. Intell. **1**(1), 132–139 (1986) (in Japanese)
13. Goebel, R.: A sketch of analogy as reasoning with equality hypotheses. In: Jantke, K.P. (ed.) AII 1989. LNCS, vol. 397, pp. 243–253. Springer, Heidelberg (1989). doi:10.1007/3-540-51734-0_65
14. Haraguchi, M.: Towards a mathematical theory of analogy. Bull. Inf. Cybern. **21**(3/4), 29–56 (1985)

Identifying Context-Dependent Modes of Reading

Miho Fuyama[1]([⊠]) and Shohei Hidaka[2]

[1] Keio University, 5322 Endo, Fujisawa-shi, Kanagawa 252-0882, Japan
miho02@sj9.so-net.ne.jp
[2] Japan Advanced Institute of Science and Technology,
1-1 Asahidai, Nomi, Ishikawa 923-1292, Japan
shhidaka@jaist.ac.jp

Abstract. Past literature has suggested that reading text as a whole cannot be reduced to merely an aggregation of sentence processing, but instead there are expected to be some context-dependent stylistic differences in the reading process. It has been, however, difficult to capture such context-dependent reading styles or modes. In this study, under the hypothesis that the statistics of reading time reflects such reading modes, we introduce a new statistical approach to capture them. Our analysis of the distributions of reading times identified two distinct modes of reading. In further analysis, we found that the temporal profiles of the two reading modes were correlated to the reader's degree of engagement. We discuss how the context dependency of the reading modes is related to dynamic construction of the reader's knowledge of narratives.

Keywords: Literary · Reading · Reading-time analysis

1 Introduction

Reading literature is not merely information processing of prose; it also evokes various feelings. Many past studies have discussed intrinsic features of readers' responses to literary works [7,9, and see also Miall [11] for a discussion about literariness]. In addition to these theoretical studies, there are also empirical studies about reading literary works [8,10]. For example, Miall and Kuiken [12] conducted four experiments to study reader response to aspects of literature known as *foregrounding* and *defamiliarization*, which are concepts in Russian formalism. They analyzed readers' emotional ratings to each word or sentence with the stylistic details of the literary text, such as alliteration, inversion, and metaphor. This study concluded that a number of these stylistic features in words or sentences were associated with an increase in reading time, higher strikingness ratings, and higher affect ratings.

M. Fuyama—Please note that the LNCS Editorial assumes that all authors have used the western naming convention, with given names preceding surnames. This determines the structure of the names in the running heads and the author index.

M. Otake et al. (Eds.): JSAI-isAI 2015 Workshops, LNAI 10091, pp. 514–527, 2017.
DOI: 10.1007/978-3-319-50953-2_38

Although there is no doubt about the effects of stylistic details in reading literature, reading experience would also be affected by the contextual structure of the story. Therefore, to approach a reader's cognitive process, which can be associated with a reader's experience *in* the story, we need to analyze the context-dependent changes in reading processing over the entire text.

In past studies, hypothetical constructs such as *story grammar* or *script*, which the reader is supposed to process in reading, were used to capture the contextual structures of stories [2,14]. Introducing the concepts of story grammar and script, Thorndyke [14] and Beaugrande [2] claimed that narratives have their own internal structure like a grammar, but at the discourse level, and these structures can be expressed by several kinds of elements (such as setting, theme, characters, goal, and so on) and associated combinatorial rules. However, using these, we can analyze only limited classes of stereotypical stories such as folk tales [2,11]. This limitation is likely to be due to their inflexibility, as literary works need to be updated to give the reader fresh interpretation. There have been other approaches to capture the context of stories, but, as far as we know, none of them have ever offered a satisfactory way to capture the context access to it.

In the present study, rather than assuming a specific story grammar, we focus on temporal changes in the reading process across different contextual structures of stories. Miall [10] analyzed the relationship between readers' affect ratings and reading times of introductory sections of novels. He analyzed readers' responses by assuming two stages of reading processes. One was called the registration stage, in which readers formed anticipations about the likely meaning of the narrative, and the other was called the interpretation stage, in which readers used the formed anticipations to comprehend the narrative. Miall assumed that the two stages could be separated according to the contextual structure of a story, and how much new information the story has. Thus, the shift across two stages would depend on both context of the story and the reader's background knowledge. From his analysis, he concluded that the reading process forms a cycle and repeatedly shifts between the two stages in the reading of a narrative.

With the working hypothesis that there are separable stages in reading processes correlated to the contextual structure, as suggested by Miall, we investigate the stage-like changes in reading process across the entire story, not only in an introductory part as analyzed by the previous study.

In this study, we investigate the "modes" of reading processes, which are supposed to be correlated to the context of the story or reader's background knowledge. The modes are operationally defined by statistical properties of reading times of each unit of text. We will further discuss modes in Sect. 1.1. Specifically, we run two experiments, in which we analyze reading time of each pair of pages for a collection of readers. A pair of pages approximately corresponds to twenty phrases, which is the unit of text analyzed for readers' responses in the previous study by Miall.

1.1   Statistical Analysis of Reading Time

We assume the reading processes are composed of several qualitatively distinct subprocesses, and we call such a subprocess a *reading mode*. The question is, given reading data, how can we infer the number of reading modes reflected in the data? In previous studies about short texts and more rapid processes, differences in reading time alone have been interpreted as a reflection of two qualitatively distinct processes [4, 10]. However, reading time may vary depending on multiple factors, such as frequency, familiarity, and the lengths of words [6, 15]. We cannot, therefore, naively interpret reading time alone as an indicator of multiple reading modes.

This observation motivates the development of a new analysis technique for reading time. The analysis we present as an alternative is based on a statistical theory of processing time [5]. In this theory, the presence of multiple different modes of processing can be detected by the statistical distribution of the processing time.

If the reading process consists of n subprocesses with the same constant processing rate over time, in other words, the process finishes only when all these subprocesses have finished, then the reading time would follow a gamma distribution with shape parameter n (Fig. 1(a)). If, on the other hand, the reading process consists of one subprocess with process rate t^k as a function of the process time t, in other words, the process finishes when at least one subprocess has finished, then the reading time would follow a Weibull distribution with shape parameter k (Fig. 1(b)).

Setting $n = 1$ in a gamma distribution or $k = 1$ in a Weibull distribution yields an exponential distribution. There is, therefore, a statistical relationship between the types of distributions of the processing time and the numbers of subprocesses.

This statistical analysis allows us to distinguish processes that have the same average speed of processes, but have different numbers of subprocesses (Fig. 2 Mode A and Mode B), and to distinguish processes that have the same number of subprocesses but have different average speeds of processes. This subprocess estimation gives an advantage over previous studies that analyzed differences in the reading time alone.

We adopt this statistical account of processing time for evaluating the number of reading modes based on reading time. If each observation in a reading time dataset follows essentially the same distribution as the others, we treat this as an indicator of a single reading mode. If, on the other hand, the data set appears to have been generated by sampling from a mixture of distributions, we treat it as an indicator of multiple reading modes (Fig. 3). Each dataset in question is composed of observations about a single subject. This technique, therefore, removes overall reading speed as a factor in the analysis.

1.2   Approach

When reading, one is generally also engaged in many other processes, such as eye movements, posture management, and so on. If one were only lightly engaged

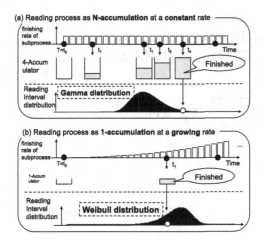

Fig. 1. Schematic illustrations of the different types of reading processes and corresponding statistical distributions.

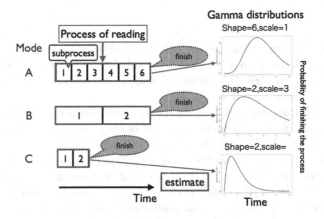

Fig. 2. Three hypothetical processing modes, A, B, and C, which have different numbers of subprocesses with different average rates. In each mode, all the subprocesses run in serial order, and the reading time follows a gamma distribution. Mode A: 6 subprocesses, each takes a short time on average; Mode B: 2 subprocesses, each takes a long time; and Mode C: 2 subprocesses, each takes a short time. The overall average of both Mode A and Mode B is the same, but their distributions (on the right-hand side of the figure) are different.

in reading and more heavily preoccupied with a number of these other activities, it is entirely possible that their preoccupation could appear as distinct reading modes in our statistical analysis. To prevent the detection of such false modes, it would be valuable to have a measure of reading engagement independent from reading time. We could then test the results of our statistical analysis based on their correlation with that measure.

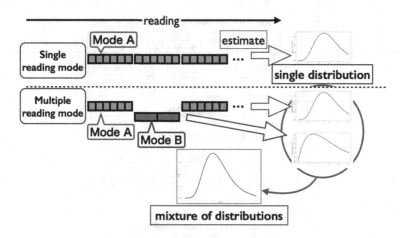

Fig. 3. (top) If one type of reading mode is repeated across multiple pages, it would result in a gamma distribution. (bottom) If both types of reading modes, A and B, take place across different pages, it would result in a mixture of two gamma distributions.

Since the analytic technique we will use is statistical in nature, it requires relatively large datasets to produce meaningful results. To this end, and although this is not typical of existing studies of reading, we use entire prose narratives as the texts in our experiments.

Given the burden that reading such long texts places on the subjects of our experiments, our first experiment consisted of only one subject: Miho Fuyama, the first author of this paper. She is an avid reader, which suggests that she is generally easily engaged in reading as an activity. In Experiment 1, we studied her reading time and the degree of engagement in reading across two books in order to empirically establish the validity of our analysis. We then analyzed data generated during her readings of 18 additional novels in order to test whether her reading process had a single or multiple reading modes.

Having validated our statistical analysis, we adopted it in our second experiment to a cross-sectional study of multiple subjects. In Experiment 2, we asked ten subjects to read a short story and introductory part of a longer story. The subjects were also asked to evaluate their degrees of reading engagement each two pages after the reading session. This experiment was designed to evaluate whether our findings from Experiment 1 hold in general. We also evaluated whether changes in reading modes could be related to the semantic structure of the text itself. To do so, we analyzed the consistency of the dynamics governing the change of reading modes across subjects and treated consistent dynamics as text-specific semantic effects in reading.

2   Experiment 1

The first author was the sole subject of several high-load reading tasks. We asked her to read 20 Japanese novels. Each session took one day, including breaks. The set of samples from these 20 sessions of 20 novels was submitted to statistical analysis using the scheme described in the previous section, and we estimated the statistical distribution of her reading time for each two pages for which the reader needs to turn over a page. For two of the novels (novels 17 and 18 in Table 1), she evaluated her degrees of absorption each two pages as an indicator of her engagement in reading. Specifically, we asked her how absorbed she was in reading every pair of pages in these novels. These absorption ratings were used to validate the statistical analysis.

2.1   Participant

The subject was the first author, Miho Fuyama, who was 30 years old when the experiment was conducted. She is a native Japanese speaker, is a regular reader, and has normal vision.

2.2   Material

We used 20 Japanese novels, which the first author read for the first time in this experiment. The titles, authors, and page lengths of the books are listed in Table 1. We selected as texts books written by authors who have won Japan's prestigious literature prizes, such as the Naoki Prize or Akutagawa Prize.

2.3   Procedure

In each session of the experiment, the subject was asked to read a novel. Each session lasted several hours (including breaks), but was completed in one day. The subject reported her degrees of absorption for every two pages read in novels number 17 and 18. These reports were made approximately 100 days after the reading sessions. This delay in the absorption rating is due to experimenter's procedural mistake. Her degree of absorption was measured on a five-level scale – "extremely bored", "bored", "normal", "absorbed", and "deeply absorbed". This scale was coded using the numbers $-2, -1, 0, 1$, and 2, respectively, for each of the states. As the experiment required her to focus on and become absorbed in such long texts, the subject was allowed to perform her readings at her home in order to minimize her tension. She was also allowed to have breaks whenever she wanted. The breaks were typically 5 to 15 min long, but there were also several hour-long lunch breaks. While reading, she sat at her desk and was videotaped with two small web cameras.

2.4   Analysis

From the videos, we transcribed the reading time for each pair of pages. These reading times were measured as the lengths of time between page turns, excluding time spent on breaks. Statistical analysis was performed on these transcribed

Table 1. The novels read in Experiment 1.

No	Title (Abbreviated)	Author	Page length
1	*Shikisai*	H. Murakami	370
2	*Kamisama*	H. Mori	314
3	*Nameraka*	H. Kawakami	189
4	*Tenchi*	T. Ubukata	474
5	*Chinmoku*	Y. Ogawa	308
6	*Hikari*	S. Miura	297
7	*Kuchi*	M. Banto	309
8	*Mizuumi*	B. Yashimoto	206
9	*Kogoeru*	A. Shino	401
10	*Self-Reference*	T. Enjo	308
11	*Shi no izumi*	H, Minagawa	427
12	*Kisetsu no kioku*	K. Hosaka	316
13	*Eien no deguchi*	E. Mori	313
14	*Hokanaranu hito he*	K. Shiraishi	295
15	*Shorou tomurai dou*	N. Kyogoku	498
16	*Kodoku no utagoe*	A, Tendo	312
17	*Neko*	Y. Ogawa	359
18	*Ruto 225*	C. Fujino	282
19	*Yasashii uttae*	Y. Ogawa	260
20	*Burahuman*	Y. Ogawa	146

reading times. We analyzed the aggregate of the data gathered across all the sessions of the experiment in order to increase the statistical power of our analysis.

In our analysis, we fitted mixtures of exponential distributions, those of Weibull distributions and those of gamma distributions, to the aggregate data based on the Expectation–Maximization algorithm [3]. For each mixture distribution, ranging from 1 to 5 components, we estimated the parameters by maximizing likelihood. As these statistical models have different numbers of parameters, we chose the model with the smallest Bayesian Information Criterion (BIC) statistic [13] as the one that best explains the data. The choice of the BIC is not exclusive. As a result, we found that other criteria, such as Akaike information criterion (AIC) [1] chooses the same model in our analysis.

2.5   Results and Discussion

We found that a mixture of two gamma distributions provided the best fit to the aggregate data amongst all the distributions considered. Figure 4 illustrates the differences between these various classes of distributions in explaining our data. It shows the hazard function $H(t)$ of the page-turn interval t. The hazard function $H(t)$ is the probability (density) to finish reading on condition of the reading being unfinished until t. Exponential distributions in general exhibit a constant $H(t)$, which means this random process has "no memory", that is, a constant rate of reading interval regardless of time. Weibull and gamma distributions, in contrast, have increasing hazard functions. This means that the reading becomes more and more likely to be finished as time goes on. The two classes of distributions, however, exhibit differences in the shapes of their hazard functions.

The exponential distribution, with a constant hazard function, did not fit the data well, as shown in Fig. 4 (BIC = 29421.71). Likewise, the Weibull distribution has large deviation from the data at the tails of distribution ($t < 30$ and $140 < t$) (BIC = 26146.06). The single gamma distribution fits better than the exponential and Weibull distributions (BIC = 25722.64), but the mixture of two gamma distributions provides the best fit (BIC = 25655.29). Further, mixtures of three gamma distributions (BIC = 25677.24) or more did not provide better fits than the two-component gamma distribution.

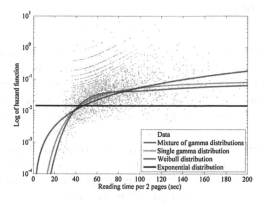

Fig. 4. The hazard functions for the sample (dots) and for the estimated probability distributions (lines) of reading time per two pages

Figure 5 shows the probability density function of empirical reading intervals and the estimated probability density function, which is a mixture of two gamma distributions. One subcomponent, Distribution 1, has shape 13.80 and scale 4.24. The other subcomponent, Distribution 2, has shape 7.58 and scale 10.67. This result suggests that the subject shows two distinct modes in her reading, with each mode involving different reading subprocesses. It is worth noting that, at this point, we have not established the relationship between the two statistically estimated modes and the putative cognitive processes for reading.

Correlation to Reading Engagement. We now address the question of whether the two distinct modes identified in our analysis are actually reflective of the text being read. In order to test this, we analyzed the correlation between the temporal change in mode and the degree of absorption reported by the reader. We obtained the reader's post-hoc report on engagement for each two pages of books 17 and 18.

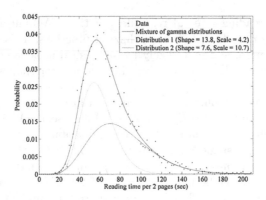

Fig. 5. Sample (dots) and estimated (solid) probability distributions of reading time per two pages. The two curves under the fitting curve show subcomponents of the gamma mixture distribution.

Taking book 17 as a representative case, Fig. 6 shows the temporal profile of the weighted-average of shape parameters (black dots) and the reader's degrees of absorption (red dots). The weights were given by the mixture of the two gamma distributions for each reading time of two pages. The corresponding moving averages of the two over five data points are shown as the black and red lines, respectively.

We performed correlation analysis for a pair of the estimated shape parameters and the degrees of absorption. For book 17 across 141 pairs of pages, we had correlation -0.284 (Pearson product-moment correlation coefficient, $p < 0.001$). For book 18 across 118 pairs of pages, we had correlation -0.283 ($p < 0.01$). This indicates that the temporal changes in the modes identified from our reading time analysis (Fig. 6) do indeed reflect changes in reading engagement.

Recall that the shape parameter can be interpreted as the number of subprocesses involved in processing a given text, and that the scale parameter can be interpreted simply as the inverse of reading speed (Fig. 1). Taking this theory into account, we conclude that the two modes estimated in this analysis are likely to represent a fast reading mode (Distribution 1) with a larger number of subprocesses, and a slow reading mode (Distribution 2) with a smaller number of subprocesses.

3   Experiment 2

In Experiment 1, our statistical analysis detected two different modes of behavior in the reading data generated by the experiment. We further showed that the change in mode over time had a statistically significant correlation to the levels of engagement with the text reported by the subject. Our goal for Experiment 2 was to establish whether or not these findings are consistent across multiple subjects and, if so, to identify the various factors governing the reading modes

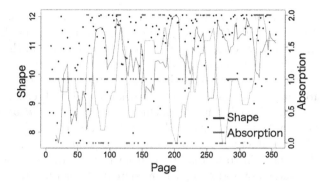

Fig. 6. Page-based temporal profile of the statistical property (shape parameter) of reading time and the absorption ratings of book 17. (Color figure online)

detected in Experiment 1. In order to answer these questions, we designed a short experiment for the other subjects. In our second experiment, we asked different subjects to read a short prose narrative but kept the rest of the procedure the same as it was in case 17 and 18 of Experiment 1. Namely, subjects were asked to read a short story or a part of a novel, and then they were asked to report their degrees of absorption for each two pages. The story itself took less than an hour to read.

We expected two possible cases:

1. We may observe individual variance in reading time across subjects, which would reflect that different subjects exhibit very different ways of processing the text.
2. The reading time may depend on the contextual structure of the text, and different subjects may exhibit similar mode changes in reading the same text.

The major factor dictating reading modes would be a subject's reading strategy in the first case. In the second case, it would be the contextual structure of the text itself.

3.1 Participants

In Experiment 2, we employed ten participants. Each participated in two reading sessions. In one session they read one short but complete story, and in the other session, they read an introductory part of a different, longer story. The order of sessions was counterbalanced across participants, and all of them participated in their second session two weeks or longer after the first session. The subjects were five male and five female undergraduate and graduate students at Keio University. Most of these subjects were not regular readers.

3.2 Procedure

The procedure was the same except for the length of the text and the environment in which the reading took place. During each session, one participant read a 49-page short story or a 39-page introductory part of a long story in a room reserved specifically for the experiment. Right after the reading session, the participant was asked to report their degrees of absorption in the same scale as Experiment 1 for each two pages. Five participants read the short story first, and the other five participants read the introductory part first. The participants took part again at a greater than two week interval, and read another text. The short story was "Kino", which is included in *Onna no inai otoko tachi*, a commercially available part of an omnibus authored by Haruki Murakami. The other text is the introductory part of *Chinmoku Hakubutukan* written by Yoko Ogawa. For the chosen introductory part, from pages 3 to 40, this particular text does not include any major change in context. After their reading session, each subject was asked to report his/her degrees of absorption for each two pages using the same five-point scale used in Experiment 1.

3.3 Analysis

For consistent comparison, we analyzed the aggregate of the reading time data across subjects by fitting to it a two-component mixture of gamma distributions. We fixed the class of distributions instead of identifying it from data. This was largely due to the small sample size of our data at this point. Each participant provided reading time data for only 23 or 18 pairs of pages, which did not provide sufficient statistical power to be conclusive even for aggregation across subjects. Thus, we employed the statistical distribution estimated in Experiment 1.

3.4 Results

The data of two readers were excluded from the analysis for their irregular method of reading back and forth many times, giving a completed dataset of 18 readers in total.

Each panel of Fig. 7 shows the page-based temporal profile of the modes estimated from reading time. In each panel, a dot shows the estimated shape parameter for each reading time data point, and the line indicates its moving average. As in Experiment 1, we found that the temporal changes in modes were significantly correlated to the reported degree of reading engagement. The trend correlations were, however, opposite to each other: the readers of "Kino" showed positive correlation ($R = 0.3, p < 0.01$), while those of *Chinmoku Hakubutukan* showed negative correlation ($R = -0.16, p < 0.04$). We have multiple possible explanations for this seemingly conflicting finding, which we will discuss later.

Further, the results shown in Fig. 7 exhibit inter-subject consistency in temporal changes in reading modes. Each panel in Fig. 7 shows the estimated shape parameter for each reader. The top panels show those of "Kino", and the bottom ones show those of *Chinmoku Hakubutukan*. We found within-story similarity in the shape parameter profiles across pages; the readers of "Kino" showed similar U-shape profiles, and those of *Chinmoku Hakubutukan* had similar flat profiles.

We performed correlation analyses on all the pairs of subjects in order to test whether readers of the same story showed correlated temporal profiles of the shape parameters. The average correlation across all the reader pairs of "Kino" was 0.67 (from 0.46 to 0.87, $p < 0.02$ for every pair of readers), and that of *Chinmoku Hakubutukan* was 0.51 (from 0.05 to 0.86, $p < 0.05$ for 23 out of 36 pairs of readers). Thus, this result suggests that each story had an effect on the reading-mode profiles, for which the readers exhibited similar profiles, while individual readers exhibited little effects on their own profiles.

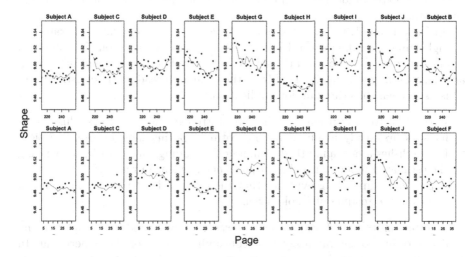

Fig. 7. The page-based temporal profiles of the estimated shape parameters for each subject. The top row shows the results of reading "Kino", and the bottom row shows the results of reading *Chinmoku Hakubutukan*.

3.5 Discussion

The results above suggest that there is a story-specific effect on the statistical properties of reading time, which we can interpret as mode-switching profiles of the reading process. As the page-based temporal profiles within the same story were similar, this suggests that the contextual structure of each story had a major impact on the switching of reading modes. In addition, the temporal profiles could be interpreted as a reflection of the semantic structure of each story. "Kino" is a short but complete story, while the selected text of *Chinmoku Hakubutukan* was an introductory part of the novel. According to Miall's theory, a full novel is expected to involve both formation and exploitation of anticipation. Consistent with this theory, we found U-shaped patterns in reading mode switching for "Kino" with its complete story, but we found flat patterns for the introductory part of *Chinmoku Hakubutukan*, which indicated no clear distinct modes. The finding of no clear modes in reading an introductory part is not well-consistent with Miall's theory, but we interpret that an introductory part of a full novel is likely to include only the formation of anticipation. In summary, these findings could be treated as supportive evidence that temporal profiles in reading distributions reflect context-dependent reading modes.

4   General Discussion

Reading is an essentially mental and subjective experience. Its cognitive under-pinnings have been difficult to characterize directly, and reading time is a major tool for drawing inferences about the underlying cognitive mechanism behind reading. This study offers a new approach to the analysis of reading time, an approach capable of identifying different modes of reading behavior from reading time data.

In Experiment 1, we collected and analyzed reading time data generated by a single subject reading several full novels in a natural situation. We observed significant correlation between the subject's report of her engagement in reading, and her reading modes as inferred from the estimated reading time distribution. This experiment has three major implications:

1. In contrast to conventional studies on controlled, short readings, this is per-haps the first study involving reading entire books in a more natural situation.
2. It establishes a new analytical technique for reading time data by associating the estimated modes with the subject's engagement in reading.
3. It provides supporting evidence that there are at least two distinct reading modes in the reading of whole novels.

A clear limitation of Experiment 1 was that we could not employ many subjects, owing to the intensely time-consuming nature of the experiment. In Experiment 2, each session was designed to be as minimally demanding as pos-sible. This allowed us to perform the experiment using a number of different subjects. We once again observed two distinct reading modes, and found that the mode switches across different subjects reading the same story were consis-tent with each other. This suggests that, to a large extent, the reading modes are dictated by the contextual structure of the text being read.

What is the contextual structure? We hypothesize that it is deeply related to the predictability of the story. Perhaps, we can consider the two discovered reading modes as low- and high- predictability modes. With reading of "Kino", the major shifts between two modes took place at the beginning and end of the story. At the beginning, a reader has little knowledge of the story, as discussed by Miall [10], and they need to build knowledge of the characters and the stage where they play their roles. When approaching the end, this story has a twist, which is unexpected for most readers. This is another place where the reader needs to rebuild their knowledge of the story. Therefore, the U-shape temporal profile is supposed to reflect lower predictability at the beginning and end of the story, and higher predictability in the middle of the story.

Our interpretation of the estimated reading modes is consistent, at least on a conceptual level, with Miall's; he hypothesized that reading can be separated in two different stages, the registration stage and the interpretation stage, and they may be repeated more than once across a narrative [10].

This hypothesis and theoretical understanding of the reading modes clearly requires further research to in order to be answered. For the questions raised by the present study, we are preparing for extensions of the current experimental paradigms.

Acknowledgements. The authors are grateful to Dr. Neeraj Kashyap for his proof-reading of this manuscript. The first author was supported by the Keio University Doctorate Student Grant-in-Aid Program and Mori Grants. The second author was supported by the Grant-in-Aid for Scientific Research B KAKENHI No. 23300099.

References

1. Akaike, H.: A new look at the statistical model identification. IEEE Trans. Autom. Control **19**(6), 716–723 (1974)
2. Beaugrande, R.D.: The story of grammars and the grammar of stories. J. Pragmat. **6**, 383–422 (1982)
3. Dempster, A.P., Laird, N.M., Rubin, D.B.: Maximum likelihood from incomplete data via the EM algorithm. J. Roy. Stat. Soc. **39**(1), 1–38 (1977)
4. Gernsbacher, M.A.: Two decades of structure building. Discourse Process. **23**(3), 265–304 (1997)
5. Hidaka, S.: A computational model associating learning process, word attributes, and age of acquisition. PLOS ONE **8**(10), e76242 (2013)
6. Inhoff, A.W., Rayner, K.: Parafoveal word processing during eye fixations in reading: effects of word frequency. Percept. Psychophys. **40**(6), 431–439 (1986)
7. Iser, W.: Der Akt des Lesens: Theorie ästhetischer Wirkung. Wilhelm Fink, München (1976)
8. Mar, R.A., Oatley, K., Djikic, M., Mullin, J.: Emotion and narrative fiction: Interactive influences before, during, and after reading. Cogn. Emot. **25**(5), 818–833 (2011)
9. Miall, D., Kuiken, D.: Beyond text theory: understanding literary response. Discourse Process. **17**(3), 337–352 (1994)
10. Miall, D.S.: Affect and narrative: a model of response to stories. Poetics **17**, 259–272 (1988)
11. Miall, D.S.: Beyond the schema given: affective comprehension of literary narratives. Cogn. Emot. **3**(1), 55–78 (1989)
12. Miall, D.S., Kuiken, D.: Foregrounding, defamiliarization, and affect: response to literary stories. Poetics **22**, 389–407 (1994)
13. Schwarz, G.: Estimating the dimension of a model. Ann. Stat. **6**(2), 461–464 (1978)
14. Thorndyke, P.W.: Cognitive structures in comprehension and memory of narrative discourse. Cogn. Psychol. **9**(1), 77–110 (1977)
15. White, S.J.: Eye movement control during reading: effects of word frequency and orthographic familiarity. J. Exp. Psychol. Hum. Percept. Perform. **34**(1), 205–223 (2008)

Whole-Body Coordination Skill for Dynamic Balancing on a Slackline

Kentaro Kodama[1]([✉]), Yusuke Kikuchi[2], and Hideo Yamagiwa[3]

[1] Kanagawa University, 3-27-1, Rokkakubashi, Kanagawa-ku, Yokohama-shi, Kanagawa, Japan
kkodama@kanagawa-u.ac.jp
[2] Future University Hakodate, 116-2, Kameda Nakano-cho, Hakodate-shi, Hokkaido, Japan
ykikuchi@fun.ac.jp
[3] Tokyo Metropolitan Tobu Medical Center, 3-3-25, Shinsuna, Koto-ku, Tokyo, Japan
somatolearning@gmail.com

Abstract. The purpose of the present study is to reveal the fundamental skills for slacklining. A slackline is a flat belt tightly spanned between two anchor points. Because it bounces and swings in all directions, maintaining balance on it is difficult. In the practical field of slackline training, instructors share their skills based on personal experience. In a basic slackline course, they begin by teaching a fundamental skill, such as single-leg standing on a slackline, by explaining how they do it. However, such first-person perspectives on slacklining skills have not been scientifically investigated. According to instructors' knowledge based on personal experience, we hypothesize the skills for single-leg standing on the slackline. The present study examines current hypotheses by comparing performances at different skill level (i.e., experienced vs. novice). This article introduces our pilot study, including current hypotheses and data from preliminary experiment, and discusses them.

Keywords: Slackline · Balance sport · Dynamic stability · Embodiment · Synergy

1 Introduction

1.1 Slackline

A slackline is a flat belt made from nylon or polyester, tightly spanned between two anchor points (Fig. 1). Because it bounces and swings in all directions, maintaining balance on it is very difficult. Slacklining has two aspects: as a balance sport, and as a balance training. Since slacklining started as a balance sport in around 2007 in Europe, it has spread over the world, and international contests have been held, wherein competitors demonstrate various acrobatic skills such as jumping on a line. Slacklining has also received much attention as a method of balance training for not only top-level athletes (e.g., Olympic ski jumpers), but also for the elderly or those with motor disorders participating in rehabilitation (one of the authors has been involved in applying slackline training to rehabilitation). We consider that slackline training is enjoyable compared to

© Springer International Publishing AG 2017
M. Otake et al. (Eds.): JSAI-isAI 2015 Workshops, LNAI 10091, pp. 528–546, 2017.
DOI: 10.1007/978-3-319-50953-2_39

traditional balance training using a balance ball or balance beam, if so slacklining is expected to be a sustainable lifelong activity.

Fig. 1. Slacklining [1]

Although slacklining has progressively spread as both a balance sport and a balance training, research on it is relatively sparse. In addition, most research has focused on slacklining in terms of balance training, through examining the effect of slackline training on balance abilities [1–3]. Although these studies are important for obtaining evidence on the effects of slackline training, the skills for slacklining remain unclear. This means that beginners of slackline training confront the problem on how to start slacklining and how to acquire the skill. Although a few studies have approached the slacklining skill, they conducted a case study that investigated the limited situation of balance recovery after perturbation [4]. The fundamental skills for slacklining have not yet been investigated and are not clear in the academic literature.

The present study hypothesized that a single-leg standing task in slacklining contains the element of the fundamental skill for slacklining. It is a first step of learning slack-lining in the practical field when being instructed in slackline training [2, 5]. In practical courses, after mastering the first step of single-leg standing without support, beginners are encouraged to proceed to the next step, such as walking on a line. We presume that starting with a single-leg standing task is better, to improve slacklining in a better way to acquiring the fundamental skills for slacklining. Therefore, the present study chose to investigate the single-standing task as an experimental task to reveal the fundamental skill for slacklining, and to describe how a single-leg standing task of maintaining balance on an unstable line is achieved.

One might argue that a static balance strategy, assuming absolute straining of muscles and fixing the joints like an inanimate object consisted of rigid materials, is one possible balance strategy. Such a static balance model, however, does not agree with the fact that living systems such as animals, including humans, always fluctuate in various time scales at different levels from the microscopic level (e.g., cell) to the macroscopic level (e.g., skeleton). Such an intrinsic fluctuation can be observed even in a quiet standing postural task [6]. Current studies on human postural balance have demonstrated the relevance of dynamic balancing [7, 8]. In these studies, human postural sway is regarded not as meaningless random noise at the muscular skeletal level, but as having meaningful structure emerging from interaction among components in the body-environment system. While balancing in an unstable environment (i.e., slacklining), dynamic balancing is more critical because the fluctuation generated by one's own body movement can easily cause amplification of the line's fluctuation and loss of stability of the body-line system. This may result in falling down from the line. Explaining how one maintains balance on a slackline by a static balance strategy is difficult, when considering the intrinsic fluctuation of an embodied system consisting of so many non-rigid components at different timescales.

1.2   Dynamical Systems Approach

The human body has many multifarious degrees of freedom (DoF), from microscopic cell level (10^{14}) to macroscopic joint level (10^2) [9, 10]. The DoF problem suggested that the large number of independently controllable DoF poses a computational burden to the central nervous system [10]. This indicated the difficulty of the unidirectional top-down motor control model by computer/information processing metaphor. It also pointed out the limitations of computing or determining one solution by solving insufficient equations (called ill-posed problem; [11]). In addition, considering whole-body movement, solving the DoF problem is more difficult because many DoFs need to be considered in a whole-body system.

Bernstein, who proposed the DoF problem, supposed that each component (DoF) is coordinated and coupled with other components to organize a functional structure/unit (i.e., synergy, not an anatomical unit) rather than being controlled separately [9]. With the idea of synergy Bernstein tried to solve the DoF problem in terms of the large number of DoFs. However, such a softly assembled temporal coupling among components is not organized in non-directional or random ways. The DoFs are supposed to regulate each other flexibly in order to achieve a specific task in specific situation/environment [12].

After Bernstein's suggestion of synergy, the self-organization theory [13, 14] was applied to human movement studies. It is called the dynamical systems approach, and has been widely applied to human movement science areas such as inter-limb coordination [15, 16], motor/coordination development [17, 18], skill acquisition [19], brain-behavior dynamics [15, 20], interpersonal coordination [21, 22]. Compared to the traditional approach to motor behavior assuming internal computation and prescription in the brain, dynamical systems approach focuses more on interactions among the body (including brain), environment, and task. Movement or coordination patterns are then supposed to emerge through interactions among several constraints from organism, environment and task [17] (Fig. 2). The large number of DoFs in an embodied system can be reduced in order to satisfy these constraints. Thus, the functional structure to achieve the task (i.e., synergy) emerges as a particular coordination pattern. The individual organism regulates its behavior in order to satisfy task demands performed in a specific environment. In this sense, the organization of movement is constrained by the individual organism, the environment, and the task. In the dynamical systems approach framework, motor behavior cannot be separated from these constraints.

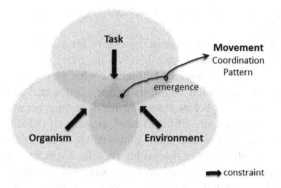

Fig. 2. Constraints on movement

Indeed, in the framework of Synergetics [13], a self-organization theory, a system's low-dimensional spatial-temporal patterns at the macroscopic level emerge through interactions among components (DoF) at the microscopic level under certain constraints from the system's environment or embedded context. The macroscopic pattern also constrains the components' behavior at the microscopic level to keep the pattern stable [15]. A representative work is the modeling of bimanual coordination through applying Synergetics to human motor behavior [23]. The model can describe the qualitative pattern change of a system with the concepts of Synergetics (e.g., an order parameter that indexes the macroscopic order or pattern of the system, a control parameter that determines the macroscopic state, and its spontaneous change of the system). Rhythmic coordinated behaviors, such as inter-limb coordination, can be modeled as a motion equation using a control parameter (i.e., movement frequency) and an order parameter (i.e., relative phase between limbs). This predicts the behavior of a system, composed of numerous mutually interacting components (DoF), as the dynamics of few order parameters [23]. Recently, dynamical systems approach has provided not only a theoretical framework but also evidence obtained by analytical tools (e.g., fractal analysis, recurrence analysis) based on nonlinear dynamics theory [24, 25]. The present study also applies these framework and techniques to investigate dynamic balancing skill in slacklining.

1.3 Dynamic Embodied Adaptability (DEA)

In the present study, we suppose that slacklining improves *dynamic embodied adaptability* (DEA). Here, DEA is defined as the ability to organize an embodied system itself rapidly and flexibly to adapt to dynamic changes in the environment and one's own body. For example, DEA can be seen in postural adjustment after trips or slips during walking. When tripping on an obstacle or slipping on a snowy road, one must regulate whole-body posture quickly and flexibly to avoid falling down. In general, however, DEA is supposed to decline with aging. Because slacklining requires one to maintain balance on a very unstable and dynamic environment (the line), it is expected to improve

DEA. Therefore, slacklining is applied to not only young people but also older people as a balance training.

DEA is supposed to be a characteristic property for living systems differing from classical artificial intelligence (AI) or robots. As the *frame problem* [26] pointed out, classical AI or robots need to describe beforehand what they will execute in the environment (i.e., internal simulation within the frame). In the real world, however, it is difficult to describe all things before executing any behaviors, because there are diverse situations and environments. In addition, situations and environments change dynamically as time proceeds. For these reason it is impossible to collect all knowledge about the world and predict all things in the world. Therefore, it is necessary to adapt to its dynamical changes depending on the individual situation and environment. After the *frame problem* pointed out the limitation of classical AI and robots, researchers in AI and robotics have been paying attention on *embodiment* and interactions between the body and environment in the real world [27, 28]. The body and environment are considered as the constraints on the system's behavior. Living system can realize such dynamic adaptations through interactions between the body and environment by active and exploratory perception-action cycle [29, 30]. Here we can find DEA in living systems.

How is DEA trained in slacklining? We suppose that specific task constraints in slacklining affect one's behavior. As explained above, in current frameworks of human movement science, movement is supposed to emerge through interactions among constraints from the organism, environment and task [17] (Fig. 2). In slacklining, the specific task constraint that one must keep balanced on a very unstable environment (i.e., line) affects one's motor behavior. As the result of such a specific constraint, muscle tone and inter-limb coordination patterns emerge through training at the skilled level. We assume that a single-leg standing task on a slackline is one of efficient training tasks for improving DEA. Because it is often used as a first step task in the practical field of slackline training [2], it contains an element of slacklining as a fundamental skill to prepare for other tasks such as walking and turning on a line.

Indeed, DEA is supposed to involve softly assembled synergy throughout the body [10, 16, 31, 32]. To improve DEA, one must acquire an adequate state of muscle tone and inter-/intra-limb coordination pattern. At the muscular level, to perform the slacklining task, muscle tone must be kept moderately relaxed because tense muscle tone causes fixation of joints and decline of movability. It also causes decreased sensitivity of muscle sense or dynamic touch for perceiving one's own body and the line [33–35]. Thus, if one strains the muscles, fluctuations of the line and the body cannot be absorbed or compensated, resulting in fluctuations being amplified and increasing the risk of falling off the line. At the joint skeletal level, to perform the slacklining task a whole-body coordination strategy is better than a lower body strategy using only the ankles or hips, because whole-body coordination allows the upper body parts (e.g., both hands, torso, neck) to participate in maintaining posture. This indicates that whole-body coordination allows an embodied system to recruit redundant degrees of freedom (DoF) in a task, leading to a more stable and flexible organization of the system [9, 36]. As suggested above, we suppose that slacklining improves DEA under the specific task constraints and an adequate muscle tone and inter-limb coordination patterns emerge through training. However skills for slacklining have not been scientifically investigated and how skilled players can achieve the task is not clear.

1.4 Interactive Approach to Skill Science

Although the skills for slacklining is not clear in the academic literature, in the practical field of slacklining (e.g., slackline teaching school or rehabilitation), instructors impart their knowledge of the skill based on their personal experience to beginners or patients. Such knowledge from the first-person perspective, however, is not always explicit or verbalized. Rather it is implicit and acquired or improved in unconscious processes through personal embodied experience. The present study focuses on such first-person perspectives, including the personal viewpoint of authors who have experience of slacklining, to generate hypotheses about the skill.

To test these hypotheses, we also conduct objective measuring and quantification of slacklining performance. As shown in the next chapter in detail, performance of a fundamental slacklining task (i.e., single-leg standing) is measured using a 3D motion capture system, and compared between participants at different skill levels (i.e., experts vs. novices) in terms of hypothesized variables (details of them will be explained in 2.1 later). If those variables differ between experts and novices, this would signify that experts control these variables to perform the task of remaining balanced on a line. Revealing these variables can lead to easier and safer learning of slacklining in the practical field.

Even though defining the first/third person perspective is difficult and its definition may be debatable, here we regard them as follows (Fig. 3). The first-person perspective involves an implicit, subjective, and qualitative approach. Therefore, it is open to be tested and has not been examined by any other approach (i.e., the third-person perspective). The third-person perspective involves an explicit, objective, and quantitative approach. If a hypothesis is generated qualitatively from the first-person perspective, then it can be tested from the third-person perspective by objective measuring and quantitative analysis. Furthermore, if a hypothesis is supported by general experimental testing, it can provide evidence to a practical field as generalized knowledge. Even if a hypothesis is not supported, it can motivate further generation of new hypotheses from different viewpoints or it can lead to understanding the individualities of the phenomenon itself that cannot be generalized or described. Thus, these two approaches can be connected through a cyclic relationship in which they motivate each other (Fig. 3). This process might be considered as rather spiral than cyclic in the long run, because a cycle can motivate a new insight, which differs from the initial question, resulting in further and endless exploration. We consider both approaches to be scientific and important for the understanding of embodied skills (i.e., *Shintai-chi*). Such a process can also make it possible to investigate knowledge or questions obtained in practical fields in academic research frameworks and to provide evidence for them, and to apply insights obtained in academic research fields to practical issues for solving problems or improving practice, and vice versa.

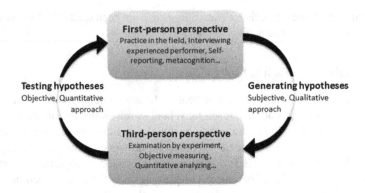

Fig. 3. Interactive approach (cycling the first-person perspective into the third-person perspective)

As described in this chapter, the present study investigates the skills for single-leg standing on a slackline to reveal the fundamental skills for slacklining. Slacklining requires ones to regulate whole-body coordination dynamically. Under constraints such as maintaining posture on an unstable line, behavioral patterns are supposed to emerge through slacklining at the skilled level. In the practical situation, instructors describe their knowledge of the skill based on their personal experience (i.e., first-person perspective). Such kinds of skill (i.e., *Shintai-chi*), however, are not always acquired or improved consciously. Therefore, to investigate these kinds of skills, an interactive approach is necessary, cycling the first-person perspective into the third-person perspective. We apply the first-person perspective for generating hypotheses and the third-person perspective for testing hypotheses by comparing performances between players at different skill level through quantification of hypothesized behavioral variables.

2   Method

The current article reports the result of our pilot study comparing single-leg standing performances of an experienced and novice player.

2.1   Current Hypotheses

Based on knowledge from slackline instructors' personal experience and the authors' personal experience, here we propose three hypotheses on the skills for single-leg standing on the slackline. To maintain whole-body balance, the overall task is regarded as keeping the center of mass (COM) above the base of support (i.e., the line area where one's foot contacts). To do so:

1. in the mediolateral direction, one should raise both hands high and coordinate them in parallel (i.e., anti-phase) to regulate the COM position above the line;
2. in the vertical direction, one should flexibly bend his/her knee of the standing leg to reduce the line's fluctuations;

3. in the anteroposterior direction, one should maintain a straight back to keep the center of gravity vertically balanced over the heel of the standing leg and the line.

Each hypothesis is portrayed in Fig. 4(1, 2, and 3 respectively). In the present study, we consider the following variables as hypothesized variables to test the above hypotheses respectively:

1. bimanual coordination stability: skilled players move their hands in a specific relation as coupled each other not control individually to regulate their COM mediolateral position;
2. knee joint flexibility of the standing leg: skilled players keep their knee joint flexibly (not fixed) so as to absorb fluctuations of the line;
3. stability of positional relation between the center of gravity (COG) and the sole of foot (contact point between the body and line): skilled players adjust the COG position to the contact point to perceive dynamic relation between the body and line constantly.

Fig. 4. Current hypotheses (regulating whole-body posture in (1) mediolateral, (2) vertical, (3) anteroposterior direction)

The current article reports partial results testing hypotheses (1) and revealing characteristic features of the experienced player's performance by preliminary experiment.

2.2 Preliminary Experiment: A Case Study

Participants. Two participants, an experienced player (age 40, male, 175 cm tall) and a novice player (age 30, male, 174.5 cm tall), were recruited to compare between slacklining performances at different skill levels. The experienced player had more than three years' experience of slacklining, while the novice player had just started slacklining. The experimental procedures were approved by the research ethics committee of Kanagawa University (No. 2014-2), where the experiment was conducted. Each participant provided informed consent to participate in this study.

Apparatus. Slacklining was performed on SLACKRACK300 (GIBBON SLACK-LINES, 300 cm length, 30 cm height). A 3D motion capture system (OptiTrack V120: Trio, NaturalPoint, Inc.) was used to measure participants' body movement (sampling frequency: 120 Hz). Nine reflective markers ware attached to the top of the head (vertex), the front of the head (frontal bone), the tips of the index fingers of both hands, the center of mass (COM, the front of the second sacral vertebra), the tops of the knees (patella), and the tips of toe (ossa digitorum manus). In the present study, the COM marker is regarded as a collective marker of whole-body behavior. Three dimensional time series data of each marker were smoothed by second order Butterworth low-pass filter with a 30 Hz cutoff frequency.

Procedure. The experimental task was to perform the single-leg standing on a line. Participants were required to keep balanced on a line for as long as possible. Participants conducted the task with their preferred leg. Twenty seconds of data while each partici-pant was balanced on a line were analyzed in terms of the hypothesized variables as shown below in detail.

Data Analysis. In the current study, three analyses were conducted. First, *cross recur-rence quantification analysis* (CRQA; [37]) was applied to evaluate first hypothesized variable, bimanual coordination stability, because CRQA is supposed to be an adequate method for analyzing nonlinear time series and it has been commonly used in bimanual coordination studies [38, 39]. Second, we applied *detrended fluctuation analysis* (DFA; [40, 41]) to assess the entire behavioral adaptability of the players' performance by means of analyzing the COM position's fluctuation considered as a collective marker position of whole-body behavior, because DFA is relatively free of assumptions about the stationarity and signal model [42], and data size [43]. Since DFA has been used to assess the system's flexibility/adaptability [44], and exploratory behavior [45], we chosen it for assessment of the entire behavioral adaptability of the players' performance to compare it between the experienced and novice players. Third, we quantified the variabilities by means of calculating the total trajectory length of the markers (of knee, COM and line) to assess how much they moved in three dimensional space. In terms of hypothesized variables mentioned above 2.1, CRQA is the only direct method to test the hypothesis (1), and the other analyses, DFA and variability analysis, are not direct methods to test our hypotheses. However we conducted them as pilot analyses to obtain basic knowledge about the players' behaviors. Details of each method were described below.

To quantify the movement stability (i.e., bimanual coordination), CRQA was applied. This is a nonlinear method that captures the recurring properties and patterns of a dynamical system, which results from two streams of information interacting over time [37, 38], and quantifies how similarly the two observed data series unfold over time [46]. Recurrence quantification analysis was originally developed to uncover subtle time correlations and repetitions of patterns, and is relatively free of assumptions about data size and distribution [47]. In CRQA, two time-delayed copies of the original time series were used for embedding the data in higher dimensional space, reconstructing the phase space of the dynamical system, to analyze the recurrent structure between them [37].

For inter-limb rhythmic coordination, two CRQA measures are regarded as significant indexes of the movement stability [39, 46]. The percent recurrence (%REC) in CRQA corresponds to the ratio of the number of shared locations relative to the number of possible shared locations in phase space [46]. In other words, %REC reflects shared locations in phase space [39]. It provides an index of the magnitude of noise in the system [39]; a higher %REC indexes lower noise in the system. The other is related to the line structure calculated from the recurrence plot (e.g., Maxline (MAXL) is the longest shared trajectory in phase space and the length of maximum diagonal line on the plot) [46, 48]. MAXL is a measure of the stability of the shared activity [46]. In other words, MAXL reflects stability of shared activity (convergence of nearby trajectories over time) [39]. It provides an index of the system's sensitivity to perturbations (i.e., the strength of the attractor against perturbations) [39]. We also calculated the average of the diagonal line (i.e., average line, AVGL) as a measure of the movement stability during a trial [38]. While %REC reflects a stochastic stability like how often two time series recur within the specific radius in phase space, metrics based on the diagonal line (MAXL and AVGL) reflect a persistent stability like how long two time series maintain recurrence state. The present study chosen these metrics to evaluate bimanual coordination stability from various perspectives.

We performed CRQA using the R package 'crqa' (version 1.0.6) [38] after determining the optimal values for the input parameters (e.g., time delay, embedding dimensions, radius) using the MATLAB toolbox 'CROSS RECURRENCE PLOT TOOLBOX' (version 5.17) [49] and referring to the standard guidelines of the RQA method [48]. As a result, we chose time delay, embedding dimensions, Euclidean distance (radius) in phase space for each analysis, (320, 3, 1.1 respectively) for bimanual coordination analysis.

To evaluate behavioral adaptability, DFA [40, 41] was applied to time series analysis of COM data. DFA is a nonlinear time series analysis method for quantification of the statistical self-affinity and evaluation for fractal properties (e.g., 1/f pink noise) of the time series [50]. It was introduced to quantify long-range temporal correlations [40], and is relatively free of assumptions about the stationarity and signal model [42], and data size of time series [43]. It has been applied to various physiological and behavioral time series, for example, heart rate variation [41], gait stride [51], and postural sway [7]. Here, according to a brief summary in Hardstone et al. (2012), the DFA procedures are described as follows [50]:

1. Compute the cumulative sum of the time series to create the signal profile.
2. Define a set of window sizes, T, which are equally spaced on a logarithmic scale between the lower bound of four samples and the length of the signal.
3. Plot the fluctuation function for all window sizes, T, on logarithmic axes.
4. The DFA exponent, α, is the slope of the trend line in the range of time-scales of interest and can be estimated using linear regression.

(Hardstone et al. 2012)

By the above procedure, the DFA exponent is calculated and is interpreted as an estimation of the Hurst exponent. If $\alpha = 0.5$, then the time series is uncorrelated (white noise). If $\alpha < 0.5$ then the time series is anti-correlated, which means that fluctuations are smaller in larger time windows than expected by chance. If $0.5 < \alpha < 1$, then positive correlations are present in the time series as you get larger fluctuations on longer time-scales than expected by chance. If $\alpha = 1$, then it is considered as 1/f pink noise. As α is closer to 1, it is regarded as stronger fractality [52]. The concept of fractality has been correlated with health/pathology [53], flexibility/adaptability [44], and exploratory behavior [45].

Knee and COM variability were quantified as the total trajectory length of each marker 3D position as follows:

$$Total\ trajectory\ length = \sqrt{(x_{t+1} - x_t)^2 + (y_{t+1} - y_t)^2 + (z_{t+1} - z_t)^2}$$

x_t, y_t and z_t indicate the mediolateral, vertical and anteroposterior position of markers of the knee, COM or line at time t respectively.

3   Results and Discussion

3.1   Bimanual Coordination

Figure 5 shows time series of both hands in the mediolateral direction for each participant (top: experienced player, bottom: novice player). X-axis represents time [s], and Y-axis represents position of mediolateral axis from the center of slackline [m]. Rigid and dashed lines represent left and right hand motion respectively. As shown in time series of both hands (Fig. 5), it seems that the experienced player can maintain a more stable bimanual coordination pattern in parallel (i.e., anti-phase coordination pattern) than the novice player. The novice player could not keep a particular pattern between hands consistently, such that he crossed both hands two times.

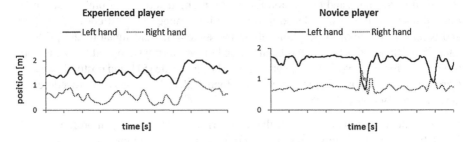

Fig. 5. Time series of both hands (20 s)

Figure 6 represents cross recurrence plots for each participant (left: experienced player, right: novice player). In each panel, x-axis and y-axis represent left and right hand time series respectively. Figures 7, 8 and 9 show %REC, MAXL, and AVGL for each participant in bimanual coordination respectively. %REC was higher in experienced player (3.65%) than in novice player (0.65%). MAXL was longer in experienced player (97) than in novice player (67). AVGL was longer in experienced player (48.64) than in novice player (26.7). As a result of CRQA for bimanual coordination,

%REC, MAXL, and AVGL were greater in the experienced player than in the novice player. These results indicate that bimanual coordination during the single-leg standing task was more stable in the experienced player than in the novice player in terms of both the magnitude of noise in the system, the system's sensitivity to perturbations, and temporally persistent stability. These facts support hypothesis (1) bimanual coordination stability is correlated with the skill level. We consider that such a bimanual coordination pattern emerges for dynamic balancing to keep the COM mediolateral position on the line. Further experiments should be conducted to obtain more samples and test the hypothesis statistically. It is also required to quantify the coupling between bimanual coordination and line as well as the actual COM mediolateral position in the future.

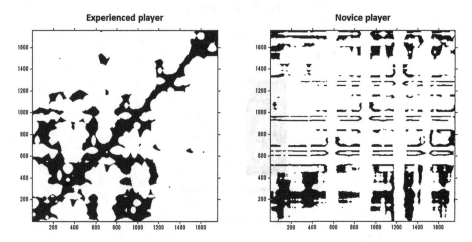

Fig. 6. Cross recurrence plot of bimanual coordination

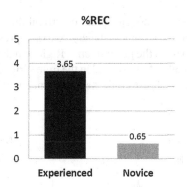

Fig. 7. Percent recurrence

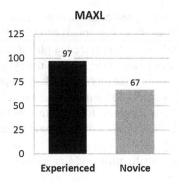

Fig. 8. Maxline

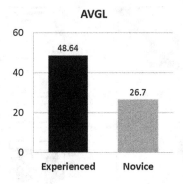

Fig. 9. Averageline

3.2 Knee Variability

Figure 10 shows time series of knee and line in vertical direction for each participant (top: experienced player, bottom: novice player). X-axis represents time [s], and Y-axis represents position (height) from the ground [cm]. Rigid and dashed lines represent knee and line variability respectively. As observed in the time series of knee and line (Fig. 10), it seems that the experienced player could keep the knee vertical position more stable

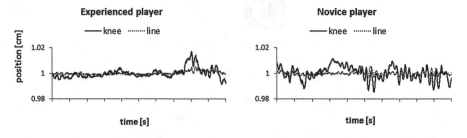

Fig. 10. Time series of knee and line

than the novice player. The novice player's knee position seems to fluctuate in the vertical direction with larger amplitudes than the experienced player's one.

Figure 11 indicates variabilities of knee and line defined as three dimensional total trajectory length [m] during 20 s performance. While line variabilities were not so different between players, knee variability was more than twice greater in the novice player than in the experienced player (Fig. 11). This result suggests that knee position of the experienced player was more stable than that of the novice player.

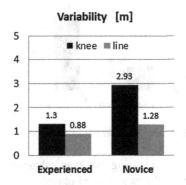

Fig. 11. Variability of knee and line

3.3   COM Variability and Fluctuation

Figure 12 shows time series of COM (around hip) and line in the vertical direction for each participant (top: experienced player, bottom: novice player). X-axis represents time [s], and Y-axis represents position (height) from the ground [cm]. Rigid and dashed lines represent COM and line variability respectively. As observed in the time series of COM and line (Fig. 12), it seems that the experienced player could keep the COM vertical position more stable than the novice player. In contrast, the novice player's COM position seemed to fluctuate in vertical direction with larger amplitudes than that of the experienced player.

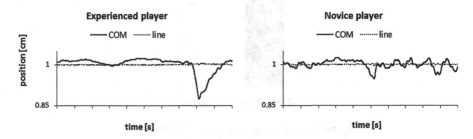

Fig. 12. Time series of COM and line

Figure 13 indicates variabilities of COM and line defined as three-dimensional total trajectory length [m] during 20 s performance. While line variabilities were not so different between players, COM variability was more than tripled greater in the novice player than in the experienced player (Fig. 13). This result suggests that COM position of the experienced player was more stable than that of the novice player.

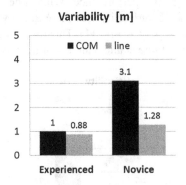

Fig. 13. Variability of COM and line

Figure 14 presents the fractal scaling exponent calculated by DFA. It is higher and closer to 1 (1/f, pink noise) in the experienced player (0.86) than in the novice player (0.66). The fractal scaling exponent calculated by DFA, was higher and closer to 1 in the experienced player than in the novice player. This result indicates that the experienced player's behavior was more flexible and exploratory than that of the novice player. As reported in previous studies on exploratory movement [35, 54], fluctuations in exploratory behaviors are fractal. If the experienced player's COM represents his whole-body behavior, the result suggests his exploration for a stable relation between his own body and the line. This possibility should be examined in future experiments quantitatively with more data from experienced players.

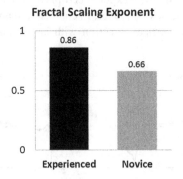

Fig. 14. Fractal scaling exponent of COM (right)

3.4   Future Directions

As a result of our pilot study, the current hypotheses seemed to be partially supported: Hypothesis (1) about bimanual coordination was tested and supported with comparison between two participants; hypothesis (2) and (3) should be tested in the future study. Even though hypothesis (1) should be also tested quantitatively with more samples, the fact seems to be meaningful that bimanual coordination was more stable in the experienced player than in the novice player. This fact suggests that a particular bimanual coordination pattern (i.e., anti-phase pattern) emerge from the interaction among components of the embodied system under certain constraints. The positional relationship among bimanual coordination and the COM-line in the mediolateral direction is expected to be revealed in future experiments. Examining how experienced players acquire skill and how the pattern emerges is also required. Variabilities observed at the knee and the COM in the vertical direction was smaller in the experienced player than in the novice player. This implies higher stability of the experienced player than that of the novice player. In the future, the coupling relationship between the knee-line and the COM-line should be investigated.

Further examination by quantitative study is expected to reveal the fundamental skills for slacklining. We will conduct this experiment comparing two groups at different skill levels (i.e., expert vs. novice) with more participants. In the future, a longitudinal study focused on the skill acquiring process of novice players can also clarify the variables that one should regulate to maintain balance and the fundamental skills for slacklining. These quantitative studies from the third-person perspective will allow us to test the hypotheses. Even if the hypothesized variables do not differ between skill levels or during the process, the results could motivate further research on the skill. Additionally, we will also reconsider current hypotheses through the qualitative approach from the first-person perspective, such as interviewing top-level slackline players. We suggest that such an interactive approach could lead to a thorough understanding of embodied skills (i.e., *Shintai-chi*).

4   Conclusion

The present article reported the results of our pilot study on single-leg standing performance to reveal the fundamental skills for slacklining. Current hypotheses were generated by personal experience of instructors and researchers, who have experience of slacklining, from the first-person perspective. The preliminary experiment was conducted to investigate how slackline players maintain balance on an unstable environment by comparing the single-leg standing performances of two participants at different skill levels (i.e., experienced vs. novice). As a result of analyzing motion-captured data in terms of our current hypotheses, highly coordinated behavior in bimanual coordination, less variability in knee and COM, and more flexible and exploratory behavior in COM were observed in the experienced player. Although these results obtained by the pilot study were from only a few participants, they necessitate further examination, and are expected to be supported by quantitative analyses from the third-person perspective in the future.

To understand such knowledge of embodied skills, applying both the first- and third-person perspectives and connecting them is important. Such an interactive approach to skill science can provide benefits to not only academic research fields such as human movement science, embodied cognitive science, robotics, sports science, but also practical fields such as balance sports/training, physical education, and rehabilitation.

References

1. Granacher, U., Iten, N., Roth, R., Gollhofer, A.: Slackline training for balance and strength promotion. Int. J. Sports Med. **31**(10), 717–723 (2010)
2. Keller, M., Pfusterschmied, J., Buchecker, M., Müller, E., Taube, W.: Improved postural control after slackline training is accompanied by reduced H-reflexes. Scand. J. Med. Sci. Sports **22**(4), 471–477 (2012)
3. Pfusterschmied, J., Buchecker, M., Keller, M., Wagner, H., Taube, W., Müller, E.: Supervised slackline training improves postural stability. Eur. J. Sport Sci. **13**, 49–57 (2011)
4. Huber, P., Kleindl, R.: A case study on balance recovery in slacklining. In: ISBS-Conference Proceedings Archive, 1990, pp. 1–4 (2010)
5. Pfusterschmied, J., Stöggl, T., Buchecker, M., Lindinger, S., Wagner, H., Müller, E.: Effects of 4-week slackline training on lower limb joint motion and muscle activation. J. Sci. Med. Sport **16**(6), 562–566 (2013)
6. Balasubramaniam, R., Wing, A.M.: The dynamics of standing balance. Trends Cogn. Sci. **6**(12), 531–536 (2002)
7. Delignières, D., Torre, K., Bernard, P.L.: Transition from persistent to anti-persistent correlations in postural sway indicates velocity-based control. PLoS Comput. Biol. **7**(2), e1001089 (2011)
8. Riley, M.A., Turvey, M.T.: Variability of determinism in motor behavior. J. Mot. Behav. **34**(2), 99–125 (2002)
9. Bernstein, N.A.: The co-ordination and regulation of movements. Pergamon Press Ltd., Oxford (1967)
10. Turvey, M.T.: Coordination. Am. Psychol. **45**(8), 938–953 (1990)
11. Latash, M.L.: The Bernstein problem: how does the central nervous system make its choices. In: Latash, M.L., Turvey, M.T. (eds.) Dexterity and Its Development. Psychology Press, London (1996)
12. Bernstein, N.A.: Dexterity and Its Development. Psychology Press, London (1996)
13. Haken, H.: Synergetics: An Introduction: Nonequilibrium Phase Transitions and Self-Organization in Physics, Chemistry and Biology. Springer, Heidelberg (1978)
14. Nicolis, G., Prigogine, I.: Self-organization in Nonequilibrium Systems: From Dissipative Structures to Order Through Fluctuations. Wiley, New York (1977)
15. Kelso, J.A.S.: Dynamic Patterns: The Self-organization of Brain and Behavior. MIT Press, Cambridge (1995)
16. Kugler, P.N., Turvey, M.T.: Information, Natural Law, and the Self-assembly of Rhythmic Movement. L. Erlbaum Associates, New Jersey (1987)
17. Newell, K.M.: Constraints on the development of coordination. In: Wade, M.G., Whiting, H.T.A. (eds.) Motor Development in Children: Aspects of Coordination and Control. Springer, Heidelberg (1986)
18. Thelen, E., Smith, L.B.: A Dynamic Systems Approach to the Development of Cognition and Action. The MIT Press, Cambridge (1996)

19. Davids, K., Button, C., Bennett, S.: Dynamics of Skill Acquisition: A Constraints-led Approach. Human Kinetics, Champaign (2008)
20. Tognoli, E., Lagarde, J., DeGuzman, G.C., Kelso, J.A.S.: The phi complex as a neuromarker of human social coordination. Proc. Natl. Acad. Sci. U.S.A. **104**(19), 8190–8195 (2007)
21. Riley, M.A., Richardson, M.J., Shockley, K.D., Ramenzoni, V.C.: Interpersonal synergies. Front. Psychol. **2**, 1–7 (2011)
22. Schmidt, R.C., Fitzpatrick, P., Caron, R., Mergeche, J.: Understanding social motor coordination. Hum. Mov. Sci. **30**(5), 834–845 (2011)
23. Haken, H., Kelso, J.A.S., Bunz, H.: A theoretical model of phase transitions in human hand movements. Biol. Cybern. **51**(5), 347–356 (1985)
24. Holden, J.G., Riley, M.A., Gao, J., Torre, K. (Eds.): Fractal Analyses: Statistical and Methodological Innovations and Best Practices (2013)
25. Van Orden, G.C., Riley, M.A. (eds.): Tutorials in Contemporary Nonlinear Methods for the Behavioral Sciences. National Science Foundation, Arlington (2005)
26. McCarthy, J., Hayes, P.J.: Some philosophical problems from the standpoint of artificial intelligence. Mach. Intell. **4**, 463–502 (1969)
27. Brooks, R.A.: Intelligent robot theory of Brooks - why the MIT robot or continue to move forward? Omusha (2006)
28. Scheier, C., Pfeifer, R.: Understanding Intelligence. MIT Press, Cambridge (2001)
29. Gibson, J.J.: The Senses Considered as Perceptual Systems. Praeger, New York (1966)
30. Gibson, J.J.: The Ecological Approach to Visual Perception: Classic Edition, vol. 20. Psychology Press, London (1979)
31. Turvey, M.T., Fonseca, S.T.: The medium of haptic perception: a tensegrity hypothesis. J. Mot. Behav. **46**(3), 143–187 (2014)
32. Turvey, M.T.: Action and perception at the level of synergies. Hum. Mov. Sci. **26**(4), 657–697 (2007)
33. Carello, C., Silva, P.L., Kinsella-Shaw, J., Turvey, M.T.: Muscle-based perception: theory, research and implications for rehabilitation. Revista Brasileira de Fisioterapia **12**(5), 339–350 (2008)
34. Carello, C., Turvey, M.T.: Physics and psychology of the muscle sense. Curr. Dir. Psychol. Sci. **13**(1), 25–28 (2004)
35. Turvey, M.T., Carello, C.: Obtaining information by dynamic (effortful) touching. Philos. Trans. R. Soc. Lond. B Biol. Sci. **366**(1581), 3123–3132 (2011)
36. Latash, M.L.: Synergy. Oxford University Press, Oxford (2008)
37. Zbilut, J.P., Giuliani, A., Webber, C.L.: Detecting deterministic signals in exceptionally noisy environments using cross-recurrence quantification. Phys. Lett. A **246**(1–2), 122–128 (1998)
38. Coco, M.I., Dale, R.: Cross-recurrence quantification analysis of categorical and continuous time series: an R package. Front. Psychol. **5**, 510 (2014)
39. Pellecchia, G.L., Shockley, K.D., Turvey, M.T.: Concurrent cognitive task modulates coordination dynamics. Cogn. Sci. **29**(4), 531–557 (2005)
40. Peng, C.-K., Buldyrev, S.V., Havlin, S., Simons, M., Stanley, H.E., Goldberger, A.L.: Mosaic organization of DNA nucleotides. Phys. Rev. E **49**(2), 1685–1689 (1994)
41. Peng, C.-K., Havlin, S., Stanley, H.E., Goldberger, A.L.: Quantification of scaling exponents and crossover phenomena in nonstationary heartbeat time series. Chaos (Woodbury, N.Y.) **5**(1), 82–87 (1995)
42. Eke, A., Herman, P., Kocsis, L., Kozak, L.R.: Fractal characterization of complexity in temporal physiological signals. Physiol. Meas. **23**(1), R1–R38 (2002)

43. Delignières, D., Ramdani, S., Lemoine, L., Torre, K., Fortes, M., Ninot, G.: Fractal analyses for "short" time series: a re-assessment of classical methods. J. Math. Psychol. **50**(6), 525–544 (2006)

44. Hausdorff, J.M.: Gait dynamics in Parkinson's disease: common and distinct behavior among stride length, gait variability, and fractal-like scaling. Chaos Interdisc. J. Nonlinear Sci. **19**(2), 026113 (2009)

45. Arzamarski, R., Isenhower, R.W., Kay, B.A., Turvey, M.T., Michaels, C.F.: Effects of intention and learning on attention to information in dynamic touch. Atten. Percept. Psychophys. **72**(3), 721–735 (2010)

46. Shockley, K.D.: Cross recurrence quantification of interpersonal postural activity. In: Riley, M., Van Orden, G. (eds.) Tutorials in Contemporary Nonlinear Methods for the Behavioral Sciences, pp. 142–177. National Science Foundation, Arlington (2005)

47. Zbilut, J.P., Webber, C.L.: Embeddings and delays as derived from quantification of recurrence plots. Phys. Lett. A **171**(3–4), 199–203 (1992)

48. Webber, C.L., Zbilut, J.P.: Recurrence quantification analysis of nonlinear dynamical systems. In: Riley, M., Van Orden, G. (eds.) Tutorials in Contemporary Nonlinear Methods for the Behavioral Sciences, pp. 26–94. National Science Foundation, Arlington (2005)

49. Marwan, N.: Cross recurrence plot toolbox (2013). http://tocsy.pik-potsdam.de/CRPtoolbox

50. Hardstone, R., Poil, S.-S., Schiavone, G., Jansen, R., Nikulin, V.V., Mansvelder, H.D., Linkenkaer-Hansen, K.: Detrended fluctuation analysis: a scale-free view on neuronal oscillations. Front. Physiol. **3**, 1–13 (2012)

51. Goldberger, A.L., Amaral, L.A.N., Hausdorff, J.M., Ivanov, P.C., Peng, C.-K., Stanley, H.E.: Fractal dynamics in physiology: alterations with disease and aging. Proc. Natl. Acad. Sci. **99**(Supplement 1), 2466–2472 (2002)

52. Van Orden, G.C., Holden, J.G., Turvey, M.T.: Human cognition and 1/f scaling. J. Exp. Psychol. Gen. **134**(1), 117–123 (2005)

53. Lipsitz, L.A., Goldberger, A.L.: Loss of "complexity" and aging. JAMA **267**(13), 1806 (1992)

54. Dixon, J.A., Stephen, D.G., Boncoddo, R., Anastas, J.R.: The self-organization of cognitive structure. Psychol. Learn. Motiv. **52**, 343–384 (2010)

Author Index

Abrusán, Márta 41
Arisaka, Ryuta 204
Asaka, Yusuke 485

Bailey, James 367
Bartolini, Cesare 233
Bekki, Daisuke 58
Bellodi, Elena 187

Caminada, Martin 317
Carvalho, Danilo S. 295
Chakraborty, Basabi 397
Chakraborty, Goutam 418
Cota, Giuseppe 187
Cupi, Loredana 217

Fukuda, Ken 380
Furuhashi, Takeshi 454
Furukawa, Koichi 499
Fuyama, Miho 514

Gavanelli, Marco 187
Goebel, Randy 282

Hara, Yurie 27
Haraguchi, Makotoc 499
Hidaka, Shohei 514
Hirata, Kouichi 409
Hou, Lei 485
Humphreys, Llio 217
Hunter, Julie 41

Imaizumi, Kazuya 380
Ishiguro, Hiroshi 447, 462, 470

Jimenez, Felix 454

Kamiyama, Takuya 418
Kanoh, Masayoshi 454
Katsura, Yuki 265
Kaufmann, Stefan 3
Kawazoe, Ai 58
Kido, Hiroyuki 348
Kikuchi, Yusuke 528

Kim, Mi-Young 282
Kinjo, Keita 499
Kiselyov, Oleg 66
Kodama, Kentaro 528
Koizumi, Masatoshi 109
Konosu, Hisae 380
Kubota, Yusuke 79
Kulik, Lars 367
Kumazaki, Hirokazu 447, 470

Lamma, Evelina 187
Levine, Robert 79
Li, Yuxuan 367
Liang, Zilu 367, 380
Liu, Wanyu 367

Matsumoto, Yoshio 447, 470
Matsuoka, Kimiko 409
McCready, Eric 27
Mimura, Masaru 447, 470
Mineshima, Koji 58
Miwa, Hiroyasu 380
Miyao, Masutomo 447, 470
Muramatsu, Taro 470
Muthuri, Robert 217, 233

Nagao, Tomoka 380
Nagayama, Kazuki 409
Nakadoi, Yoshihiro 478
Nakajima, Kanako 380
Nakamura, Makoto 249
Nakamura, Tsuyoshi 454
Nemoto, Sakiko 447
Nguyen, Minh-Le 295
Nguyen, Minh-Tien 295
Nishimura, Satoshi 380
Nishimura, Takuichi 380
Nishina, Kei 265
Nitta, Katsumi 265

Ogawa, Yasuhiro 249
Ohno, Tomohiro 249
Okada, Shogo 265

Okubo, Satoko   380
Ozaki, Tomonobu   499

Ploderer, Bernd   367

Ramamonjisoa, David   428
Riccardo, Zese   187
Rieser, Lukas   93
Riguzzi, Fabrizio   187
Robaldo, Livio   217
Rushby, John   331

Sakai, Tomohiro   15
Sakama, Chiaki   317
Sakamoto, Satomi   249
Santos, Cristiana   217, 233
Sauerland, Uli   109
Sawada, Osamu   122
Stojanovic, Isidora   138
Sumioka, Hidenobu   462
Sun, Xin   217

Tamura, Ayaka   109
Tanaka, Ribeka   58
Tomlinson Jr., John M.   109
Toyama, Katsuhiko   249
Tran, Chien-Xuan   295

Wada, Yasuo   462
Wakita, Yujin   447
Wang, Linton   151
Watanuki, Keiichi   485

Xu, Ying   282

Yamagiwa, Hideo   528
Yatabe, Shunsuke   166
Yokoyama, Shigeki   409
Yoshida, Sho   397
Yoshida, Yasuyuki   380
Yoshikawa, Tomohiro   454
Yoshikawa, Yuichiro   447, 462, 470

Printed in the United States
By Bookmasters